Guang Gong
Kishan Chand Gupta (E

Progress in Cryptology - INDOCRYPT 2010

11th International Conference on Cryptology in India
Hyderabad, India, December 12-15, 2010
Proceedings

Volume Editors

Guang Gong
University of Waterloo
Department of Electrical and Computer Engineering
Waterloo, ON, Canada N2L 3G1
E-mail: ggong@uwaterloo.ca

Kishan Chand Gupta
Indian Statistical Institute
Applied Statistics Unit
Kolkata 700108, India
E-mail: kishan@isical.ac.in

Library of Congress Control Number: 2010939793

CR Subject Classification (1998): E.3, K.6.5, D.4.6, C.2, J.1, G.2.1

LNCS Sublibrary: SL 4 – Security and Cryptology

ISSN 0302-9743
ISBN-10 3-642-17400-0 Springer Berlin Heidelberg New York
ISBN-13 978-3-642-17400-1 Springer Berlin Heidelberg New York

springer.com

Printed in Germany

Typesetting: Camera-ready by author, data conversion by Scientific Publishing Services, Chennai, India
Printed on acid-free paper 06/3180

Lecture Notes in Computer Science 6498

Message from the General Chairs

The Cryptology Research Society of India (CRSI) has been successfully coordinating Indocrypt, the International Conference series of Cryptology in India, since 2000. Indocrypt 2010 was the 11th event in this series that was organized by the C.R. Rao Advanced Institute of Mathematics, Statistics and Computer Science (AIMSCS) and the University of Hyderabad, India, under the aegis of the Cryptology Research Society of India.

Indocrypt 2010 was held during December 12–15, 2010, with a one-day workshop followed by three-day conference. Living up to its own reputation, it has stood up to become a leading forum for the dissemination of the latest research results in cryptology. It brought together eminent researchers from across the world to present and discuss a wide spectrum of research results and their applications in cryptology and information security.

The hard work of all the members of the Organizing Committee was one of the most significant factors leading to the success of the event. We express our special thanks to Guang Gong and Kishan Chand Gupta (Program Co-chairs) for coordinating and leading the effort of the Program Committee.

We are thankful to Rajat Tandon, Arun Agarwal and K. Nageswara Rao (Organizing Co-chairs) along with all other members of the Organizing Committee who coordinated all the local arrangements with ardor.

We express our heartfelt thanks to DIT, DRDO, DST and MSRI for being sponsors of the event. Last but not the least, we extend our most sincere thanks to the authors, the reviewers and the participants for their vital contributions to the success of the event.

December 2010

Siddani Bhaskara Rao
Bimal K. Roy

Message from the Technical Program Chairs

Indocrypt 2010, the 11th International Conference on Cryptology in India, took place during December 12–15, 2010 at Hyderabad, India. The General Chairs, S.B. Rao and Bimal Roy, did an excellent job in keeping all strands together and making this conference successful. The main program was preceded by a day of tutorial presentations by Sanjit Chatterjee and Guang Gong.

This year, the conference received 72 submissions. Each paper was assigned to at least 3 and on an average 3.9 Program Committee members. During 2 weeks of discussions, 240 comments were produced by 36 Program Committee members and many more reviews were added. After rigorous reviews and detailed discussions, the Program Committee accepted 22 papers out of which 3 papers were accepted under anonymous shepherding.

Our sincere thanks to the eminent invited speakers, Neal Koblitz and Bart Preneel. Neal Koblitz spoke on "Getting a Few Things Right and Many Things Wrong" and Bart Preneel gave a talk on "Cryptographic Hash Functions: Theory and Practice."

It would have been impossible to make this event successful without the contribution and support we received from several corners. We would like to express our gratitude to all authors around the world for submitting their papers to this conference. Our special thanks to the Program Committee members and external reviewers who spared their valuable time and gave their expert comments on the submitted papers.

We would like to give our heartfelt thanks to Honggang Hu and Sumit Kumar Pandey for their constant support. We extend our gratitude to the Organizing Committee, volunteers and participants. We would also like to thank Qi Chai for the website design, and easychair.org, a nice and easy-to-use conference management system.

We are grateful to the Department of Electrical and Computer Engineering, University of Waterloo, Canada and Applied Statistics Unit, Indian Statistical Institute, Kolkata, India for providing infrastructural support. Last but not the least, we are grateful to those whose names are missed here, but have contributed in some way or the other to Indocrypt 2010.

It is a pleasure to see Indocrypt as a well-accepted cryptology conference where new results are submitted. We wish you a happy reading.

December 2010

Guang Gong
Kishan Chand Gupta

Organization

Indocrypt 2010 was organized by the C.R. Rao Advanced Institute of Mathematics, Statistics and Computer Science (AIMSCS), and the University of Hyderabad, India, under the aegis of Cryptology Research Society of India (CRSI), India.

General Chairs

S.B. Rao	C.R. Rao AIMSCS, Hyderabad, India
Bimal K. Roy	Indian Statistical Institute, Kolkata, India

Organizing Co-chairs

Rajat Tandon	University of Hyderabad, India
Arun Agarwal	University of Hyderabad, India
K. Nageswara Rao	C.R. Rao AIMSCS, India

Organizing Committee

V.C. Venkaiah	C.R. Rao AIMSCS, India
Chakravarthy Bhagvathi	University of Hyderabad, India
Y.V. Subba Rao	University of Hyderabad, India
Mahabir Prasad Jhanwar	C.R. Rao AIMSCS, India
M. Prem Laxman Das	C.R. Rao AIMSCS, India
Prabal Paul	C.R. Rao AIMSCS, India
N. Anil Kumar	C.R. Rao AIMSCS, India
Neelima Jampala	C.R. Rao AIMSCS, India
G. Uma Devi	C.R. Rao AIMSCS, India

Program Chairs

Guang Gong	University of Waterloo, Canada
Kishan Chand Gupta	Indian Statistical Institute, Kolkata, India

Program Committee

Daniel J. Bernstein	University of Illinois at Chicago, USA
Colin Boyd	Queensland University of Technology, Australia
Anne Canteaut	INRIA Paris-Rocquencourt, France
Claude Carlet	University of Paris 8, France
Debrup Chakraborty	CINVESTAV-IPN, Mexico City, Mexico
Sanjit Chatterjee	University of Waterloo, Canada
Lily Chen	NIST, USA

Sherman Chow	New York University, USA
Abhijit Das	IIT Kharagpur, India
Steven Galbraith	Auckland University, New Zealand
Sugata Gangopadhyay	IIT Roorkee, India
Praveen Gauravaram	Technical University of Denmark
Vipul Goyal	Microsoft Research, India
Tor Helleseth	University of Bergen, Norway
Honggang Hu	University of Waterloo, Canada
Shaoquan Jiang	University of Electronic Science and Technology, China
Xuejia Lai	Shanghai Jiao Tong University, China
Tanja Lange	Technische Universiteit Eindhoven, The Netherlands
C. E. Veni Madhavan	IISC Bangalore, India
Willi Meier	FHNW, Switzerland
Alfred Menezes	University of Waterloo, Canada
Miodrag Mihaljevic	RCIS-AIST, Tokyo, and Mathematical Institute SANU, Belgrade
Mridul Nandi	The George Washington University, USA
Goutam Paul	Jadavpur University, Kolkata, India
Manoj Prabhakaran	University of Illinois, Urbana-Champaign, USA
C. Pandu Rangan	IIT Chennai, India
Vincent Rijmen	K.U.Leuven, Belgium and TU Graz, Austria
Sushmita Ruj	University of Ottawa, Canada
Rei Safavi-Naini	University of Calgary, Canada
Kouichi Sakurai	Kyushu University, Japan
P. K. Saxena	SAG, Delhi, India
Nicolas Sendrier	INRIA, France
Nicolas Theriault	Universidad de Talca, Chile
Ayineedi Venkateswarlu	CRRAO AIMSCS, Hyderabad, India
Huaxiong Wang	Nanyang Technological University, Singapore
Amr Youssef	Concordia University, Canada

External Reviewers

Abdel Alim Kamal
Aleksandar Kircanski
Alexander May
Arpita Patra
Ashish Choudhary
Ashraful Tuhin
Avradip Mandal
Berkant Ustaoglu
Bing Wang
Bo Zhu
Bo-Yin Yang
C. Mancillas-Lopez
Carles Padro
Chen-Mou Cheng
Cheng-Kang Chu
Chunhua Su
Cline Blondeau
Crystal Lee Clough
Daesung Kwon
Damien Stehlé
Daniel Augot
Daniel Smith
Debdeep Mukhopadhyay
Dhananjoy Dey
Dmitry Khovratovich
Dustin Moody
Emmanuel Thomé
F.-X. Standaert
Fumihiko Sano
Gilles Millerioux

Grigori Kabatianski
H.V. Kumar Swamy
Hadi Ahmadi
Hongsong Shi
Huseyin Demirci
Håvard Raddum
Igor Semaev
Indivar Gupta
Jian Guo
Julia Borghoff
Kazuhide Fukushima
Kerry McKay
K.-K. Raymond Choo
Kjell Wooding
Kohtaro Tadaki
Koichiro Akiyama
Kris Narayan
Lei Wei
M. Prem Laxman Das
Mahabir Prasad Jhanwar
Maria Naya-Plasencia
Martin Gagne
Meena Kumari
Mehdi Hassanzadeh
Ming Duan
Morshed U. Chowdhury
N. Rajesh Pillai
Nadia Heninger
Nashad Safa
Nasoor Bagheri
Ning Ding
Olivier Pereira
P.R. Mishra
Peter Schwabe
S.K. Tiwari
S.S. Bedi
Saket Srivastava
Sanjam Garg
Serdar Pehlivanoglu
Shiho Moriai
Shinsaku Kiyomoto
Siamak Shahandashti
Somitra Kr Sanadhya
Sondre Rønjom
Souradyuti Paul
Srimanta Bhattacharya
Srinivas Vivek Venkatesh
Stefan Popoveniuc
Stphane Manuel
Suresh V.
Takashi Nishide
Thomas Peyrin
Thomas Roche
Xiaorui Sun
Yannick Seurin
Yannis Rouselakis
Yasufumi Hashimoto
Yiyuan Luo
Zheng Gong

Table of Contents

Hash Functions

Attacks on Block Ciphers and Stream Ciphers

Fast Cryptographic Computation

Cryptanalysis of AES

Efficient Implementation

Getting a Few Things Right and Many Things Wrong

Neal Koblitz

Department of Mathematics
University of Washington
Seattle, Washington 98195-4350, Seattle, USA
koblitz@math.washington.edu

Abstract. The history of cryptography from ancient times to the present is full of tales of blunders and oversights, typically occurring when an over-confident encryptor is outwitted by a patient and clever cryptanalyst. In contrast, mathematics (if properly peer-reviewed) is perfect. There is never error, because by definition one cannot prove a theorem if it is false. So in order to remove the contingent and subjective elements from cryptography there have been concerted efforts in recent years to transform the field into a branch of mathematics, or at least a branch of the exact sciences. In my view, this hope is misguided, because in its essence cryptography is as much an art as a science.

I will start by describing a setting (taken from a recent paper written with Alfred Menezes and Ann Hibner Koblitz) in which the conventional wisdom about parameter selection might (or might not) be wrong. Then I will illustrate the pitfalls of working in cryptography by giving a (far from exhaustive) survey of the many misjudgments I have made and erroneous beliefs I have had over the course of 25 years working in this field. I will then describe a few of the embarrassing moments in the history of "provable security", which is the name of an ambitious program that aims to transform cryptography into a science.

G. Gong and K.C. Gupta (Eds.): INDOCRYPT 2010, LNCS 6498, p. 1, 2010.

Partial Key Exposure Attack on RSA – Improvements for Limited Lattice Dimensions

Santanu Sarkar, Sourav Sen Gupta, and Subhamoy Maitra

Applied Statistics Unit, Indian Statistical Institute,
203 B T Road, Kolkata 700 108, India
sarkar.santanu.bir@gmail.com, sg.sourav@gmail.com, subho@isical.ac.in

Abstract. Consider the RSA public key cryptosystem with the parameters $N = pq$, $q < p < 2q$, public encryption exponent e and private decryption exponent d. In this paper, cryptanalysis of RSA is studied given that some amount of the Most Significant Bits (MSBs) of d is exposed. In Eurocrypt 2005, a lattice based attack on this problem was proposed by Ernst, Jochemsz, May and de Weger. In this paper, we present a variant of their method which provides better experimental results depending on practical lattice parameters and the values of d. We also propose a sublattice structure that improves the experimental results significantly for smaller decryption exponents.

Keywords: Cryptanalysis, Factorization, Lattice Reduction, Public Key Cryptosystem, RSA, Sublattice.

1 Introduction

The RSA [15] public key cryptosystem can be briefly described as follows:

- primes p, q, (generally considered of same bit size, i.e., $q < p < 2q$);
- $N = pq$, $\phi(N) = (p-1)(q-1)$;
- e, d are such that $ed = 1 + k\phi(N)$, $k \geq 1$;
- N, e are public and plaintext $M \in \mathbb{Z}_N$ is encrypted as $C \equiv M^e \bmod N$;
- secret key d needed to decrypt ciphertext $C \in \mathbb{Z}_N$ as $M \equiv C^d \bmod N$.

One important model of cryptanalysis in the field of RSA is side channel attacks such as fault attacks, timing attacks, power analysis etc. [3,12,13], by which an adversary may obtain some bits of the private key d.

Boneh et al. [3] studied how many bits of d need to be known to factor the RSA modulus N. The constraint in [3] was the upper bound on e, that had been $\sqrt{N}$. The idea of [3] has been improved by Blömer and May [2] where the bound on e was increased upto $N^{0.725}$. Then the work by Ernst et al. [8] improved the result for full size public exponent e. Sarkar and Maitra [16] extended the work of [8] by guessing few bits of one prime. Recently, the work by Aono [1] improved the results of [8] when some portion of Least Significant Bits (LSBs) of d are exposed and $d < N^{0.5}$. In this paper, we propose a variant of the idea presented in [8] to make the results more practical when some portion of Most Significant

G. Gong and K.C. Gupta (Eds.): INDOCRYPT 2010, LNCS 6498, pp. 2–16, 2010.

Bits (MSBs) of d are exposed and $d < N^{0.6875}$. One may argue that exposing LSBs and MSBs pose two different scenarios. But if we compare the two methods by the total number of bits of d that one needs to know for cryptanalysis, our method improves the results of [8] for a larger range of d than [1].

In this paper we consider the case when some MSBs of d are exposed. So one can write $d = d_0 + d_1$ where the attacker knows d_0. Attacker can also find an approximation $k_0 = \lfloor \frac{ed_0 - 1}{N} \rfloor$ of k. Let $k_1 = k - k_0$ and d be of bitsize $\delta \log_2 N$. Ernst et al. [8] considered the polynomial $f_1(x, y, z) = ed_0 - 1 + ex - Ny - yz$ and it is clear that (d_1, k, s) is a root of f_1 where $s = 1 - p - q$. Further, the approximation k_0 of k has been used to consider the polynomial $f_2(x, y, z) = ed_0 - 1 - k_0 N + ex - Ny - yz - k_0 z$. In this case, (d_1, k_1, s) is the root of f_2. If one can find the root of either f_1 or f_2, N can be factored.

Given $(\delta - \gamma) \log_2 N$ many MSBs of d, one can find the root of f_1 or f_2 in poly$(\log N)$ time if any of the following holds [8][1]:

$$\gamma < \tfrac{5}{6} - \tfrac{1}{3}\sqrt{1 + 6\delta},$$
$$\gamma < \tfrac{3}{16} \text{ and } \delta \leq \tfrac{11}{16},$$
$$\gamma < \tfrac{1}{3} + \tfrac{1}{3}\delta - \tfrac{1}{3}\sqrt{4\delta^2 + 2\delta - 2} \text{ and } \delta \geq \tfrac{11}{16}.$$

In this paper we consider the polynomial $f_e(x, y) = 1 + (k_0 + x)(N + y)$ over $\mathbb{Z}_e$ where the terms k_0, k_1, d_0, d_1, s are same as mentioned before. Clearly (k_1, s) is the root of f_e. If one gets s, then immediately the factorization of N is possible. We use Coppersmith's [5] method for roots of modular polynomials to find such a root. As predicted in [8, Section 5], this leads to lattices of smaller dimension and hence better practical results for fixed lattice parameters within certain range of d. However, [8] does not precisely analyze this situation, and so we provide a comprehensive treatment of such an analysis in Section 2.

Though the theoretical bounds on γ, as given in [8], work for $d < \phi(N)$, the experimental results could only be achieved for the range $d \leq N^{0.7}$ with lattice dimension upto 50. Our experimental results are better than that of [8] for $d \leq N^{0.64}$ with smaller lattice dimensions. The results explaining experimental advantage are presented in Section 4.

Although the practical attacks are mounted using lattices with small dimension, where the lattice parameters are generally predetermined, the results of this kind are often compared in asymptotic sense in literature. In this direction, we show that the root of f_e can be obtained in poly$(\log N)$ time if $\lambda < \frac{3}{16}$, where $\lambda = \max\{\gamma, \delta - \frac{1}{2}\}$. Our results are as good as [8] for $N^{0.4590} \leq d < N^{0.6875}$ in terms of asymptotic bound.

The reader may note that our results are not better than those in [8] if we consider the asymptotic performance. It is only better in practical experimental scenario, where we obtain results of the same quality as in [8] by using lattices with comparatively smaller dimension. The reason is that we use Coppersmith's [5] idea for the modular polynomial, while [8] used Coron's [6] version for ease of presentation. Thus, the lattice dimension obtained in [8] was a cubic in a certain parameter while we obtain a quadratic, hence smaller. This idea has

[1] The terms δ, β in [8] are denoted as γ, δ respectively in our analysis.

already been pointed out in [8] itself, but rigorous analysis for limited lattice dimensions is studied here.

Further, in Section 3, we propose the construction of a sublattice by deleting certain rows of the above mentioned lattice. This provides significantly improved results for smaller decryption exponents. Once again, experimental evidences in Section 4 support our claim.

1.1 Preliminaries

Let us start with some basic concepts on lattice reduction techniques. Consider a set of linearly independent vectors $u_1, \ldots, u_\omega \in \mathbb{Z}^n$, with $\omega \leq n$. The lattice L, spanned by $\{u_1, \ldots, u_\omega\}$, is the set of all integer linear combinations of the vectors $u_1, \ldots, u_\omega$. The number of vectors ω is the dimension of the lattice. Such a lattice is called full rank when $\omega = n$. By $u_1^*, \ldots, u_\omega^*$, we denote the vectors obtained by applying the Gram-Schmidt process [4, Page 81] to $u_1, \ldots, u_\omega$. The determinant of L is defined as $\det(L) = \prod_{i=1}^{\omega} ||u_i^*||$, where $||.||$ denotes the Euclidean norm on vectors. Given a bivariate polynomial $g(x, y) = \sum a_{i,j} x^i y^j$, the Euclidean norm is defined as $\| g(x, y) \| = \sqrt{\sum_{i,j} a_{i,j}^2}$ and the infinity norm is defined as $\| g(x, y) \|_\infty = \max_{i,j} |a_{i,j}|$. We shall follow these notation in this paper.

Fact 1. *Given a basis $u_1, \ldots, u_\omega$ of a lattice L, the LLL algorithm [14] generates a new basis $b_1, \ldots, b_\omega$ of L with the following properties.*

1. $\| b_i^* \|^2 \leq 2 \| b_{i+1}^* \|^2$, *for* $1 \leq i < \omega$.
2. *For all i, if $b_i = b_i^* + \sum_{j=1}^{i-1} \mu_{i,j} b_j^*$ then $|\mu_{i,j}| \leq \frac{1}{2}$ for all j.*
3. $\| b_i \| \leq 2^{\frac{\omega(\omega-1)+(i-1)(i-2)}{4(\omega-i+1)}} det(L)^{\frac{1}{\omega-i+1}}$ *for* $i = 1, \ldots, \omega$.

Here $b_1^*, \ldots, b_\omega^*$ denote the vectors obtained by applying Gram-Schmidt process to $b_1, \ldots, b_\omega$.

In [5], Coppersmith discusses lattice based techniques to find small integer roots of univariate polynomials $\bmod n$, and of bivariate polynomials over the integers. The idea of [5] can also be extended to more than two variables, but the method becomes heuristic. Lemma 1 is relevant to the idea of [5] for finding roots of bivariate polynomials over integers.

Lemma 1. *Let $g(x_1, x_2)$ be a polynomial which is the sum of ω many monomials. Suppose $g(y_1, y_2) \equiv 0 \bmod n$, where $|y_1| < Y_1$ and $|y_2| < Y_2$. If $\| g(x_1 Y_1, x_2 Y_2) \| < \frac{n}{\sqrt{\omega}}$, then $g(y_1, y_2) = 0$ holds over integers.*

We apply Gröbner Basis based techniques to solve for the roots of bivariate polynomials. Though our technique works in practice as noted from the experiments we perform, theoretically this may not always happen. Thus we formally state the following heuristic assumption, that we will require for our theoretical results.

Assumption 1. *Suppose that one constructs a lattice using the idea of Coppersmith [5] in order to find the root of a bivariate modular equation. Further, consider that the lattice reduction is executed using the LLL algorithm. Let the polynomials corresponding to the first two basis vectors of the lattice after LLL reduction be* $\{f_1, f_2\}$ *and they share the common root of the form* $(x_1^{(0)}, x_2^{(0)})$. *If* J *be the ideal generated by* $\{f_1, f_2\}$, *then one can efficiently collect the root by computing the Gröbner Basis of* J.

Note that the time complexity of the Gröbner Basis computation is in general double-exponential in the degree of the polynomials [7].

2 The Lattice Based Technique

We start with the following theorem.

Theorem 1. *Consider the RSA equation* $ed \equiv 1 \pmod{\phi(N)}$. *Let* $d = N^{\delta}$ *and* e *be* $\Theta(N)$. *Suppose we know an integer* d_0 *such that* $|d - d_0| < N^{\gamma}$. *Then, under Assumption 1, one can factor* N *in poly*$(\log N)$ *time when*

$$\lambda < \frac{\frac{1}{12}m^3 - \frac{13}{12}m + \frac{1}{4}m^2t + \frac{1}{4}mt}{\frac{1}{2}m^3 + m^2 + \frac{1}{2}m + \frac{1}{2}t^2 + \frac{1}{2}t + m^2t + \frac{1}{2}mt^2 + \frac{3}{2}mt}$$

where $\lambda = \max\{\gamma, \delta - \frac{1}{2}\}$ *and* m, t *are non-negative integers.*

Proof. From the RSA equation, we have $ed = 1 + k(N + 1 - p - q)$. When MSBs of d are known, we can write $d = d_0 + d_1$, where d_0, corresponding the MSBs of d, is known to the attacker, but d_1 is not. The attacker can also calculate $k_0 = \lfloor \frac{ed_0 - 1}{N} \rfloor$ as an approximation of k and set $k_1 = k - k_0$.

We can write $ed = 1 + (k_0 + k_1)(N + s)$, where $s = 1 - p - q$. Thus $1 + k_0N + k_0s + k_1N + k_1s \equiv 0 \pmod{e}$ and we are interested in finding the solution $(x_0, y_0) = (k_1, s)$ of

$$f_e(x, y) = \overline{1 + k_0N} + k_0y + Nx + xy$$

in $\mathbb{Z}_e$. Note that we are considering the polynomial $f_e(x, y)$ reduced modulo e, and hence the modified constant term $\overline{1 + k_0N}$ is actually equivalent to $1 + k_0N - ed_0$, which is much smaller than the original. This helps in reducing the bit size of some elements in the matrix corresponding to the lattice we describe below.

Following results by Blömer and May [2, Proof of Theorem 6], and Ernst et al. [8, Section 2], it can be shown that $|k_1| < 4N^{\lambda}$, for $\lambda = \max\{\gamma, \delta - \frac{1}{2}\}$. We also have $|s| \leq 2N^{0.5}$, by definition. Now, let us take $X = N^{\lambda}$ and $Y = N^{0.5}$. One may note that X, Y are the upper bounds of the roots $(x_0, y_0) = (k_1, s)$ of $f_e(x, y)$, neglecting the respective small constants 4 and 2 respectively. For integers $m, t \geq 0$, we define two sets of polynomials

$$g_{i,j}(x, y) = x^i f_e^j(x, y) e^{m-j} \quad \text{where } j = 0, \ldots, m,\ i = 0, \ldots, m - j + t,$$
$$h_{i,j}(x, y) = y^i f_e^j(x, y) e^{m-j} \quad \text{where } j = 0, \ldots, m,\ i = 1, \ldots, m - j.$$

Note that $g_{i,j}(k_1, s) \equiv 0 \pmod{e^m}$ and $h_{i,j}(k_1, s) \equiv 0 \pmod{e^m}$. We call $g_{i,j}$ the x-shift and $h_{i,j}$ the y-shift polynomials, as per construction.

Next, we form a lattice L by taking the coefficient vectors of the shift polynomials $g_{i,j}(xX, yY)$ and $h_{i,j}(xX, yY)$ as basis. One can verify that the dimension of the lattice L is $\omega = (m+1)^2 + t(m+1)$. The matrix L_M, containing the basis vectors of L, is lower triangular and has diagonal entries of the form

$$\begin{aligned} &X^{i+j}Y^j e^{m-j} \text{ for } j = 0, \ldots, m \text{ and } i = 0, \ldots, m-j+t, \text{ and} \\ &X^j Y^{i+j} e^{m-j} \text{ for } j = 0, \ldots, m \text{ and } i = 1, \ldots, m-j, \end{aligned}$$

coming from $g_{i,j}$ and $h_{i,j}$ respectively. Thus, one can calculate the determinant of L as

$$\det(L) = \left[\prod_{j=0}^{m} \prod_{i=0}^{m-j+t} X^{i+j}Y^j e^{m-j}\right] \left[\prod_{j=0}^{m} \prod_{i=1}^{m-j} X^j Y^{i+j} e^{m-j}\right] = X^{s_1} Y^{s_2} e^{s_3}$$

where

$$\begin{aligned} s_1 &= \frac{1}{2}m^3 + m^2 + \frac{1}{2}m + \frac{1}{2}t^2 + \frac{1}{2}t + m^2 t + \frac{1}{2}mt^2 + \frac{3}{2}mt, \\ s_2 &= \frac{1}{2}m^3 + m^2 + \frac{1}{2}m + \frac{1}{2}m^2 t + \frac{1}{2}mt, \text{ and} \\ s_3 &= \frac{2}{3}m^3 + \frac{1}{2}m^2 t + \frac{3}{2}m^2 + \frac{1}{2}mt + \frac{5}{6}m. \end{aligned}$$

To utilize Gröbner basis techniques and Assumption 1, we need two polynomials $f_1(x, y)$, $f_2(x, y)$ which share the roots (k_1, s) over integers. From Lemma 1 and Fact 1, we know that one can find such $f_1(x, y)$, $f_2(x, y)$ using LLL lattice reduction algorithm over L when

$$2^{\frac{\omega(\omega-1)}{4}} (det(L))^{\frac{1}{\omega-1}} < \frac{e^m}{\sqrt{\omega}}.$$

Now, for large N, e we have $(det(L))^{\frac{1}{\omega-1}}, e$ is much larger than $2^{\frac{\omega(\omega-1)}{4}}$, $\sqrt{\omega}$. Hence we approximate the required condition by $det(L) < e^{m(\omega-1)}$. Given the values of $det(L)$ and ω obtained above, we get the required condition as $X^{s_1} Y^{s_2} e^{s_3} < e^{m\left((m+1)^2 + t(m+1) - 1\right)}$, i.e., $X^{s_1} Y^{s_2} < e^{s_0}$, where

$$\begin{aligned} s_0 &= m\left((m+1)^2 + t(m+1) - 1\right) - s_3 \\ &= \frac{1}{3}m^3 + \frac{1}{2}m^2 - \frac{5}{6}m + \frac{1}{2}m^2 t + \frac{1}{2}mt. \end{aligned}$$

Now putting the values of the bounds $X = N^{\lambda}, Y = N^{0.5}$ in $X^{s_1} Y^{s_2} < e^{s_0}$, and considering e to be $\Theta(N)$, we get the condition as

$$\begin{aligned} &\lambda \left(\frac{m^3}{2} + m^2 + \frac{m}{2} + \frac{t^2}{2} + \frac{t}{2} + m^2 t + \frac{mt^2}{2} + \frac{3mt}{2}\right) + \\ &\frac{1}{2}\left(\frac{m^3}{2} + m^2 + \frac{m}{2} + \frac{m^2 t}{2} + \frac{mt}{2}\right) < \frac{m^3}{3} + \frac{m^2}{2} - \frac{5m}{6} + \frac{m^2 t}{2} + \frac{mt}{2}. \end{aligned} \quad (1)$$

From Equation (1) we get the required bound for λ as follows:

$$\lambda < \frac{\frac{1}{12}m^3 - \frac{13}{12}m + \frac{1}{4}m^2t + \frac{1}{4}mt}{\frac{1}{2}m^3 + m^2 + \frac{1}{2}m + \frac{1}{2}t^2 + \frac{1}{2}t + m^2t + \frac{1}{2}mt^2 + \frac{3}{2}mt}.$$

Now, one can find the root (k_1, s) from f_1, f_2 under Assumption 1. The claimed time complexity of poly$(\log N)$ can be achieved because

- the time complexity of the LLL lattice reduction is poly$(\log N)$; and
- given a fixed lattice dimension of small size, we get constant degree polynomials and the Gröbner Basis calculation is in general double-exponential in the degree of the polynomial.

This completes the proof of Theorem 1. □

Let us illustrate the lattice generation technique for $m = 3, t = 0$. We use the shift polynomials e^3, xe^3, ye^3, fe^2, x^2e^3, xfe^2, x^2e^3, xfe^2, x^3e^3, x^2fe^2, y^2e^3, yfe^2, f^2e, xf^2e, y^3e^3, y^2fe^2, yf^2e, f^3 and build the following lattice L with the basis elements coming from the coefficients of these shift polynomials, as discussed before. In this case, the lattice dimension turns to be $(m+1)^2+t+mt = 16$. The '−' marked places contain non-zero elements, but we do not write those as those elements do not contribute in the calculation of the determinant.

poly	1	x	y	xy	x^2	x^2y	x^3	x^3y	y^2	xy^2	x^2y^2	x^3y^2	y^3	xy^3	x^2y^3	x^3y^3
e^3	e^3															
xe^3		Xe^3														
ye^3			Ye^3													
fe^2	−	−	−	XYe^2												
x^2e^3					X^2e^3											
xfe^2		−		−	−	X^2Ye^2										
x^3e^3							X^3e^3									
x^2fe^2					−	−	−	X^3Ye^2								
y^2e^3									Y^2e^3							
yfe^2			−	−					−	XY^2e^2						
f^2e	−	−	−	−	−	−			−	−	X^2Y^2e					
xf^2e		−		−	−	−	−	−		−	−	X^3Y^2e				
y^3e^3													Y^3e^3			
y^2fe^2									−	−			−	XY^3e^2		
yf^2e			−	−		−			−	−	−		−	−	X^2Y^3e	
f^3	−	−	−	−	−	−	−	−	−	−	−	−	−	−	−	X^3Y^3

The technique of Ernst et al. [8] as well as our strategy explained in the proof of Theorem 1 fall under the generalized strategy presented in Jochemsz and May [10].

In [8], Ernst et al. present two methods for lattice based cryptanalysis of RSA with partial key exposure. In Method I, dimension of the proposed lattice is $\omega_1 = (\frac{m^2}{2} + \frac{5m}{2} + 3)t + \frac{m^3}{6} + \frac{3}{2}m^2 + \frac{13}{3}m + 4$ and the technique will be successful for

$$\gamma < \frac{\left(\frac{1}{12} - \frac{1}{6}\delta\right)m^3 + \frac{1}{4}m^2t - \frac{1}{4}mt^2 + \left(\frac{1}{2} - \delta\right)m^2 + \frac{1}{2}mt - \frac{1}{2}t^2 + \left(\frac{5}{12} - \frac{17}{6}\delta\right)m - \frac{1}{2}t - 2\delta - \frac{1}{2}}{\frac{1}{6}m^3 + \frac{1}{2}m^2t + m^2 + \frac{3}{2}mt + \frac{17}{6}m + t + 1} \qquad (2)$$

In case of Method II of [8], dimension of the corresponding lattice is $\omega_2 = (\frac{1}{2}m^2 + \frac{5}{2}m + 3)t + \frac{1}{3}m^3 + \frac{5}{2}m^2 + \frac{37}{6}m + 5$ and the required condition for success, with $\delta \leq \frac{11}{16}$, is

$$\gamma < \frac{\frac{1}{12}m^3 + \frac{1}{4}m^2t + \frac{1}{4}m^2 + \frac{3}{4}mt - \frac{1}{3}m + \frac{1}{2}t - 1}{\frac{1}{2}m^3 + m^2t + \frac{1}{2}mt^2 + \frac{5}{2}m^2 + \frac{7}{2}mt + t^2 + 6m + 4t + 3} \tag{3}$$

In case of our method, the corresponding lattice dimension of $\omega = mt + t + m^2 + 2m + 1$ produces equivalent results if $\lambda = \max\{\gamma, \delta - \frac{1}{2}\}$, and

$$\lambda < \frac{\frac{1}{12}m^3 - \frac{13}{12}m + \frac{1}{4}m^2t + \frac{1}{4}mt}{\frac{1}{2}m^3 + m^2 + \frac{1}{2}m + \frac{1}{2}t^2 + \frac{1}{2}t + m^2t + \frac{1}{2}mt^2 + \frac{3}{2}mt}. \tag{4}$$

At this point, let us present some numerical values of m, t, as in Table 1, that clearly show that theoretical bound presented in Theorem 1 is better than that of Ernst et al. [8] for similar lattice dimension. Larger values of γ in our case indicate that we need to know less amount of MSBs of the decryption exponent d. Moreover, the negative values of γ in case of Method I of [8] suggests that it is not theoretically possible to get desired results for the corresponding values of (m, t).

Table 1. Comparison of our theoretical results with that of [8] for some specific m, t

δ	Our			Method I of [8]			Method II of [8]		
	γ	(m, t)	LD	γ	(m, t)	LD	γ	(m, t)	LD
0.45	0.158	(10, 4)	165	0.082	(6, 2)	192	0.118	(5,1)	168
0.45	0.160	(11, 4)	192	0.099	(7, 3)	300	0.132	(5,2)	196
0.5	0.143	(7, 3)	88	-0.007	(4, 2)	98	0.107	(4,1)	112
0.55	0.162	(11, 5)	204	-0.012	(7, 1)	210	0.126	(6,1)	240

In view of the above data, our method proves to be considerably efficient in terms of the lattice dimension as well. One can observe that our method offers same or better values of γ compared to [8, Method I] or [8, Method II] with a considerably lower lattice dimension. The reason, as already mentioned in the Introduction, is that we use Coppersmith's [5] idea for solving the modular polynomial, while [8] used Coron's [6] version. Thus, the lattice dimension they obtained was a cubic in m whereas we obtain a quadratic in m (as t is linear in m).

Note that the maximum bit size of an entry corresponding to x shift is $X^{m+t}N^m$ and the maximum bit size of an entry corresponding to y shift is $Y^m e^m$ in our lattice. These bounds are of the same (or lower) size as those in case of the lattice constructed by Ernst et al. [8] in most of the cases. Hence, a smaller lattice dimension in our case will automatically imply better efficiency. It is worth noticing that comparatively smaller lattice dimension for same values of m, t allows us to tune these parameters to higher values and obtain better results at the same cost.

As it is generally studied in cryptanalytic materials, we also obtain the asymptotic bounds for our technique as follows.

Corollary 1. *Consider the RSA equation* $ed \equiv 1 \pmod{\phi(N)}$. *Let* $d = N^{\delta}$ *and* e *be* $\Theta(N)$. *Suppose we know an integer* d_0 *such that* $|d - d_0| < N^{\gamma}$. *Then one can factor* N *in poly*$(\log N)$ *time under Assumption 1 when* $\lambda < \frac{3}{16}$, *where* $\lambda = \max\{\gamma, \delta - \frac{1}{2}\}$.

Proof. Putting $t = \tau m$ and neglecting $o(m^3)$ terms in Equation (1), we get

$$\frac{1}{2}\tau^2\lambda + \left(\lambda - \frac{1}{4}\right)\tau + \left(\frac{1}{2}\lambda - \frac{1}{12}\right) < 0.$$

Substituting the optimal value of $\tau = \frac{1}{\lambda}\left(\frac{1}{4} - \lambda\right)$, we get the required condition as $\lambda < \frac{3}{16}$. □

The corresponding asymptotic bounds for the methods proposed by Ernst et al. [8] are

- Method I: $\gamma < \frac{5}{6} - \frac{1}{3}\sqrt{1 + 6\delta}$,
- Method II (1st result): $\gamma < \frac{3}{16}$ and $\delta \leq \frac{11}{16}$,
- Method II (2nd result): $\gamma < \frac{1}{3} + \frac{1}{3}\delta - \frac{1}{3}\sqrt{4\delta^2 + 2\delta - 2}$ and $\delta \geq \frac{11}{16}$.

On the other hand, cryptanalysis using our method is possible when $\lambda < \frac{3}{16}$, with $\lambda = \max\{\gamma, \delta - \frac{1}{2}\}$. As $\lambda < \frac{3}{16}$, we have $\gamma < \frac{3}{16}$ and $\delta - \frac{1}{2} < \frac{3}{16}$, that is, $\delta < \frac{11}{16}$.

Thus our result and Method II (1st result) are of same quality in terms of asymptotic bound when $\delta < \frac{11}{16} = 0.6875$. However, when $\delta < 0.4590$, then the bound on γ using Method I of [8] is $\geq \frac{3}{16}$, and our result is worse than that of [8] in this case. Hence, our asymptotic results are of the same quality as the work of Ernst et al. [8] for $N^{0.4590} \leq d < N^{0.6875}$.

But in experimental situations, our result is better than that of [8] for $d \leq N^{0.64}$. These experimental advantages are detailed in Section 4.

3 Further Improvement Using Sublattice

From the experimental results of Ernst et al. [8, Method II], one may note that for small values of δ (e.g., $\delta = 0.3$), the experimental results are better than the theoretical bounds. This happens in case of our experiments as well. Our method suggests the theoretical bound

$$\lambda < \frac{\frac{1}{12}m^3 - \frac{13}{12}m + \frac{1}{4}m^2t + \frac{1}{4}mt}{\frac{1}{2}m^3 + m^2 + \frac{1}{2}m + \frac{1}{2}t^2 + \frac{1}{2}t + m^2t + \frac{1}{2}mt^2 + \frac{3}{2}mt}.$$

When, $t = 0$, we have $\lambda < \frac{\frac{1}{12}m^3 - \frac{13}{12}m}{\frac{1}{2}m^3 + m^2 + \frac{1}{2}m} < \frac{1}{6} \approx 0.167$, for all m. But the experimental evidences for $t = 0$ in the range $\delta = 0.3$ and $\delta = 0.35$ are clearly better. This is because, for these parameters, the shortest vectors may belong to some sub-lattice. However, the theoretical calculation in [8] as well as in our Theorem 1 cannot capture that. Further, identifying such optimal sub-lattice seems to be difficult as pointed out by Jochemsz and May [11, Section 7.1]. In this section, we propose a strategy to obtain better experimental results using a special structure of the sublattice.

Our strategy: Recall our construction of the lattice L in Section 2. The rows of the matrix L_M corresponding to L came from the coefficients of $g_{i,j}(xX, yY)$ and $h_{i,j}(xX, yY)$, where

$$g_{i,j}(x, y) = x^i f_e^j(x, y)e^{m-j} \quad \text{with } j = 0, \ldots, m,\ i = 0, \ldots, m - j + t,$$
$$h_{i,j}(x, y) = y^i f_e^j(x, y)e^{m-j} \quad \text{with } j = 0, \ldots, m,\ i = 1, \ldots, m - j.$$

The strategy for constructing a sublattice is to keep the x-shift portion of L_M unchanged and judiciously delete a few rows from the y-shift portion of L_M to produce a new matrix $L_M^{'}$. We propose deleting the rows generated by $h_{i,j} = y^i f_e^j e^{m-j}$, where $j = 0, \ldots, m$ and $i = 2, \ldots, m - j$. In other words, the new matrix $L_M^{'}$ can be constructed from the shift polynomials $g_{i,j}(xX, yY)$ and $h_{i,j}(xX, yY)$, where

$$g_{i,j}(x, y) = x^i f_e^j(x, y)e^{m-j} \quad \text{with } j = 0, \ldots, m,\ i = 0, \ldots, m - j + t,$$
$$h_{i,j}(x, y) = y f_e^j(x, y)e^{m-j} \quad \text{with } j = 0, \ldots, m - 1.$$

Let $L^{'}$ be the lattice defined by $L_M^{'}$. As all the rows of $L_M^{'}$ come from L_M, $L^{'}$ is a sublattice of L and we propose $L^{'}$ to be our chosen sublattice.

One may easily calculate that the number of rows of the sublattice is $\omega_R^{'} = \frac{1}{2}(m + 1)(m + 2) + m + t(m + 1)$. Hence, we obtain a substantial reduction of $\frac{1}{2}m(m - 1)$ in terms of lattice dimension, which makes the LLL operation considerably faster. Experiments show that applying LLL to $L^{'}$ (with lower lattice dimension) yield results of same quality as those in case of L as shown in Table 4 in Section 4.

Let us illustrate the strategy for choosing a sublattice in case of $m = 3, t = 0$. Please refer back to Section 2 for our original lattice having 16 rows. Here, following our strategy, we delete rows 9, 11, 12 from top and obtain the following sublattice. The reduction in number of rows in this case is $\frac{1}{2}m(m - 1) = 3$, as expected. This reduction produces considerably better results in practice as higher values of m can be used. For example, number of rows reduces to 43 from 64 in case of $m = 7, t = 0$.

poly	1	x	y	xy	x^2	x^2y	x^3	x^3y	y^2	xy^2	x^2y^2	x^3y^2	y^3	xy^3	x^2y^3	x^3y^3
e^3	e^3															
xe^3		Xe^3														
ye^3			Ye^3													
fe^2	–	–	–	XYe^2												
x^2e^3					X^2e^3											
xfe^2		–		–	–	X^2Ye^2										
x^3e^3							X^3e^3									
x^2fe^2					–	–	–	X^3Ye^2								
yfe^2			–	–					–	XY^2e^2						
f^2e	–	–	–	–	–	–			–	–	X^2Y^2e					
xf^2e		–		–	–	–	–	–		–	–	X^3Y^2e				
yf^2e			–	–		–			–	–	–		–	–	X^2Y^3e	
f^3	–	–	–	–	–	–	–	–	–	–	–	–	–	–	–	X^3Y^3

It is also worth noting that this reduction in dimension allows us some extra x-shifts by increasing the value of t, which improve our results even further.

Note that our choice of sublattice is purely heuristic at this point and it will be interesting if one can furnish the theoretical justification for this strategy. We have noted that the idea of [9] cannot be immediately exploited to theoretically capture the sublattice structure.

The main motivation of exploring the idea of sublattice is the observation that experimental results perform better than theoretical bounds. This happens for low values of d. During experimentation, we indeed observed that improved results are obtained for $d = N^{0.3}, N^{0.35}$ using sublattices. However, for $d \geq N^{0.4}$, we could not achieve any improvement using the sublattice based technique over our lattice based technique.

4 Experimental Results

We have implemented the code in SAGE 4.1 on a Linux Ubuntu 8.10, Dual CORE Intel(R) Pentium(R) D CPU 1.83 GHz, 2 GB RAM, 2 MB Cache machine. Let us present two examples to explain our improvements.

Example 1. We consider 500 bits p, q, i.e., 1000 bits $N = pq$. The exponent e is of 1000 bits and d is of 300 bits. The details of p, q, e, d are available in Appendix A. The idea of [8, Method I] has been implemented on our platform and we get the following comparison which shows that our method is more efficient. By LD, we mean the Lattice Dimension.

Method	m, t, LD	MSBs of d to be known	Time (seconds)
Method I of [8]	2, 2, 40	95	30.22
Our (Lattice)	5, 0, 36	75	6.15
Method I of [8]	3, 1, 50	75	451.42
Our (Lattice)	6, 0, 49	66	26.72
Method I of [8]	4, 2, 98	66	9101.23
Our (Lattice)	7, 0, 64	63	104.57
Our (Sublattice)	7, 0, 43	63	49.66

For lattice dimension 98, using [8, Method I], successful result could not be achieved when 63 MSBs are available. □

Example 2. We take the same p, q as in Example 1, and consider 1000-bit e and 600-bit d. The details of p, q, e, d are given in Appendix A. We implemented the idea of [8, Method II] on our platform to get the following comparison, which shows the efficiency of our method.

Method	m, t, LD	MSBs of d to be known	Time (seconds)
Method II of [8]	2, 2, 50	491	82.47
Method II of [8]	3, 1, 70	477	618.86
Our method	5, 0, 36	467	12.34
Our method	6, 0, 49	459	67.02
Our method	6, 1, 56	451	197.68

In these cases, we have checked that our sublattice based technique does not provide any improvement over the general method. This is because there are probably no sublattice structures to improve the bound of γ. □

Table 2. Experimental results for Method I (left) and Method II (right) of [8] for 1000 bit N in our implementation. LLL time is presented in seconds.

δ	γ	γ (expt.), $m = 1$			γ (expt.), $m = 2$		
	asym.	$t = 0$	$t = 1$	$t = 2$	$t = 0$	$t = 1$	$t = 2$
0.30	0.28	0.194	0.195	0.199	0.209	0.209	0.210
0.35	0.25	0.136	0.148	0.153	0.142	0.159	0.158
0.40	0.22	0.097	0.117	0.114	0.096	0.140	0.139
0.45	0.19	0.048	0.100	0.098	0.047	0.117	0.117
0.50	0.17	0	0.083	0.083	0	0.098	0.111
0.55	0.14	0	0.081	0.083	0	0.086	0.108
0.60	0.12	0	0.045	0.048	0	0.061	0.105
0.638	0.10	0	0	0	0	0.013	0.069
0.65	0.10	0	0	0	0	0	0.055
0.70	0.07	0	0	0	0	0	0
0.75	0.05	0	0	0	0	0	0
0.80	0.03	0	0	0	0	0	0
0.85	0.01	0	0	0	0	0	0
Lattice dim.		10	16	22	20	30	40
LLL time		<1	1	3	5	11	29

δ	γ	γ (expt.), $m = 1$				γ (expt.), $m = 2$		
	asym.	$t = 0$	$t = 1$	$t = 2$	$t = 3$	$t = 0$	$t = 1$	$t = 2$
0.30	0.19	0.197	0.197	0.198	0.194	0.192	0.193	0.201
0.35	0.19	0.147	0.147	0.146	0.143	0.158	0.159	0.158
0.40	0.19	0.116	0.119	0.120	0.124	0.139	0.140	0.140
0.45	0.19	0.101	0.109	0.117	0.115	0.120	0.129	0.135
0.50	0.19	0.084	0.111	0.120	0.118	0.109	0.123	0.133
0.55	0.19	0.081	0.110	0.116	0.118	0.109	0.122	0.134
0.60	0.19	0.052	0.109	0.115	0.121	0.112	0.124	0.132
0.638	0.19	0	0.074	0.078	0.082	0.076	0.110	0.123
0.65	0.19	0	0.058	0.058	0.060	0.060	0.096	0.106
0.70	0.18	0	0	0	0	0	0.048	0.051
0.75	0.14	0	0	0	0	0	0	0
0.80	0.11	0	0	0	0	0	0	0
0.85	0.08	0	0	0	0	0	0	0
0.90	0.05	0	0	0	0	0	0	0
0.95	0.03	0	0	0	0	0	0	0
Lattice dim.		14	20	26	32	30	40	50
LLL time		< 1	2	4	37	9	50	415

We present the results for 1000 bit N here because RSA moduli of this order are used in practice. We also detail the comparison of results of our method with that of [8] for 256 bit N in Appendix B. We experiment with Methods I, II of [8] on our platform as results for 1000 bits N are not available in the paper [8]. These results are presented in Table 2. We present the experimental results of our lattice based technique for 1000 bit N in Table 3.

Our results (presented in Table 3) are better than that of [8] (presented in Table 2) for $\delta \leq 0.638$. In these cases the experiments we performed were always successful. Beyond that bound, not every attempt with the lattice dimensions mentioned in Table 2 was successful. However, we successfully reached the range $\delta = 0.64$ in some of our experiments. The results can be further improved with higher lattice dimensions.

In Table 4 we present the improvements using our sublattice based technique for small values of d. Improved results are obtained only for $\delta = 0.3, 0.35$, as we discussed before.

Table 3. Experimental result of our lattice based method for 1000 bit N

δ	γ	$m = 4, t = 0$	$m = 5, t = 0$	$m = 6, t = 0$
	asympt.	expt.	expt.	expt.
0.30	0.19	0.211	0.226	0.232
0.35	0.19	0.178	0.191	0.194
0.40	0.19	0.152	0.162	0.169
0.45	0.19	0.135	0.145	0.154
0.50	0.19	0.124	0.134	0.144
0.55	0.19	0.125	0.134	0.141
0.60	0.19	0.127	0.133	0.141
0.638	0.19	0	0	0.142
Lattice dimension		25	36	49
LLL time (in sec)		3	14	100

Table 4. Experimental result of our sublattice based method for 1000 bit N

δ	$m = 4, t = 0$	$m = 5, t = 0$	$m = 6, t = 0$	$m = 7, t = 0$	$m = 8, t = 0$
0.30	0.211	0.226	0.232	0.235	0.237
0.35	0.178	0.191	0.193	0.196	0.199
Sublattice dimension	19	26	34	43	53
LLL time (in sec)	< 1	3	14	50	140

5 Conclusion

In this paper we consider the partial key exposure attack on RSA and provide better results than what were obtained by Ernst et al. [8], for certain parameters. We present experimental evidences to show how our technique improves those of [8] in the following ways:

- we provide better efficiency at smaller lattice dimensions in practice,
- our method offers similar asymptotic results for certain range of δ,
- we propose a strategy for constructing sublattice to improve the efficiency even further.

We would like to clarify that the practical advantages we obtain over [8] are due to using Coppersmith's techniques (for modular polynomials) instead of Coron's idea (for integer polynomials), as predicted in [8].

Our work puts forward two natural open problems. The first is to improve the range of δ for our improvements over the work of Ernst et al. [8]. The second open problem would be to provide a theoretical model for constructing the sublattice or a formal justification of our heuristic sublattice strategy. This will further improve the bounds of γ within a certain range of δ, as expected from our experimental observations.

Acknowledgments. The authors are grateful to the anonymous reviewers and the shepherd for their invaluable comments and suggestions that helped in improving the technical as well as the editorial quality of the paper.

References

1. Aono, Y.: A New Lattice Construction for Partial Key Exposure Attack for RSA. In: Jarecki, S., Tsudik, G. (eds.) Public Key Cryptography – PKC 2009. LNCS, vol. 5443, pp. 34–53. Springer, Heidelberg (2009)
2. Blömer, J., May, A.: New Partial Key Exposure Attacks on RSA. In: Boneh, D. (ed.) CRYPTO 2003. LNCS, vol. 2729, pp. 27–43. Springer, Heidelberg (2003)
3. Boneh, D., Durfee, G., Frankel, Y.: Exposing an RSA Private Key Given a Small Fraction of its Bits. In: Ohta, K., Pei, D. (eds.) ASIACRYPT 1998. LNCS, vol. 1514, pp. 25–34. Springer, Heidelberg (1998)
4. Cohen, H.: A Course in Computational Algebraic Number Theory. Springer, Heidelberg (1996)
5. Coppersmith, D.: Small Solutions to Polynomial Equations and Low Exponent Vulnerabilities. Journal of Cryptology 10(4), 223–260 (1997)

6. Coron, J.-S.: Finding Small Roots of Bivariate Integer Equations Revisited. In: Cachin, C., Camenisch, J.L. (eds.) EUROCRYPT 2004. LNCS, vol. 3027, pp. 492–505. Springer, Heidelberg (2004)
7. Cox, D., Little, J., O'Shea, D.: Ideals, Varieties, and Algorithms: An Introduction to Computational Algebraic Geometry and Commutative Algebra, 3rd edn. Springer, New York (2007)
8. Ernst, M., Jochemsz, E., May, A., de Weger, B.: Partial Key Exposure Attacks on RSA up to Full Size Exponents. In: Cramer, R. (ed.) EUROCRYPT 2005. LNCS, vol. 3494, pp. 371–386. Springer, Heidelberg (2005)
9. Herrmann, M., May, A.: Maximizing Small Root Bounds by Linearization and Applications to Small Secret Exponent RSA. In: Nguyen, P.Q., Pointcheval, D. (eds.) Public Key Cryptography – PKC 2010. LNCS, vol. 6056, pp. 53–69. Springer, Heidelberg (2010)
10. Jochemsz, E., May, A.: A Strategy for Finding Roots of Multivariate Polynomials with new Applications in Attacking RSA Variants. In: Lai, X., Chen, K. (eds.) ASIACRYPT 2006. LNCS, vol. 4284, pp. 267–282. Springer, Heidelberg (2006)
11. Jochemsz, E., May, A.: A Polynomial Time Attack on RSA with Private CRT-Exponents Smaller Than $N^{0.073}$. In: Menezes, A. (ed.) CRYPTO 2007. LNCS, vol. 4622, pp. 395–411. Springer, Heidelberg (2007)
12. Kocher, P.: Timing attacks on implementations of Diffie-Hellman, RSA, DSS and other systems. In: Koblitz, N. (ed.) CRYPTO 1996. LNCS, vol. 1109, pp. 104–113. Springer, Heidelberg (1996)
13. Kocher, P., Jaffe, J., Jun, B.: Differential power analysis. In: Wiener, M. (ed.) CRYPTO 1999. LNCS, vol. 1666, pp. 388–397. Springer, Heidelberg (1999)
14. Lenstra, A.K., Lenstra, H.W., Lovász, L.: Factoring Polynomials with Rational Coefficients. Mathematische Annalen 261, 513–534 (1982)
15. Rivest, R.L., Shamir, A., Adleman, L.: A Method for Obtaining Digital Signatures and Public Key Cryptosystems. Communications of ACM 21(2), 158–164 (1978)
16. Sarkar, S., Maitra, S.: Improved Partial Key Exposure Attacks on RSA by Guessing a Few Bits of One of the Prime Factors. In: Lee, P.J., Cheon, J.H. (eds.) ICISC 2008. LNCS, vol. 5461, pp. 37–51. Springer, Heidelberg (2009)

Appendix A

Details of Example 1

We consider 500 bits p, q, i.e., 1000 bits $N = pq$. The primes p, q are

2912084397987488530101897161744729347962392317939825896973727031332611750392993689176153198751509218175826108298590364493248615130368371602851843098873, and

2657783139024632223171522538547082247641650337318532802844728917890624461019916484333889619460842058436768987401997469575864127623067945908748061533789.

The exponents e (1000 bit) and d (300 bit) are respectively

6336592072697636916143183091126456156249373448914681680770114707307345673704169135132242996825746665043587681813217139957921752851107827093765230755870784287989301300080332740805893387704531628007459742805317603659777158458744950023018031526899032667410842692995720687649380410410817145354934370336339, and

1814845758605508107926070456170619634431455166454166878249390985115060612463635344515223099.

Details of Example 2

We take the same p, q as in Example 1, and e (1000 bit), d (600 bit) are

7558665224136375374055690448131132089893236520631697389659469251681205924949978104059590671041756882393215403400574000254028912282454283048431737459758911375883732281559627012563046500414822556109522044727622686588774322310244883084236536267052528517810444320170108273107293503245751392242116148600457, and

310467162954524537052338316072300982788100901813501555561309089978891346487969392904811057700697506260195168811102476929658811788095093743188030863035429129416894638056811082358191 3.

Appendix B

Here we present the experimental results for 256 bit N in tabular form to compare our results with that of [8]. In Table 5, we reproduce the results of [8, Fig. 5, 6] when N is of 256 bits. We add one extra row of data containing the run time of the program to show how the implementation of the techniques of [8] works on our platform.

Table 5. Experimental results for the techniques of [8] for 256 bit N. In the table on the left, LLL time A is the data given for Method I in [8, Fig. 6] and LLL time B is the data from our implementation for Method II of [8]. In the table on the right, LLL time A is the data given for Method II in [8, Fig. 6] and LLL time B is the data from our implementation for Method II of [8]. All the LLL times are given in seconds.

δ	γ	γ (expt.), $m=1$			γ (expt.), $m=2$		
	asym.	$t=0$	$t=1$	$t=2$	$t=0$	$t=1$	$t=2$
0.30	0.28	0.19	0.19	0.19	0.19	0.21	0.21
0.35	0.25	0.13	0.14	0.14	0.14	0.16	0.16
0.40	0.22	0.09	0.11	0.11	0.09	0.14	0.15
0.45	0.19	0.04	0.10	0.10	0.05	0.12	0.12
0.50	0.17	0	0.08	0.09	0	0.10	0.11
0.55	0.14	0	0.08	0.08	0	0.09	0.11
0.60	0.12	0	0.04	0.04	0	0.06	0.10
0.65	0.10	0	0	0	0	0	0.06
0.70	0.07	0	0	0	0	0	0.01
0.75	0.05	0	0	0	0	0	0
0.80	0.03	0	0	0	0	0	0
0.85	0.01	0	0	0	0	0	0
Lattice dim.		10	16	22	20	30	40
LLL time A		1	2	8	3	25	100
LLL time B		<1	<1	<1	<1	2	4

δ	γ	γ (expt.), $m=1$				γ (expt.), $m=2$		
	asym.	$t=0$	$t=1$	$t=2$	$t=3$	$t=0$	$t=1$	$t=2$
0.30	0.19	0.19	0.20	0.20	0.20	0.19	0.19	0.19
0.35	0.19	0.15	0.16	0.16	0.16	0.16	0.16	0.16
0.40	0.19	0.12	0.12	0.12	0.12	0.14	0.15	0.15
0.45	0.19	0.10	0.11	0.12	0.12	0.12	0.13	0.13
0.50	0.19	0.08	0.11	0.12	0.12	0.12	0.13	0.13
0.55	0.19	0.08	0.11	0.11	0.11	0.11	0.12	0.13
0.60	0.19	0.05	0.11	0.11	0.11	0.11	0.12	0.13
0.65	0.19	0	0.05	0.06	0.06	0.05	0.08	0.10
0.70	0.18	0	0	0	0	0	0.04	0.05
0.75	0.14	0	0	0	0	0	0	0
0.80	0.11	0	0	0	0	0	0	0
0.85	0.08	0	0	0	0	0	0	0
0.90	0.05	0	0	0	0	0	0	0
0.95	0.03	0	0	0	0	0	0	0
LLL time A		1	7	17	32	30	40	50
LLL time B		< 1	< 1	< 1	5	< 1	6	43

Table 6. Experimental results for our method for 256 bit N

δ	γ asympt.	$m = 4, t = 0$ expt.	$m = 5, t = 0$ expt.	$m = 6, t = 0$ expt.
0.30	0.19	0.211	0.219	0.227
0.35	0.19	0.172	0.184	0.195
0.40	0.19	0.145	0.160	0.164
0.45	0.19	0.133	0.141	0.156
0.50	0.19	0.121	0.129	0.137
0.55	0.19	0.117	0.133	0.137
0.60	0.19	0.117	0.129	0.145
0.625	0.19	0	0.109	0.137
Lattice dimension		25	36	49
LLL time (in sec)		<1	1	5

Next we present our results when N is of 256 bits in Table 6. One may note that our results are better than that of [8] (presented in Table 5) for $\delta \leq 0.625$. The experimental data till $\delta \leq 0.625$ is presented based on that fact that we are always successful to factorize N in experiments following the idea of Theorem 1.

Towards Provable Security of the Unbalanced Oil and Vinegar Signature Scheme under Direct Attacks

Stanislav Bulygin[1], Albrecht Petzoldt[2], and Johannes Buchmann[1,2]

[1] Center for Advanced Security Research Darmstadt - CASED
Mornewegstraße 32, 64293 Darmstadt, Germany
{johannes.buchmann,Stanislav.Bulygin}@cased.de

[2] Technische Universität Darmstadt, Department of Computer Science
Hochschulstraße 10, 64289 Darmstadt, Germany
{apetzoldt,buchmann}@cdc.informatik.tu-darmstadt.de

Abstract. In this paper we show that solving systems coming from the public key of the Unbalanced Oil and Vinegar (UOV) signature scheme is on average at least as hard as solving a certain quadratic system with completely random quadratic part. In providing lower bounds on direct attack complexity we rely on the empirical fact that complexity of solving a non-linear polynomial system is determined by the homogeneous part of this system of the highest degree. Our reasoning explains, in particular, the results on solving the UOV systems presented by J.-C. Faugere and L. Perret at the SCC conference in 2008.

Keywords: Multivariate Cryptography, UOV Signature Scheme, provable security, security reduction, semi-regular sequence.

1 Introduction

Multivariate public key cryptography is one of the alternatives for the post-quantum era, i.e. when a large enough quantum computer is built and the public key cryptosystems used today (RSA, ECC, El Gamal) are broken. Other than resistance to quantum computer attacks, multivariate public key cryptosystems (MPKCs) enjoy other useful properties. In particular, they are quite fast compared to conventional schemes and require only very moderate resources. This makes MPKCs excellent candidates for use in resource constraint devices, like RFIDs and smart cards. Still there are two issues that pose obstacles on the way of using MPKCs. The first one is the issue of key sizes. The second problem is that MPKC proposals are being broken on a regular basis, which weakens believe in a possibility of constructing both secure and efficient MPKC.

Quite a few attempts have been undertaken in order to tackle the first problem. Mainly, the researchers concentrated on reducing the secret key size. In the recent paper [23] the authors undertook an attempt to reduce the public key size, based on yet unbroken (under proper parameter choice) UOV scheme [15].

G. Gong and K.C. Gupta (Eds.): INDOCRYPT 2010, LNCS 6498, pp. 17–32, 2010.

There has been no lack in proposals of MPKCs, see [7,10] for an overview. On the down side for the designers, the cryptanalytic progress has also been substantial. New proposals aim mainly at fixing problems exposed by the cryptanalysis, but then it often happens that "fixed" proposals get broken again (observe for example the sequence Matsumoto-Imai scheme [17] → its cryptanalysis [20] → HFE [21] → cryptanalysis of HFE challenge 1 [13] or a less known sequence MFE cryptosystem [26] → its SOLE cryptanalysis [8] → improved MFE [25] → cryptanalysis of the improved versions [5]; there are many more such "sequences"). There is a need in theoretical backing of design principles used in constructions of MPKCS. Note that for the classical cryptosystems one has some empirical certainty in the security of these systems. Namely, it is believed that breaking RSA in the classical computational model is as hard as factoring. For the ECC there is a believe that there exists no sub-exponential algorithm for solving the discrete logarithm problem in a group of points of an elliptic curve. Even better arguments are provided in the lattice-base and hash-based cryptography. For example, rigorous security reductions are provided for the cryptosystems based on random lattices as well as more compactly representable ideal lattices. Some attempts on providing "provability" or "reducibility" for MPKCs were undertaken by N. Courtois in his note [6]. There for providing security proofs he assumed strong properties of certain multivariate constructions themselves. It would be desirable instead to anchor security to some known problem(s) on which MPKCs are built. In this paper this will be a weaker version of the MQ-problem. Surprisingly enough, the methods used by the authors in [23] are also applicable to tackle the problem of "provable security" in the case of MPKCs as we will show.

The object of this paper is the Unbalanced Oil and Vinegar (UOV) scheme proposed in [15]. Note that for suitably chosen parameters (in particular, $v > o$, e.g. $v = 2o$, see Section 2) the progress in cryptanalysis of this scheme is connected mainly with the progress in solving generic quadratic systems over a finite field [4,14], which is one of the underlying hard problems the UOV is based on. Despite some progress, the above problem is still considered to be hard on average. Considering lack of structural attacks on the UOV for carefully chosen parameters, the system remains unbroken for more than ten years now. In this paper we show that breaking a UOV system directly is on average at least as hard as solving a quadratic system with a random quadratic part. We would like to be careful here on what we mean. What is meant is that using only direct (or general) attacks on the UOV, i.e. attacks based on Gröbner bases/XL-like, it is not possible to break the UOV if the parameters are large enough (and it is in principle possible to compute these). Still, our approach says nothing about structural attacks on the UOV. In particular, our approach says that the balanced Oil and Vinegar is secure against direct attacks, but it is a matter of the polynomial time algorithm to find an equivalent secret key [16]. So our claim is related to the direct solving attacks only. One, of course, should also be careful with "provably secure" in this context. If it is possible to solve a random quadratic system with certain parameters, this implies a jeopardy for a UOV scheme which anchors to such a system. In particular, in [14,4] it was shown that it is possible to forge signatures of the UOV with certain parameters. Still, moderate increase of parameters would render such an attack inefficient, due to high complexity of the anchoring problem.

The paper is organized as follows. In Section 2 we review the UOV scheme. Then in Section 3 we present the idea of [23] and show how it can be used for inserting a random (rather than partially cyclic) matrix in a UOV public key. In the following section we gather important points necessary for the further exposition. Our reduction arguments follow in Section 5 where we present our main result in Theorem 1. Section 6 provides some lower bounds on direct attacks using results of the previous section. We conclude in Section 7.

2 The UOV Signature Scheme

The idea of the Oil and Vinegar trapdoor was first proposed by J. Patarin in [22] and stems from his cryptanalysis of the Matsumoto-Imai scheme [20].

Let K be a finite field. Let o and v be two integers and set $n = o + v$. Patarin suggested to choose $o = v$. The original scheme was broken by Kipnis and Shamir in [16], and it was recommended in [15] to choose $v > o$ (Unbalanced Oil and Vinegar (UOV)). Next we describe the idea of the UOV scheme.

The UOV scheme is a single field construction, so we work solely in the polynomial ring $K[X]$, where $X = \{x_1, \ldots, x_n\}$. We divide the variable set X into two sets: vinegar variables $(x_i)_{i \in V}, V = \{1, \ldots, v\}$ and oil variables $(x_i)_{i \in O}, O = \{v+1, \ldots, n\}$. Here $|V| = v, |O| = o$ and $v + o = n$. We define o quadratic polynomials $q_k(X) = q_k(x_1, \ldots, x_n)$ by

$$q_k(X) = \sum_{i \in V,\ j \in O} \alpha_{ij}^{(k)} x_i x_j + \sum_{i,j \in V,\ i \leq j} \beta_{ij}^{(k)} x_i x_j + \sum_{i \in V \cup O} \gamma_i^{(k)} x_i + \eta^{(k)}, k = 1, \ldots, o \tag{1}$$

Note that oil and vinegar variables are not fully mixed, just like oil and vinegar in a salad dressing.

The map $\mathcal{Q} = (q_1(X), \ldots, q_o(X))$ can be easily inverted. First, we choose the values of the v vinegar variables $x_1, \ldots, x_v$ at random. Therewith we get a system of o linear equations in the o variables $x_{v+1}, \ldots, x_n$ which can be solved by Gaussian elimination. If the system does not have a solution, choose other values of $x_1, \ldots, x_v$ and try again.

The public key $\mathcal{P}$ of the UOV scheme consists of o quadratic polynomials in n variables.

$$\begin{aligned} P &= (p^{(1)}, \ldots, p^{(o)}) \\ &= \left(\sum_{i=1}^{n} \sum_{j=i}^{n} p_{ij}^{(1)} x_i x_j + \sum_{i=1}^{n} p_i^{(1)} x_i + p_0^{(1)}, \ldots, \sum_{i=1}^{n} \sum_{j=i}^{n} p_{ij}^{(o)} x_i x_j + \sum_{i=1}^{n} p_i^{(o)} x_i + p_0^{(o)} \right) \end{aligned} \tag{2}$$

After having chosen an ordering on monomials, we can write down the public coefficients into an $o \times \frac{(n+1)\cdot(n+2)}{2}$-matrix M_P.

$$M_P = \begin{pmatrix} p_{11}^{(1)} & p_{12}^{(1)} & \dots & p_{nn}^{(1)} & p_1^{(1)} & \dots & p_n^{(1)} & p_0^{(1)} \\ \vdots & & & & & & & \vdots \\ p_{11}^{(o)} & p_{12}^{(o)} & \dots & p_{nn}^{(o)} & p_1^{(o)} & \dots & p_n^{(o)} & p_0^{(o)} \end{pmatrix}.$$

In the case of UOV, the public key is given as

$$\mathcal{P} = \mathcal{Q} \circ \mathcal{T}, \tag{3}$$

with an affine invertible map $\mathcal{T}$ and the central map $\mathcal{Q}$ as defined in (1).

Remark 1. In contrast to other multivariate schemes the second affine map $\mathcal{S}$ is not needed for the security of UOV and therefore is left out. So we indeed use $\mathcal{P} = \mathcal{Q} \circ \mathcal{T}$ and not $\mathcal{P} = \mathcal{S} \circ \mathcal{Q} \circ \mathcal{T}$.

Other than the attack on the balanced version [16] , there exists a number of attacks on different parameter choices, see e.g. [4,2]. Essentially, the UOV scheme remains unbroken. For example the parameter choice: $v = 2o, o = 26$ over the field $GF(2^8)$ is considered to be secure [2,4,7].

3 Inserting a Random Matrix in the UOV Public Key

Let $q_{ij}^{(k)}$ be the coefficients of quadratic terms of the central map polynomials from (1). Due to equations (2) and (3), we get the following equations for the coefficients of the quadratic terms of the public key:

$$p_{ij}^{(r)} = \sum_{k=1}^{n} \sum_{l=k}^{n} \alpha_{kl}^{ij} \cdot q_{kl}^{(r)} = \sum_{k=1}^{v} \sum_{l=k}^{n} \alpha_{kl}^{ij} \cdot q_{kl}^{(r)} \ (1 \le i \le j \le n, \ r = 1, \dots, o) \tag{4}$$

with

$$\alpha_{kl}^{ij} = \begin{cases} t_{ki} \cdot t_{li} & (i = j) \\ t_{ki} \cdot t_{lj} + t_{kj} \cdot t_{li} & (i \ne j) \end{cases} \tag{5}$$

Note that the right hand side of equation (4) only contains coefficients of the quadratic terms of $\mathcal{Q}$ and coefficients of $\mathcal{T}$ and is linear in the former ones. The second "=" in equation (4) is due to the fact that all the q_{ij} $(i, j \in O)$ are zero.

Denote $D := \frac{v \cdot (v+1)}{2} + o \cdot v$. Let the monomials $x_i x_j, 1 \le i, j \le n$ be ordered w.r.t the given degree monomial ordering [1]. The given monomial ordering $<_{ord}$ also induces an ordering on the set of pairs $Pr = \{(i, j) | 1 \le i \le j \le n\}$, namely $(i', j') > (i'', j'')$ iff $x_{i'} x_{j'} >_{ord} x_{i''} x_{j''}$.

We define Q to be the $o \times D$ matrix containing the non-zero coefficients of the central polynomials with respect to the monomial ordering defined above.

[1] In fact we do not need a monomial ordering as is used in computer algebra; we just need some ordering of monomials. Still we prefer to work with monomial orderings in this paper.

Additionally, we define a $D \times D$ matrix A containing the coefficients of the equations (4):

$$A = \left(\alpha_{kl}^{ij} \right),$$

where indices (k, l) are taken according to the monomial ordering as above and $1 \leq i \leq v,\ i \leq j \leq n$. Thus equation (4) yields

$$M' = Q \cdot A, \tag{6}$$

where M' is a submatrix of M_P composed of the first D columns.

In order to obtain a UOV scheme, we assign random values from K to the coefficients of $\mathcal{T}$. Then the entries of the matrix A can be computed by equation (5). Equation (6) yields a linear relation between the coefficients of $\mathcal{P}$ and $\mathcal{Q}$, given $\mathcal{T}$. To use this relation properly, we need the matrix A to be invertible. Practically for large enough fields (e.g. $K = GF(2^8)$) this property is satisfied in an overwhelming number of cases. Assuming A is invertible, we can prove the following proposition:

Proposition 1. *Given an $o \times D$ matrix B and an affine invertible map $\mathcal{T}$ such that the corresponding matrix A is invertible, it is possible to construct a UOV scheme with the secret key $(\mathcal{Q}, \mathcal{T})$ and the public key $\mathcal{P}$ with $M_P = (B|C)$, where C is a $(o \times ((n+1)(n+2)/2 - D))$-matrix.*

Proof. Under the assumption of A being invertible, equation (6) yields a bijection between the entries of M' and the quadratic coefficients of $\mathcal{Q}$. Therefore, if we assign the entries of M' the values of the matrix B, we get a uniquely determined quadratic part of the central map $\mathcal{Q}$. Since the linear part and constant terms of $\mathcal{Q}$ do not have any influence on the quadratic part of the public key, they can be chosen arbitrarily.

4 Preparation

In this section we gather some points which are useful for understanding the core part that follows.

We have already mentioned and will be mentioning the notion of a quadratic system with a completely random quadratic part.

Definition 1. *Let $\{f_1, \ldots, f_m\}$ be a set of quadratic polynomials from the polynomial ring $K[x_1, \ldots, x_n]$, K a finite field, of the form $f_l = \sum_{i<j} a_{ij} x_i x_j + \sum_i b_i x_i + c_l, 1 \leq l \leq m$. We say that the quadratic system $f_1 = \cdots = f_m = 0$ has a completely random quadratic part, if a_{ij}'s are chosen from K uniformly and independently at random and b_i's and c_l's are arbitrary.*

So in this sense, we require coefficients of quadratic monomials to be random and we do not impose any restrictions on the affine part.

All the following reasoning will be based on one assumption that appears to be quite reasonable due to considerable empirical evidence gathered by the community of polynomial system solving.

Assumption. *Solving a random quadratic system with m equations and n variables is as hard as solving a quadratic system with m equations and n variables with a completely random quadratic part.*

This assumption deals mainly with Gröbner bases techniques and other general techniques for polynomial system solving. Since these techniques are general, it is quite plausible to assume that they are not able to catch any peculiarities that may be present in an affine part, and thus the complexity is determined mainly by the quadratic part. See more discussion on this issue after Corollary 1.

We will proceed as follows. First, by inserting a random matrix in the Macauley matrix of the public key as in Section 3, we will show that solving the system coming from the public key yields a solution to a certain system with a completely random quadratic part (Proposition 2). Still this is not enough, since additional information about that system is known, which renders the problem of finding a solution simple. Therefore, we modify a bit the key generation procedure to vaporize this additional information. Then in Theorem 1 we show that solving the system coming from the public key yields a solution to a certain system with a completely random quadratic part, where one has no additional information about the system in the sense of the above assumption. Using the assumption we conclude average hardness of direct solving the public system and then derive some (although quite conservative) estimates on complexity.

Let us make a remark about "security reduction" we are doing here. It is not a security reduction in a usual cryptographic sense. In particular, we do not have any probabilistic models, advantages, etc. What we mean is that if one is able to solve the public system, then one is able (given some values of the solution vector!) solve a system with a completely random quadratic part, which we assume to be hard by the assumption we made here.

5 Security Reduction

Lemma 1. *Let $\mathcal{T}$ be an invertible affine map which leads to an invertible transformation matrix A. Then, every UOV-scheme that has the affine map $\mathcal{T}$ as a part of the secret key can be obtained via the construction of Proposition 1.*

Proof. Let $(\mathcal{Q}, \mathcal{T})$ be a UOV scheme, such that $\mathcal{T}$ leads, via equation (5), to an invertible transformation matrix A. Let Q be the $o \times D$ matrix containing the non-zero quadratic coefficients of $\mathcal{Q}$ and L be the $o \times (n+1)$-matrix containing its linear coefficients and constant terms. Let B be defined as $B = Q \cdot A$. Since A was assumed to be invertible, this is an 1:1 relation between the matrices B and Q. We start with $(B, \mathcal{T})$ and follow the construction described in the Proposition. If the linear and constant part is chosen to be L, we will end up with the UOV scheme $(\mathcal{Q}, \mathcal{T})$.

In the classical UOV key generation we start with a random UOV central map $\mathcal{Q}$, random invertible affine $\mathcal{T}$ and then obtain a public key $\mathcal{P}$, which can be written with a matrix $M_P = (B|C)$. According to Lemma 1 we may equivalently start with B, the same $\mathcal{T}$ and end up with the same $\mathcal{Q}$ up to linear terms, which may be assigned arbitrarily. In this sense both constructions are equivalent. In order to provide a security reduction we will need the latter construction.

Let $(\mathcal{P}, \mathcal{Q}, \mathcal{T})$ be a UOV scheme obtained via Proposition 1. We impose the following monomial ordering. Let $Y, Z \subset X = \{1, \dots, n\}$ be two disjoint subsets of X (note that X is now a set on indexes not to be confused with the variables set in Section 2) such that $X = Y \cup Z$. The sets Y and Z have cardinalities v and o resp., so that $o + v = n$. The monomial ordering with $x_1 > \dots > x_n$ is then an ordering chosen in such a way that the following holds:

$$x_i x_j > x_k x_l > x_m x_p > x_u \ \forall i, j, k \in Y \ \forall l, m, p \in Z \ \forall u \in X. \tag{7}$$

One example of such an ordering is a weighted-degree ordering where each variable of Y has weight 3 and each variable of Z has weight 2. With this ordering quadratic monomials composed of Y-variables will be the largest, then follow "mixed" with variables from Y and Z, and finally those composed of Z-variables. We will need this ordering later in the proof of Proposition 2, in the follow-up procedure of the key generation, and in Theorem 1.

Now let $P(x)$ be polynomials of the public key $\mathcal{P}$. Let h be a hash value of the given document. The task of an attacker that wants to attack UOV directly is to find a solution of $P(x) = h$. Any solution to $P(x) = h$ provides a valid signature, therefore enables signature forgery.

Proposition 2. *Let $P(x)$ be polynomials of the public key $\mathcal{P}$ of a UOV scheme that is constructed by choosing completely random B and $\mathcal{T}$, following Proposition 1, and using an ordering satisfying (7). If it is possible to get a solution $x' = (x'_1, \dots, x'_n)$ of $P(x) = h$, then it is possible to get a solution of a quadratic system of o equations and v variables with a completely random quadratic part.*

Proof. The public key $\mathcal{P}$ is represented by a matrix $M_P = (B|C)$, where the columns are ordered according to the chosen ordering that satisfies (7). Due to this ordering the public key polynomials may be written as

$$p^{(k)} = \sum_{i,j \in Y} a_{ij}^{(k)} x_i x_j + \sum_{i \in Y, j \in Z} b_{ij}^{(k)} x_i x_j + \sum_{i,j \in Z} c_{ij}^{(k)} x_i x_j + \sum_{i \in X} d_i^{(k)} x_i + e^{(k)}, \tag{8}$$

for $k = 1, \dots, o$. Again according to the monomial ordering we have chosen, coefficients $a_{ij}^{(k)}$ and $b_{ij}^{(k)}$ are elements of the matrix B and therefore are chosen completely at random. Now a solution $x' = (x'_1, \dots, x'_n)$ may be seen as $x' = ((x'_i)_{i \in Y}, (x'_j)_{j \in Z})$. Plug in values $(x'_j)_{j \in Z}$ for variables $(x_j)_{j \in Z}$ in (8). Therewith one obtains a quadratic system with o equations and v variables $(x_i)_{i \in Y}$ of the form

$$\tilde{p}^{(k)} = \sum_{i,j \in Y} a_{ij}^{(k)} x_i x_j + \sum_{i \in Y} \tilde{d_{ij}}^{(k)} x_i + \tilde{e}^{(k)}, k = 1, \dots, o.$$

Note that coefficients $a_{ij}^{(k)}$ are completely random and are taken from the initial construction in (8). The system we need to solve is therefore

$$\tilde{P}((x_i)_{i \in Y}) = h. \tag{9}$$

So finding a solution x' of $P(x) = h$ provides a solution to (9), where $\tilde{P}$ has completely random quadratic part.

Proposition 2 seemingly provides a reduction for the problem of direct solving of $P(x) = h$ to the problem of solving a quadratic system with a completely random quadratic part. The way we presented the public key $\mathcal{P}$ this is not really true. Indeed, if the attacker knows the variable sets Y and Z he may simply fix the variables from Y ending up with a "non-random" system with variables from Z. In fact, if we suppose that the coefficients $c_{ij}^{(k)}$ are zero, then by fixing Y-variables the attacker ends up with a linear system as is the case for UOV maps. So the problem here is that having additional information about Y and Z the system (9) is not as hard as it is supposed to be.

Note that when the attacker uses Gröbner methods for solving, he/she would usually fix v variables first in order to end up with an $o \times o$ system. This is due to the fact that a random quadratic system with o equations and v variables over $GF(q)$ is expected to have q^{v-o} solutions. In order to be able to compute a solution it is preferable to "cut down" the solution space. By assigning values to some $v - o$ variables, the system still has o equations, but o variables, and is expected to have a unique solution, which is the found with Gröbner basis techniques. The idea of our reduction is to disguise the monomial ordering that was used and, in particular, the sets Y and Z. Below we show that after the process of fixing the attacker, at least on average, is intrinsically faced with solving a quadratic system which is at least as hard as a certain "random" one. For the reduction we need the following key generation procedure:

Key generation procedure

1. Choose a $o \times D$ matrix B completely (and uniformly) at random.
2. Choose an affine map $\mathcal{T}$ at random. If it is not invertible, choose again.
3. Choose $Y \subset X, |Y| = v$ at random. Set $Z := X \setminus Y$.
4. Use the construction of Proposition 1 with a monomial ordering satisfying (7), obtain the central map $\mathcal{Q}$ and the matrix C. The secret key is $(\mathcal{Q}, \mathcal{T})$.
5. Let $M = (B|C)$ be a matrix with the columns indexed by monomials with degree up to 2 ordered with the ordering chosen in the previous step. Let M' be the matrix M whose columns are permuted according to the graded lexicographic ordering. The public key is the set of polynomials with the matrix $M_P = M'$.

Now the attacker observing the matrix M_P does not know which monomial ordering was used and what are the sets Y and Z. Therefore he is not able to figure out where the random part $\sum_{i,j \in Y} a_{ij}^{(k)} x_i x_j$ is (see proof of Proposition 2). Note also that except of a minor modification of the key generation procedure, the scheme stays essentially the same. In particular, one does not need to store the set Y: after obtaining a key pair, signature generation and verification are independent on a monomial ordering used.

Remark 2. A legitimate question here is whether an attacker observing the matrix M_P is able to figure out monomials $x_i x_j, i, j \in Z$ and therefore the sets Y and Z. In principle, he must be able to do so, since the coefficients of monomials $x_i x_j, i, j \in Z$ are not completely random, but obtained via the "reverse" computation, after computing $\mathcal{Q}$ from B and $\mathcal{T}$. It may be shown that the coefficients of the matrix C in the construction satisfy certain quadratic relations. So, in principle, by choosing a subset $S \subset X$ of cardinality o and checking if coefficients $x_i x_j, i, j \in S$ satisfy these quadratic relations, it may be possible to distinguish Z from other subsets of X. In order to do so one has to go through all o-subsets of n and this has complexity dominated by $\mathcal{O}(\binom{n}{o})$. It can be shown to be worse than one can do with system solving, as we proceed below.

Moreover, in the monomial ordering as in the above construction, we may also choose an arbitrary order of variables, unknown to the attacker. Then it is not really clear for the attacker, how to apply the quadratic relations that exist for the matrix C.

All in all, it seems that it is computationally impossible for the attacker to figure out the partition of X into Y and Z. It is a future research point to confirm this statement more rigorously.

The next theorem shows that the attacker applying the fixing+solving technique has to face some random system, at least on average.

Theorem 1. *Let* $v = \alpha o, \alpha \geq 1$. *Let*

$$P(x) = h \tag{10}$$

be an $o \times (o + v)$ *system of public equations for a UOV scheme obtained with the procedure above. Suppose that the system (10) is solved by first fixing* v *variables (variables are chosen at random, as well as the values fixed) and then solving the* $o \times o$ *system, which is obtained after plugging in the fixed values in (10). Then solving (10) is on average at least as hard as solving an* $o \times \frac{\alpha}{\alpha+1} o$ *quadratic system with a completely random quadratic part.*

Proof. Let Y and Z be disjoint variable sets as in the construction. So the public key is given by equations (8), where Y and Z are unknown to the attacker. The attacker fixes v variables to concrete values. Since $v = \alpha o$, we expect on average $\frac{\alpha}{\alpha+1} v$ variables to be fixed in Y and $\frac{1}{\alpha+1} v$ in Z. So there remains a set $Y_F \subset Y$ of variables in Y that are not fixed, $|Y_F| = v - \frac{\alpha}{\alpha+1} v = \frac{1}{\alpha+1} v = \frac{\alpha}{\alpha+1} o$. Denote the non-fixed variables in Z by Z_F. After plugging in the fixed values the attacker obtains a system

$$\tilde{P}(x_i | i \in Y_F \cup Z_F) = h. \tag{11}$$

Again note that the sets Y_F and Z_F are not known to the attacker. Let $x' = (x'_i | i \in Y_F \cup Z_F)$ be a solution of (11). Suppose the attacker is given the values of $x'_i, i \in Z_F$. After plugging in these values in (11) he/she obtains a system

$$\tilde{\tilde{P}}(x_i | i \in Y_F) = h \tag{12}$$

with a completely random quadratic part (all quadratic terms in (8) are "killed" except the ones with $i, j \in Y_F$). The system (12) has o equations and $|Y_F| = \frac{\alpha}{\alpha+1} o$ variables. Since the values $x'_i, i \in Z_F$ were given to the attacker, the actual solving of (11) is at least as hard as solving (12).

The corollary below specifies the above theorem to the choice of α that is used in the UOV to avoid the structural attack of [15].

Corollary 1. *If $v = 2o$, then the quadratic system with a completely random quadratic part from Theorem 1 has o equations and $\frac{2}{3}o$ variables. In other words, the ratio $\#eqs/\#vars = 3/2$ in this case.*

Let us discuss the above results. It may seem surprising that we have such a reduction to a hard problem, considering that the legitimate signer is able to get the solution x'. The catch here is that the system (12) is not completely random, only its quadratic part is. So it is not surprising that the signer, knowing the decomposition $\mathcal{P} = \mathcal{Q} \circ \mathcal{T}$ is able to get a solution which also yields a solution to (12). Similarly, the attacker, who via some structural attack is able to get the decomposition $\mathcal{P} = \mathcal{Q} \circ \mathcal{T}$ or an equivalent one, is also able to solve (12). If we consider only direct attacks, though, the situation is different. There is a quite strong experimental evidence that the complexity of solving a non-linear system is determined by its homogeneous part of the highest degree, e.g. [14,4]. In particular, in [3] an affine sequence of polynomials is defined to be semi-regular (practically speaking random) if its homogeneous part of the highest degree is semi-regular. Complexity estimates in [3] rely on the domination of the homogeneous highest degree part. If we take this domination assumption, we may state that from the point of view of Gröbner basis algorithms (in particular F5 [12]) complexity of solving (12) and a completely random quadratic system of o equations in $\frac{\alpha}{\alpha+1} o$ variables is roughly the same. Figuratively speaking, Gröbner basis algorithms are not able to see peculiarities that are hidden in the linear part of (12). The above reasoning is compliant with the assumption we made in Section 4.

In [14] J.-C. Faugère and L. Perret discuss security of the UOV scheme under direct attacks. In particular, they apply their implementation of the F5 algorithm to solve UOV systems with $o = 16$ and $v = 16, 32$ over $GF(2^4)$. Based on their experimental data, they conclude in particular: "These experiments suggest that the systems obtained when mounting a specify[2]+solve signature forgery attack against UOV behave like semi-regular systems". It is now clear why they came to such a conclusion: they intrinsically face solving a "random" system (12). In fact as we have seen in the proof of Theorem 1 a system the attacker faces is a $o \times o$ system (11), which is (much) harder than (12). We made a reduction to (12) to use a "random" quadratic system as a "provably secure" anchor. Also the following known observation is noticeable. We see from Theorem 1 that the

[2] We called it "fix".

larger α is, the more variables we expect in an "underlying" random system (and the number of equations stays the same). Therefore, as α (and thus v) increases, we expect this system to be "more random" and thus harder to solve. This is confirmed in [2], where the authors say "From experiments, we could conclude that the time complexity increases exponentially with increasing v. This fact can be understood intuitively by the observation that for increasing v, the scheme becomes more random, which makes it more difficult to solve.". Theorem 1 provides a theoretical explanation of this intuition. Note that although we are mentioning some previously known work here, clearly the key generation construction from Section 4 was not used there. Still, due to Lemma 1 the usual key generation and the modified one yield essentially the same result. Therefore, the reduction results naturally explain the older experimental results.

As a result, we have a theoretical argument for security of the UOV schemes under direct attacks. Note that such security is not that common for MPKCs. Many proposals may actually be broken already by direct attacks: Matsumoto-Imai, some instances of HFE [13], MQQ scheme [18], and many others. Some others although not broken by direct methods, show their distinction from random systems. In particular, solving succeeds at degrees lower than one would expect from a random system.

6 Expected Lower Bounds on Direct Attacks Complexity

In this section we present lower bounds of attacker's complexity, when using direct solving methods. These lower bounds are based on average hardness the attacker has to face as is described in Theorem 1 and Corollary 1. In order to provide concrete formulas for lower bounds we use complexity estimates for the F5 algorithm that exist for semi-regular sequences [3]. These complexity estimates are also in accordance with the assumption we made in the previous section. Similar complexity estimates exist for the XL algorithm, see [27].

In order to formalize the notion of a random system, the notion of a (semi-)regular system was introduced in [3]. The definition of a semi-regular system is as follows.

Definition 2 ([3]). *Let $f_1, \ldots, f_m \in K[X]$ be a sequence of homogeneous polynomials. This sequence of polynomials is* semi-regular *if*

- $\langle f_1, \ldots, f_m \rangle \neq K[X]$,
- *for all* $1 \leq i \leq m$ *and* $g \in K[X] : \deg(g \cdot f_i) < d_{reg}$ *and* $g \cdot f_i \in \langle f_1, \ldots, f_{i-1} \rangle \Rightarrow g \in \langle f_1, \ldots, f_{i-1} \rangle$.

Here d_{reg} is the degree of regularity *defined in [3] and it determines the degree at which a Gröbner basis algorithm like F5 terminates.*

The definition above is for homogeneous systems only. A sequence $f_1, \ldots, f_m$ of affine polynomials is called *semi-regular* if the sequence $f_1^h, \ldots, f_m^h$ is semi-regular, where f_i^h is the homogeneous part of f_i of the highest degree.

Further, there is a result saying what is the asymptotic complexity of solving a semi-regular system.

Proposition 3 ([3]). *Let $f_1, \ldots, f_m$ be an affine semi-regular sequence. Then the total number of arithmetic operations in K performed by the F5 algorithm is bounded by*

$$\mathcal{O}\Big(m \cdot d_{reg}^h \binom{n + d_{reg}^h - 1}{d_{reg}^h}\Big)^{\omega}.$$

Here d_{reg}^h is the degree of regularity of the corresponding homogeneous semi-regular sequence, and ω is the exponent of linear algebra elimination procedure, $2 < \omega \leq 3$.

For our results we also need explicit formulas for the degree of regularity that plays a role in the proposition above. Namely, the following result from [3] is of interest.

Theorem 2 ([3]). *With the notation as above, let $m = kn$, k is a constant $k > 1$. Then the degree of regularity of a homogeneous quadratic semi-regular sequence in m polynomials and n variables behaves asymptotically like:*

$$d_{reg} = (k - \frac{1}{2} - \sqrt{k(k-1)})n + \mathcal{O}(n^{1/3}), n \to \infty.$$

Note that the above results are asymptotic. Therefore, one has to be careful when applying these to concrete instances. Now having all this machinery we may state the main result of this section.

Theorem 3. *We use the same notation as in previous sections. Let $v = \alpha o, \alpha \geq 1$ and o is large enough. Let (10) be an $o \times (o+v)$ system of public equations for a UOV scheme obtained with the procedure of Section 3. Suppose that the system (10) is solved by first fixing v variables and then solving the $o \times o$ system with the F5 algorithm, which is obtained after plugging in the fixed values in (10). The complexity of this approach is lower bounded by*

$$\mathcal{O}\Big(o^2 \cdot DR(\alpha) \cdot \binom{\frac{\alpha}{\alpha+1}o + DR(\alpha)o - 1}{DR(\alpha)o}^{\omega}\Big), \tag{13}$$

where

$$DR(\alpha) = \Big(1 - \frac{\alpha}{2(\alpha+1)} - \frac{1}{\sqrt{\alpha+1}}\Big).$$

Proof. Due to Theorem 1 complexity of solving (10) is bounded from below by complexity of solving a $o \times \frac{\alpha}{\alpha+1}o$ affine quadratic semi-regular system. Now the result is obtained by setting $n = \frac{\alpha}{\alpha+1}o, m = o, k = \frac{\alpha+1}{\alpha}$ in Theorem 2.

The above lower bound is dominated by the binomial coefficient. The following result gives a simplified lower bound on the logarithm of complexity necessary for the direct attack

Proposition 4. *Using notation as above, if we denote by Compl the lower bound on complexity as in Theorem 3, then for large enough o we have*

$$\log Compl \geq \omega \cdot \frac{3\alpha+2-2\sqrt{\alpha+1}}{2(\alpha+1)} \cdot H\Big(\frac{\alpha+2-2\sqrt{\alpha+1}}{3\alpha+2-2\sqrt{\alpha+1}}\Big) \cdot o,$$

where log *is the binary logarithm and* $H(x) = -x\log x - (1-x)\log(1-x)$ *is the binary entropy function.*

Proof. Let us rewrite the binomial coefficient in (13) in terms of o and α:

$$\binom{\frac{\alpha}{\alpha+1}o + DR(\alpha)o - 1}{DR(\alpha)o}^{\omega} = \binom{\frac{3\alpha+2-2\sqrt{\alpha+1}}{2(\alpha+1)}o}{\frac{\alpha+2-2\sqrt{\alpha+1}}{2(\alpha+1)}o}^{\omega}.$$

Using Stirling's approximation for large n and $0 < \lambda < 1$: $\log\binom{n}{\lambda n} \approx nH(\lambda)$ and the fact that (13) is dominated by the binomial coefficient, we have the result by a direct computation.

The following corollary shows the above results specified for the case $\alpha = 2$.

Corollary 2. *Using the notation above and assuming $\alpha = 2$ we have that complexity of direct F5-approach is lower bounded by*

$$\mathcal{O}\Big(o^2 \cdot \frac{2-\sqrt{3}}{3} \cdot \binom{\frac{4-\sqrt{3}}{3}o}{\frac{2-\sqrt{3}}{3}o}^{\omega}\Big),$$

The lower bound on the logarithm of complexity is

$$\log Comp \geq \omega \cdot \frac{4-\sqrt{3}}{3} \cdot H\Big(\frac{2-\sqrt{3}}{4-\sqrt{3}}\Big) \cdot o \approx 0.4\,\omega \cdot o. \tag{14}$$

In practice the lower bound (14) is pretty bad: by setting $\omega = 3$ one needs to have o around 70 to guarantee the security level of 80 bits. So for practical security tighter bounds are needed.

Remark 3. The main reason why our lower bound is so bad is the use of the oracle in the proof of Theorem 1. This oracle gives an attacker on average the values of $\frac{o}{\alpha+1}$ variables and therefore makes the system much easier to solve.

Note that we assumed that the attacker in the direct attack proceeds by first fixing v variables to concrete values and then solving an $o \times o$ system. Here we implicitly assumed that the attacker solves this $o \times o$ system "directly". There are other possibilities. For example, it has been shown to be a good practice (especially when the underlying coefficient field is not too large) first to guess at a couple of variables and then proceed with solving, e.g. with a Gröbner basis algorithm. Recent results, also in context of the UOV, on a "hybrid" approach [4] indicate that one may actually improve a bit on the complexity estimates above[3].

[3] In the sense that the attack is more efficient

Still, we believe that the complexity estimates above grasp the essence of the problem, namely that on average one deals with an exponential-time algorithm. Therefore, we do not use the improved strategies here to derive more accurate lower bounds. Potentially, one may even try to proceed without the initial fixing of v variables. This may be possible if one uses e.g. a SAT-solver approach, see e.g. [1]. By this approach we do not need to cut down our variety to make things work; a SAT-algorithm is able to find one solution of a system directly. SAT-solvers may be quite efficient for sparse systems over $GF(2)$. Note that here we are dealing with larger fields, rather than $GF(2)$, and there methods of SAT-solving are not so well understood. Moreover, complexity of such algorithms is hard to estimate due to rich heuristics employed there. Therefore, we do not attempt to include analysis based on SAT-solver in this paper.

7 Conclusion and Future Work

In this paper we presented a theoretical reasoning on why breaking UOV systems directly is on average at least as hard as solving quadratic systems with a random quadratic part. This reasoning is based on the assumption that the complexity of solving an affine system is determined by its homogeneous part of the highest degree, which we believe to be a very plausible assumption. It would be interesting to test this assumption further, e.g. by using the mutant concept, [19].

As an immediate future work we see investigating the question whether similar results may be obtained for other trapdoors, e.g. Rainbow [9] and enSTS [24]. A far more reaching question for the UOV systems would be to see under which assumptions (if any) finding a decomposition of the form $\mathcal{P} = \mathcal{Q} \circ \mathcal{T}$ can be reduced to some problem that is believed to be hard. Existence of an efficient decomposition finding for the balanced variant makes funding such a reduction a very challenging task. Moreover, it is of interest to develop a more formal approach which would enable security reduction in a classical cryptographic sense.

Acknowledgements

The first two authors would like to thank Enrico Thomae and Christopher Wolf for fruitful discussions and helpful comments. The authors are grateful to anonymous referees for valuable comments that helped to improve exposition in the paper.

References

1. Bard, A.: Algebraic Cryptanalysis. Springer, Heidelberg (2009)
2. Braeken, A., Wolf, C., Preneel, B.: A Study of the Security of Unbalanced Oil and Vinegar Signature Schemes. Topics in Cryptology CT-RSA (2005)
3. Bardet, M., Faugère, J.-C., Salvy, B., Yang, B.-Y.: Asymptotic Behaviour of the Degree of Regularity of Semi-Regular Polynomial Systems. In: Proceedings of MEGA 2005, Eighth International Symposium on Effective Methods in Algebraic Geometry (2005)

4. Bettale, L., Faugère, J.-C., Perret, L.: Hybrid approach for solving multivariate systems over finite fields. Journal of Math. Cryptology, 177–197 (2009)
5. Cao, W., Niw, X., Hu, L., Tang, X., Ding, J.: Cryptanalysis of Two Quartic Encryption Schemes and One Improved MFE Scheme. In: Sendrier, N. (ed.) PQCrypto 2010. LNCS, vol. 6061, pp. 41–60. Springer, Heidelberg (2010)
6. Courtois, N.: Short Signatures, Provable Security, Generic Attacks and Computational Security of Multivariate Polynomial Schemes such as HFE, Quartz and Sflash, available at eprint 2004/143 (2004)
7. Ding, J., Gower, J.E., Schmidt, D.: Multivariate Public Key Cryptosystems. Springer, Heidelberg (2006)
8. Ding, J., Hu, L., Nie, X., Li, J., Wagner, J.: High Order Linearization Equation (HOLE) Attack on Multivariate Public Key Cryptosystems. In: Okamoto, T., Wang, X. (eds.) PKC 2007. LNCS, vol. 4450, pp. 233–248. Springer, Heidelberg (2007)
9. Ding, J., Schmidt, D.: Rainbow, a new multivariate polynomial signature scheme. In: Ioannidis, J., Keromytis, A.D., Yung, M. (eds.) ACNS 2005. LNCS, vol. 3531, pp. 164–175. Springer, Heidelberg (2005)
10. Ding, J., Yang, B.-Y.: Multivariate Public Key Cryptography. In: Bernstein, D.J., Buchmann, J., Dahmen, E. (eds.) Post-Quantum Cryptography, Springer, Heidelberg (2009)
11. Ding, J., Yang, B.-Y., Chen, C.-H.O., Chen, M.-S., Cheng, C.M.: New Differential-Algebraic Attacks and Reparametrization of Rainbow. In: Bellovin, S.M., Gennaro, R., Keromytis, A.D., Yung, M. (eds.) ACNS 2008. LNCS, vol. 5037, pp. 242–257. Springer, Heidelberg (2008)
12. Faugère, J.-C.: A new efficient algorithm for computing Gröbner bases without reduction to zero (F5). In: Mora, T. (ed.) Proceedings of the 2002 International Symposium on Symbolic and Algebraic Computation ISSAC, pp. 75–83. ACM Press, New York (July 2002)
13. Faugère, J.-C., Joux, A.: Algebraic Cryptanalysis of Hidden Field Equations (HFE) using Gröbner Bases. In: Boneh, D. (ed.) CRYPTO 2003. LNCS, vol. 2729, pp. 44–60. Springer, Heidelberg (2003)
14. Faugère, J.-C., Perret, L.: On the Security of UOV. In: Proceedings of SCC 2008, pp. 103–109 (2008)
15. Kipnis, A., Patarin, L., Goubin, L.: Unbalanced Oil and Vinegar Schemes. In: Stern, J. (ed.) EUROCRYPT 1999. LNCS, vol. 1592, pp. 206–222. Springer, Heidelberg (1999)
16. Kipnis, A., Shamir, A.: Cryptanalysis of the Oil and Vinegar Signature scheme. In: Krawczyk, H. (ed.) CRYPTO 1998. LNCS, vol. 1462, pp. 257–266. Springer, Heidelberg (1998)
17. Matsumoto, T., Imai, H.: Public quadratic polynomial-tuples for efficient signature verification and message-encryption. In: Günther, C.G. (ed.) EUROCRYPT 1988. LNCS, vol. 330, pp. 419–453. Springer, Heidelberg (1988)
18. Mohamed, S.E.M., Ding, J., Buchmann, J., Werner, F.: Algebraic Attack on the MQQ Public Key Cryptosystem. In: Garay, J.A., Miyaji, A., Otsuka, A. (eds.) CANS 2009. LNCS, vol. 5888, pp. 392–401. Springer, Heidelberg (2009)
19. Mohamed, S.E.M., Cabarcas, D., Ding, J., Buchmann, J., Bulygin, S.: MXL3: An efficient algorithm for computing Gröbner bases of zero-dimensional ideals. In: Lee, D., Hong, S. (eds.) ICISC 2009. LNCS, vol. 5984, pp. 87–100. Springer, Heidelberg (2010)

20. Patarin, J.: Cryptanalysis of the Matsumoto and Imai public key scheme of Eurocrypt 88. In: Coppersmith, D. (ed.) CRYPTO 1995. LNCS, vol. 963, pp. 248–261. Springer, Heidelberg (1995)
21. Patarin, J.: Hidden Field Equations (HFE) and Isomorphism of Polynomials (IP): two new families of asymmetric algorithms. In: Maurer, U.M. (ed.) EUROCRYPT 1996. LNCS, vol. 1070, pp. 33–48. Springer, Heidelberg (1996)
22. Patarin, J.: The oil and vinegar signature scheme, presented at the Dagstuhl Workshop on Cryptography (September 1997)
23. Petzoldt, A., Bulygin, S., Buchmann, J.: A Multivariate Signature Scheme with a Partially Cyclic Public Key. In: Proceedings of SCC 2010, pp. 229–235 (2010)
24. Tsujii, S., Gotaishi, M., Tadaki, K., Fujita, R.: Proposal of a Signature Scheme based on STS Trapdoor. In: Sendrier, N. (ed.) PQCrypto 2010. LNCS, vol. 6061, pp. 201–217. Springer, Heidelberg (2010)
25. Wang, X., Feng, F., Wang, X., Wang, Q.: A More Secure MFE Multivariate Public Key Encryption Scheme. International Journal of Computer Science and Applications 6(3), 1–9 (2009)
26. Wang, L., Yang, B., Hu, Y., Lai, F.: A Medium-Field Multivariate Public Key Encryption Scheme. In: Pointcheval, D. (ed.) CT-RSA 2006. LNCS, vol. 3860, pp. 132–149. Springer, Heidelberg (2006)
27. Yang, B.-Y., Chen, J.-M.: All in the XL family: Theory and practice. In: Park, C., Chee, S. (eds.) ICISC 2004. LNCS, vol. 3506, pp. 67–86. Springer, Heidelberg (2005)

CyclicRainbow – A Multivariate Signature Scheme with a Partially Cyclic Public Key

Albrecht Petzoldt[1], Stanislav Bulygin[2], and Johannes Buchmann[1,2]

[1] Technische Universität Darmstadt, Department of Computer Science
Hochschulstraße 10, 64289 Darmstadt, Germany
{apetzoldt,buchmann}@cdc.informatik.tu-darmstadt.de

[2] Center for Advanced Security Research Darmstadt - CASED
Mornewegstraße 32, 64293 Darmstadt, Germany
{johannes.buchmann,Stanislav.Bulygin}@cased.de

Abstrct. Multivariate Cryptography is one of the alternatives to guarantee the security of communication in the post-quantum world. One major drawback of such schemes is the huge size of their keys. In [PB10] Petzoldt et al. proposed a way how to reduce the public key size of the UOV scheme by a large factor. In this paper we extend this idea to the Rainbow signature scheme of Ding and Schmidt [DS05]. By our construction it is possible to reduce the size of the public key by up to 62 %.

Keywords: Multivariate Cryptography, Rainbow Signature Scheme, Key Size Reduction.

1 Introduction

Besides lattice-, code- and hash-based cryptosystems, multivariate cryptography is one of the main alternatives to guarantee the security of communication in the post-quantum world [BB08]. Multivariate schemes are fast and efficient and seem especially suitable for signatures on low cost devices like RFIDs or smart cards.

Since the invention of multivariate cryptography in the 1980's, a huge variety of schemes both for encryption and signatures have been proposed. On the one hand, we have the so called BigField-Schemes like Matsumoto-Imai [MI88] and HFE [Pa96]. On the other hand, we have the SingleField-Schemes like UOV [KP99] and Rainbow [DS05]. In between, there are the so called MiddleField schemes like ℓ-IC [DW07] and MFE [WY06]. For all of these schemes there exist many variations and improvements, like the minus variation [PG98] [PC01], Internal Perturbation [Di04] and Projection [DY07]. One common drawback of all multivariate schemes is the large size of their public and private keys. Therefore, the question of key size reduction for multivariate schemes is an important area of research.

In the last years, a lot of work was done to look at possibilities to reduce the key sizes. Most researchers hereby concentrated on the reduction of the private key. We mention here the proposals of Yang and Chen for creating schemes with

G. Gong and K.C. Gupta (Eds.): INDOCRYPT 2010, LNCS 6498, pp. 33–48, 2010.

sparse central maps [YC05] and approaches with so called equivalent keys of Hu et al. [HW05]. In [PB10] Petzoldt et al. presented an idea how to reduce the public key size of the UOV signature scheme by a large factor. The principle idea is, to compute the coefficients of the central map in such a way, that the corresponding public key gets a compact structure.

In this paper we show how to extend this idea to the Rainbow signature scheme, which was proposed by J. Ding and D. Schmidt in 2005 [DS05]. The result is a Rainbow scheme, whose public key belongs to a certain subset of the set of all valid Rainbow public keys. By doing so it is possible to reduce the size of the public key by up to 62 %. Furthermore, we can reduce the number of field multiplications needed during the verification process by 30 %.

The structure of this paper is as follows:

In Section 2 we describe the Rainbow signature scheme of Ding and Schmidt. Section 3 gives an overview on the approach of [PB10] to create a UOV scheme with partially cyclic public key. Section 4 deals with notations and definitions we need for our construction in Section 5. Section 6 looks at security aspects of the new scheme and Section 7 gives concrete parameter sets and compares it with other multivariate schemes of the UOV family. Finally, Section 8 concludes the paper.

2 Multivariate Public Key Cryptography

Multivariate Public Key Cryptography is one of the main approaches for secure communication in the post-quantum world. The principle idea is to choose a multivariate system $\mathcal{F}$ of quadratic polynomials which can be easily inverted (central map). After that one chooses two affine linear invertible maps $\mathcal{S}$ and $\mathcal{T}$ to hide the structure of the central map. The public key of the cryptosystem is the composed map $\mathcal{P} = \mathcal{S} \circ \mathcal{F} \circ \mathcal{T}$ which is difficult to invert. The private key consists of $\mathcal{S}$, $\mathcal{F}$ and $\mathcal{T}$ and therefore allows to invert $\mathcal{P}$.

2.1 The Principle of Oil and Vinegar (OV)

One way to create easily invertible multivariate quadratic systems is the principle of Oil and Vinegar, which was first proposed by J. Patarin in [Pa97].

Let K be a finite field. Let o and v be two integers and set $n = o + v$. Patarin suggested to choose $o = v$. After this original scheme was broken by Kipnis and Shamir in [KS98], it was recommended in [KP99] to choose $v > o$ (Unbalanced Oil and Vinegar (UOV)). In the following we describe the more general approach UOV.

We set $V' = \{1, \ldots, v\}$ and $O = \{v + 1, \ldots, n\}$. Of the n variables $x_1, \ldots, x_n$ we call $x_1, \ldots, x_v$ the Vinegar variables and $x_{v+1}, \ldots, x_n$ Oil variables. We define o quadratic polynomials $f_k(\mathbf{x}) = f_k(x_1, \ldots, x_n)$ by

$$f_k(\mathbf{x}) = \sum_{i \in V',\ j \in O} \alpha_{ij}^{(k)} x_i x_j + \sum_{i,j \in V',\ i \leq j} \beta_{ij}^{(k)} x_i x_j + \sum_{i \in V' \cup O} \gamma_i^{(k)} x_i + \eta^{(k)} \ (k \in O)$$

Note that Oil and Vinegar variables are not fully mixed, just like oil and vinegar in a salad dressing.

The map $\mathcal{F} = (f_{v+1}(\mathbf{x}), \ldots, f_n(\mathbf{x}))$ can be easily inverted. First, we choose the values of the v Vinegar variables $x_1, \ldots, x_v$ at random. Therewith we get a system of o linear equations in the o variables $x_{v+1}, \ldots, x_n$ which can be solved by Gaussian Elimination. (If the system doesn't have a solution, choose other values of $x_1, \ldots, x_v$ and try again).

2.2 The Rainbow Signature Scheme

In [DS05] J. Ding and D. Schmidt proposed a signature scheme called Rainbow, which is based on the idea of (Unbalanced) Oil and Vinegar [KP99].

Let K be a finite field and V be the set $\{1, \ldots, n\}$. Let $v_1, \ldots, v_{u+1}, u \geq 1$ be integers such that $0 < v_1 < v_2 < \cdots < v_u < v_{u+1} = n$ and define the sets of integers $V_i = \{1, \ldots, v_i\}$ for $i = 1, \ldots, u$. We set $o_i = v_{i+1} - v_i$ and $O_i = \{v_i + 1, \ldots, v_{i+1}\}$ $(i = 1, \ldots, u)$. The number of elements in V_i is v_i and we have $|O_i| = o_i$. For $k = v_1 + 1, \ldots, n$ we define multivariate quadratic polynomials in the n variables $x_1, \ldots, x_n$ by

$$f_k(\mathbf{x}) = \sum_{i \in O_l,\ j \in V_l} \alpha_{i,j}^{(k)} x_i x_j + \sum_{i,j \in V_l,\ i \leq j} \beta_{i,j}^{(k)} x_i x_j + \sum_{i \in V_l \cup O_l} \gamma_i^{(k)} x_i + \eta^{(k)},$$

where l is the only integer such that $k \in O_l$. Note that these are Oil and Vinegar polynomials with x_i, $i \in V_l$ being the Vinegar variables and x_j, $j \in O_l$ being the Oil variables.

The map $\mathcal{F}(\mathbf{x}) = (f_{v_1+1}(\mathbf{x}), \ldots, f_n(\mathbf{x}))$ can be inverted as follows. First, we choose $x_1, \ldots, x_{v_1}$ at random. Hence we get a system of o_1 linear equations (given by the polynomials f_k $(k \in O_1)$) in the o_1 unknowns $x_{v_1+1}, \ldots, x_{v_2}$, which can be solved by Gaussian Elimination. The so computed values of x_i $(i \in O_1)$ are plugged into the polynomials $f_k(\mathbf{x})$ $(k > v_2)$ and a system of o_2 linear equations (given by the polynomials f_k $(k \in O_2)$) in the o_2 unknowns x_i $(i \in O_2)$ is obtained. By repeating this process we can get values for all the variables x_i $(i = 1, \ldots, n)$[1].

The Rainbow signature scheme is defined as follows:

Key Generation. The private key consists of two invertible affine maps $\mathcal{S} : K^m \rightarrow K^m$ and $\mathcal{T} : K^n \rightarrow K^n$ and the map $\mathcal{F}(\mathbf{x}) = (f_{v_1+1}(\mathbf{x}), \ldots, f_n(\mathbf{x})) : K^n \rightarrow K^m$. Here, $m = n - v_1$ is the number of components of $\mathcal{F}$.

The public key consists of the field K and the composed map $\mathcal{P}(\mathbf{x}) = \mathcal{S} \circ \mathcal{F} \circ \mathcal{T}(\mathbf{x}) : K^n \rightarrow K^m$.

Signature Generation. To sign a document d, we use a hash function $\mathbf{h} : K^* \rightarrow K^m$ to compute the value $\mathbf{h} = \mathbf{h}(d) \in K^m$. Then we compute recursively $\mathbf{x} = \mathcal{S}^{-1}(\mathbf{h})$, $\mathbf{y} = \mathcal{F}^{-1}(\mathbf{x})$ and $\mathbf{z} = \mathcal{T}^{-1}(\mathbf{y})$. The signature of the document is $\mathbf{z} \in K^n$. Here, $\mathcal{F}^{-1}(\mathbf{x})$ means finding one (of the possibly many) pre-image of $\mathbf{x}$.

[1] It may happen, that one of the linear systems does not have a solution. If so, one has to choose other values of $x_1, \ldots x_{v_1}$ and try again.

Verification. To verify the authenticity of a signature, one simply computes $\mathbf{h}' = P(\mathbf{z})$ and the hashvalue $\mathbf{h} = \mathbf{h}(d)$ of the document. If $\mathbf{h}' = \mathbf{h}$ holds, the signature is accepted, otherwise rejected.

The size of the public key is

$$m \cdot \left(\frac{n \cdot (n+1)}{2} + n + 1 \right) = m \cdot \frac{(n+1) \cdot (n+2)}{2} \text{ field elements,} \quad (1)$$

the size of the private key

$$m \cdot (m+1) + n \cdot (n+1) + \sum_{l=1}^{u} o_l \cdot \left(v_l \cdot o_l + \frac{v_l \cdot (v_l + 1)}{2} + v_{l+1} + 1 \right) \text{ field elements.} \quad (2)$$

The length of the needed hash value is m field elements, the length of the signature is n field elements.

The scheme is denoted by Rainbow$(v_1, o_1, \ldots, o_u)$. For $u = 1$ we get the original UOV scheme.

Rainbow over $GF(2^8)$ is commonly believed to be secure for at least 26 equations [BF09], [PB1a]. The actual design of the Rainbow layers is thereby not so important, as long as the following four items are taken into consideration [PB1a]:

- to defend the scheme against the Rainbow-Band-Separation attack (see subsection 6.2) one must have $n \geq \lceil \frac{5}{3} \cdot (m-1) \rceil$.
- to defend the scheme against the MinRank attack (see subsection 6.3) one must have $v_1 \geq 9$.
- to defend the scheme against the HighRank attack (see subsection 6.4) one must have $o_u \geq 10$.
- to defend the scheme against the UOV attack (see subsection 6.5) one must have $n - 2 \cdot o_u \geq 11$.

In particular, $(v_1, o_1, o_2) = (17, 13, 13)$ is a good choice for the parameters of Rainbow over $GF(2^8)$.

3 The Approach of [PB10]

In this section we describe briefly the approach of [PB10] to create a UOV-based scheme with a partially cyclic public key.

For SingleField schemes, both the public key $\mathcal{P}$ and the central map $\mathcal{F}$ are quadratic maps from K^n to K^o and therefore can be written as

$$\mathcal{P}(\mathbf{x}) = \sum_{i=1}^{n} \sum_{j=i}^{n} p_{ij}^{(k)} x_i x_j + \sum_{i=1}^{n} p_i^{(k)} x_i + p_0^{(k)} \quad \text{resp.}$$

$$\mathcal{F}(\mathbf{x}) = \sum_{i=1}^{n} \sum_{j=i}^{n} f_{ij}^{(k)} x_i x_j + \sum_{i=1}^{n} f_i^{(k)} x_i + f_0^{(k)} \ (k = 1, \ldots, o)$$

In the special case of the unbalanced Oil and Vinegar Signature Scheme [KP99] $\mathcal{P}$ is given as a concatenation of the central UOV-map $\mathcal{F}$ and an affine invertible map $\mathcal{T} = ((t_{ij})_{i,j=1}^{n}, c_T)$, i.e. $\mathcal{P} = \mathcal{F} \circ \mathcal{T}$.

The authors of [PB10] observed, that this equation leads (after fixing the affine map $\mathcal{T}$) to a linear relation between the coefficients of $\mathcal{P}$ and those of $\mathcal{F}$ of the form

$$p_{ij}^{(k)} = \sum_{i=1}^{n} \sum_{j=i}^{n} \alpha_{ij}^{rs} \cdot f_{rs}^{(k)}, \tag{3}$$

where the coefficients α_{ij}^{rs} are given as

$$\alpha_{ij}^{rs} = \begin{cases} t_{ri} \cdot t_{si} & (i = j) \\ t_{ri} \cdot t_{sj} + t_{rj} \cdot t_{si} & \text{otherwise} \end{cases}. \tag{4}$$

The relation (3) can be written in the form

$$\mathbf{p}^{(k)} = A' \cdot \mathbf{f}^{(k)}, \tag{5}$$

with two vectors containing the coefficients of the quadratic monomials of the k-th components of $\mathcal{P}$ resp. $\mathcal{F}$ and a matrix

$$A' = \left(\alpha_{ij}^{rs}\right) \ (1 \leq i \leq v,\ i \leq j \leq n \text{ for the rows, } 1 \leq r \leq v,\ r \leq s \leq n \text{ for the columns}). \tag{6}$$

By fixing the vectors $\mathbf{p}^{(i)}$ $i = 1, \ldots, o$ and inverting this relation, the authors of [PB10] were able to compute the central map $\mathcal{F}$ of a UOV scheme (with invertible affine map $\mathcal{T}$), whose public key has a coefficient matrix M_P of the form

$$M_P = (B|C),$$

where the rows of B are given by the vectors $\mathbf{p}^{(i)}$ $i = 1, \ldots, o$ and C is a matrix without apparent structure. By choosing the matrix B as a partially circulant matrix, they were able to reduce the public key size of the UOV scheme by a large factor.

4 Preliminaries

In this section we introduce some notations and definitions we need for the construction of our scheme in the next section. We restrict ourselves to the case of two Rainbow layers.

4.1 Notations

We denote
$D_1 = \frac{v_1 \cdot (v_1+1)}{2} + v_1 \cdot o_1$ the number of quadratic terms in the central polynomials of the first layer.

$D_2 = \frac{v_2 \cdot (v_2+1)}{2} + v_2 \cdot o_2$ the number of quadratic terms in the central polynomials of the second layer.
$D = \frac{n \cdot (n+1)}{2}$ the number of quadratic terms in the public polynomials.
For the invertible affine map $\mathcal{S} = (S, c_S)$ we divide the $m \times m$ matrix S into four parts:
$S = \begin{pmatrix} S_{11} & S_{12} \\ S_{21} & S_{22} \end{pmatrix}$, where S_{11} is the upper left $o_1 \times o_1$ submatrix of S.

4.2 The Monomial Ordering

To make the description of our construction easier, we use a special "blockwise" ordering of monomials:

- The first block (consisting of D_1 monomials) contains the monomials which appear in the first Rainbow layer (i.e. the monomials $x_i x_j$ $(1 \leq i \leq v_1,\ i \leq j \leq v_2)$).
- The second block (consisting of $D_2 - D_1$ monomials) contains the monomials which appear in the second but not in the first Rainbow layer (i.e. the monomials $x_i x_j$ $((1 \leq i \leq v_1,\ v_2+1 \leq j \leq n) \vee (v_1+1 \leq i \leq v_2,\ i \leq j \leq n))$.
- The third block contains the remaining quadratic monomials (i.e. the monomials $x_i x_j$ $(v_2 + 1 \leq i \leq j \leq n)$).
- The fourth and last block consists of the linear and constant monomials.

Inside the blocks we use the lexicographical ordering.

Example. For $(v_1, v_2, n) = (2, 4, 6)$ we get the following ordering of monomials $x_1^2 > x_1x_2 > x_1x_3 > x_1x_4 > x_2^2 > x_2x_3 > x_2x_4 > x_1x_5 > x_1x_6 > x_2x_5 > x_2x_6 > x_3^2 > x_3x_4 > x_3x_5 > x_3x_6 > x_4^2 > x_4x_5 > x_4x_6 > x_5^2 > x_5x_6 > x_6^2 > x_1 > x_2 > x_3 > x_4 > x_5 > x_6 > 1$.

5 The Scheme

In this section we describe how to construct a Rainbow scheme with a partially cyclic key. We restrict here to the case of two Rainbow layers.[2]

5.1 Properties of the Rainbow Public Key

For the Rainbow signature scheme the public key is given as a the concatenation of three maps

$$\mathcal{P} = \mathcal{S} \circ \mathcal{F} \circ \mathcal{T}.$$

We denote the concatenated map $\mathcal{F} \circ \mathcal{T}$ by $\mathcal{Q}$ and get

$$\mathcal{P} = \mathcal{S} \circ \mathcal{Q}.$$

[2] With a similar idea it is possible to create a partially cyclic public key for a Rainbow scheme with u layers. We don't handle it here.

Note that the relation between the maps $\mathcal{Q}$ and $\mathcal{F}$ has the same form as the relation between public key and central map in the UOV case. Therefore we get exactly the same equations as in section 3.

$$q_{ij}^{(k)} = \sum_{r=1}^{n} \sum_{s=i}^{n} \alpha_{ij}^{rs} \cdot f_{rs}^{(k)} \ (1 \leq k \leq m), \tag{7}$$

where the coefficients α_{ij}^{rs} are given as

$$\alpha_{ij}^{rs} = \begin{cases} t_{ri} \cdot t_{si} & (i = j) \\ t_{ri} \cdot t_{sj} + t_{rj} \cdot t_{si} & \text{otherwise} \end{cases}. \tag{8}$$

Due to the special structure of the central map $\mathcal{F}$, we can reduce the number of terms in equation (7). We get

$$q_{ij}^{(k)} = \sum_{r=1}^{v_1} \sum_{s=r}^{v_2} \alpha_{ij}^{rs} \cdot f_{rs}^{(k)} \ (1 \leq k \leq o_1)$$
$$q_{ij}^{(k)} = \sum_{r=1}^{v_2} \sum_{s=r}^{n} \alpha_{ij}^{rs} \cdot f_{rs}^{(k)} \ (o_1 + 1 \leq k \leq m), \tag{9}$$

Analogous to the case of the UOV we want to write equation (9) in a compact form. To do this, we define a quadratic $D_2 \times D_2$ matrix A containing the coefficients α_{ij}^{rs}

$$A = \left(a_{ij}^{rs}\right) \ (1 \leq i \leq v_2,\ i \leq j \leq n \text{ for the rows, } 1 \leq r \leq v_2,\ r \leq s \leq n \text{ for the columns}). \tag{10}$$

The order in which the α_{ij}^{rs} appear in the matrix, is thereby given by the monomial ordering defined in subsection 4.2 (for both rows and columns). We divide the matrix A into the four parts

$$A = \begin{pmatrix} A_{11} & A_{12} \\ A_{21} & A_{22} \end{pmatrix},$$

where A_{11} is the upper left $D_1 \times D_1$ submatrix of A.

We write down the coefficients of $\mathcal{P}$, $\mathcal{Q}$ and $\mathcal{F}$ (according to the monomial ordering defined above) into three matrices P', Q' and F' and divide these matrices as follows

We define the matrices P, Q and F to be the matrices consisting of the first D_2 columns of P', Q', resp. F'. With these definitions we get the following relations between the three matrices P, Q and F:

$$P = S \cdot Q \text{ or } \begin{pmatrix} B_1 & C_1 \\ & B_2 \end{pmatrix} = \begin{pmatrix} S_{11} & S_{12} \\ S_{21} & S_{22} \end{pmatrix} \cdot \begin{pmatrix} Q_{11} & Q_{12} \\ Q_{21} & Q_{22} \end{pmatrix} \tag{11}$$

$$Q = F \cdot A^T \text{ or } \begin{pmatrix} Q_{11} & Q_{12} \\ Q_{21} & Q_{22} \end{pmatrix} = \begin{pmatrix} F_1 & 0 \\ & F_2 \end{pmatrix} = \begin{pmatrix} A_{11}^T & A_{21}^T \\ A_{12}^T & A_{22}^T \end{pmatrix} \tag{12}$$

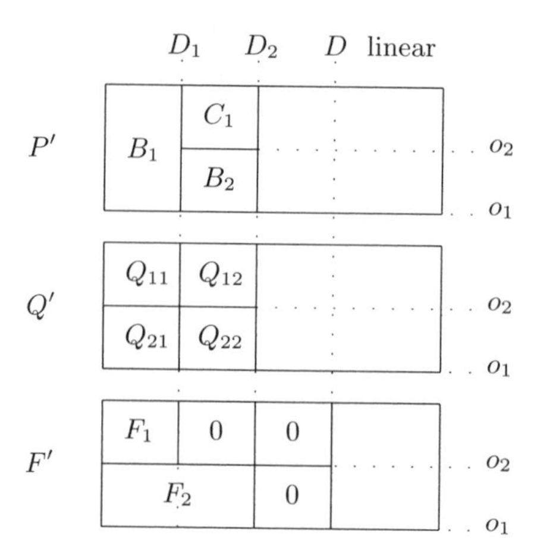

Fig. 1. Layout of the matrices P', Q' and F'

5.2 Construction

Additionally to the requirement that S and T are invertible, which is needed for the correctness of the scheme, we need the following assumptions to be true:

- The lower right $o_2 \times o_2$ submatrix S_{22} of S must be invertible.
- The transformation matrix A must be invertible.
- The upper left $D_1 \times D_1$ submatrix A_{11} of A must be invertible.

To justify these assumptions we carried out a number of experiments. For each of the values of (v_1, o_1, o_2) listed in Table 1 we created 1000 matrices S and A and observed how many of them were invertible.

Table 1. Percentage of invertible (sub-)matrices

Rainbow$(256, v_1, o_1, o_2)$	(4,2,2)	(9,6,6)	(11,9,9)	(14,11,11)	(17,13,13)
invertible matrices S_{22}	99.6	99.8	99.5	99.4	99.6
invertible matrices A	99.8	99.7	99.6	99.8	99.7
invertible matrices A_{11}	99.5	99.6	99.4	99.5	99.4

At the beginning of our construction we assign the elements of B_1 and B_2 elements of K, so that they get a compact structure.

For this we choose two vectors $\mathbf{a}^{(1)} = (a_1^{(1)}, \ldots, a_{D_1}^{(1)}) \in K^{D_1}$ and $\mathbf{a}^{(2)} = (a_1^{(2)}, \ldots, a_{D_2-D_1}^{(2)}) \in K^{D_2-D_1}$ at random. Then we set

$$b_{ij}^{(1)} = a^{(1)}_{((j-i) \bmod D_1)+1} \tag{13}$$

for the elements of the $m \times D_1$ matrix B_1 and

$$b_{ij}^{(2)} = a^{(2)}_{((j-i) \bmod (D_2-D_1))+1} \tag{14}$$

for the elements of the $o_2 \times (D_2 - D_1)$ matrix B_2.

Our goal is to compute the coefficients of the central map F in such a way that B_1 and B_2 appear in the matrix P representing the public key as shown in figure 1. From equations (11) and (12) we get

$$\begin{pmatrix} Q_{11} \\ Q_{21} \end{pmatrix} = S^{-1} \cdot B_1 \tag{15}$$

$$F_1 = Q_{11} \cdot (A_{11}^{-1})^T \tag{16}$$

$$Q_{12} = F_1 \cdot A_{21}^T \tag{17}$$

$$Q_{22} = S_{22}^{-1} \cdot (B_2 - S_{21} \cdot Q_{12}) \tag{18}$$

$$F_2 = (Q_{21}||Q_{22}) \cdot (A^{-1})^T \tag{19}$$

5.3 Key Generation and Key Sizes

Key Generation

1. Choose randomly two vectors $\mathbf{a}^{(1)} \in K^{D_1}$ and $\mathbf{a}^{(2)} \in K^{D_2 - D_1}$. Compute the entries of the matrices B_1 and B_2 by formulas (9) and (10).
2. Choose randomly two affine invertible maps $\mathcal{S} = (S, c_S) : K^m \rightarrow K^m$ and $\mathcal{T} = (T, c_T) : K^n \rightarrow K^n$. If the matrix S_{22} (see subsection 4.1) is not invertible, choose another map $\mathcal{S}$.
3. Compute for $\mathcal{T}$ the corresponding transformation matrix A using (10) and (8). Both A and its upper left $D_1 \times D_1$ submatrix A_{11} have to be invertible. If this is not the case, choose another map $\mathcal{T}$.
4. Compute the matrix $\begin{pmatrix} Q_{11} Q_{21} \end{pmatrix}$ using (15) .
5. Compute the quadratic coefficients of the central polynomials of the first layer by formula (16).
6. Compute the entries of the matrices Q_{12} and Q_{22} by formulas (17) and (18).
7. Compute the quadratic coefficients of the central polynomials of the second Rainbow layer by formula (19).
8. Choose the coefficients of the linear and constant terms of the central polynomials at random.
9. Compute the public key of the scheme by $\mathcal{P} = \mathcal{S} \circ \mathcal{F} \circ \mathcal{T}$.

The resulting public key has the form shown in figure 1.

The *public key* consists of the vectors $\mathbf{a}^{(1)}$ and $\mathbf{a}^{(2)}$, the matrix $C_1 = S_{11} \cdot Q_{12} + S_{12} \cdot Q_{22}$ and the last $\frac{(n+1)\cdot(n+2)}{2} - D_2$ columns of the matrix P.

The *private key* consists of the maps $\mathcal{S}$, $\mathcal{Q}$ and $\mathcal{T}$.

Note that both public and private keys of our scheme are from a subset of all valid Rainbow public resp. private keys. So, each instance of our scheme can be seen as a Rainbow scheme.

The *size of the public key* is

$$D_1+(D_2-D_1)+o_1\cdot(D_2-D_1)+m\cdot\left(\frac{(n+1)\cdot(n+2)}{2}-D_2\right)=m\cdot\frac{(n+1)\cdot(n+2)}{2}-o_1\cdot D_1-(o_2-1)\cdot D_2 \quad (20)$$

field elements, the *size of the private key*

$$m\cdot(m+1)+n\cdot(n+1)+\sum_{l=1}^{2}o_l\cdot\left(v_l\cdot o_l+\frac{v_l\cdot(v_l+1)}{2}+v_{l+1}+1\right)\text{ field elements.} \quad (21)$$

Signature generation and *verification* work as for the standard Rainbow scheme.

5.4 Efficiency of the Verification Process

Besides the considerable reduction of the public key size, the number of multiplications needed in the verification process is decreased by about 30 %.

This can be seen as follows: To evaluate an arbitrary public key, for every quadratic term two K-multiplications are needed. Together with the n multiplications for the linear terms, one needs $n\cdot(n+2)$ multiplications for each polynomial. Hence, to evaluate the whole public key, one needs

$$m\cdot n\cdot(n+2)\ K\text{−multiplications} \quad (22)$$

When evaluating our partially cyclic public key, some of the multiplications can be used several times (For example, $a_1^{(1)}\times x_1$ appears in every of the m public polynomials.) Thus, we do not have to carry out all the multiplications one by one. A close analysis shows, that by using this strategy we can reduce the number of K-multiplications needed in the verification process to

$$m\cdot n\cdot(n+2)-\left(\frac{m}{2}\cdot(2\cdot v_1\cdot v_2-v_1^2-v_1)+\frac{o_2}{2}\cdot(v_1^2-2v_1v_2-v_1+2v_2v_3-v_2^2-v_2)\right), \quad (23)$$

which, for $(v_1,o_1,o_2)=(17,13,13)$, leads to a reduction of 30 %.

6 Security

In this section we look at known attacks against the Rainbow signature scheme and study their effects against our scheme.

6.1 Direct Attacks [BB08], [YC07]

The most straightforward way for an attacker to forge a signature for a message h is to solve the public system $P(\mathbf{x})=h$ by an algorithm like XL or a Gröbner Basis method. To study the security of our scheme against direct attacks, we carried out experiments with MAGMA [BC97], which contains an efficient implementation of Faugeres F_4-algorithm [Fa99] for computing Gröbner Bases. Table 2 shows the results of our experiments against random systems, the standard Rainbow scheme and our partially cyclic version.

As the table shows, F_4 cannot solve our systems significantly faster than those of the standard Rainbow scheme.

Definition 1. *Let $p(\mathbf{x}) = p(x_1, \ldots, x_n)$ be a quadratic multivariate polynomial and*

$$dp(\mathbf{x}, \mathbf{c}) = p(\mathbf{x} + \mathbf{c}) - p(\mathbf{x}) - p(\mathbf{c}) + p(\mathbf{0})$$

its discrete differential. For p we define a matrix H_p by

$$dp = \mathbf{x}^T \cdot H_p \cdot \mathbf{c}$$

For the matrix H_{p_i} representing the quadratic part of the i-th public polynomial we write in short H_i.

6.2 Rainbow-Band-Separation [DY08]

The goal of this attack is to find an equivalent private key by which one can forge signatures for arbitrary messages. One tries to find a basis change of variables which transforms the matrices H_i into "Rainbow form" (see figure 2)

To achieve this, one has to solve several overdetermined systems of quadratic equations. The complexity of the attack is determined by the complexity of its first step, which consists of solving an overdetermined system of $m + n - 1$ quadratic equations in n variables. Table 3 shows the results of our experiments with the Rainbow Band Separation attack. The quadratic systems were again solved with MAGMA. As the table shows, the RBS attack can not take an advantage out of the special structure of our public key.

Table 2. Results of the experiments with direct attacks

(v_1, o_1, o_2)	(8,5,6)	(9,6,6)	(10,6,7)	(11,7,7)
cyclicRainbow	406 s	3135 s	23528 s	220372 s
Rainbow	405 s	3158 s	23560 s	222533 s
random system	408 s	3178 s	23621 s	221372 s

$*_{v_1 \times v_1}$	$*_{v_1 \times o_1}$	$0_{v_1 \times o_2}$
$*_{o_1 \times v_1}$	$0_{o_1 \times o_1}$	$0_{o_1 \times o_2}$
$0_{o_2 \times v_1}$	$0_{o_2 \times o_1}$	$0_{o_2 \times o_2}$

$1 \leq i \leq o_1$

$*_{v_1 \times v_1}$	$*_{v_1 \times o_1}$	$*_{v_1 \times o_2}$
$*_{o_1 \times v_1}$	$*_{o_1 \times o_1}$	$*_{o_1 \times o_2}$
$*_{o_2 \times v_1}$	$*_{o_2 \times o_1}$	$0_{o_2 \times o_2}$

$o_1 + 1 \leq i \leq m$

Fig. 2. Matrices H_i in the Rainbow form

Table 3. Results of our experiments with the Rainbow Band Separation attack

$(256, v_1, o_1, o_2)$	(8,5,6)	(9,6,6)	(10,6,7)	(11,7,7)
cyclicRainbow	403 s	3163 s	23583 s	223726 s
Rainbow	412 s	3152 s	23652 s	224273 s

6.3 MinRank Attack [GC00], [BG06]

In the MinRank attack one tries to find linear combinations $H = \sum_{i=1}^{m} \alpha_i H_i$ of the matrices representing the homogeneous quadratic parts of the public polynomials such that $\text{rank}(H) \leq v_2$. This linear combination are with high probability linear combinations of the central polynomials of the first Rainbow layer.

These linear combinations can be found by choosing randomly a vector $\mathbf{v} \in K^n$ and trying to solve the system $(\sum_{i=1}^{m} \alpha_i H_i) \cdot \mathbf{v} = \mathbf{0}$ for the α_i $(i = 1, \ldots, m)$. After having found o_1 linear combinations of this form, the attacker is able to extract the first Rainbow layer. After that, it is possible to recover the other layers one by one and therefore to find an equivalent private key. The complexity of the MinRank attack is determined by the complexity of finding the linear combinations, which is about $o_1 \cdot q^{v_1+1} \cdot m^3$.

Table 4 shows the results of our experiments with the MinRank attack. For every parameter set listed in the table we created 100 Rainbow schemes and attacked each of these schemes by the MinRank attack. The table shows the average number of vectors $\mathbf{v}$ we had to test until finding o_1 linear combinations of rank $\leq v_2$.

As the table shows, linear combinations with rank $\leq v_2$ can not be found easier for our scheme than for the standard Rainbow scheme. Furthermore, for our scheme these linear combinations do not show any visible structure. Note that the parameters listed in the table are far below those actually used for Rainbow. For the parameters proposed in subsection 2.2 the complexity of the attack is much higher than 2^{80}.

6.4 HighRank Attack [GC00], [DY08]

In the HighRank attack one tries to identify the variables appearing the lowest number of times in the central equations. These are the variables of the last Rainbow layer.

To do this, one forms random linear combinations H of the matrices H_i. If H has nontrivial kernel, one checks if the solution set of $(\sum_{i=1}^{m} \lambda_i H_i) \cdot \ker H = \mathbf{0}$ has dimension $n - o_2$. Then, with probability q^{-o_2}, we have

$$\ker(H) \subseteq \mathcal{T}(\mathcal{O}) \text{ with } \mathcal{O} = \{\mathbf{x} \in K^n | x_1 = \cdots = x_{n-o_2} = 0\}.$$

Table 4. Results of experiments with the MinRank attack

(q, v_1, o_1, o_2)	(8,3,2,2)	(8,4,3,3)	(16,3,2,2)	(16,4,3,3)
cyclicRainbow	7635	83534	124174	2982618
Rainbow	7724	84676	125463	3028357

After having found a basis of $\mathcal{T}^{-1}(\mathcal{O})$, one extends this basis to a basis of the whole space K^n. This enables an attacker to forge signatures the same way as a legitimate user. The complexity of the attack is determined by the complexity of finding a basis of $\mathcal{T}^{-1}(\mathcal{O})$, which is about $q^{o_u} \cdot m^3$.

For each of the parameter sets listed in Table 5 we created 100 Rainbow schemes. The table shows the average number of linear combinations we had to test until finding a basis of $\mathcal{T}^{-1}(\mathcal{O})$.

Table 5. Results of our experiments with the HighRank attack

(q, v_1, o_1, o_2)	(8,3,2,2)	(8,4,3,3)	(16,3,2,2)	(16,4,3,3)
cyclicRainbow	64.2	511.5	257.3	4093.7
Rainbow	65.1	512.3	256.8	4097.8

As the table shows, for both the Rainbow and the cyclic Rainbow scheme we have to test nearly the same number of linear combinations to find a basis of $\mathcal{T}^{-1}(\mathcal{O})$. Note that the parameters listed in the table are far below those actually used for Rainbow. For the parameters proposed in subsection 2.2 the complexity of the attack is much higher than 2^{80}.

6.5 UOV Attack [KP99]

Since a Rainbow scheme can be seen as a UOV scheme with v_u vinegar and o_u oil variables, it can be attacked by the UOV attack of Kipnis and Shamir [KP99]. The goal of this attack is to find the pre-image $\mathcal{T}^{-1}(\mathcal{O})$ of the Oil-subspace $\mathcal{O} = \{\mathbf{x} \in K^n | x_1 = \cdots = x_{n-o_2} = 0\}$ under the affine invertible map $\mathcal{T}$. One chooses randomly a linear combination H of the matrices $H_1, \ldots, H_m$ and sets $G := H \cdot H_j^{-1}$ for some $j \in \{1, \ldots, m\}$. After that, one computes all the minimal invariant subspaces of G. With high probability, these invariant subspaces are also subspaces of $\mathcal{T}^{-1}(\mathcal{O})$. After having found a basis of $\mathcal{T}^{-1}(\mathcal{O})$, one extends this basis to a basis of the whole space K^n. This enables an attacker to forge signatures for arbitrary messages. The complexity of the attack is determined by the complexity of finding a basis of $\mathcal{T}^{-1}(\mathcal{O})$, which is about $q^{n-2 \cdot o_u} \cdot m^3$.

For each of the parameter sets listed in the table we created 100 instances of both schemes. Then we attacked these instances by the UOV-attack. Table 6 shows the average number of matrices G we had to test until finding a basis of $\mathcal{T}^{-1}(\mathcal{O})$.

Table 6. Results of the experiments with the UOV attack

$(16, v_1, o_1, o_2)$	(3,2,2)	(5,3,3)	(9,6,6)	(12,10,10)
cyclicRainbow	1734	531768	852738	1183621
Rainbow	1728	532614	847362	1146382

As the table shows, for both schemes we have to test nearly the same number of matrices G to find a basis of $\mathcal{T}^{-1}(\mathcal{O})$. Note that the parameters listed in the table are far below those actually used for Rainbow. For the parameters proposed in subsection 2.2 the complexity of the attack is much higher than 2^{80}.

6.6 Summary

As the previous five subsections showed, known attacks against the Rainbow signature scheme do not work significantly better in our case, which means that they can not use the special structure of our public key. So, in this sense our scheme seems to be secure and we do not have to increase our parameter sets.

However, in the future we are going to study the security of our scheme under other attacks, e.g. decomposition attacks [FP09]. It might also be possible that some dedicated attacks against our scheme exist.

7 Parameters

Based on the security analysis in the previous section we propose for our scheme the same parameters as suggested for the standard Rainbow Scheme (see section 2), namely

$$(q, v_1, o_1, o_2) = (256, 17, 13, 13).$$

Table 7 compares our scheme with others from the UOV family. Additionally to the parameters proposed above, the table contains key- and signature sizes for a more conservative parameter set for $m = 28$.

For 26 equations, we get a key size reduction of $\frac{25.9-10.2}{25.9} = 62\%$, for 28 equations $\frac{32.2-12.9}{32.2} = 60\%$.

Table 7. Comparison of different UOV-based signature schemes

Scheme	public key size (kB)	private key size (kB)	hash size (bit)	signature size (bit)
UOV(256,26,52)	80.2	76.1	208	624
cyclicUOV(256,26,52)	14.5	76.1	208	624
Rainbow(256,17,13,13)	25.9	19.1	208	344
cyclicRainbow(256,17,13,13)	10.2	19.1	208	344
UOV(256,28,56)	99.9	92.8	224	672
cyclicUOV(256,28,56)	16.5	92.8	224	672
Rainbow(256,19,14,14)	32.2	24.3	224	376
cyclicRainbow(256,19,14,14)	12.9	24.3	224	376

8 Conclusion

In this paper we showed a way how to extend the approach of [PB10] to the Rainbow signature scheme. The result is a Rainbow-like scheme, which reduces the size of the public key by 62 % and the number of field multiplications needed during the verification process by 30 %. We believe that our idea might be a good approach for implementing the Rainbow scheme on low cost devices, e.g. smartcards. Furthermore, it's a quite general idea, which should be applicable to a number of other SingleField Scheme, for example enSTS [TG10].

Points of research for the future are in particular security issues of the scheme as well as the use of PRNG's to construct the public key.

Acknowledgements

We thank Enrico Thomae and Christopher Wolf for fruitful discussions and helpful comments. Furthermore we want to thank the anonymous reviewers for their valuable comments which helped to improve the paper.

References

[BB08] Bernstein, D.J., Buchmann, J., Dahmen, E. (eds.): Post Quantum Cryptography. Springer, Heidelberg (2009)

[BC97] Bosma, W., Cannon, J., Playoust, C.: The Magma algebra system. I. The user language. J. Symbolic Comput. 24(3-4), 235–265 (1997)

[BF09] Bettale, L., Faugere, J.-C., Perret, L.: Hybrid approach for solving multivariate systems over finite fields. Journal of Math. Cryptology, 177–197 (2009)

[BG06] Billet, O., Gilbert, H.: Cryptanalysis of Rainbow. In: De Prisco, R., Yung, M. (eds.) SCN 2006. LNCS, vol. 4116, pp. 336–347. Springer, Heidelberg (2006)

[DS05] Ding, J., Schmidt, D.: Rainbow, a new multivariate polynomial signature scheme. In: Ioannidis, J., Keromytis, A.D., Yung, M. (eds.) ACNS 2005. LNCS, vol. 3531, pp. 164–175. Springer, Heidelberg (2005)

[Di04] Ding, J.: A new variant of the Matsumoto-Imai cryptosystem through perturbation. In: Bao, F., Deng, R., Zhou, J. (eds.) PKC 2004. LNCS, vol. 2947, pp. 266–281. Springer, Heidelberg (2004)

[DY08] Ding, J., Yang, B.-Y., Chen, C.-H.O., Chen, M.-S., Cheng, C.M.: New Differential-Algebraic Attacks and Reparametrization of Rainbow. In: Bellovin, S.M., Gennaro, R., Keromytis, A.D., Yung, M. (eds.) ACNS 2008. LNCS, vol. 5037, pp. 242–257. Springer, Heidelberg (2008)

[DW07] Ding, J., Wolf, C., Yang, B.-Y.: ℓ-invertible Cycles for Multivariate Quadratic Public Key Cryptography. In: Okamoto, T., Wang, X. (eds.) PKC 2007. LNCS, vol. 4450, pp. 266–281. Springer, Heidelberg (2007)

[DY07] Ding, J., Yang, B.-Y., Cheng, C.-M., Chen, O., Dubois, V.: Breaking the symmetry: a Way to Resist the new Differential Attack, eprint 366/2007

[Fa99] Faugére, J.C.: A new efficient algorithm for computing Groebner bases (F4). Journal of Pure and Applied Algebra 139, 61–88 (1999)

[FP09] Faugére, J.C., Perret, L.: An efficient algorithm for decomposing multivariate polynomials and its applications to cryptography. Journal of Symbolic Computation 44(12), 1676–1689 (2009)

[GC00] Goubin, L., Courtois, N.T.: Cryptanalysis of the TTM cryptosystem. In: Okamoto, T. (ed.) ASIACRYPT 2000. LNCS, vol. 1976, pp. 44–57. Springer, Heidelberg (2000)

[HW05] Hu, Y.-H., Wang, L.-C., Chou, C.-P., Lai, F.: Similar Keys of Multivariate Public Key Cryptosystems. In: Desmedt, Y.G., Wang, H., Mu, Y., Li, Y. (eds.) CANS 2005. LNCS, vol. 3810, pp. 211–222. Springer, Heidelberg (2005)

[KP99] Kipnis, A., Patarin, L., Goubin, L.: Unbalanced Oil and Vinegar Schemes. In: Stern, J. (ed.) EUROCRYPT 1999. LNCS, vol. 1592, pp. 206–222. Springer, Heidelberg (1999)

[KS98] Kipnis, A., Shamir, A.: Cryptanalysis of the Oil and Vinegar Signature scheme. In: Krawczyk, H. (ed.) CRYPTO 1998. LNCS, vol. 1462, pp. 257–266. Springer, Heidelberg (1998)

[MI88] Matsumoto, T., Imai, H.: Public Quadratic Polynomial-Tuples for efficient Signature-Verification and Message-Encryption. In: Günther, C.G. (ed.) EUROCRYPT 1988. LNCS, vol. 330, pp. 419–453. Springer, Heidelberg (1988)

[Pa96] Patarin, J.: Hidden Field equations (HFE) and Isomorphisms of Polynomials (IP). In: Maurer, U.M. (ed.) EUROCRYPT 1996. LNCS, vol. 1070, pp. 38–48. Springer, Heidelberg (1996)

[Pa97] Patarin, J.: The oil and vinegar signature scheme, presented at the Dagstuhl Workshop on Cryptography (September 1997)

[PB10] Petzoldt, A., Bulygin, S., Buchmann, J.: A Multivariate Signature Scheme with a partially cyclic public key. In: Proceedings of SCC 2010, pp. 229–235 (2010)

[PB1a] Petzoldt, A., Bulygin, S., Buchmann, J.: Selecting Parameters for the Rainbow Signature Scheme. In: Sendrier, N. (ed.) Post-Quantum Cryptography. LNCS, vol. 6061, pp. 218–240. Springer, Heidelberg (2010)

[PC01] Patarin, J., Courtois, N., Goubin, L.: Flash, a fast multivariate signature algorithm. In: Naccache, D. (ed.) CT-RSA 2001. LNCS, vol. 2020, pp. 298–307. Springer, Heidelberg (2001)

[PG98] Patarin, J., Goubin, L., Courtois, N.: C*-+ and HM: Variations around two schemes of T. Matsumoto and H. Imai. In: Ohta, K., Pei, D. (eds.) ASIACRYPT 1998. LNCS, vol. 1514, pp. 35–50. Springer, Heidelberg (1998)

[TG10] Tsuji, S., Gotaishi, M., Tadaki, K., Fujita, R.: Proposal of a Signature Scheme based on STS Trapdoor. In: Sendrier, N. (ed.) Post-Quantum Cryptography. LNCS, vol. 6061, pp. 201–217. Springer, Heidelberg (2010)

[WY06] Wang, L.C., Yang, B.Y., Hu, Y.H., Lai, F.: A medium field multivariate public-key encryption scheme. In: Pointcheval, D. (ed.) CT-RSA 2006. LNCS, vol. 3860, pp. 132–149. Springer, Heidelberg (2006)

[YC05] Yang, B.-Y., Chen, J.-M.: Building secure tame like multivariate public-key cryptosystems: The new TTS. In: Boyd, C., González Nieto, J.M. (eds.) ACISP 2005. LNCS, vol. 3574, pp. 518–531. Springer, Heidelberg (2005)

[YC07] Yang, B.-Y., Chen, J.-M.: All in the XL family: Theory and practice. In: Park, C.-s., Chee, S. (eds.) ICISC 2004. LNCS, vol. 3506, pp. 67–86. Springer, Heidelberg (2005)

Combined Security Analysis of the One- and Three-Pass Unified Model Key Agreement Protocols

Sanjit Chatterjee[1], Alfred Menezes[1], and Berkant Ustaoglu[2]

[1] Department of Combinatorics & Optimization, University of Waterloo
{s2chatte,ajmeneze}@uwaterloo.ca
[2] NTT Information Sharing Platform Laboratories, Tokyo, Japan
bustaoglu@cryptolounge.net

Abstract. The unified model (UM) is a family of key agreement protocols that has been standardized by ANSI and NIST. The NIST standard explicitly permits the reuse of a static key pair among the one-pass and three-pass UM protocols. However, a recent study demonstrated that such reuse can lead to security vulnerabilities. In this paper we revisit the security of the one- and three-pass UM protocols when static key pairs are reused. We propose a shared security model that incorporates the individual security attributes of the two protocols. We then show, provided appropriate measures are taken, that the protocols are secure even when static key pairs are reused.

1 Introduction

The unified model (UM) is a family of two-party key agreement protocols of the Diffie-Hellman variety. This well-known family of protocols has been standardized in ANSI X9.42 [1], ANSI X9.63 [2] and NIST SP 800-56A [12]. The NIST SP 800-56A standard, which includes several discrete logarithm based key agreement protocols along with the UM family, explicitly permits the reuse of static (long-term) key pairs among different key agreement protocols.

Recent work studied the effect of reusing static key pairs among different key agreement protocols [4]. The paper described a plausible scenario where the three-pass UM protocol may become insecure if parties reuse static key pairs with the one-pass UM protocol as permitted in NIST SP 800-56A. The paper also proposed a shared security model for key agreement protocols having *identical* security attributes. Protocols having different security attributes, such as the one-pass and three-pass UM protocols, are not covered in that security model.

In this work, we revisit the security of the one- and three-pass UM protocols when parties are allowed to share their static keys among the two protocols. Our aim is to investigate whether it is possible to obtain meaningful assurances about the security of the individual members of this protocol pair in the combined setting. The adversary in this combined model is afforded considerable additional strength since she can use information learned in protocol sessions of one-pass UM to attack protocol sessions of three-pass UM and vice versa.

G. Gong and K.C. Gupta (Eds.): INDOCRYPT 2010, LNCS 6498, pp. 49–68, 2010.

While a reductionist security argument for three-pass UM can be found in the literature [11], no such formal argument has been presented for one-pass UM. We illustrate by way of example in §2 that a formal security argument for a multi-pass key agreement protocol does not guarantee that the corresponding "natural" one-pass variant is also secure. This leads us to formally study the security of one-pass UM. In §3 a security model is presented for one-pass UM that is motivated by the corresponding model of three-pass UM. This is followed by a reductionist security argument for one-pass UM in the proposed model.

In §4 we propose a combined security model for one- and three-pass UM that incorporates the security attributes of the individual protocols. We then show in §5 that it is possible to maintain the individual security guarantees of one- and three-pass UM in the combined model. This is achieved by adding appropriate protocol identifiers in the key derivation functions and without any negative impact on the efficiency of the protocols.

Notation and terminology. Let $\mathcal{G} = \langle g \rangle$ denote a multiplicatively-written cyclic group of prime order q, and let $\mathcal{G}^* = \mathcal{G} \setminus \{1\}$. The *Computational Diffie-Hellman (CDH) assumption* in $\mathcal{G}$ is that computing $\mathrm{CDH}(U, V) = g^{uv}$ is infeasible given $U = g^u$ and $V = g^v$ where $u, v \in_R [1, q-1]$. The *Decisional Diffie-Hellman (DDH) assumption* in $\mathcal{G}$ is that distinguishing DH triples (g^a, g^b, g^{ab}) from random triples (g^a, g^b, g^c) is infeasible. The *Gap Diffie-Hellman (GDH) assumption* in $\mathcal{G}$ is that the CDH assumption holds even when a CDH solver is given a DDH oracle that distinguishes DH triples from random triples. Let MAC denote a message authentication code algorithm such as HMAC.

The Unified Model (UM) family of key agreement protocols are of the Diffie-Hellman variety where the two communicating parties $\hat{A}$ and $\hat{B}$ exchange static public keys. Party $\hat{A}$'s static private key is an integer $a \in_R [1, q-1]$, and her corresponding static public key is $A = g^a$. Similarly, party $\hat{B}$ has a static key pair (b, B), and so on. A certifying authority (CA) issues certificates that binds a party's identifier to its static public key. A party $\hat{A}$ called the *initiator* commences the protocol by selecting an ephemeral (one-time) key pair and then sends the ephemeral public key (and possibly other data) to the second party. The ephemeral private key is a randomly selected integer $x \in [1, q-1]$ and the corresponding ephemeral public key is $X = g^x$. Upon receipt of X, the *responder* $\hat{B}$ selects an ephemeral private key y and sends $Y = g^y$ (and possibly other data) to $\hat{A}$; this step is omitted in one-pass protocols. The parties may exchange some additional messages, after which they accept a session key.

2 (In)Security of One-Pass Key Agreement Protocols

In a multi-pass key agreement protocol both parties have their own ephemeral secrets which play distinct roles in the derivation of the session key. However, there are situations, e.g., email systems, where one of the parties may be off-line and hence not available for immediate response. One-pass key agreement protocols can be useful in such scenarios.

One-pass protocols are by their very nature non-interactive and proceed without any ephemeral secret contributed by the responder. Such a protocol can be obtained by appropriately modifying a multi-pass key agreement protocol. In fact, most (if not all) of the one-pass protocols available in the literature (including those standardized by different standards bodies) are derived from a two- or three-pass key agreement protocol. Some well known examples are the UM [12], MQV [9], HMQV [7] and CMQV [13] families of protocols.

A natural method to derive a one-pass key agreement protocol from a multi-pass protocol is to set the responder $\hat{B}$'s ephemeral public key Y to be equal to his static public key B. Examples are one-pass UM [12], one-pass MQV [9], and one-pass HMQV [7]. Another natural method is to set Y to be the identity element of $\mathcal{G}$. The latter approach was followed in the case of one-pass CMQV [13]. However, given any provably secure multi-pass protocol, application of one of these natural conversions does not necessarily yield a *secure* one-pass variant.

As an illustrative example, consider the following version of the two-pass MTI/C0 protocol that was proven secure in [8]. The initiator $\hat{A}$ chooses an ephemeral secret $x \in_R [1, q-1]$ and sends the message B^x to the responder $\hat{B}$. The responder, in turn, chooses an ephemeral secret $y \in_R [1, q-1]$ and sends the message A^y to $\hat{A}$. The shared secret is g^{xy} which $\hat{A}$ computes as $(A^y)^{x/a}$ and $\hat{B}$ computes as $(B^x)^{y/b}$. It is easy to see that setting $y = b$ or $y = 0$ yields a trivially insecure one-pass protocol. In particular, the shared secret is B^x in case $y = b$ and B^0 in case $y = 0$. So the adversary can simply choose x and impersonate any party $\hat{A}$ to $\hat{B}$. As another example, consider Protocol 2 (in the public key setting) as proposed by Boyd et al. in [3]. Here $\hat{A}$ (resp. $\hat{B}$) encapsulates keying material K_A (resp. K_B) under the public key of $\hat{B}$ (resp. $\hat{A}$) and sends the encapsulation along with an ephemeral public key g^x (resp. g^y) to $\hat{B}$ (resp. $\hat{A}$). The session key is then derived from K_A, K_B and g^{xy} as per the protocol specification. Like the MTI/C0 protocol, the one-pass variant will be trivially insecure as the receiver $\hat{B}$ has no means of determining who sent him the message.

The above two examples are only illustrative and there is no suggestion in the literature to use them as *secure* one-pass protocols. However, consider the case of the one-pass HMQV protocol which at first glance appears to be secure and in fact was claimed to be provably secure in [7]. It was later shown [10] that the protocol succumbs to an unknown-key share attack [5]. On the other hand, not all one-pass key agreement protocols are necessarily insecure. The one-pass UM and MQV protocols appear to resist all known attacks and one-pass CMQV comes with a reductionist security argument in a security model appropriately designed for the one-pass protocol environment.

3 One-Pass UM Protocol

In this section we provide a security model for one-pass key agreement protocols, followed by a formal description of the one-pass UM protocol and its reductionist security argument in the given security model.

3.1 Security Model

The security model for one-pass key agreement protocols as defined here is primarily motivated by the corresponding definition of one-pass CMQV in the full version of [13]. However, our ultimate aim is to provide a combined security analysis of the one- and three-pass UM protocols. Hence our definition is specifically crafted with one-pass UM in mind (as specified in NIST SP 800-56A [12]) and the security model for three-pass UM as defined in [11].

The model assumes that there are n parties each having a static key pair and a certificate that binds the corresponding public key to that party. There is no requirement for the certifying authority (CA) to obtain a proof of possession of the static private key from the respective party. However, the CA must perform a validation check of the public key to ascertain that it belongs to $\mathcal{G}^*$.[1]

Protocol sessions. A party $\hat{A}$ can be activated via an incoming message to create a session. The incoming message has the form $(\hat{A}, \hat{B})$ or $(\hat{A}, \hat{B}, X)$. If $\hat{A}$ was activated with $(\hat{A}, \hat{B})$ then $\hat{A}$ is the session *initiator*; otherwise $\hat{A}$ is the session *responder*. If $\hat{A}$ is the session initiator then $\hat{A}$ creates a separate session state and selects an ephemeral public key X. The session is labeled *active* and identified via a session identifier $(\hat{A}, \hat{B}, X)$. The initiator then computes the session key as per the protocol specification, sends the message $(\hat{B}, \hat{A}, X)$, and completes the session. If $\hat{A}$ is the session responder then $\hat{A}$ creates a separate session state that is identified via a session identifier $(\hat{A}, \hat{B}, X)$, computes the session key as per the protocol specification, and completes the session. Note that the responder does not send a message to the initiator in a one-pass protocol. Suppose that $s = (\hat{A}, \hat{B}, X)$ is a session owned by $\hat{A}$. A session $s^* = (\hat{C}, \hat{D}, Y)$ is said to be *matching* to s if $\hat{C} = \hat{B}$, $\hat{D} = \hat{A}$ and $Y = X$.

The model mandates that the responder validate that $X \in \mathcal{G}^*$ upon receiving the message $(\hat{A}, \hat{B}, X)$. The session is *aborted* in case the verification fails and the party deletes all information specific to that session.

Adversary. The adversary $\mathcal{M}$ is modeled as a probabilistic Turing machine and controls all communications. Parties submit outgoing messages to $\mathcal{M}$, who makes decisions about their delivery. The adversary presents parties with incoming messages via *Send*(message), thereby controlling the activation of parties. The adversary does not have immediate access to a party's private information. However, in order to capture possible leakage of private information, $\mathcal{M}$ is allowed to make the following queries:

- *SessionStateReveal*(s): This query allows $\mathcal{M}$ to obtain all the secret information available in the session state of s. We assume that the query is only made at an initiator and that $\mathcal{M}$ can learn the ephemeral secret x through this query.
- *Expire*(s): If s has completed then its session key is deleted and the session is labeled expired. We assume that $\mathcal{M}$ only issues this query to sessions that have completed but are unexpired. At any point in time a session is in exactly one of the following states: active, completed, aborted, expired.

[1] It is easy to mount a small-subgroup attack on the UM family of protocols if the static public key is not validated.

• *SessionKeyReveal*(s): If the session is completed but unexpired then $\mathcal{M}$ obtains the corresponding session key. Like the previous query, we assume that $\mathcal{M}$ only issues this query to sessions that have completed but are unexpired.

• *Corrupt*(party): This query allows $\mathcal{M}$ to obtain complete control over the party. In particular, $\mathcal{M}$ obtains that party's static private key, the contents of all active sessions owned by the party, and the session keys of the completed but unexpired sessions owned by the party. (However, $\mathcal{M}$ cannot obtain the session key of an already expired session.) In addition, $\mathcal{M}$ is allowed to select a new static key pair for the party. Parties against whom the adversary issued this query are called *corrupt*, while a party that is not corrupt is called *honest*.

Fresh session. As in a multi-pass key agreement protocol, the goal of the adversary $\mathcal{M}$ in the one-pass protocol is to distinguish the session key held by a 'fresh' session from a random key. However, the definition of a fresh session in a one-pass protocol differs from that in a multi-pass protocol. This is due to the intrinsic asymmetric nature of the protocol where only the initiator contributes an ephemeral secret. Formally, we define a fresh session as follows.

Definition 1 (one-pass fresh session). Let s be the identifier of a completed session, owned by party $\hat{A}$ with peer $\hat{B}$. Let s^* be the identifier of a matching session of s, if it exists. Define s to be *fresh* if none of the following hold:

1. $\mathcal{M}$ issued a *SessionKeyReveal*(s) query or a *SessionKeyReveal*(s^*) query.
2. $\hat{A}$ is the initiator and one of the following holds: (a) $\mathcal{M}$ issued *Corrupt*($\hat{A}$) before *Expire*(s); (b) $\mathcal{M}$ issued *SessionStateReveal*(s) and *Corrupt*($\hat{A}$); (c) $\mathcal{M}$ issued *Corrupt*($\hat{B}$).
3. $\hat{A}$ is the responder and one of the following holds: (a) $\mathcal{M}$ issued *Corrupt*($\hat{A}$); (b) s^* exists and $\mathcal{M}$ issued either *Corrupt*($\hat{B}$) before *Expire*(s^*) or both *Corrupt*($\hat{B}$) and *SessionStateReveal*(s^*); (c) s^* does not exist and $\mathcal{M}$ issued *Corrupt*($\hat{B}$).

Adversary's goal. To capture indistinguishability, $\mathcal{M}$ is allowed to make a special *Test* query to a fresh session s. In response, $\mathcal{M}$ is given with equal probability either the session key held by s or a random key. If $\mathcal{M}$ guesses correctly whether the key is random or not, then $\mathcal{M}$ is said to be successful. Note that $\mathcal{M}$ can continue interacting with the parties after issuing the *Test* query, but must ensure that the test session remains fresh throughout the experiment.

Definition 2 (one-pass security). The one-pass UM protocol is said to be *secure* if the following conditions hold:

1. If two honest parties complete matching sessions then, except with negligible probability, they both compute the same session key.
2. No polynomially-bounded adversary $\mathcal{M}$ can distinguish the session key of a fresh session from a randomly chosen session key with probability greater than $\frac{1}{2}$ plus a negligible fraction.

Remark 1. By replaying messages from $\hat{A}$ to $\hat{B}$ an adversary $\mathcal{M}$ can force multiple sessions owned by $\hat{B}$ to hold the same session key κ. Let S_κ be the set of all such sessions. Since all sessions in S_κ have the same session identifier, $\mathcal{M}$ cannot compromise a single session in S_κ without compromising all sessions in S_κ.

Remark 2. The definition of fresh session in one-pass UM is more restrictive than the corresponding definition for three-pass UM. In particular, a session in the one-pass protocol is no longer considered as fresh if the adversary just issues a *Corrupt* query to the responder. On the other hand, whereas three-pass UM does not provide key-compromise impersonation (KCI) resilience [6] (for either initiator or the responder), KCI resilience is provided for the initiator in one-pass UM. This is by virtue of the asymmetric nature of the one-pass protocol which also accounts for (weak) forward secrecy when the initiator is compromised. Our security model for one-pass protocols captures weak forward secrecy with resepct to the initiator but does not capture KCI resilience.

3.2 Protocol Descriptions

In the one-pass UM protocol, denoted UM_1 and depicted in Figure 1, the responder's ephemeral public key is set to be the static public key of that party. The protocol is formally described in Definition 3. In UM_1, Λ denotes optional public information that can be included in the key derivation function H.[2]

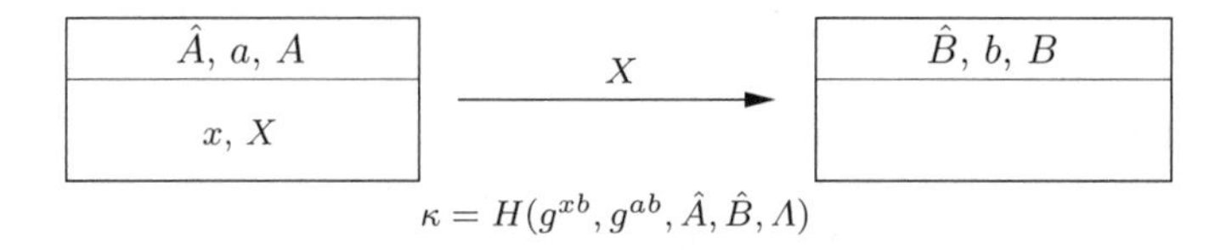

Fig. 1. UM_1: The one-pass UM protocol (simplified)

Definition 3 (UM_1). Each party obtains the certified static public key of its peer along with assurance that the key is valid. The protocol proceeds as follows.

1. Upon activation $(\hat{A}, \hat{B})$, party $\hat{A}$ (the initiator) does the following:
   (a) Select an ephemeral private key $x \in_R [1, q-1]$ and compute the ephemeral public key $X = g^x$.
   (b) Compute $\sigma_e = B^x$ and $\sigma_s = B^a$.
   (c) Compute $\kappa = H(\sigma_e, \sigma_s, \hat{A}, \hat{B}, \Lambda)$ and destroy x, σ_e, σ_s.
   (d) Send $(\hat{B}, \hat{A}, X)$ to $\hat{B}$ and complete session $(\hat{A}, \hat{B}, X)$ with session key κ.

[2] The key derivation function in the one- and three-pass UM protocols also includes an integer keydatalen that indicates the bitlength of the secret keying material to be generated, and a bit string AlgorithmID that indicates how the derive keying material will be parsed and for which algorithm(s) it will be used. We will henceforth omit keydatalen and AlgorithmID since they are not relevant to our security analysis.

2. Upon activation $(\hat{B}, \hat{A}, X)$, party $\hat{B}$ (the responder) does the following:
   (a) Verify that $X \in \mathcal{G}^*$.
   (b) Compute $\sigma_e = X^b$ and $\sigma_s = A^b$.
   (c) Compute $\kappa = H(\sigma_e, \sigma_s, \hat{A}, \hat{B}, \Lambda)$ and destroy σ_e, σ_s.
   (d) Complete session $(\hat{B}, \hat{A}, X)$ with session key κ.

Three-pass UM. The three-pass UM protocol, denoted UM_3, is depicted in Figure 2 (and formally described in [11]). In Figure 2, $\mathcal{I}$ and $\mathcal{R}$ denote the constant strings "KC_2_U" and "KC_2_V", and Λ_1 and Λ_2 are optional public strings. The session key is κ, whereas κ' is an ephemeral secret key used to authenticate the exchanged ephemeral public keys and the identifiers.

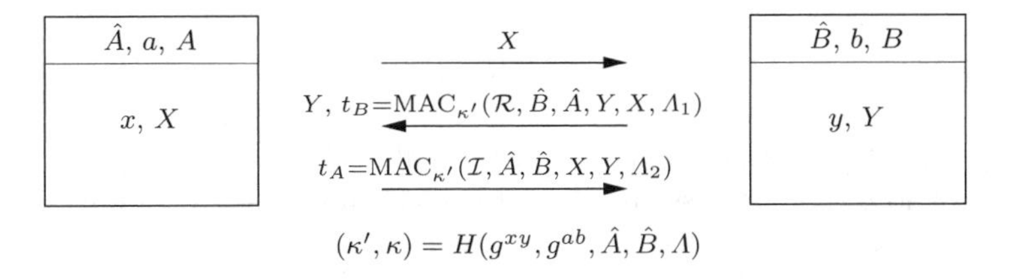

Fig. 2. UM_3: The three-pass UM protocol (simplified)

3.3 Security Argument

For simplicity we consider $\Lambda = X$ where X is the initiator's ephemeral public key.[3] Furthermore, we do not allow parties to initiate sessions with themselves; this restriction is relaxed at the end of §4.

Theorem 1. *If H is modeled as a random oracle and $\mathcal{G}$ is a group where the GDH assumption holds, then one-pass UM is a secure key agreement protocol.*

Proof. It is easy to see that matching sessions produce the same session key. We will verify that for a security parameter λ, no polynomially-bounded adversary $\mathcal{M}$ can distinguish the session key of a fresh session from a randomly chosen session key with probability $\frac{1}{2} + p(\lambda)$ for some non-negligible function $p(\lambda)$.

Let M denote the event that $\mathcal{M}$ succeeds in the distinguishing game, and suppose that $\Pr(M) = \frac{1}{2} + p(\lambda)$ where $p(\lambda)$ is non-negligible. We assume that $\mathcal{M}$ operates in an environment with n parties, and where each party is activated at most t times to create a new session. We will show how $\mathcal{M}$ can be used to construct a polynomial-time algorithm $\mathcal{S}$ that, with non-negligible probability of success, solves a CDH instance (U, V).

Since H is a modeled as a random function, $\mathcal{M}$ has only two strategies for winning the distinguishing game with probability significantly greater than $\frac{1}{2}$:

[3] If Λ is empty, then maintaining consistency in the responses to *SessionKeyReveal* and H queries in Theorem 1's simulation requires further checks. More precisely, $\mathcal{S}$ has to identify relations among H queries, session keys, and ephemeral public keys by using the DDH oracle for each possible combination.

(i) induce a non-matching session s' to establish the same session key as the test session s, and thereafter issue a *SessionKeyReveal*(s') query; or
(ii) query the random oracle H with $(g^{xb}, g^{ab}, \hat{A}, \hat{B}, X)$ where $s = (\hat{A}, \hat{B}, X)$ is the test session or its matching session.

Since the input to the key derivation function includes the identities of the communicating parties and the exchanged ephemeral public key, non-matching completed sessions produce different session keys except with negligible probability of H collisions. This rules out strategy (i).

Now, let H^* denote the event that $\mathcal{M}$ queries H with $(g^{xb}, g^{ab}, \hat{A}, \hat{B}, X)$ where $(\hat{A}, \hat{B}, X)$ is the test session or its matching session. Since H is a random function, we have $\Pr(M|\overline{H^*}) = \frac{1}{2}$ where negligible terms are ignored. Hence

$$\Pr(M) = \Pr(M \wedge H^*) + \Pr(M|\overline{H^*}) \Pr(\overline{H^*}) \leq \Pr(M \wedge H^*) + \frac{1}{2},$$

so $\Pr(M \wedge H^*) \geq p(\lambda)$. The event $M \wedge H^*$ will henceforth be denoted by M^*.

Let s^t denote the test session selected by $\mathcal{M}$, and let s^m denote its matching session (if it exists). Consider the following events:

1. Event $E1$: $\mathcal{M}$ issues neither *Corrupt*$(\hat{A})$ nor *Corrupt*$(\hat{B})$.
2. Event $E2$: The event is comprised of the following sub-events:
   - the owner of s^t is the session initiator and $\mathcal{M}$ does not issue *SessionStateReveal*(s^t);
   - the owner of s^t is the session responder, the session matching to s^t exists, and $\mathcal{M}$ does not issue *SessionStateReveal*(s^m).

It is easy to see that $M^* = (M^* \wedge E1) \vee (M^* \wedge E2)$. Since $\Pr(M^*)$ is non-negligible, it must be the case that either $p_1 = \Pr(M^* \wedge E1)$ or $p_2 = \Pr(M^* \wedge E2)$ is non-negligible. The events $E1$ and $E2$ are considered separately.

We will show how to construct a solver $\mathcal{S}$ that takes as input a CDH challenge (U, V), has access to adversary $\mathcal{M}$ and a DDH oracle, and produces a solution to the CDH challenge. We use the following conventions: the DDH oracle on input (g^a, g^b, g^c) returns the bit 0 if $g^c \neq g^{ab}$, and the bit 1 if $g^c = g^{ab}$. Also, $\xi : \mathcal{G} \times \mathcal{G} \to \mathcal{G}$ is a random function known only to $\mathcal{S}$ and such that $\xi(X, Y) = \xi(Y, X)$ for all $X, Y \in \mathcal{G}$. The algorithm $\mathcal{S}$, which simulates $\mathcal{M}$'s environment, will use $\xi(S, T)$ to 'represent' CDH(S, T) in certain situations where $\mathcal{S}$ does not know $\log_g S$ or $\log_g T$. Except with negligible probability, $\mathcal{M}$ will not detect that $\xi(S, T)$ is being used instead of CDH(S, T).

Event $M^* \wedge E1$. Suppose that $M^* \wedge E1$ occurs with non-negligible probability. In this case $\mathcal{S}$ establishes n parties. Two of these parties denoted by $\hat{U}$ and $\hat{V}$ are assigned static public keys U and V, respectively. The remaining $n-2$ parties are assigned random static key pairs. The simulation of $\mathcal{M}$'s environment proceeds as follows:

1. *Send*$(\hat{A}, \hat{B})$. $\mathcal{S}$ follows protocol UM_1. However, if $\hat{A} \in \{\hat{U}, \hat{V}\}$, then $\mathcal{S}$ deviates from the protocol description by setting $\sigma_s = \xi(A, B)$.

2. *Send*$(\hat{B}, \hat{A}, X)$. $\mathcal{S}$ follows protocol UM_1. However, if $\hat{B} \in \{\hat{U}, \hat{V}\}$, then $\mathcal{S}$ deviates from the protocol description by setting $\sigma_s = \xi(A, B)$, $\sigma_e = \xi(X, B)$.
3. *SessionStateReveal*(s). $\mathcal{S}$ responds faithfully to the query.
4. *Expire*(s). $\mathcal{S}$ responds faithfully to the query.
5. *SessionKeyReveal*(s). $\mathcal{S}$ responds faithfully to the query.
6. *Corrupt*$(\hat{A})$. If $\hat{A} \in \{\hat{U}, \hat{V}\}$ then $\mathcal{S}$ aborts with failure. Otherwise, $\mathcal{S}$ responds faithfully to the query.
7. $H(\sigma_e, \sigma_s, \hat{A}, \hat{B}, X)$.
   (a) If $\hat{A} \in \{\hat{U}, \hat{V}\}$ and $\sigma_s \neq \xi(A, B)$, then $\mathcal{S}$ obtains $\tau = \text{DDH}(A, B, \sigma_s)$. If $\tau = 1$ and $\{\hat{A}, \hat{B}\} = \{\hat{U}, \hat{V}\}$, then $\mathcal{S}$ aborts with success and outputs $\text{CDH}(U, V) = \sigma_s$. If $\tau = 1$ and $\{\hat{A}, \hat{B}\} \neq \{\hat{U}, \hat{V}\}$, then $\mathcal{S}$ returns $H(\sigma_e, \xi(A, B), \hat{A}, \hat{B}, X)$.
   (b) If $\hat{B} \in \{\hat{U}, \hat{V}\}$ and either $\sigma_e \neq \xi(X, B)$ or $\sigma_s \neq \xi(A, B)$, then $\mathcal{S}$ sets $\tau_e = 1$ if either $\text{DDH}(X, B, \sigma_e) = 1$ or $\sigma_e = \xi(X, B)$. Similarly, $\mathcal{S}$ sets $\tau_s = 1$ if either $\text{DDH}(A, B, \sigma_s) = 1$ or $\sigma_s = \xi(A, B)$. If $\tau_e = 1$ and $\tau_s = 1$, then $\mathcal{S}$ returns $H(\xi(X, B), \xi(A, B), \hat{A}, \hat{B}, X)$.
   (c) $\mathcal{S}$ simulates a random oracle in the usual way[4].
8. *Test*(s). If $\hat{U}$ and $\hat{V}$ are not the session peers then $\mathcal{S}$ aborts with failure. Otherwise, $\mathcal{S}$ answers the query faithfully.

Analysis of $M^* \wedge E1$. The simulation of $\mathcal{M}$'s environment is perfect except with negligible probability. The probability that $\hat{U}$ and $\hat{V}$ are the communicating parties of the test session is at least $2/n^2$, in which case $\mathcal{S}$ does not abort as in Step 8. Suppose this is indeed the case and suppose that event $M^* \wedge E1$ occurs. In event $E1$ the adversary does not corrupt the test session's communicating peers and therefore failure as in Step 6 does not occur. Under event M^* the adversary queries H with $\sigma_s = \text{CDH}(U, V)$ and thus $\mathcal{S}$ is successful as in Step 7a. The probability of $\mathcal{S}$'s success in this event is therefore bounded by $\Pr(S) \geq 2p_1/n^2$.

Event $M^* \wedge E2$. Suppose that $M^* \wedge E2$ occurs with non-negligible probability. In this case $\mathcal{S}$ establishes n parties. One of these parties, denoted by $\hat{V}$, is assigned the static public V. The remaining parties are assigned random static key pairs. Further, $\mathcal{S}$ selects a random integer $r \in_R [1, nt]$. The rth session created will be called s^U. The simulation of $\mathcal{M}$'s environment proceeds as follows:

1. *Send*$(\hat{A}, \hat{B})$. $\mathcal{S}$ follows protocol UM_1. However, if $\hat{A} = \hat{V}$, then $\mathcal{S}$ deviates from the protocol description by setting $\sigma_s = \xi(A, B)$. Further, if the session activated is s^U, then $\mathcal{S}$ deviates from the protocol description by setting $X = U$ and $\sigma_e = \xi(X, B)$.
2. *Send*$(\hat{B}, \hat{A}, X)$. $\mathcal{S}$ follows protocol UM_1. However, if $\hat{B} = \hat{V}$, then $\mathcal{S}$ deviates from the protocol description by setting $\sigma_s = \xi(A, B)$ and $\sigma_e = \xi(X, B)$.
3. *SessionStateReveal*(s). If $s = s^U$ then $\mathcal{S}$ aborts with failure. Otherwise, $\mathcal{S}$ responds faithfully to the query.
4. *Expire*(s). $\mathcal{S}$ responds faithfully to the query.

[4] i.e., $\mathcal{S}$ returns random values for new queries and replays answers if the queries were previously made.

5. *SessionKeyReveal*(s). If s is either s^U or s^U's matching session then $\mathcal{S}$ aborts with failure. Otherwise, $\mathcal{S}$ responds faithfully to the query.
6. *Corrupt*$(\hat{A})$. If $\hat{A} = \hat{V}$ then $\mathcal{S}$ aborts with failure. Otherwise, $\mathcal{S}$ responds faithfully to the query.
7. $H(\sigma_e, \sigma_s, \hat{A}, \hat{B}, X)$.
   (a) If $\hat{A} = \hat{V}$ and $\sigma_s \neq \xi(A, B)$, then $\mathcal{S}$ obtains $\tau = \mathrm{DDH}(A, B, \sigma_s)$. If $\tau = 1$, then $\mathcal{S}$ returns $H(\sigma_e, \xi(A, B), \hat{A}, \hat{B}, X)$.
   (b) If $X = U$ and $\sigma_e \neq \xi(X, B)$, then $\mathcal{S}$ obtains $\tau = \mathrm{DDH}(X, B, \sigma_e)$. If $\tau = 1$ and $\hat{B} = \hat{V}$, then $\mathcal{S}$ aborts with success and outputs $\mathrm{CDH}(U, V) = \sigma_e$. If $\tau = 1$ and $\hat{B} \neq \hat{V}$, then $\mathcal{S}$ returns $H(\xi(X, B), \sigma_s, \hat{A}, \hat{B}, X)$.
   (c) If $\hat{B} = \hat{V}$ and either $\sigma_e \neq \xi(X, B)$ or $\sigma_s \neq \xi(A, B)$, then $\mathcal{S}$ sets $\tau_e = 1$ if either $\mathrm{DDH}(X, B, \sigma_e) = 1$ or $\sigma_e = \xi(X, B)$. Similarly, $\mathcal{S}$ sets $\tau_s = 1$ if either $\mathrm{DDH}(A, B, \sigma_s) = 1$ or $\sigma_s = \xi(A, B)$. If $\tau_e = 1$ and $\tau_s = 1$, then $\mathcal{S}$ returns $H(\xi(X, B), \xi(A, B), \hat{A}, \hat{B}, X)$.
   (d) $\mathcal{S}$ simulates a random oracle in the usual way.
8. *Test*(s). If the responder $\hat{B}$ is not $\hat{V}$ and the ephemeral public key X is not U then $\mathcal{S}$ aborts with failure. Otherwise, $\mathcal{S}$ answers the query faithfully.

Analysis of $M^* \wedge E2$. The simulation of $\mathcal{M}$'s environment is perfect except with negligible probability. Suppose that event $E2$ occurs; then the probability that $\hat{B} = \hat{V}$ and that either the test session or its matching session is s^U (that is $X = U$) is at least $2/tn^2$. Suppose this is the case, whence $\mathcal{S}$ does not abort as in Step 8. Note that in this case the test session's responder is $\hat{V}$ and ephemeral public key is U. Suppose also that event $M^* \wedge E2$ occurs. Since the adversary is successful, the test session is fresh and therefore $\mathcal{S}$ does not abort as in Steps 3, 5 and 6. Under event M^* the adversary queries H with σ_e such that $\sigma_e = \mathrm{CDH}(U, V)$, and thus $\mathcal{S}$ is successful as in Step 7b. The probability of $\mathcal{S}$'s success in this event is therefore bounded by $\Pr(S) \geq 2p_2/tn^2$.

Overall analysis. During the simulation, $\mathcal{S}$ performs group exponentiations, simulates oracle queries, and accesses the DDH oracle, all of which take polynomial time. If $\mathcal{M}$'s running is also polynomial then $\mathcal{S}$ is a polynomially-bounded algorithm that succeeds in solving the CDH challenge with probability at least

$$\Pr(S) \geq \frac{2}{n^2} \max\left(p_1, \frac{1}{t} p_2\right). \tag{1}$$

Thus $\mathcal{S}$ is a polynomially-bounded CDH solver, contradicting the GDH assumption in $\mathcal{G}$. □

4 Combined Security Model for UM_1 and UM_3

In this section we propose a combined security model for UM_1 and UM_3. The aim is to capture the security assurances guaranteed by the individual protocols even when parties use the same static key in the two protocols. Note that such reuse is explicitly allowed by the NIST standard [12], but can lead to a security vulnerability [4]. The description of the combined model closely follows the description of the individual models of one-pass (see §3) and three-pass UM (see [11]) and is essentially a "union" of the two models. The crucial addition is that of an identifier Π in the protocol description. For the one-pass protocol this identifier Π is set to UM_1 and for the three-pass protocol it is set to UM_3.

In the following, we assume that messages are represented as vectors of binary strings. If m is a vector then $\#m$ denotes the number of its components. Two vectors m_1 and m_2 are said to be *matched*, written $m_1 \sim m_2$, if the first $t = \min\{\#m_1, \#m_2\}$ components of the vectors are pairwise equal as binary strings.

Session creation. A party $\hat{A}$ can be activated via an incoming message to create a session. The message has the form $(\Pi, \hat{A}, \hat{B})$ or $(\Pi, \hat{A}, \hat{B}, Y)$, where $\Pi \in \{\mathsf{UM}_1, \mathsf{UM}_3\}$ identifies which protocol is activated. If $\hat{A}$ was activated with $(\Pi, \hat{A}, \hat{B})$ then $\hat{A}$ is the session *initiator*; otherwise $\hat{A}$ is the session *responder.*

Session initiator. If $\hat{A}$ is the session initiator then $\hat{A}$ creates a separate session state where session-specific short-lived data is stored and selects an ephemeral public key X. The session is labeled *active.* If $\Pi = \mathsf{UM}_1$ the session is identified via a session identifier $s = (\mathsf{UM}_1, \hat{A}, \hat{B}, X)$ and a session key is computed as per the protocol specification. $\hat{A}$ sends the message $(\mathsf{UM}_1, \hat{B}, \hat{A}, X)$ and then completes the session. If $\Pi = \mathsf{UM}_3$ then the session is identified via a (temporary and incomplete) session identifier $s = (\mathsf{UM}_3, \hat{A}, \hat{B}, X, *, *, *)$. In this case the outgoing message is $(\mathsf{UM}_3, \hat{B}, \hat{A}, X)$.

Session responder. If $\hat{A}$ is the session responder and $\Pi = \mathsf{UM}_1$, then $\hat{A}$ creates a separate session state that is identified by a session identifier $s = (\mathsf{UM}_1, \hat{A}, \hat{B}, Y)$, computes the session key as per the one-pass UM protocol, and completes the session. If $\hat{A}$ is the session responder and $\Pi = \mathsf{UM}_3$, then $\hat{A}$ creates a separate session state and prepares an ephemeral public key X and key confirmation tag t_A. The session is labeled active and identified via a (temporary and incomplete) session identifier $s = (\mathsf{UM}_3, \hat{A}, \hat{B}, Y, X, t_A, *)$. The outgoing message is $(\mathsf{UM}_3, \hat{B}, \hat{A}, Y, X, t_A)$.

Session update. This is applicable only in the case of UM_3. A party $\hat{A}$ can be activated to update an active session via an incoming message of the form $(\mathsf{UM}_3, \hat{A}, \hat{B}, X, Y, t_B)$ or $(\mathsf{UM}_3, \hat{A}, \hat{B}, Y, X, t_A, t_B)$. If the message is of the former type then upon receipt of this message, $\hat{A}$ checks whether she owns an active session with identifier $s = (\mathsf{UM}_3, \hat{A}, \hat{B}, X, *, *, *)$; except with negligible probability, $\hat{A}$ can own at most one such session. If no such session exists then the message is rejected; otherwise, $\hat{A}$ prepares a key confirmation tag t_A, updates the identifier to $s = (\mathsf{UM}_3, \hat{A}, \hat{B}, X, Y, t_B, t_A)$, sends the message $(\mathsf{UM}_3, \hat{B}, \hat{A}, X, Y, t_B, t_A)$, and completes the session by accepting a session key. On the other hand, if the incoming message is $(\mathsf{UM}_3, \hat{A}, \hat{B}, Y, X, t_A, t_B)$, then $\hat{A}$ first checks whether she owns an active session with identifier $s = (\mathsf{UM}_3, \hat{A}, \hat{B}, Y, X, t_A, *)$. If not the message is rejected; otherwise $\hat{A}$ updates the identifier to $s = (\mathsf{UM}_3, \hat{A}, \hat{B}, Y, X, t_A, t_B)$ and completes the session by accepting a session key.

Aborted sessions. Both protocols require that parties perform some checks on incoming messages. For example, in both UM_1 and UM_3 the parties need to perform some form of public-key validation while in UM_3 parties are required to verify a tag. If in a protocol a party is activated to create a session with an incoming message that does not meet the protocol specifications, then that message is rejected and no session is created. If a party is activated in UM_3 to update an active session with an incoming message that does not meet the protocol specifications, then the party deletes all information specific to that session (including the session state and the session key if it has been computed) and *aborts* the session. Abortion occurs before the session identifier is updated.

Matching sessions. A session s with identifier $(\mathsf{UM}_i, \ldots)$, $i \in \{1, 3\}$, is called a UM_i*-session*. For a session $(\mathsf{UM}_i, \hat{A}, \hat{B}, \ldots)$, we call $\hat{A}$ the session *owner*, $\hat{B}$ the session *peer*, and together $\hat{A}$ and $\hat{B}$ are referred to as the *communicating parties*. Let $s = (\mathsf{UM}_i, \hat{A}, \hat{B}, Comm_A)$ be a session owned by $\hat{A}$. A session $s = (\mathsf{UM}_j, \hat{C}, \hat{D}, Comm_C)$ is said to be *matching* to s if $i = j$, $\hat{A} = \hat{D}$, $\hat{B} = \hat{C}$ and $Comm_A \sim Comm_C$. It can be seen that the session s, except with negligible probability, can have more than one matching session if and only if $Comm_A$ has exactly one component, i.e., is comprised of a single outgoing message.

Adversary. As in §3.1, the adversary $\mathcal{M}$ is a probabilistic Turing machine, controls all communications, and presents messages to parties via the *Send* query. In addition, $\mathcal{M}$ can issue *SessionStateReveal*, *Expire*, *SessionKeyReveal*, *Corrupt* and *Test* queries as described in §3.1. In the case of UM_1, the *SessionStateReveal* query can only be issued at an initiator, while in UM_3 the query can be issued to initiators and responders.

Remark 3. The adversary in this combined model has the same power that she enjoys against UM_1 or UM_3 when run in isolation. Nevertheless, the model allows $\mathcal{M}$ considerable additional strength, namely using the information learned in protocol sessions of UM_1 to attack protocol sessions of UM_3 and vice versa. In fact, this is precisely the attack scenario that was considered in [4].

Definition 4 (fresh session). Let s be the identifier of a completed Π-session, owned by party $\hat{A}$ with peer $\hat{B}$. Let s^* be the identifier of the matching session of s, if the matching session exists. If $\Pi = \mathsf{UM}_1$, define s to be *fresh* if none of the following conditions hold:

1. $\mathcal{M}$ issued a *SessionKeyReveal*(s) query or a *SessionKeyReveal*(s^*) query.
2. $\hat{A}$ is the initiator and one of the following holds: (a) $\mathcal{M}$ issued *Corrupt*$(\hat{A})$ before *Expire*(s); (b) $\mathcal{M}$ issued *SessionStateReveal*(s) and *Corrupt*$(\hat{A})$; (c) $\mathcal{M}$ issued *Corrupt*$(\hat{B})$.
3. $\hat{A}$ is the responder and one of the following holds: (a) $\mathcal{M}$ issued *Corrupt*$(\hat{A})$; (b) s^* exists and $\mathcal{M}$ issued either *Corrupt*(B) before *Expire*(s^*) or both *Corrupt*$(\hat{B})$ and *SessionStateReveal*(s^*); (c) s^* does not exist and $\mathcal{M}$ issued *Corrupt*$(\hat{B})$.

If $\Pi = \mathsf{UM}_3$, define s to be *fresh* if none of the following conditions hold:

1. $\mathcal{M}$ issued a *SessionKeyReveal*(s) query or a *SessionKeyReveal*(s^*) query.
2. $\mathcal{M}$ issued *Corrupt*$(\hat{A})$ before *Expire*(s).

3. $\mathcal{M}$ issued *SessionStateReveal*(s) and either *Corrupt*$(\hat{A})$ or *Corrupt*$(\hat{B})$.
4. s^* exists and $\mathcal{M}$ issued one of the following:
   (a) *Corrupt*$(\hat{B})$ before *Expire*(s^*).
   (b) *SessionStateReveal*(s^*) and either *Corrupt*$(\hat{A})$ or *Corrupt*$(\hat{B})$.
5. s^* does not exist and $\mathcal{M}$ issued *Corrupt*$(\hat{B})$ before *Expire*(s).

Definition 5 (security in the combined model). UM_1 and UM_3 are said to be *secure* in the combined model if the following conditions hold:

1. For $i \in \{1,3\}$ if two honest parties complete matching UM_i-sessions then, except with negligible probability, they both compute the same session key.
2. For $i \in \{1,3\}$ no polynomially-bounded adversary $\mathcal{M}$ can distinguish the session key of a fresh UM_i-session from a randomly chosen session key with probability greater than $\frac{1}{2}$ plus a negligible fraction.

To circumvent the protocol interference attack of [4] on one- and three-pass UM in the combined model, one-pass UM (see §3.2) is modified by including the protocol identifier UM_1 (in addition to the ephemeral public key X) in the optional input Λ to the key derivation function. Similarly, the protocol identifier UM_3 is included in the optional input Λ (in addition to the ephemeral public keys X and Y) to the key derivation function in three-pass UM (see §3.2).

5 Security Argument for Shared Reuse

This section provides a reductionist security argument for UM_1 and UM_3 in the combined model.

Theorem 2. *If H is a random oracle, the MAC scheme is secure, and $\mathcal{G}$ is a group where the GDH assumption holds, then UM_1 and UM_3 are secure in the combined model.*

Proof. It is easy to see that matching sessions compute the same session key. We will verify that no adversary $\mathcal{M}$ can distinguish the session key of a fresh session from a randomly chosen session key.

Let M denote the event that $\mathcal{M}$ is successful, and suppose that $\Pr(M) = \frac{1}{2} + p(\lambda)$ where $p(\lambda)$ is non-negligible. Let H^* denote the event that $\mathcal{M}$ queries H with $(\sigma_e, \sigma_s, \hat{A}, \hat{B}, \Lambda)$, where $s = (\Pi, \hat{A}, \hat{B}, *)$ is the test session or its matching session. Since H is a random function, we have $\Pr(M|\overline{H^*}) = \frac{1}{2}$ where negligible terms are ignored. It follows that $\Pr(M \wedge H^*) \geq p(\lambda)$. The event $M \wedge H^*$ will henceforth be denoted by M^*.

Let $s^t = (\Pi, \hat{A}, \hat{B}, *)$ denote the test session selected by $\mathcal{M}$, and let $s^m = (\Pi, \hat{B}, \hat{A}, *)$ denote its matching session (if it exists). Consider the following events:

1. Event E_1: $\Pi = \mathsf{UM}_3$, s^m exists, and $\mathcal{M}$ issues neither *SessionStateReveal*(s^t) nor *SessionStateReveal*(s^m).
2. Event E_2: $\mathcal{M}$ issues neither *Corrupt*$(\hat{A})$ nor *Corrupt*$(\hat{B})$.

3. Event E_3: $\Pi = \mathsf{UM}_1$ and either
   - the owner of s^t is the session initiator and $\mathcal{M}$ does not issue *SessionStateReveal*(s^t);
   - the owner of s^t is the session responder, s^m exists, and $\mathcal{M}$ does not issue *SessionStateReveal*(s^m).
4. Event E_4: $\Pi = \mathsf{UM}_3$, s^m does not exist, and $\mathcal{M}$ does not issue *SessionStateReveal*(s^t).

It can be see that $M^* = (M^* \wedge E_1) \vee (M^* \wedge E_2) \vee (M^* \wedge E_3) \vee (M^* \wedge E_4)$. Since $\Pr(M^*)$ is non-negligible, it must be the case that either $p_1 = \Pr(M^* \wedge E_1)$, $p_2 = \Pr(M^* \wedge E_2)$, $p_3 = \Pr(M^* \wedge E_3)$, or $p_4 = \Pr(M^* \wedge E_4)$ is non-negligible in λ. These events are considered separately.

We will show how to construct a solver $\mathcal{S}$ that takes as input a CDH challenge (U, V), has access to adversary $\mathcal{M}$, a DDH oracle, and a MAC oracle, and produces a solution to the CDH challenge or a MAC forgery. We adopt the same conventions for the DDH oracle and the ξ function as presented in the security argument for the one-pass protocol.

Event $M^* \wedge E_1$. Suppose that $M^* \wedge E_1$ occurs with non-negligible probability. In this case $\mathcal{S}$ establishes n parties, who are assigned random static key pairs, and selects $s_1, s_2 \in_R [1, nt]$. The s_1'th and s_2'th sessions created will be called s^U and s^V, respectively. The simulation of $\mathcal{M}$'s environment proceeds as follows:

1. *Send*$(\mathsf{UM}_1, \hat{A}, \hat{B})$. $\mathcal{S}$ follows protocol UM_1. However, if the session being created is the s_1'th or s_2'th session, then $\mathcal{S}$ aborts with failure.
2. *Send*$(\mathsf{UM}_1, \hat{A}, \hat{B}, X)$. $\mathcal{S}$ follows protocol UM_1. However, if the session being created is the s_1'th or s_2'th session, then $\mathcal{S}$ aborts with failure.
3. *Send*$(\mathsf{UM}_3, \hat{A}, \hat{B})$. $\mathcal{S}$ follows protocol UM_3. However, if the session being created is the s_1'th or s_2'th session, then $\mathcal{S}$ deviates from the protocol description by setting the ephemeral public key X to be U or V, respectively.
4. *Send*$(\mathsf{UM}_3, \hat{B}, \hat{A}, X)$. $\mathcal{S}$ follows protocol UM_3. However, if the session being created is the s_1'th or s_2'th session, then $\mathcal{S}$ deviates from the protocol description by setting the ephemeral public key Y to be U or V, respectively, and setting $\sigma_e = \xi(Y, X)$.
5. *Send*$(\mathsf{UM}_3, \hat{A}, \hat{B}, X, Y, t_B)$. $\mathcal{S}$ follows protocol UM_3. However, if $X \in \{U, V\}$, then $\mathcal{S}$ deviates from the protocol description by setting $\sigma_e = \xi(X, Y)$.
6. *Send*$(\mathsf{UM}_3, \hat{B}, \hat{A}, X, Y, t_B, t_A)$. $\mathcal{S}$ follows protocol UM_3.
7. *SessionStateReveal*(s). $\mathcal{S}$ responds faithfully to the query except if $s \in \{s^U, s^V\}$ in which case $\mathcal{S}$ aborts with failure.
8. *Expire*(s). $\mathcal{S}$ responds faithfully to the query.
9. *SessionKeyReveal*(s). $\mathcal{S}$ responds faithfully to the query except if $s \in \{s^U, s^V\}$ in which case $\mathcal{S}$ aborts with failure.
10. *Corrupt*$(\hat{A})$. If $\hat{A}$ owns session s^U or s^V, and that session is not expired, then $\mathcal{S}$ aborts with failure. Otherwise, $\mathcal{S}$ responds faithfully to the query.
11. $H(\sigma_e, \sigma_s, \hat{A}, \hat{B}, X, \mathsf{UM}_1)$. $\mathcal{S}$ simulates a random oracle in the usual way.
12. $H(\sigma_e, \sigma_s, \hat{A}, \hat{B}, X, Y, \mathsf{UM}_3)$.

(a) If $X \in \{U, V\}$ and $\sigma_e \neq \xi(X, Y)$, then $\mathcal{S}$ obtains $\tau = \text{DDH}(X, Y, \sigma_e)$. If $\tau = 1$ and $Y \in \{U, V\}$ and $Y \neq X$, then $\mathcal{S}$ aborts with success and outputs $\text{CDH}(U, V) = \sigma_e$. If $\tau = 1$ and either $Y \notin \{U, V\}$ or $Y = X$, then $\mathcal{S}$ returns $H(\xi(X, Y), \sigma_s, \hat{A}, \hat{B}, X, Y, \mathsf{UM}_3)$.
(b) If $Y \in \{U, V\}$ and $\sigma_e \neq \xi(X, Y)$, then $\mathcal{S}$ obtains $\tau = \text{DDH}(X, Y, \sigma_e)$. If $\tau = 1$, then $\mathcal{S}$ returns $H(\xi(X,Y), \sigma_s, \hat{A}, \hat{B}, X, Y, \mathsf{UM}_3)$.
(c) $\mathcal{S}$ simulates a random oracle in the usual way.

13. *Test*(s). If $s \notin \{s^U, s^V\}$ or if s^U and s^V are non-matching, then $\mathcal{S}$ aborts with failure. Otherwise, $\mathcal{S}$ answers the query faithfully.

Analysis. $\mathcal{S}$'s simulation of $\mathcal{M}$'s environment is perfect except with negligible probability. Suppose that event $M^* \wedge E_1$ occurs. The probability that $\mathcal{M}$ selects one of s^U, s^V as the test session and the other as its matching session is at least $2/(nt)^2$. In this case $\mathcal{S}$ does not abort as described in Steps 1, 2, 7 and 13. Since the test session is fresh, $\mathcal{S}$ does not abort as described in Steps 9 and 10. Except with negligible probability of guessing $\xi(U, V)$, a successful $\mathcal{M}$ queries H with $(\text{CDH}(U, V), \text{CDH}(A, B), \hat{A}, \hat{B}, X, Y, \mathsf{UM}_3)$ where $\{X, Y\} = \{U, V\}$, in which case $\mathcal{S}$ is successful as described in Step 12a. The probability that $\mathcal{S}$ is successful is bounded by $\Pr(\mathcal{S}) \geq 2p_1/(nt)^2$, where negligible terms are ignored.

Event $M^* \wedge E_2$. Suppose that $M^* \wedge E_2$ occurs with non-negligible probability. In this case $\mathcal{S}$ establishes n parties, two of which, denoted $\hat{U}$ and $\hat{V}$, are assigned static public keys U and V, respectively. The remaining $n-2$ parties are assigned random static key pairs. The simulation of $\mathcal{M}$'s environment proceeds as follows:

1. *Send*$(\mathsf{UM}_1, \hat{A}, \hat{B})$. $\mathcal{S}$ follows protocol UM_1. However, if $\hat{A} \in \{\hat{U}, \hat{V}\}$, then $\mathcal{S}$ deviates from the protocol description by setting $\sigma_s = \xi(A, B)$.
2. *Send*$(\mathsf{UM}_1, \hat{B}, \hat{A}, X)$. $\mathcal{S}$ follows protocol UM_1. However, if $\hat{B} \in \{\hat{U}, \hat{V}\}$, then $\mathcal{S}$ deviates from the protocol description by setting $\sigma_s = \xi(A, B)$, $\sigma_e = \xi(X, B)$.
3. *Send*$(\mathsf{UM}_3, \hat{A}, \hat{B})$. $\mathcal{S}$ follows protocol UM_3.
4. *Send*$(\mathsf{UM}_3, \hat{B}, \hat{A}, X)$. $\mathcal{S}$ follows protocol UM_3. However, if $\hat{B} \in \{\hat{U}, \hat{V}\}$, then $\mathcal{S}$ deviates from the protocol description by setting $\sigma_s = \xi(A, B)$.
5. *Send*$(\mathsf{UM}_3, \hat{A}, \hat{B}, X, Y, t_B)$. $\mathcal{S}$ follows protocol UM_3. However, if $\hat{A} \in \{\hat{U}, \hat{V}\}$, then $\mathcal{S}$ deviates from the protocol description by setting $\sigma_s = \xi(A, B)$.
6. *Send*$(\mathsf{UM}_3, \hat{B}, \hat{A}, X, Y, t_B, t_A)$. $\mathcal{S}$ follows protocol UM_3.
7. *SessionStateReveal*(s). $\mathcal{S}$ responds faithfully to the query.
8. *Expire*(s). $\mathcal{S}$ responds faithfully to the query.
9. *SessionKeyReveal*(s). $\mathcal{S}$ responds faithfully to the query.
10. *Corrupt*$(\hat{A})$. If $\hat{A} \in \{\hat{U}, \hat{V}\}$ then $\mathcal{S}$ aborts with failure. Otherwise, $\mathcal{S}$ responds faithfully to the query.
11. $H(\sigma_e, \sigma_s, \hat{A}, \hat{B}, X, \mathsf{UM}_1)$.
    (a) If $\hat{A} \in \{\hat{U}, \hat{V}\}$ and $\sigma_s \neq \xi(A, B)$, then $\mathcal{S}$ obtains $\tau = \text{DDH}(A, B, \sigma_s)$. If $\tau = 1$ and $\{\hat{A}, \hat{B}\} = \{\hat{U}, \hat{V}\}$, then $\mathcal{S}$ aborts with success and outputs $\text{CDH}(U, V) = \sigma_s$. If $\tau = 1$ and $\{\hat{A}, \hat{B}\} \neq \{\hat{U}, \hat{V}\}$, then $\mathcal{S}$ returns $H(\sigma_e, \xi(A, B), \hat{A}, \hat{B}, X, \mathsf{UM}_1)$.

   (b) If $\hat{B} \in \{\hat{U}, \hat{V}\}$ and either $\sigma_e \neq \xi(X, B)$ or $\sigma_s \neq \xi(A, B)$, then $\mathcal{S}$ sets $\tau_e = 1$ if either $\text{DDH}(X, B, \sigma_e) = 1$ or $\sigma_e = \xi(X, B)$. Similarly, $\mathcal{S}$ sets $\tau_s = 1$ if either $\text{DDH}(A, B, \sigma_s) = 1$ or $\sigma_s = \xi(A, B)$.
   If $\tau_e = 1$ and $\tau_s = 1$, then $\mathcal{S}$ returns $H(\xi(X, B), \xi(A, B), \hat{A}, \hat{B}, X, \mathsf{UM}_1)$.
   (c) $\mathcal{S}$ simulates a random oracle in the usual way.
12. $H(\sigma_e, \sigma_s, \hat{A}, \hat{B}, X, Y, \mathsf{UM}_3)$.
   (a) If $\hat{A} \in \{\hat{U}, \hat{V}\}$ and $\sigma_s \neq \xi(A, B)$, then $\mathcal{S}$ obtains $\tau = \text{DDH}(A, B, \sigma_s)$. If $\tau = 1$ and $\hat{B} \in \{\hat{U}, \hat{V}\}$, then $\mathcal{S}$ aborts with success and outputs $\text{CDH}(U, V) = \sigma_s$. If $\tau = 1$ and $\hat{B} \notin \{\hat{U}, \hat{V}\}$, then $\mathcal{S}$ returns $H(\sigma_e, \xi(A, B), \hat{A}, \hat{B}, X, Y, \mathsf{UM}_3)$.
   (b) If $\hat{B} \in \{\hat{U}, \hat{V}\}$ and $\sigma_s \neq \xi(A, B)$, then $\mathcal{S}$ obtains $\tau = \text{DDH}(A, B, \sigma_s)$. If $\tau = 1$, then $\mathcal{S}$ returns $H(\sigma_e, \xi(A, B), \hat{A}, \hat{B}, X, Y, \mathsf{UM}_3)$.
   (c) $\mathcal{S}$ simulates a random oracle in the usual way.
13. *Test*(s). If $\hat{U}$ and $\hat{V}$ are not the session peers then $\mathcal{S}$ aborts with failure. Otherwise, $\mathcal{S}$ answers the query faithfully.

Analysis. The simulation of $\mathcal{M}$'s environment is perfect except with negligible probability. The probability that $\hat{U}$ and $\hat{V}$ are the communicating parties of the test session is at least $2/n^2$, in which case $\mathcal{S}$ does not abort as in Step 13. Suppose this is indeed the case and suppose that event $M^* \wedge E_2$ occurs. In this event $\mathcal{M}$ does not corrupt the test session communicating peers and therefore failure as in Step 10 does not occur. Under event M^* the adversary queries H with $\sigma_s = \text{CDH}(U, V)$, and thus $\mathcal{S}$ is successful as in Step 11a or 12a. The probability of $\mathcal{S}$'s success in this event is therefore bounded by $\Pr(\mathcal{S}) \geq 2p_2/n^2$, where negligible terms are ignored.

Event $M^* \wedge E_3$. Suppose that $M^* \wedge E_3$ occurs with non-negligible probability. In this case $\mathcal{S}$ establishes n parties. One of these parties, denoted by $\hat{V}$, is assigned the static public V. The remaining parties are assigned random static key pairs. Furthermore, $\mathcal{S}$ selects a random integer $r \in_R [1, nt]$. The rth session created will be called s^U. The simulation of $\mathcal{M}$'s environment proceeds as follows:

1. *Send*$(\mathsf{UM}_1, \hat{A}, \hat{B})$. $\mathcal{S}$ follows protocol UM_1. However, if $\hat{A} = \hat{V}$, then $\mathcal{S}$ deviates from the protocol description by setting $\sigma_s = \xi(A, B)$. Further, if the session activated is s^U, then $\mathcal{S}$ deviates from the protocol description by setting $X = U$ and $\sigma_e = \xi(X, B)$.
2. *Send*$(\mathsf{UM}_1, \hat{B}, \hat{A}, X)$. $\mathcal{S}$ follows protocol UM_1. However, if $\hat{B} = \hat{V}$, then $\mathcal{S}$ deviates from the protocol description by setting $\sigma_s = \xi(A, B)$ and $\sigma_e = \xi(X, B)$.
3. *Send*$(\mathsf{UM}_3, \hat{A}, \hat{B})$. $\mathcal{S}$ follows protocol UM_3.
4. *Send*$(\mathsf{UM}_3, \hat{B}, \hat{A}, X)$. $\mathcal{S}$ follows protocol UM_3. However, if $\hat{B} = \hat{V}$, then $\mathcal{S}$ deviates from the protocol description by setting $\sigma_s = \xi(A, B)$.
5. *Send*$(\mathsf{UM}_3, \hat{A}, \hat{B}, X, Y, t_B)$. $\mathcal{S}$ follows protocol UM_3. However, if $\hat{A} = \hat{V}$, then $\mathcal{S}$ deviates from the protocol description by setting $\sigma_s = \xi(A, B)$.
6. *Send*$(\mathsf{UM}_3, \hat{B}, \hat{A}, X, Y, t_B, t_A)$. $\mathcal{S}$ follows protocol UM_3.
7. *SessionStateReveal*(s). If $s = s^U$ then $\mathcal{S}$ aborts with failure. Otherwise, $\mathcal{S}$ responds faithfully to the query.

8. *Expire*(s). $\mathcal{S}$ responds faithfully to the query.
9. *SessionKeyReveal*(s). If s is either s^U or s^U's matching session then $\mathcal{S}$ aborts with failure. Otherwise, $\mathcal{S}$ responds faithfully to the query.
10. *Corrupt*$(\hat{A})$. If $\hat{A} = \hat{V}$ then $\mathcal{S}$ aborts with failure. Otherwise, $\mathcal{S}$ responds faithfully to the query.
11. $H(\sigma_e, \sigma_s, \hat{A}, \hat{B}, X, \mathsf{UM}_1)$.
    (a) If $\hat{A} = \hat{V}$ and $\sigma_s \neq \xi(A, B)$, then $\mathcal{S}$ obtains $\tau = \mathrm{DDH}(A, B, \sigma_s)$. If $\tau = 1$, then $\mathcal{S}$ returns $H(\sigma_e, \xi(A, B), \hat{A}, \hat{B}, X, \mathsf{UM}_1)$.
    (b) If $X = U$ and $\sigma_e \neq \xi(X, B)$, then $\mathcal{S}$ obtains $\tau = \mathrm{DDH}(X, B, \sigma_e)$. If $\tau = 1$ and $\hat{B} = \hat{V}$, then $\mathcal{S}$ aborts with success and outputs $\mathrm{CDH}(U, V) = \sigma_e$. If $\tau = 1$ and $\hat{B} \neq \hat{V}$, then $\mathcal{S}$ returns $H(\xi(X, B), \sigma_s, \hat{A}, \hat{B}, X, \mathsf{UM}_1)$.
    (c) If $\hat{B} = \hat{V}$ and either $\sigma_e \neq \xi(X, B)$ or $\sigma_s \neq \xi(A, B)$, then $\mathcal{S}$ sets $\tau_e = 1$ if either $\mathrm{DDH}(X, B, \sigma_e) = 1$ or $\sigma_e = \xi(X, B)$. Similarly, $\mathcal{S}$ sets $\tau_s = 1$ if either $\mathrm{DDH}(A, B, \sigma_s) = 1$ or $\sigma_s = \xi(A, B)$. If $\tau_e = 1$ and $\tau_s = 1$, then $\mathcal{S}$ returns $H(\xi(X, B), \xi(A, B), \hat{A}, \hat{B}, X, \mathsf{UM}_1)$.
    (d) $\mathcal{S}$ simulates a random oracle in the usual way.
12. $H(\sigma_e, \sigma_s, \hat{A}, \hat{B}, X, Y, \mathsf{UM}_3)$.
    (a) If $\hat{A} = \hat{V}$ or $\hat{B} = \hat{V}$, and $\sigma_s \neq \xi(A, B)$, then $\mathcal{S}$ obtains $\tau = \mathrm{DDH}(X, Y, \sigma_e)$. If $\tau = 1$, then $\mathcal{S}$ returns $H(\sigma_e, \xi(A, B), \hat{A}, \hat{B}, X, Y, \mathsf{UM}_3)$.
    (b) $\mathcal{S}$ simulates a random oracle in the usual way.
13. *Test*(s). If the responder $\hat{B}$ is not $\hat{V}$ and the ephemeral public key X is not U then $\mathcal{S}$ aborts with failure. Otherwise, $\mathcal{S}$ answers the query faithfully.

Analysis. The simulation of $\mathcal{M}$'s environment is perfect except with negligible probability. Suppose that event $M^* \wedge E_3$ occurs. The probability that $\hat{B} = \hat{V}$ and that either the test session or its matching session is s^U (i.e., $X = U$) is at least $2/n^2t$. Suppose this is the case, whence $\mathcal{S}$ does not abort as in Step 13. Since $\mathcal{M}$ is successful, the test session is fresh and therefore $\mathcal{S}$ does not abort as in Steps 7, 9 and 10. Under event M^* the adversary queries H with $\sigma_e = \mathrm{CDH}(U, V)$ and thus $\mathcal{S}$ is successful as in Step 11b. The probability of $\mathcal{S}$'s success in this event is therefore bounded by $\Pr(S) \geq 2p_3/n^2t$, where negligible terms are ignored.

Event $M \wedge E_4$. Suppose that $M \wedge E_4$ occurs with non-negligible probability. In this case $\mathcal{S}$ establishes n parties. Two of these parties, denoted by $\hat{U}$ and $\hat{V}$, are assigned static public keys U and V, respectively. The remaining $n-2$ parties are assigned random static key pairs. Furthermore, $\mathcal{S}$ is given a MAC oracle with key $\tilde{k}$ that is unknown to $\mathcal{S}$. $\mathcal{S}$ selects $r \in_R [1, nt]$; the rth session created will be called s^r. The simulation of $\mathcal{M}$'s environment proceeds as follows:

1. *Send*$(\mathsf{UM}_1, \hat{A}, \hat{B})$. $\mathcal{S}$ follows protocol UM_1. However, if $\hat{A} \in \{\hat{U}, \hat{V}\}$, then $\mathcal{S}$ deviates from the protocol description by setting $\sigma_s = \xi(A, B)$. If the session created is the rth session, then $\mathcal{S}$ aborts with failure.
2. *Send*$(\mathsf{UM}_1, \hat{B}, \hat{A}, X)$. $\mathcal{S}$ follows protocol UM_1 However, if $\hat{B} \in \{\hat{U}, \hat{V}\}$, then $\mathcal{S}$ deviates from the protocol description by setting $\sigma_s = \xi(A, B)$, $\sigma_e = \xi(X, B)$. If the session created is the rth session, then $\mathcal{S}$ aborts with failure.
3. *Send*$(\mathsf{UM}_3, \hat{A}, \hat{B})$. $\mathcal{S}$ follows protocol UM_3. If the session created is the rth session and $\{\hat{A}, \hat{B}\} \neq \{\hat{U}, \hat{V}\}$, then $\mathcal{S}$ aborts with failure.

4. *Send*($\mathsf{UM}_3, \hat{B}, \hat{A}, X$). $\mathcal{S}$ follows protocol UM_3. However, if $\hat{B} \in \{\hat{U}, \hat{V}\}$, then $\mathcal{S}$ deviates from the protocol description by setting $\sigma_s = \xi(A, B)$. If the created session is the rth session, then $\mathcal{S}$ deviates from the protocol description as follows. If $\{\hat{A}, \hat{B}\} \neq \{\hat{U}, \hat{V}\}$ then $\mathcal{S}$ aborts with failure. Otherwise, $\mathcal{S}$ selects a random session key κ and sets the MAC key κ' equal to the (unknown) key $\tilde{k}$ of the MAC oracle. $\mathcal{S}$ queries the MAC oracle with $(\mathcal{R}, \hat{B}, \hat{A}, Y, X)$ and sets t_B equal to the oracle response.
5. *Send*($\mathsf{UM}_3, \hat{A}, \hat{B}, X, Y, t_B$). $\mathcal{S}$ follows protocol UM_3. However, if $\hat{A} \in \{\hat{U}, \hat{V}\}$, then $\mathcal{S}$ deviates from the protocol description by setting $\sigma_s = \xi(A, B)$. If $\hat{A}$ was activated to update s^r, then $\mathcal{S}$ selects a random session key κ and sets the MAC key κ' equal to the (unknown) key $\tilde{k}$ of the MAC oracle. $\mathcal{S}$ queries the MAC oracle with $(\mathcal{I}, \hat{A}, \hat{B}, X, Y)$ and sets t_A equal to the oracle response.
6. *Send*($\mathsf{UM}_3, \hat{B}, \hat{A}, X, Y, t_B, t_A$). $\mathcal{S}$ follows protocol UM_3. If $\hat{B}$ was activated to update s^r, then $\mathcal{S}$ completes the session without verifying the received t_A.
7. *SessionStateReveal*(s). $\mathcal{S}$ answers the query faithfully. However, if $s = s^r$ and the owner of s^r is the session responder, then $\mathcal{S}$ aborts with failure.
8. *Expire*(s). $\mathcal{S}$ answers the query faithfully. However, if $s = s^r$ and s^r does not have a matching session, then $\mathcal{S}$ aborts with success and outputs as its MAC forgery the key confirmation tag received by s (and the associated message).
9. *SessionKeyReveal*(s). $\mathcal{S}$ answers the query faithfully.
10. *Corrupt*($\hat{A}$). If $\hat{A} \in \{\hat{U}, \hat{V}\}$ then $\mathcal{S}$ aborts with failure. Otherwise, $\mathcal{S}$ answers the query faithfully.
11. $H(\sigma_e, \sigma_s, \hat{A}, \hat{B}, X, \mathsf{UM}_1)$.
    (a) If $\hat{A} \in \{\hat{U}, \hat{V}\}$ and $\sigma_s \neq \xi(A, B)$, then $\mathcal{S}$ obtains $\tau = \mathrm{DDH}(X, Y, \sigma_s)$. If $\tau = 1$ and $\{\hat{A}, \hat{B}\} = \{\hat{U}, \hat{V}\}$, then $\mathcal{S}$ aborts with success and outputs $\mathrm{CDH}(U, V) = \sigma_s$. If $\tau = 1$ and $\{\hat{A}, \hat{B}\} \neq \{\hat{U}, \hat{V}\}$, then $\mathcal{S}$ returns $H(\sigma_e, \xi(A, B), \hat{A}, \hat{B}, X, \mathsf{UM}_1)$.
    (b) If $\hat{B} \in \{\hat{U}, \hat{V}\}$ and either $\sigma_e \neq \xi(X, B)$ or $\sigma_s \neq \xi(A, B)$, then $\mathcal{S}$ sets $\tau_e = 1$ if either $\mathrm{DDH}(X, B, \sigma_e) = 1$ or $\sigma_e = \xi(X, B)$. Similarly, $\mathcal{S}$ sets $\tau_s = 1$ if either $\mathrm{DDH}(A, B, \sigma_s) = 1$ or $\sigma_s = \xi(A, B)$. If $\tau_e = 1$ and $\tau_s = 1$, then $\mathcal{S}$ returns $H(\xi(X, B), \xi(A, B), \hat{A}, \hat{B}, X, \mathsf{UM}_1)$.
    (c) $\mathcal{S}$ simulates a random oracle in the usual way.
12. $H(\sigma_e, \sigma_s, \hat{A}, \hat{B}, X, Y, \mathsf{UM}_3)$.
    (a) If $\hat{A} \in \{\hat{U}, \hat{V}\}$ and $\sigma_s \neq \xi(A, B)$, then $\mathcal{S}$ obtains $\tau = \mathrm{DDH}(A, B, \sigma_s)$. If $\tau = 1$ and $\hat{B} \in \{\hat{U}, \hat{V}\}$ and $\hat{B} \neq \hat{A}$, then $\mathcal{S}$ aborts with success and outputs $\mathrm{CDH}(U, V) = \sigma_s$. If $\tau = 1$ and either $\hat{B} \notin \{\hat{U}, \hat{V}\}$ or $\hat{B} = \hat{A}$, then $\mathcal{S}$ returns $H(\sigma_e, \xi(A, B), \hat{A}, \hat{B}, X, Y, \mathsf{UM}_3)$.
    (b) If $\hat{B} \in \{\hat{U}, \hat{V}\}$ and $\sigma_s \neq \xi(A, B)$, then $\mathcal{S}$ obtains $\tau = \mathrm{DDH}(A, B, \sigma_s)$. If $\tau = 1$, then $\mathcal{S}$ returns $H(\sigma_e, \xi(A, B), \hat{A}, \hat{B}, X, Y, \mathsf{UM}_3)$.
    (c) $\mathcal{S}$ simulates a random oracle in the usual way.
13. *Test*(s). If $s \neq s^r$ or if s^r has a matching session, then $\mathcal{S}$ aborts with failure. Otherwise, $\mathcal{S}$ answers the query faithfully.

Analysis. $\mathcal{S}$'s simulation of $\mathcal{M}$'s environment is perfect except with negligible probability. Suppose that event $M^* \wedge E_4$ occurs. The probability that the test session is the rth session, and $\hat{U}$ and $\hat{V}$ are its communicating parties, is at least $2/(n^3 t)$. Suppose that this is indeed the case (so $\mathcal{S}$ does not abort in Steps 1, 2, 3 and 4). Since event E_4 has occurred, $\mathcal{S}$ does not abort in Steps 7 and 13. Also by definition of a fresh session, $\mathcal{M}$ is only allowed to corrupt either $\hat{U}$ or $\hat{V}$ after expiring the test session. Therefore before aborting as in Step 10, $\mathcal{S}$ is successful as in Step 8. Except with negligible probability of guessing $\xi(U,V)$, a successful $\mathcal{M}$ must query H with $(\mathrm{CDH}(X,Y), \mathrm{CDH}(U,V), \hat{A}, \hat{B}, X, Y)$ where $\{\hat{A}, \hat{B}\} = \{\hat{U}, \hat{V}\}$, in which case $\mathcal{S}$ is successful as described in Step 11a. The probability that $\mathcal{S}$ is successful is bounded by $\Pr(S) \geq 2p_4/n^3 t$, where negligible terms are ignored.

Overall analysis. During the simulation $\mathcal{S}$ performs group exponentiations, simulates oracle queries, and accesses the DDH oracle, all of which take polynomial time. If $\mathcal{M}$'s running is also polynomial then $\mathcal{S}$ is a polynomially-bounded algorithm that produces a MAC forgery or succeeds in solving the CDH challenge with probability at least

$$\Pr(S) \geq \max\left(\frac{2p_1}{(nt)^2}, \frac{2p_2}{n^2}, \frac{2p_3}{n^2 t}, \frac{2p_4}{n^3 t}\right). \tag{2}$$

□

Reflections. In the simulations of events $M^* \wedge E_2$ and $M^* \wedge E_4$, it was implicitly assumed (in the success events in the hash queries) that $\hat{U}$ and $\hat{V}$ are distinct parties. More precisely, if a party is allowed to initiate a session with itself then $\mathcal{S}$ may fail as $\mathcal{M}$ may produce $\mathrm{CDH}(U,U)$ or $\mathrm{CDH}(V,V)$ instead of $\mathrm{CDH}(U,V)$. The case $\hat{U} = \hat{V}$ can be encompassed by a reduction from the Gap Square Problem (GSP), which is the problem of computing g^{u^2} given g^u and a DDH oracle. $\mathcal{S}$'s actions are modified as follows: given $U = g^u$, $\mathcal{S}$ selects $v \in_R [1, q-1]$ and computes $V = U^v$. The output produced by $\mathcal{S}$ in events $M^* \wedge E_1$ and $M^* \wedge E_3$ is $\sigma_e^{v^{-1}}$. In events $M^* \wedge E_2$ and $M^* \wedge E_4$, $\mathcal{S}$'s output is $\sigma_s^{v^{-1}}$ if the communicating parties are $\hat{U}$ and $\hat{V}$, σ_s if $\hat{U}$ is both the owner and peer of the test session, and $\sigma_s^{v^{-2}}$ if $\hat{V}$ is both the owner and peer of the test session.

6 Concluding Remarks

We revisited the question of security of the one- and three-pass UM protocols where parties are allowed to reuse their static keys among the two protocols. Our work shows that it is possible to formally argue the security of key agreement protocols having different security attributes in a combined setting.

References

1. ANSI X9.42, Agreement of Symmetric Keys Using Discrete Logarithm Cryptography, American National Standards Institute (2003)
2. ANSI X9.63, Key Agreement and Key Transport Using Elliptic Curve Cryptography, American National Standards Institute (2001)

3. Boyd, C., Cliff, Y., Nieto, J., Paterson, K.: Efficient one-round key exchange in the standard model. In: Mu, Y., Susilo, W., Seberry, J. (eds.) ACISP 2008. LNCS, vol. 5107, pp. 69–83. Springer, Heidelberg (2008) http://eprint.iacr.org/2008/007
4. Chatterjee, S., Menezes, A., Ustaoglu, B.: Reusing static keys in key agreement protocols. In: Roy, B., Sendrier, N. (eds.) INDOCRYPT 2009. LNCS, vol. 5922, pp. 39–56. Springer, Heidelberg (2009)
5. Diffie, W., van Oorschot, P., Wiener, M.: Authentication and authenticated key exchanges. Design. Code. Cryptogr. 2(2), 107–125 (1992)
6. Just, M., Vaudenay, S.: Authenticated multi-party key agreement. In: Kim, K.-c., Matsumoto, T. (eds.) ASIACRYPT 1996. LNCS, vol. 1163, pp. 36–49. Springer, Heidelberg (1996)
7. Krawczyk, H.: HMQV: A high-performance secure Diffie-Hellman protocol. In: Shoup, V. (ed.) CRYPTO 2005. LNCS, vol. 3621, pp. 546–566. Springer, Heidelberg (2005), http://eprint.iacr.org/2005/176
8. Kunz-Jacques, S., Pointcheval, D.: About the security of MTI/C0 and MQV. In: De Prisco, R., Yung, M. (eds.) SCN 2006. LNCS, vol. 4116, pp. 156–172. Springer, Heidelberg (2006)
9. Law, L., Menezes, A., Qu, M., Solinas, J., Vanstone, S.: An efficient protocol for authenticated key agreement. Design. Code. Cryptogr. 28(2), 119–134 (2003)
10. Menezes, A.: Another look at HMQV. J. Math. Cryptology 1(1), 47–64 (2007)
11. Menezes, A., Ustaoglu, B.: Security arguments for the UM key agreement protocol in the NIST SP 800-56A standard. In: Proceedings of the 2008 ACM Symposium on Information, Computer and Communications Security, pp. 261–270. ACM Press, New York (2008)
12. SP 800-56A, Recommendation for Pair-Wise Key Establishment Schemes Using Discrete Logarithm Cryptography (Revised), National Institute of Standards and Technology (March 2007)
13. Ustaoglu, B.: Obtaining a secure and efficient key agreement protocol from (H)MQV and NAXOS. Design. Code. Cryptogr. 46(3), 329–342 (2008), http://eprint.iacr.org/2007/123

Indifferentiability beyond the Birthday Bound for the Xor of Two Public Random Permutations

Avradip Mandal[1], Jacques Patarin[2], and Valerie Nachef[3]

[1] University of Luxembourg, Luxembourg
avradip.mandal@uni.lu
[2] PRISM, Université de Versailles, France
jacques.patarin@prism.uvsq.fr
[3] UMR CNRS 8088, University of Cergy-Pontoise, France
valerie.nachef@u-cergy.fr

Abstract. Xoring two permutations is a very simple way to construct pseudorandom functions from pseudorandom permutations. The aim of this paper is to get precise security results for this construction when the two permutations on n bits f and g are public. We will first prove that $f \oplus g$ is indifferentiable from a random function on n bits when the attacker is limited with q queries, with $q \ll \sqrt{2^n}$. This bound is called the "birthday bound". We will then prove that this bound can be improved to $q^3 \ll 2^{2n}$. We essentially instantiate length preserving random functions, starting from fixed key ideal cipher with high security guarantee.

Keywords: Indifferentiability, Luby-Rackoff Backwards with public permutations, Building random oracles from ideal block ciphers.

1 Introduction

The thema of this paper is to prove some security bounds about the indifferentiability of the Xor of two public random permutations on n bits from one random public function on n bits. We will look for security bounds "beyond the birthday bound" and smaller than the "information bound", i.e. when the number of queries q is $q \ll 2^n$, but we may have $q \gg \sqrt{2^n}$. Therefore, this paper is in relation with previous work about the Xor of two permutations, about previous work dealing with security proofs "beyond the birthday bound" for various ideal cryptographic constructions, and, of course, about previous work on indifferentiability.

Luby-Rackoff Backwards

The problem to construct pseudorandom functions (PRF) from pseudorandom permutations (PRP) is called "Luby-Rackoff Backwards". This problem was first considered in [3]. This problem is obvious if the pseudorandom permutations are secret and if we are interested in an asymptotical polynomial versus non polynomial model (since a PRP is then a PRF). However, this problem is not obvious if we want security beyond the birthday bound, or if the permutations are

G. Gong and K.C. Gupta (Eds.): INDOCRYPT 2010, LNCS 6498, pp. 69–81, 2010.

public. When the permutations are secret, Lucks ([11]) has proved that the Xor of k independent pseudorandom permutations gives security when $q \ll 2^{\frac{k}{k+1}n}$. (For $k = 2$ this gives $O(2^{\frac{2}{3}n})$). This bound was improved in [2,19] and [20] where proofs of security for $q \ll 2^n$ are given. (However, Lucks proof is much simpler). When $q = 2^n$, as pointed out in these papers, it is easy to distinguish $\pi_1 \oplus \pi_2$ from a random functions R since $\oplus_{x \in \{0,1\}^n}(\pi_1 \oplus \pi_2)(x) = 0$. In this paper π_1 and π_2 will be public, and therefore our bounds are necessary smaller than the bounds obtained when π_1 and π_2 are secret. We will in fact match the original bound proven by Lucks, i.e. $q \ll 2^{\frac{2}{3}n}$.

Security proofs beyond the birthday bound
Many papers have been published with security proofs beyond the birthday bound for various ideal cryptographic constructions. For example Aiello and Verkatesan [1] for doubling the length of pseudorandom functions with the Benes construction, or Maurer and Prietrzak [12] or Patarin [16,17] for Feistel schemes.

Indifferentiability
However the main topic of this paper is related to indifferentiability theory since, again, π_1 and π_2 will be public. The notion of indifferentiability was introduced by Maurer, Renner and Holenstein [13]. Since then, a lot of works has been done about indifferentiability. For example, in [6], Coron et al have shown how to construct a random oracle from an ideal block cipher. Their proved security bound is in $q \ll \sqrt{2^n}$, where n is the number of bits of the output of the ideal cipher. With our construction we will also be able to construct random oracles from ideal block ciphers. The other direction (constructing an ideal cipher from an oracle model) was proved in [7]. Their construction uses a 6-round Feistel scheme, and the security is proved for $q^{16} \ll 2^n$. This is below the birthday bound ($q \ll \sqrt{2^n}$). In his PhD, [22], Seurin has obtained a better bound, $q^4 \ll 2^n$, but for more rounds: 10 rounds instead of 6. This bound is better but still below the birthday bound. Our problem is different and simpler, this is why we will be able to obtain better security bounds.

Related Works
Recently Dodis *et al* [8], has shown XOR of a random permutation and its inverse is actually indifferentiable from a random function. However, they achieved a birthday security bound. Whereas, we show with XOR of two independent random permutations one can get a security proof with beyond birthday security guarantee.

2 Bounding Distinguisher's Advantage

A *distinguisher* (attacker) D, for two oracles $\mathcal{F}$ and $\mathcal{G}$, is an algorithm which has access to either oracle $\mathcal{F}$ or oracle $\mathcal{G}$ and outputs either 0 or 1 after making queries to the given oracle. The *advantage* $\mathsf{Adv}_D(\mathcal{F}, \mathcal{G})$ or simply Adv_D of the distinguisher D is defined as

$$\mathsf{Adv}_D = |\Pr[D^F \to 1] - \Pr[D^G \to 1]|.$$

The *view* V of the distinguisher is nothing but the list of the queries made by and the responses it received from the given oracle. $\mathcal{V}^F$ and $\mathcal{V}^G$ be the random variables corresponding to the distinguisher's view, when D is interacting with $\mathcal{F}$ and $\mathcal{G}$ respectively. $\mathcal{V}$ be the set of all possible views. One can actually easily show [4,5,15],

$$\mathsf{Adv}_D \leq \frac{1}{2} \sum_{V \in \mathcal{V}} |\Pr[\mathcal{V}^F = V] - \Pr[\mathcal{V}^G = V]|.$$

Below, we state two well known theorems [4,5,15] regarding upper bounds on Adv_D.

Theorem 1. *If for all* $V \in \mathcal{V}$, *we have*

$$\Pr[\mathcal{V}^G = V] \geq (1 - \varepsilon) \Pr[\mathcal{V}^F = V],$$

then $\mathsf{Adv}_D \leq \varepsilon$.

Theorem 2. *If for all* $V \in \mathcal{M} \subseteq \mathcal{V}$, *we have*

$$\Pr[\mathcal{V}^G = V] \geq (1 - \varepsilon_1) \Pr[\mathcal{V}^F = V]$$

and $\Pr[\mathcal{V}^F \notin \mathcal{M}] \leq \varepsilon_2$, *then* $\mathsf{Adv}_D \leq \varepsilon_1 + \varepsilon_2$.

3 Indifferentiability

The notion of indifferentiability was introduced by Maurer, Renner and Holenstein in [13]. This is an extension of the classical notion of indistinguishability, where one or more oracles are publicly available, such as random oracles, random permutations, or ideal ciphers. This notion of indifferentiability is used to show that an ideal primitive $\mathcal{G}$ (for example a random function) can be replaced by a construction C that is based on some other ideal primitive $\mathcal{F}$ (for example, C is the Xor of two random permutations).

Definition 1. Indifferentiability [13]
A Turing machine C *with oracle access to an ideal primitive* $\mathcal{F}$ *is said to be* $(t, q_C, q_{\mathcal{F}}, \varepsilon)$ *indifferentiable from an ideal primitive* $\mathcal{G}$ *if there exists a simulator* S *with an oracle access to* $\mathcal{G}$ *and running time at most* t, *such that for any distinguisher* D, *it holds that*

$$\mathsf{Adv}_D((C^{\mathcal{F}}, \mathcal{F}), (\mathcal{G}, S^{\mathcal{G}})) < \varepsilon.$$

The distinguisher makes at most q_C *queries to* C *or* $\mathcal{G}$ *and at most* $q_{\mathcal{F}}$ *queries to* $\mathcal{F}$ *or* S. *Similarly,* $C^{\mathcal{F}}$ *is said to be (computationally) indifferentiable from* $\mathcal{G}$ *if running time of* D *is bounded above by some polynomial in the security parameter* k *and* ε *is a negligible function of* k.

The previous definition is illustrated in Figure 1. $I_n = \{0, 1\}^n$ denotes the set of all bit strings of length n. As in this paper, $R : I_n \to I_n$ is a random function,

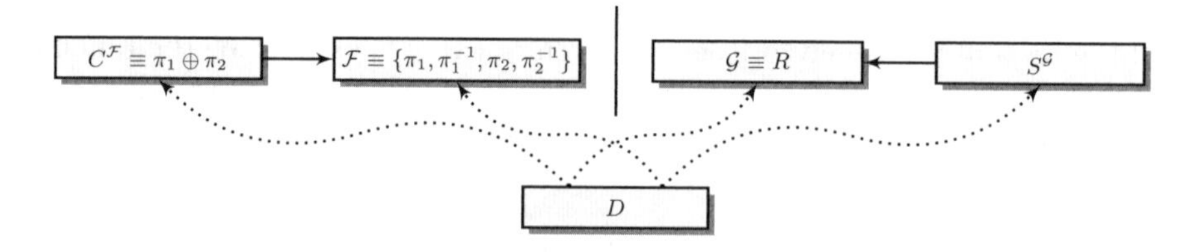

Fig. 1. The indifferentiability notion

$\pi_1, \pi_2 : I_n \to I_n$ are two random permutations, and C is $\pi_1 \oplus \pi_2$. The distinguisher has either access to the system formed by the construction C and π_1, π_2, or to the system formed by the random function R and a simulator S. In the first system (left), the construction C computes its output by making calls to π_1, π_2, π_1^{-1} and π_2^{-1}. In our case $C^{\{\pi_1,\pi_2\}}$ is just $\pi_1 \oplus \pi_2$, so in our case C does not need access to π_1^{-1}, π_2^{-1} but just π_1, π_2. The distinguisher can also make calls to π_1, π_2, π_1^{-1} and π_2^{-1} directly. In the second system (right), the distinguisher can either query the random function R, or the simulator S that can make query to R. We see that the role of the simulator is to simulate the random permutations π_1, π_2 and to simulate also π_1^{-1}, π_2^{-1}, such that no distinguisher can tell whether it is interacting with C and π_1, π_2, π_1^{-1} and π_2^{-1}, or with R and S. Notice that the simulator does not see the distinguisher's queries to π. However, it can call R directly when needed for the simulation. The output of S should be indistinguishable from that of random oracle permutations π_1, π_2, and the output of S should look consistent with what the distinguisher can obtain from R.

4 Our Simulator

S denotes the simulator, and D the distinguisher. After α queries, S maintains always the sequence (x_i, a_i, b_i), $1 \leq 1 \leq \alpha$, containing previous responses, as we will see, with $\forall i$, $1 \leq i \leq \alpha$, $a_i \oplus b_i = R(x_i)$.When D contacts R, he makes only direct queries: D gives a value x'_i to R, and obtains the value $R(x'_i)$. S does not know these values x'_i. When D contacts S, we can assume without losing generality that D can make only 3 types of queries: A direct query, or an inverse query with a_α, or an inverse query with b_α.

Direct query
In a direct query, D gives a new value x_α to S (i.e. $x_\alpha \notin \{x_1, \ldots, x_{\alpha-1}\}$) and S will give to D a value a_α to simulate $\pi_1(x_\alpha)$ and a value b_α to simulate $\pi_2(x_\alpha)$. We can assume, without losing generality that D chooses $x_\alpha \notin \{x_1, \ldots, x_{\alpha-1}\}$ because if $x_\alpha = x_i$, $i \leq \alpha - 1$, then S will always answer $a_\alpha = a_i$ and $b_\alpha = b_i$ and D will learn nothing new. Our simulator will compute a_α and b_α like this:

1. S asks for the value $R(x_\alpha)$.
2. a_α is randomly chosen with a uniform distribution in
   $I_n \setminus \{a_1, a_2, \ldots, a_{\alpha-1}, R(x_\alpha) \oplus b_1, R(x_\alpha) \oplus b_2, \ldots, R(x_\alpha) \oplus b_{\alpha-1}\}$.
3. $b_\alpha = R(x_\alpha) \oplus a_\alpha$

We will denote $V_{\alpha-1} = \{a_1, a_2, \ldots, a_{\alpha-1}\}$ and $Q_{\alpha-1} = \{R(x_\alpha) \oplus b_1, R(x_\alpha) \oplus b_2, \ldots, R(x_\alpha) \oplus b_{\alpha-1}\}$. Therefore we will have: $Q_{\alpha-1} = \{b_\alpha \oplus a_\alpha \oplus b_1, b_\alpha \oplus a_\alpha \oplus b_2, \ldots, b_\alpha \oplus a_\alpha \oplus b_{\alpha-1}\}$, and a_α is randomly chosen in $I_n \setminus (V_{\alpha-1} \cup Q_{\alpha-1})$.

Inverse query with a_α
In such inverse query, D gives a new value a_α to S (i.e. $a_\alpha \notin \{a_1, a_2, \ldots, a_{\alpha-1}\}$) and S will give to D a value x_α to simulate $\pi_1^{-1}(a_\alpha)$ and a value b_α to simulate $\pi_2(x_\alpha)$. We can assume, without losing generality that D chooses $a_\alpha \notin \{a_1, a_2, \ldots, a_{\alpha-1}\}$ because if $a_\alpha = a_i$, $i \leq \alpha - 1$, then S will always answer $x_\alpha = x_i$ and $b_\alpha = b_i$ and D will learn nothing new. Our simulator will compute x_α and b_α like this:

1. x_α is randomly chosen with a uniform distribution in $I_n \setminus \{x_1, \ldots, x_{\alpha-1}\}$.
2. S asks for the value $R(x_\alpha)$.
3. If $R(x_\alpha) \oplus a_\alpha \notin \{b_1, b_2, \ldots, b_{\alpha-1}\}$ then S gives this x_α to D and gives $b_\alpha = R(x_\alpha) \oplus a_\alpha$ to D.
4. If $R(x_\alpha) \oplus a_\alpha \in \{b_1, b_2, \ldots, b_{\alpha-1}\}$ then S goes back to 1 above, and tries with another x_α randomly chosen in $I_n \setminus \{x_1, \ldots, a_{\alpha-1}\}$.

This process continues until S has found like this a value x_α such that $R(x_\alpha) \oplus a_\alpha \notin \{b_1, b_2, \ldots, b_{\alpha-1}\}$ and then it gives this x_α to D and $b_\alpha = R(x_\alpha) \oplus a_\alpha$. If S cannot find such a x_α it does not answer, but in general this probability will be negligible if $\alpha \ll 2^n$. Therefore, when S answers, the value x_α has been randomly chosen with a uniform distribution in $I_n \setminus \{x_1, \ldots x_{\alpha-1}\} \setminus W_{\alpha-1}$ where $W_{\alpha-1} = \{x \in I_n \text{ such that } R(x) \oplus a_\alpha \in \{b_1, b_2, \ldots, b_{\alpha-1}\}\}$.

Remark. A variant would be to choose a simulator S that will abort after k failed x_α, where k is an integer. In this paper we can actually assume $k = 2$, as we are looking for a security proof when $q \ll 2^{\frac{2n}{3}}$. We do not assume that D makes "timing attacks" but only that D computes from the values given by S without using the time for S to give them. However it would not change anything to assume that D tries to use this time since when S answers a value x_α, whatever the time S has used to compute x_α, x_α is always randomly chosen in $I_n \setminus \{x_1, \ldots x_{\alpha-1}\} \setminus W_{\alpha-1}$ and D knows this set $I_n \setminus \{x_1, \ldots x_{\alpha-1}\} \setminus W_{\alpha-1}$. Therefore the time gives no more information to D.

Inverse query with b_α
In such inverse query, D gives a new value b_α to S (i.e. $b_\alpha \notin \{b_1, \ldots, b_{\alpha-1}\}$) and S will give to D a value x_α to simulate π_2^{-1} and a value a_α to simulate $\pi_1(x_\alpha)$. Our simulator will compute x_α in a symmetric way as we have just seen for inverse query with a_α. This means that S will randomly choose x_α with uniform distribution in $I_n \setminus \{x_1, \ldots x_{\alpha-1}\} \setminus W'_{\alpha-1}$ with $W'_{\alpha-1} = \{x \in I_n \text{ such that } R(x) \oplus b_\alpha \in \{a_1, a_2, \ldots, a_{\alpha-1}\}\}$ and that S will give this x_α to D and S will give a_α to D with $a_\alpha = R(x_\alpha) \oplus b_\alpha$.

Whatever the query of D is, direct, inverse with a_α, or inverse with b_α, S will store and memorize the values $(x_\alpha, a_\alpha, b_\alpha)$ generated to the sequence (x_i, a_i, b_i), $1 \leq i \leq q$.

5 Distinguisher Characterization

Distinguisher D'
For any distinguisher D, with q queries, we consider another distinguisher D' with q' queries, $q' \geq q$ such that:
1. The first q queries of D' are exactly those of D.
2. D' outputs the same decision, 0 or 1, as D.
3. For any direct query x that D makes directly to R, D' will make **at the end** a direct query with this value x to S.
We can assume that D' does not make any duplicate query, since S and R will always give the same answers on the same questions. Since D and D' always output the same decision, we have: $Adv_D = Adv_{D'}$. If q is the number of queries that D makes to S or R, with q_1 the number of queries that D makes to R and q_2 the number of queries that D makes to S, then $q = q_1 + q_2$, $q' = q + q_1$, D' makes q_1 queries to R (as D) and $q = q_1 + q_2$ queries to S.

Let, $T = ((x_1, a_1, b_1), \cdots, (x_q, a_q, b_q))$ be the view of the distinguisher D'. The ith triple (x_i, a_i, b_i) implies D' received the triple during the ith direct/inverse query to the Simulator or the Random permutations. B_n be the set of all permutations from I_n to I_n. D' tries to distinguish whether the sequence T, came from π_1 and π_2 with $\pi_1, \pi_2 \in_R B_n$ or from the simulator. p_T and p_T^* be the probabilities that D' receives the tuple T while interacting with the random permutations and simulator respectively. As D and D' output the same decision bit we have,

$$\mathsf{Adv}_D = \mathsf{Adv}_{D'} \leq \frac{1}{2} \sum_T |p_T - p_T^*|$$

Any sequence $T = \{(x_i, a_i, b_i), 1 \leq i \leq q\}$, such that $x_i, a_i, b_i \in I_n$, x_i's are pairwise distinct, a_i's are pairwise distinct and b_i's are pairwise distinct as well is called *distinct* q-sequence. For any distinct q-sequence T we have,

$$p_T = \prod_{\alpha=1}^{q} \frac{1}{(2^n - (\alpha - 1))^2},$$

because π_1, π_2 are random permutations. For any other sequence T of q-triples p_T is zero.

6 Proof of Security When $q \ll \sqrt{2^n}$

Theorem 3. *π_1, π_2 be two random permutations and R be a random function $I_n \to I_n$. S be the simulator as defined before. Then for any distinguisher D, for the systems $(\pi_1 \oplus \pi_2, \{\pi_1, \pi_2\})$ and (R, S) making at most q queries we have,* $\mathsf{Adv}_D \leq \mathcal{O}(\frac{q^2}{2^n})$

Proof. As discussed before, at first we construct the distinguisher D' starting from the distinguisher D. We know,

$$\mathsf{Adv}_D = \mathsf{Adv}_{D'}.$$

p_T and p_T^* are the probabilities that D' receives the distinct q-sequence $T = \{(x_i, a_i, b_i), 1 \leq i \leq q\}$ while interacting with the random permutations and simulator respectively. We already know,

$$p_T = \prod_{\alpha=1}^{q} \frac{1}{(2^n - (\alpha - 1))^2}.$$

Now, if we can show

$$p_T^* \geq (1 - \epsilon) p_T,$$

for all distinct q-sequence T, then by Theorem 1 that would imply

$$\mathsf{Adv}_{D'} \leq \epsilon.$$

As we are interested in the lower bound of p_T^*, we evaluate $p_T^{*'}$ where we impose one extra condition, that is while answering the inverse queries the simulator never makes a bad guess. We define G_α to be the event, that this extra condition is satisfied during $(x_1, a_1, b_1), \cdots, (x_\alpha, a_\alpha, b_\alpha)$ responses. Also, $(X_\alpha, A_\alpha, B_\alpha)$ be the random variable corresponding to α^{th} *triple* received by D'. We are interested in a lower bound for

$$P_\alpha \equiv \Pr[(X_\alpha, A_\alpha, B_\alpha) = (x_\alpha, a_\alpha, b_\alpha) \cap G_\alpha \\ |(X_i, A_i, B_i) = (x_i, a_i, b_i) \ \textit{for}\ 1 \leq i \leq \alpha - 1 \cap G_{\alpha-1}].$$

Direct query with x_α

We have $|V_{\alpha-1} \cup Q_{\alpha-1}| \geq (\alpha - 1)$. Hence,

$$P_\alpha \geq \frac{1}{2^n} \times \frac{1}{2^n - (\alpha - 1)}.$$

The term $\frac{1}{2^n}$, comes from the probability over the random function i.e. $R(x_\alpha) = a_\alpha \oplus b_\alpha$, the condition $G_{\alpha-1}$ implies $R(x_\alpha)$ was not queried beforehand.

Inverse query with a_α

Conditioned on the event $G_{\alpha-1}$, the probability that the simulator guesses x_α in the first trial and $R(x_\alpha)$ outputs $a_\alpha \oplus b_\alpha$ is $\frac{1}{2^n-(\alpha-1)} \times \frac{1}{2^n}$. For inverse query

with b_α we also get the same bound as above. So, whether α^{th} query is direct or inverse query we always have

$$P_\alpha \geq \frac{1 - \frac{\alpha-1}{2^n}}{(2^n - (\alpha - 1))^2}.$$

Hence,

$$p_T^* \geq p_T^{*'} = \prod_{\alpha=1}^{q} P_\alpha \geq \prod_{\alpha=1}^{q} \frac{1 - \frac{\alpha-1}{2^n}}{(2^n - (\alpha - 1))^2} \geq p_T (1 - \frac{q^2}{2^n}).$$

This would imply $\mathsf{Adv}_{D'} \leq \frac{q^2}{2^n}$. □

7 Proof of Security When $q \ll 2^{\frac{2}{3}n}$

Theorem 4. *π_1, π_2 be two random permutations and R be a random function $I_n \to I_n$. S be the simulator as defined before. Then for any distinguisher D, for the systems $(\pi_1 \oplus \pi_2, \{\pi_1, \pi_2\})$ and (R, S) making at most q queries we have,* $\mathsf{Adv}_D \leq \mathcal{O}(\frac{q^3}{2^{2n}})$

Proof. As discussed before, at first we construct the distinguisher D' starting from the distinguisher D. We know,

$$\mathsf{Adv}_D = \mathsf{Adv}_{D'}.$$

For any tuple $T = ((x_1, a_1, b_1), \cdots, (x_q, a_q, b_q))$ for all the values $A \in I_n$, let $N_A(T)$ be the number of (i, j), $1 \leq i \leq q$, $1 \leq j \leq q$ such that $a_i \oplus b_j = A$. M be the set of q-tuples such that $T \in M$ iff $N_A(T) \leq \frac{24q^2}{2^n - q}$ for all $A \in I_n$. p_T and p_T^* be the probabilities that D' receives the tuple $T = \{(x_i, a_i, b_i), 1 \leq i \leq q\}$ while interacting with the random permutations and simulator respectively. We already know,

$$p_T = \prod_{\alpha=1}^{q} \frac{1}{(2^n - (\alpha - 1))^2}.$$

Theorem 5 from Appendix A implies,

$$\sum_{T \notin M} p_T \leq \frac{2^n}{2^{12n}} = \frac{1}{2^{11n}}.$$

Now, if we can show

$$p_T^* \geq (1 - \epsilon) p_T$$

for all $T \in M$, then by Theorem 2 that would imply

$$\mathsf{Adv}_{D'} \leq \frac{1}{2^{11n}} + \epsilon.$$

While answering the inverse queries, it might be possible that the simulator makes some bad guess of x_α. As we are interested in the lower bound of p_T^*, we evaluate $p_T^{*'}$ where we impose some extra conditions, the simulator is allowed to make only one bad guess while answering inverse queries and the bad guess can not be same as any x_i for $1 \leq i \leq q$ or some previous bad guess. For $1 \leq \alpha \leq q$, we define G_α to be the event, that this extra condition is satisfied during $(x_1, a_1, b_1), \cdots, (x_\alpha, a_\alpha, b_\alpha)$ responses. Also, $(X_\alpha, A_\alpha, B_\alpha)$ be the random variable corresponding to α^{th} *triple* received by D'. We are interested in a lower bound for

$$P_\alpha \equiv \Pr[(X_\alpha, A_\alpha, B_\alpha) = (x_\alpha, a_\alpha, b_\alpha) \cap G_\alpha$$
$$|(X_i, A_i, B_i) = (x_i, a_i, b_i) \textit{ for } 1 \leq i \leq \alpha - 1 \cap G_{\alpha-1}].$$

Direct query
If $T \in M$, we have $|V_{\alpha-1} \cup Q_{\alpha-1}| \geq 2(\alpha-1) - \frac{24q^2}{2^n-q}$. Hence, we have

$$P_\alpha \geq \frac{1}{2^n} \times \frac{1}{2^n - 2(\alpha-1) + \frac{24q^2}{2^n-q}}.$$

The term $\frac{1}{2^n}$, comes from the probability over the random function i.e. $R(x_\alpha) = a_\alpha \oplus b_\alpha$. The condition $G_{\alpha-1}$, guarantees that x_α was not queried to the random oracle R, as a bad first guess in some previous inverse query. Assuming $q \leq 2^n/4$ we get,

$$P_\alpha \geq \frac{1 - \frac{96q^2}{2^{2n}}}{(2^n - (\alpha-1))^2}.$$

Inverse query with a_α
Conditioned on the event $G_{\alpha-1}$, the probability that the simulator guesses x_α in the first trial and $R(x_\alpha)$ outputs $a_\alpha \oplus b_\alpha$ is $\frac{1}{2^n-(\alpha-1)} \times \frac{1}{2^n}$. Conditioned on the event $G_{\alpha-1}$, the probability that the simulator guesses x_α in the second trial satisfying $R(x_\alpha) = a_\alpha \oplus b_\alpha$ and condition G_α is at least,

$$\frac{2^n - q - (\alpha-1)}{2^n - (\alpha-1)} \times \frac{\alpha-1}{2^n} \times \frac{1}{2^n - (\alpha-1)} \times \frac{1}{2^n}.$$

$\frac{2^n-q-(\alpha-1)}{2^n-(\alpha-1)}$ corresponds to the probability that the first guess does not collide with $x_1, \cdots, x_q$ and the possible bad first guesses in the previous inverse queries. As the first guess is not queried before, $\frac{\alpha-1}{2^n}$ is the probability that the first guess is bad. $\frac{1}{2^n-(\alpha-1)} \times \frac{1}{2^n}$ is the probability that the second guess is x_α and $R(x_\alpha) = a_\alpha \oplus b_\alpha$. Hence all together we have,

$$\begin{aligned} P_\alpha &\geq \frac{1}{2^n - (\alpha-1)} \times \frac{1}{2^n} + \frac{2^n - q - (\alpha-1)}{2^n - (\alpha-1)} \times \frac{\alpha-1}{2^n} \times \frac{1}{2^n - (\alpha-1)} \times \frac{1}{2^n} \\ &= \frac{1}{2^n} \times \frac{1}{2^n - (\alpha-1)} \times \left(1 + \frac{2^n - q - (\alpha-1)}{2^n - (\alpha-1)} \times \frac{\alpha-1}{2^n}\right) \end{aligned}$$

Again assuming $q \leq 2^n/4$, we can show

$$P_\alpha \geq \frac{1 - \frac{4q^2}{2^{2n}}}{(2^n - (\alpha-1))^2}.$$

For inverse queries with b_α, we also get the same lower bound as above. Hence, whether α^{th} query is direct or inverse query we always have

$$P_\alpha \geq \frac{1 - \frac{96q^2}{2^{2n}}}{(2^n - (\alpha-1))^2}.$$

Hence,

$$p_T^* \geq p_T^{*'} = \prod_{\alpha=1}^{q} P_\alpha \geq \prod_{\alpha=1}^{q} \frac{1 - \frac{96q^2}{2^{2n}}}{(2^n - (\alpha - 1))^2} \geq p_T(1 - \frac{96q^3}{2^{2n}}).$$

This would imply $\mathsf{Adv}_{D'} \leq \frac{1}{2^{11n}} + \frac{96q^3}{2^{2n}}$. □

8 Application of Our Work

Even though the problem of constructing a public random function from public random permutations are interesting in its own right, here we briefly mention some possible application of our result. There are numerous cryptographic schemes [8,14,23,24] where length preserving Random Functions are needed. Only known instantiation of those non-invertible length-preserving primitives were due to Dodis *et al* [8]. However, as stated below their instantiation does not always serve the purpose as the birthday security bound over there fails to preserve the high security of the constructions.

1. In Crypto 2007, Maurer and Tessaro [14] considered the problem of extending the domain of public random functions approaching optimal security bound, starting from length preserving random functions. With our result, if we choose to instantiate the random function as XOR of two fixed key ideal ciphers. Even though we won't be able to guarantee the optimal $\Theta(2^{n(1-\varepsilon)})$ security bound we can easily guarantee beyond birthday security bound up to $O(2^{\frac{2n}{3}})$ queries.
2. Stam in Crypto 2008 [24], Shrimpton and Stam in ICALP 2008 [23] considered the problem of building collision resistant compression functions starting from length preserving random functions. However, here whether we use our instantiation or instantiation due to Dodis *et al* [8] do not matter, because here the goal of their work is to achieve collision resistancy as close as the Birthday Barrier.

9 Conclusion

In this paper, we have proved the indifferentiability of the Xor of two random permutations on n bits from a random function on n bits when the number of queries satisfies $q \ll \sqrt{2^n}$ (birthday bound) or $q \ll 2^{2n/3}$. The simulator S used was the same in both cases. In fact, it is conjectured that for this simulator the security is probably in $q \ll 2^n$, which if true would extend Maurer and Tessaro's [14] result preserving the optimal $\Theta(2^{n(1-\varepsilon)})$ bound.

Acknowledgements. We sincerely thank Jean Sébastien Coron for his valuable comments and long discussions on initial drafts of this paper.

References

1. Aiello, W., Venkatesan, R.: Foiling Birthday Attacks in Length-Doubling Transformations - Benes: A Non-Reversible Alternative to Feistel. In: Maurer, U.M. (ed.) EUROCRYPT 1996. LNCS, vol. 1070, pp. 307–320. Springer, Heidelberg (1996)
2. Bellare, M., Impagliazzo, R.: A Tool for Obtaining Tighter Security Analyses of Pseudorandom Function Based Constructions, with Applications to PRP to PRF Conversion. ePrint Archive 1999/024: Listing for 1999 (1999)
3. Bellare, M., Krovetz, T., Rogaway, P.: Luby-Rackoff Backwards: Increasing Security by Making Block Ciphers Non-invertible. In: Nyberg, K. (ed.) EUROCRYPT 1998. LNCS, vol. 1403, pp. 266–280. Springer, Heidelberg (1998)
4. Bhattacharyya, R., Mandal, A., Nandi, M.: Security analysis of the mode of jh hash function. In: beyer, i. (ed.) FSE 2010. LNCS, vol. 6147, Springer, Heidelberg (2010)
5. Chang, D., Lee, S., Nandi, M., Yung, M.: Indifferentiable security analysis of popular hash functions with prefix-free padding. In: Lai, X., Chen, K. (eds.) ASIACRYPT 2006. LNCS, vol. 4284, pp. 283–298. Springer, Heidelberg (2006)
6. Coron, J.C., Dodis, Y., Malinaud, C., Puniya, P.: Merkle-Damgård Revisited: How to Construct a Hash Function. In: Shoup, V. (ed.) CRYPTO 2005. LNCS, vol. 3621, pp. 430–448. Springer, Heidelberg (2005)
7. Coron, J.-S., Patarin, J., Seurin, Y.: The Random Oracle Model and the Ideal Cipher Model are Equivalent. In: Wagner, D. (ed.) CRYPTO 2008. LNCS, vol. 5157, pp. 1–20. Springer, Heidelberg (2008)
8. Dodis, Y., Pietrzak, K., Puniya, P.: A new mode of operation for block ciphers and length-preserving macs. In: Smart, N.P. (ed.) EUROCRYPT 2008. LNCS, vol. 4965, pp. 198–219. Springer, Heidelberg (2008)
9. Hall, C., Wagner, D., Kelsey, J., Schneier, B.: Building PRFs from PRPs. In: Krawczyk, H. (ed.) CRYPTO 1998. LNCS, vol. 1462, pp. 370–389. Springer, Heidelberg (1998)
10. Hall Jr., M.: A Combinatorial Problem on Abelian Groups. Proceedings of the Americal Mathematical Society 3(4), 584–587 (1952)
11. Lucks, S.: The Sum of PRPs Is a Secure PRF. In: Preneel, B. (ed.) EUROCRYPT 2000. LNCS, vol. 1807, pp. 470–487. Springer, Heidelberg (2000)
12. Maurer, U., Pietrzak, K.: The Security of Many-Round Luby-Rackoff Pseudo-Random Permutations. In: Biham, E. (ed.) EUROCRYPT 2003. LNCS, vol. 2656, pp. 544–561. Springer, Heidelberg (2003)
13. Maurer, U., Renner, R., Holenstein, C.: Indifferentiability Impossibility Results on Reductions, and Applications to the Random Oracle Methodology. In: Naor, M. (ed.) TCC 2004. LNCS, vol. 2951, pp. 21–39. Springer, Heidelberg (2004)
14. Maurer, U.M., Tessaro, S.: Domain extension of public random functions: Beyond the birthday barrier. In: Menezes, A. (ed.) CRYPTO 2007. LNCS, vol. 4622, pp. 187–204. Springer, Heidelberg (2007)
15. Nandi, M.: A simple and unified method of proving indistinguishability. In: Barua, R., Lange, T. (eds.) INDOCRYPT 2006. LNCS, vol. 4329, pp. 317–334. Springer, Heidelberg (2006)
16. Patarin, J.: Luby-Rackoff: 7 Rounds are Enough for $2^{n(1-\epsilon)}$ Security. In: Boneh, D. (ed.) CRYPTO 2003. LNCS, vol. 2729, pp. 513–529. Springer, Heidelberg (2003)
17. Patarin, J.: On linear systems of equations with distinct variables and Small block size. In: Won, D.H., Kim, S. (eds.) ICISC 2005. LNCS, vol. 3935, pp. 299–321. Springer, Heidelberg (2006)

18. Patarin, J.: A proof of security in $O(2^n)$ for the Benes schemes. In: Vaudenay, S. (ed.) AFRICACRYPT 2008. LNCS, vol. 5023, pp. 209–220. Springer, Heidelberg (2008)
19. Patarin, J.: A Proof of Security in $O(2^n)$ for the Xor of Two Random Permutations. In: Safavi-Naini, R. (ed.) ICITS 2008. LNCS, vol. 5155, pp. 232–248. Springer, Heidelberg (2008)
20. Patarin, J.: A Proof of Security in $O(2^n)$ for the Xor of Two Random Permutations - Extended Version. Cryptology ePrint archive: 2008/010: Listing for 2008 (2008)
21. Salzborn, F., Szekeres, G.: A Problem in Combinatorial Group Theory. Ars Combinatoria 7, 3–5 (1979)
22. Seurin, Y.: Primitives et Protocoles cryptographics à sécurité prouvée. In: Ph. Thesis. Université de Versailles - Saint Quentin – France (2009)
23. Shrimpton, T., Stam, M.: Building a collision-resistant compression function from non-compressing primitives. In: Aceto, L., Damgård, I., Goldberg, L.A., Halldórsson, M.M., Ingólfsdóttir, A., Walukiewicz, I. (eds.) ICALP 2008, Part II. LNCS, vol. 5126, pp. 643–654. Springer, Heidelberg (2008)
24. Stam, M.: Beyond uniformity: Better security/efficiency tradeoffs for compression functions. In: Wagner, D. (ed.) CRYPTO 2008. LNCS, vol. 5157, pp. 397–412. Springer, Heidelberg (2008)

A Property of the $a_i \oplus b_j$ Values

We will assume that $q^2 \geq n \cdot 2^n$.

Theorem 5. *For all A of I_n, if the values $(a_1, a_2, \ldots a_q)$ are pairwise distinct and randomly chosen in I_n and if the values $(b_1, b_2, \ldots b_q)$ are pairwise distinct and randomly chosen in I_n, then: the number N_A of (i, j), $1 \leq i \leq q$, $1 \leq j \leq q$ such that $a_i \oplus b_j = A$ satisfies:*

$$\Pr[N_A \geq \frac{24q^2}{2^n - q}] \leq \frac{1}{2^{12n}}$$

This also means that the number U_A of $(a_1, a_2, \ldots, a_q, b_1, b_2, \ldots, b_q)$ such that the a_i are pairwise distinct, the b_i are pairwise distinct, and $N_A \geq \frac{24q^2}{2^n - q}$ satisfies: $U_A \leq \frac{1}{2^{12n}}[2^n(2^n - 1) \ldots (2^n - q + 1)]^2$.

Remark. The coefficient $\frac{1}{2^{12n}}$ here is not very important, we can easily change it to another even smaller coefficient. What is important for us here is the $O(\frac{q^2}{2^n})$ value.

Proof of Theorem 5. When new values a_α and b_α are generated, the probability that $\exists j, j \leq \alpha - 1$ such that $a_\alpha \oplus b_j = A$ is $\leq \frac{\alpha - 1}{2^n - (\alpha - 1)}$ since we have $\alpha - 1$ values b_j, and since a_α is randomly generated in $I_n \setminus \{a_1, a_2, \ldots, a_{\alpha-1}\}$. Similarly, the probability that $\exists j, j \leq \alpha - 1$ such that $a_j \oplus b_\alpha = A$ is $\leq \frac{\alpha - 1}{2^n - (\alpha - 1)}$, and the probability that $a_\alpha \oplus b_\alpha = A$ is $\leq \frac{1}{2^n - (\alpha - 1)}$. Therefore the probability that a_α and b_α will increase N_A from the values a_i, b_i, $1 \leq i \leq \alpha - 1$ is $\leq \frac{2\alpha}{2^n - (\alpha - 1)}$. Moreover, if it occurs, then N_A will increase by at maximum 2 since if it exists j such that $a_\alpha \oplus b_j = A$ then j is unique because all the b_j values are pairwise distinct. If $N_A \geq 2k$, where k is an integer, then for at least k such values α, N_A was increased by at least 1, i.e. for at least k values α we had: $\exists j \leq \alpha - 1$ such that $a_\alpha \oplus b_j = A$, or $a_j \oplus b_\alpha = A$, or $a_\alpha \oplus b_\alpha = A$. Therefore,

$$\Pr[N_A \geq 2k] \leq \binom{q}{k} \cdot \left(\frac{2q}{2^n - (q-1)}\right)^k$$

$$\Pr[N_A \geq 2k] \leq \frac{q!}{k!(q-k)!}\left(\frac{2q}{2^n - (q-1)}\right)^k$$

$$\Pr[N_A \geq 2k] \leq \frac{q^k}{k!}\left(\frac{2q}{2^n - q}\right)^k$$

$$\Pr[N_A \geq 2k] \leq \frac{1}{k!}\left(\frac{2q^2}{2^n - q}\right)^k \quad (1)$$

From Stirling formula, $k! \sim_{k \to +\infty} k^k e^{-k} \sqrt{2\pi k}$. If $k \geq \frac{12q^2}{2^n - q}$ this gives

$$k! \geq \left(\frac{12q^2}{e(2^n - q)}\right)^k \geq \left(\frac{4q^2}{2^n - q}\right)^k$$

Therefore, from (1), $\Pr[N_A \geq 2k] \leq \frac{1}{2^k}$ and since $k \geq \frac{12q^2}{2^n - q} \geq 12n$ we have: $\Pr[N_A \geq \frac{24q^2}{2^n - q}] \leq \frac{1}{2^{12n}}$ as claimed.

The Characterization of Luby-Rackoff and Its Optimum Single-Key Variants

Mridul Nandi

C.R. Rao AIMSCS Institute, Hyderabad*
mridul.nandi@gmail.com

Abstract. Luby and Rackoff provided a construction (LR) of $2n$-bit (strong) pseudorandom permutation or (S)PRP from n-bit pseudorandom function (PRF), which was motivated by the structure of DES. Their construction consists of four rounds of Feistel permutations (or three rounds, for PRP), each round involves an application of an independent PRF (i.e. with an independent round key). The definition of the LR construction can be extended by reusing round keys in a manner determined by a *key-assigning* function. So far several key-assigning functions had been analyzed (e.g. LR with 4-round keys K_1, K_2, K_2, K_2 was proved secure whereas K_1, K_2, K_2, K_1 is not secure). Even though we already know some key-assigning functions which give secure and insecure LR constructions, the exact characterization of all secure LR constructions for arbitrary number of rounds is still unknown. Some characterizations were being conjectured which were later shown to be wrong. In this paper we solve this long-standing open problem and (informally) prove the following:

> *LR is secure iff its key-assigning is* **not palindrome** (i.e. the order of key indices is not same with its reverse order).

We also study the class of LR-variants where some of its round functions can be tweaked (our previous characterization would not work for the variants). We propose a single-key LR-variant SPRP, denoted by LRv, making only four invocations of the PRF. It is exactly same as single-key, 4-round LR with an additional operation (e.g. rotation) applied to the first round PRF output. So far the most efficient single-key LR construction is due to Patarin, which requires five invocations. Moreover, we show a PRP-distinguishing attack on a wide class of single-key, LR-variants with three PRF-invocations. So,

> *4 invocations of PRF is minimum for a class of a single-key LR-variants SPRP and LRv is* **optimum** *in the class.*

Keywords: Luby-Rackoff, Feistel, PRP, SPRP, PRF, distinguisher, palindrome.

* A large part of the work has been done while working in The George Washington University.

G. Gong and K.C. Gupta (Eds.): INDOCRYPT 2010, LNCS 6498, pp. 82–97, 2010.

1 Introduction

Strong Pseudorandom permutations or SPRPs, which were introduced by Luby and Rackoff [4], formalize the well established cryptographic notion of block ciphers. They provided a construction of SPRP, well known as LR construction, which was motivated by the structure of DES [6]. The basic building block is the so called $2n$-bit *Feistel permutation* (or *LR round permutation*) LR_{F_K} based on an n-bit pseudorandom function (PRF) F_K:

$$\mathsf{LR}_{F_K}(x_1, x_2) = (F_K(x_1) \oplus x_2, x_1), \quad x_1, x_2 \in \{0,1\}^n.$$

Their construction consists (see Fig 1) of four rounds of Feistel permutations (or three rounds, for PRP), each round involves an application of an independent PRF (i.e. with independent random keys K_1, K_2, K_3, and K_4). More precisely, $\mathsf{LR}_{K_1,K_2,K_3}$ and $\mathsf{LR}_{K_1,K_2,K_3,K_4}$ are PRP and SPRP respectively where

$$\mathsf{LR}_{K_1,\ldots,K_r} := \mathsf{LR}_{F_{K_1},\ldots,F_{K_r}} := \mathsf{LR}_{F_{K_r}}(\ldots(\mathsf{LR}_{F_{K_1}}(\cdot))\ldots).$$

After this work, many results are known improving performance (reducing the number of invocations of F_K) [5] and reducing the key-sizes (i.e. reusing the round keys [7,8,10,12,11] or generate more keys from single key by using a PRF [2]). However there are some limitations. For example, we cannot use as few as single-key LR (unless we tweak the round permutation) or as few as two-round since they are not secure. Distinguishing attacks for some other LR constructions are also known [8]. We list some of the know related results (see Table 1). Here all keys $K_1, K_2, \ldots$ are independently chosen.

- $\mathsf{LR}_{K_1,K_2,K_3}$ is PRP but not SPRP. [4] and $\mathsf{LR}_{K_1,K_2,K_3,K_4}$ is SPRP. [4]
- $\mathsf{LR}_{K_1,K_2,K_2}$ is PRP. [14]
- $\mathsf{LR}_{K_1,K_2,K_1,K_2}, \mathsf{LR}_{K_1,K_2,K_2,K_2}, \mathsf{LR}_{K_1,K_2,K_1,K_1}$ and $\mathsf{LR}_{K_1,K_1,K_2,K_2}$ are SPRP. [8]

Our Contribution. In [11] author conjectured a necessary and sufficient condition for all secure LR constructions, which was later shown to be wrong [8]. So far we do not know any proven characterization. In this paper we solve it and prove the following theorem.

Theorem 1. Let F_K be a PRF, $K_1, \ldots, K_t$ be t independent keys and $\sigma = (\sigma_1, \ldots, \sigma_r)$ be an r-tuple with elements from $[1..t] := \{1, 2, \ldots t\}$, called a *key-assigning* function. The construction $\mathsf{LR}_{K_{\sigma_1},\ldots,K_{\sigma_r}}$ is (S)PRP if and only if σ is not palindrome[1] and $r \geq 3$ (or 4 respectively). As a corollary, any 4-round LR is SPRP if and only if it is PRP.

Due to the above result we now know that no single-key with any arbitrary round LR can be secure. However if one modifies the round permutation then secure single-key construction is possible. There are some known secure variants [8,11] among which the designs due to Patarin are most efficient. There are

[1] An r-tuple $\sigma = (\sigma_1, \ldots, \sigma_r)$ is called *palindrome* if $\sigma_i = \sigma_{r+1-i}, \forall i$.

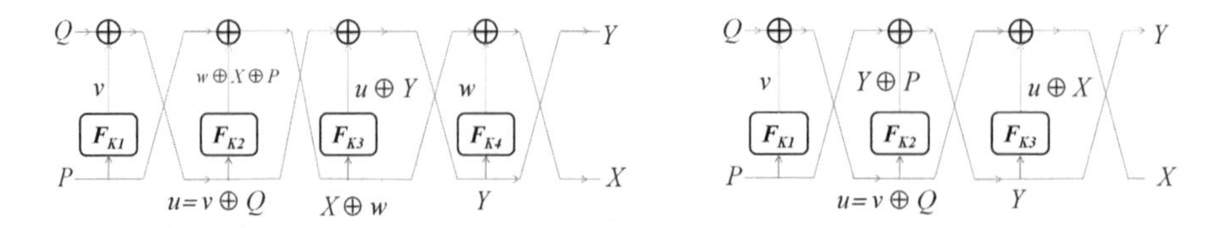

Fig. 1. The 4-round and 3-round LR constructions with independent round keys, i.e. K_1, K_2, K_3 and K_4 are chosen independently

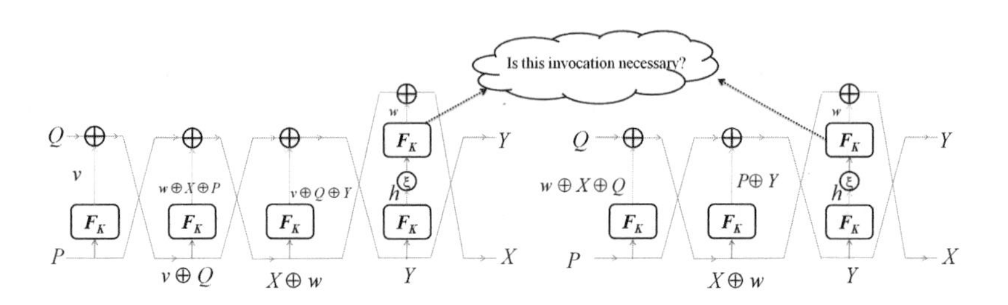

Fig. 2. 4-round (3-round for PRP) single-key LR-variant due to Patarin requires 5 (or 4) invocations of F_K. Here ξ is a simple function with low spreading number [8] e.g. one-bit rotation.

other efficient LR variants where k-wise independent universal hash function is required [5]. The SPRP design $\mathsf{LR}_{F_K \circ \xi \circ F_K}(\mathsf{LR}_{K,K,K}(\cdot))$ due to Patarin requires five invocations of the underlying PRF F_K (see Fig 2). This is almost same as 4-round single-key LR except the last round in which the composition function $F_K \circ \xi \circ F_K := F_K(\xi(F_K(\cdot)))$ is applied instead of F_K where ξ is a simple function, e.g. one-bit rotation. The similar construction is PRP for 3 rounds. The same result is true if we apply the tweak in the first round or use any ξ with small spreading number[2] [8]. In this paper we also prove the following results:

1. We first show that the PRF invocation in the tweak of 3-round Patarin construction (see right part of Fig 2) is essential. If we drop it then we have a PRP distinguishing attack. Moreover, this distinguishing attack works for many other choices of ξ (instead of rotation).
2. One may ask the same for the 4-round construction. Surprisingly, we show that the extra invocation of F_K in the tweak is redundant. In particular, $\mathsf{LR}_{K,K,K}(\mathsf{LR}_{\xi \circ F_K}(\cdot))$ is SPRP (see Fig 3).
3. Next we show that we cannot go below 4 invocations in a wide class of LR variants (with linear shuffle, defined the class and shuffle in Sec 5). In particular we show that any single-key, 3-round, Feistel encryption with linear shuffle is not PRP. So in that class, our construction is optimum. However,

[2] It is a parameter defined in [8]. Spreading of ξ is the $\max_{c \in \{0,1\}^n} \#\{x : x \oplus \xi(x) = c\}$.

we do not know the optimality when we have non-linear shuffle. This is an interesting open problem.

In summary, we prove the following result.

Theorem 2. $\mathsf{LR}_{K,K,K}(\mathsf{LR}_{\xi \circ F_K}(\cdot))$ is SPRP (see Fig 3) whenever F_K is PRF where ξ is one-bit rotation (or other simple function with low spreading number). Moreover, a single-key, 3-round LR-variant with any linear shuffle is not PRP.

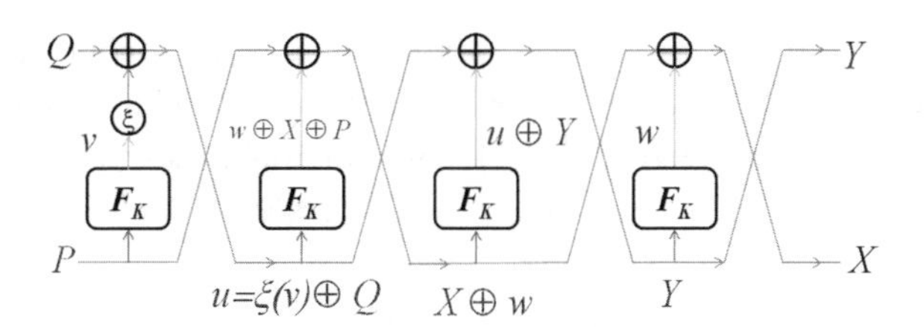

Fig. 3. Our 4-round single-key LR variant requires only 4 invocations of F_K. It is almost same as 4-round single-key LR except the rotation (or ξ) which is applied to the output of the first internal function.

Organization of the paper. We first describe notation and the proof tool, called Patarin's coefficient H-technique in Section 2. In section 3, we demonstrate our distinguishing attacks on LR and some of its variants. Then we characterize the secure LR construction in Section 4. In Section 5, we generalize LR variants and show that 3-round single-key general LR constructions are not secure. In section 6, we propose an optimum 4-round LR variant and prove its SPRP security and finally we conclude.

2 Notation and Preliminaries

A distinguishing adversary A is a probabilistic algorithm which has access to some oracles and which outputs either 0 or 1. Oracles are written as superscripts. The notation $A^{\mathcal{O}_1,\mathcal{O}_2} \Rightarrow 1$ (or $A^{\mathcal{O}} \Rightarrow 1$) denotes the event that the adversary A, interacts with the oracles $\mathcal{O}_1, \mathcal{O}_2$ (or $\mathcal{O}$), and finally outputs the bit 1. In what follows, by the notation $X \xleftarrow{\$} \mathcal{S}$, we will denote the event of choosing X uniformly at random from the finite set $\mathcal{S}$. Let RF_n be an n-bit to n-bit random function. An n-bit pseudorandom function (PRF) is a function $F : \mathcal{K} \times \{0,1\}^n \to \{0,1\}^n$, where $\mathcal{K} \neq \emptyset$ is the key space of the PRF such that the prf-advantage

$$\mathbf{Adv}_F^{\mathrm{prf}}(A) = \left| \Pr\left[K \xleftarrow{\$} \mathcal{K} : A^{F_K} \Rightarrow 1\right] - \Pr\left[A^{\mathsf{RF}_n} \Rightarrow 1\right] \right|$$

is negligible for any efficient adversary A. We denote $\mathbf{Adv}_F^{\mathrm{prf}}(q)$ (or $\mathbf{Adv}_F^{\mathrm{prf}}(q,t)$) by $\max_A \mathbf{Adv}_E^{\mathrm{prf}}(A)$ where maximum is taken over all adversaries which makes at most q queries (and runs in time t respectively). We write $F_K()$ instead of

$F(K,.)$. Let Perm(n) denote the set of all permutations on $\{0,1\}^n$. We require a blockcipher $E(K,)$, $K \in \mathcal{K}$, to be a strong pseudorandom permutation. The advantage of an adversary A in breaking the strong pseudorandomness of $E(,)$ is defined in the following manner.

$$\mathbf{Adv}_E^{\pm\mathrm{prp}}(A) = |\Pr\left[K \xleftarrow{\not c} \mathcal{K} : A^{E_K(),E_K^{-1}()} \Rightarrow 1\right] -$$

$$\Pr\left[\pi \xleftarrow{\not c} \mathrm{Perm}(n) : A^{\pi(),\pi^{-1}()} \Rightarrow 1\right]|.$$

If adversary has only access to encryption oracle $E(K,\cdot)$ then we call it prp-advantage $\mathbf{Adv}_E^{\mathrm{prp}}(A)$. Similar to prf-advantage we define $\mathbf{Adv}_E^{\pm\mathrm{prp}}(q)$ and $\mathbf{Adv}_E^{\pm\mathrm{prp}}(q,t)$ by $\max_A \mathbf{Adv}_{\mathbf{E}}^{\pm\mathrm{prp}}(A)$. Similar definition can be given for $\mathbf{Adv}_E^{\mathrm{prp}}(q)$ and $\mathbf{Adv}_E^{\mathrm{prp}}(q,t)$. In this paper we reserve q to mean the number of queries.

Pointless queries: Let M and C represent plaintext and ciphertext respectively. We assume that an adversary never repeats a query, i.e., it does not ask the encryption oracle with a particular value of M more than once and neither does it ask the decryption oracle with a particular value of C more than once. Furthermore, an adversary never queries its deciphering oracle with C if it got C in response to an encipher query M for some M and vice versa. These queries are called *pointless* as the adversary knows what it would get as responses for such queries. In this paper we assume adversaries make no pointless queries.

2.1 Patarin's Coefficient H-Technique

The following describes Patarin's coefficient H technique [9] (also known as Decorrelation theorem due to Vaudenay [13]) which would be used in our security analysis.

The view of an adversary $A^{\mathcal{O}_{+1},\mathcal{O}_{-1}}$ is the tuple $\psi := ((M_1,C_1,\delta_1), \ldots, (M_q,C_q,\delta_q))$ where A makes i^{th} query M_i or C_i and obtains responses C_i or M_i if $\delta_i = +1$ or -1 respectively. In case of $\mathcal{O}_{+1} = \mathcal{O}$ and $\mathcal{O}_{-1} = \mathcal{O}^{-1}$ we have that $\mathcal{O}(M_i) = C_i, \forall i$. Since A does not make any pointless query, all M_i (and C_i) are distinct.

Patarin's coefficient H technique says that prp-advantage of any distinguisher $A^{\mathcal{O}_{+1},\mathcal{O}_{-1}}$ making total q non-trivial queries to E is small if the followings hold for a subset $S \subseteq (\mathcal{M} \times \mathcal{M} \times \{+1,-1\})^q$ (the set S is known as set of **bad** views):

1. $\Pr[\mathsf{view}(A^{\mathsf{RP}_{\mathcal{M}},\mathsf{RP}_{\mathcal{M}}^{-1}}) \in S] \leq \epsilon_1$ i.e. the probability of bad view is small.
2. For any $\psi := ((M_1,C_1,\delta_1),\ldots,(M_q,C_q,\delta_q)) \notin S$ (ψ is called a **good** or non-bad view),

$$\Pr_K[E(K,M_i) = C_i,\ \forall i] \geq \frac{(1-\epsilon_2)}{|\mathcal{M}|^q} = (1-\epsilon_2) \times \Pr[\mathsf{RF}(M_i) = C_i,\ \forall i].$$

So, each good view can occur with probability more than $(1-\epsilon_2)$ times the probability of the view for the random function RF. In other words, on the average probability of good view are almost same for both $E(K,)$ and random function.

More precisely, if above holds then $\mathbf{Adv}_E^{\pm\mathrm{prp}}(A) \le \epsilon_1 + \epsilon_2 + q(q-1)/2|\mathcal{M}|$. The third term arises from the well known fact [1] that $\mathbf{Adv}_{\mathsf{RF}}^{\pm\mathrm{prp}}(A') \le \frac{q(q-1)}{2|\mathcal{M}|}$ for any A' making q non-trivial queries. Thus given any encryption algorithm $E(\cdot)$ it suffices to identify a set of bad views S and the values of ϵ_1 and ϵ_2 corresponding to the bad-views set.

3 Distinguishing Attack on Luby-Rackoff and Its Variants

Given $f : \{0,1\}^n \to \{0,1\}^n$, called *internal function*, the Luby-Rackoff (LR) round function (or the Feistel permutaion) $\mathsf{LR}_f : \{0,1\}^{2n} \to \{0,1\}^{2n}$ is defined by $\mathsf{LR}_f(x_1, x_2) = (f(x_1) \oplus x_2, x_1)$ where $x_1, x_2 \in \{0,1\}^n$. Clearly, $\mathsf{LR}_f^{-1}(y_1, y_2) = (y_2, f(y_2) \oplus y_1)$ and hence we also call it LR round permutation. The r-round LR permutation is defined by the sequential composition of the r round permutations $\mathsf{LR}_{\mathbf{f}} := \mathsf{LR}_{f_r} \circ \ldots \circ \mathsf{LR}_{f_1}$ where $\mathbf{f} = (f_1, \ldots, f_r)$. If the internal functions are keyed functions, i.e. $f_i = F_{K_i}$, then we simply denote the r-round LR encryption by $\mathsf{LR}_{K_1,\ldots,K_r}$. Given a family of function-tuples $\mathcal{F}$, the induced family of LR permutations (or encryption with key $\mathbf{f} \stackrel{\$}{\leftarrow} \mathcal{F}$) is $\mathsf{LR}_{\mathcal{F}} = \{\mathsf{LR}_{\mathbf{f}} : \mathbf{f} \in \mathcal{F}\}$.

A *Key-assigning* is an r-tuple $\sigma = (i_1, \ldots, i_r)$ with elements from $[1..k]$. Now given a sequence of k function families $\mathcal{F} = \langle \mathcal{F}_1, \ldots \mathcal{F}_k \rangle$ we define the function-tuple family $\mathcal{F}^{\otimes\sigma} := \{(f_{i_1}, \ldots f_{i_r}) : f_j \in \mathcal{F}_j\}$. In practice, each function family is indexed by an independent key. The single-key (i.e. $k = 1$) r-round LR for a function family $\mathcal{F}$ is nothing but $\mathcal{F}^{\otimes 1^r}$. In case of k independent random functions $\Gamma = \langle \Gamma_1, \ldots, \Gamma_k \rangle$ mapping n-bits to n-bits, we have a random function tuple $\Gamma^{\otimes\sigma}$. It is easy to see that if σ is palindrome then $\mathcal{F}^{\otimes\sigma}$ is a family of palindrome function tuples. By using hybrid argument one can show that for any PRF F,

$$\mathbf{Adv}_{\mathsf{LR}_{K_{\sigma_1},\ldots,K_{\sigma_r}}}^{\pm\mathrm{prp}}(q) \le \mathbf{Adv}_{\mathsf{LR}_{\Gamma^{\otimes\sigma}}}^{\pm\mathrm{prp}}(q) + \mathbf{Adv}_F^{\mathrm{prf}}(rq).$$

Because of it, we always assume random function instead of PRF. The Table 1 provides some known designs which are proved (or mentioned) secure.

Table 1. Some known secure, efficient LR designs and its variants. Let $\Gamma = (\Gamma_1, \Gamma_2)$ where Γ_1 and Γ_2 are independent random functions. $\sigma' = (1,2,2)$, $\sigma \in \{(1,2,2,2), (1,2,1,1), (1,2,1,2), (1,1,2,2)\}$. The author of [8] did not provide any proof, only mentioned that H technique can be applied to prove these SPRP. However in this paper we provide a general proof which covers the proof of these constructions.

Construction	$(\Gamma_1 \circ \xi \circ \Gamma_1, \Gamma_1, \Gamma_1)$	$(\Gamma_1 \circ \xi \circ \Gamma_1, \Gamma_1, \Gamma_1, \Gamma_1)$	$\Gamma^{\otimes\sigma}$	$\Gamma^{\otimes\sigma'}$
Security	PRP [8]	SPRP [8]	SPRP [8]	PRP [14]

3.1 Distinguishing Attack on Luby-Rackoff Encryptions with Palindrome Key-Assigning

Let $\mathsf{sw} : \{0,1\}^n \times \{0,1\}^n \to \{0,1\}^n \times \{0,1\}^n$ be the swap-function i.e. $\mathsf{sw}(x_1, x_2) = (x_2, x_1)$. Note that for any function f and palindrome function-tuple $\mathbf{f} = (f_1, \ldots, f_r)$ we have $(\mathsf{LR}_f \circ \mathsf{sw} \circ \mathsf{LR}_f)(x_1, x_2) = \mathsf{LR}_f(x_1, f(x_1) \oplus x_2) = (x_2, x_1)$ and hence we have the following (see Fig 4)

$$\begin{aligned}\mathsf{LR}_{\mathbf{f}} \circ \mathsf{sw} \circ \mathsf{LR}_{\mathbf{f}} &= \mathsf{LR}_{f_1} \circ \ldots \circ (\mathsf{LR}_{f_r} \circ \mathsf{sw} \circ \mathsf{LR}_{f_r}) \circ \ldots \circ \mathsf{LR}_{f_1} \quad \text{(since } \mathbf{f} \text{ is palindrome)}\\ &= \mathsf{LR}_{f_1} \circ \ldots \circ (\mathsf{LR}_{f_{r-1}} \circ \mathsf{sw} \circ \mathsf{LR}_{f_{r-1}}) \circ \ldots \circ \mathsf{LR}_{f_1}\\ &= \ldots = \mathsf{LR}_{f_1} \circ \mathsf{sw} \circ \mathsf{LR}_{f_1} = \mathsf{sw}\end{aligned}$$

So, if $\mathcal{F}$ is any palindrome family then $\Pr_{\mathbf{f} \xleftarrow{\$} \mathcal{F}}[\mathsf{LR}_{\mathbf{f}} \circ \mathsf{sw} \circ \mathsf{LR}_{\mathbf{f}}(\mathbf{0}, \mathbf{0}) = (\mathbf{0}, \mathbf{0})] = 1$. So, $\mathsf{LR}_{\mathcal{F}}$ can be distinguished from the random permutation RP by making two adaptive queries $(Y_1, Y_2) := \mathcal{O}(\mathbf{0}, \mathbf{0})$ and $(Z_1, Z_2) := \mathcal{O}(Y_2, Y_1)$. The probability that $Z_1 = Z_2 = \mathbf{0}$ is one when $\mathcal{O} = \mathsf{LR}_{\mathcal{F}}$ (this also gives ciphertext-forging attack). When the distinguisher is interacting with RP, the probability is almost $1/2^{2n}$. Hence $\mathbf{Adv}^{\mathrm{prp}}_{\mathsf{LR}_{r \otimes \sigma}}(2, t) \geq 1 - 1/2^{2n}$. As a corollary single-key LR with any number of rounds is not PRP.

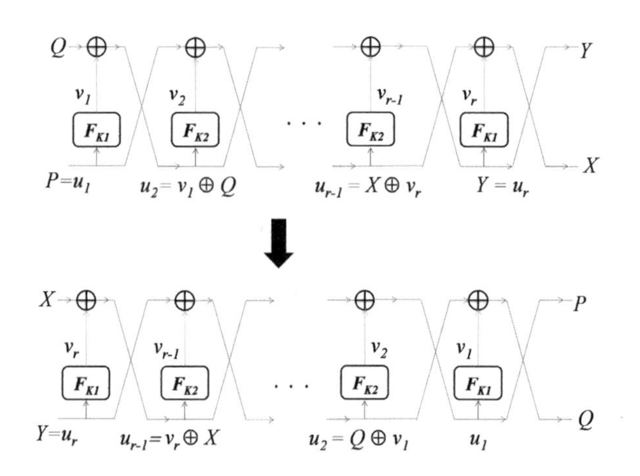

Fig. 4. It illustrates how the distinguishing attacks works for palindrome key-assigning

3.2 Distinguishing Attack on Some Variant of Single-Key 3-Round Luby-Rackoff Encryptions

From the previous section we now know that the any r-round single-key (or 3-round, double-key with key assigning $\sigma = \langle 1, 2, 1 \rangle$) LR is not PRP. The best known PRP, single-key LR variant is due to Patarin [8]. In this variant (see Fig 2) the last round (or the first round) internal function is defined as $f \circ \xi \circ f$ where ξ is the one-bit left rotation. As mentioned in Fig 2, we want to study whether we can simply use rotation tweak without using the extra invocation of f (to

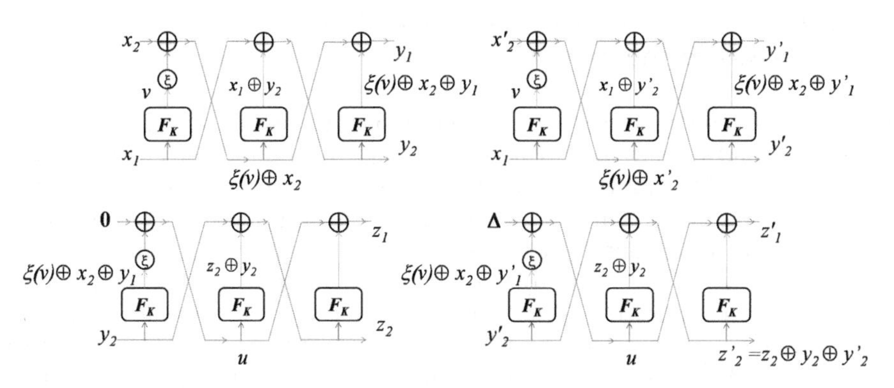

Fig. 5. It gives an idea how our 3-round LR variant (simpler version of Patarin's 3-round by dropping one invocation) distinguishing attack works. Here $u = v^{<<2} \oplus (x_2 \oplus y_1)^{<<1}$ and $\Delta = (x_2 \oplus x_2' \oplus y_1 \oplus y_1')^{<<1}$.

save one extra invocation of f). Here we show that we cannot do that for three round constructions. We provide a distinguishing attack where the rotation is applied to the first round instead of the last round. Same analysis would work for the last round, too. The function family of this modification can be described as $\mathcal{F}^{(2)} := \{(\xi \circ f, f, f) : f \in \mathcal{F}\}$. Our attack on $\mathsf{LR}_\xi := \mathsf{LR}_{\mathcal{F}^{(2)}}$ requires four encryption queries.

1. 1st *and* 2nd *Query*: (1) $\mathsf{LR}_\xi(x_1, x_2) = (y_1, y_2)$ and (2) $\mathsf{LR}_\xi(x_1, x_2') = (y_1', y_2')$. Call this event by E_1 and let $\Delta = (x_2 \oplus x_2' \oplus y_1 \oplus y_1')^{<<1}$.

   Lemma 1. $\Pr_{f \xleftarrow{\$} \mathcal{F}}[f(x_1)^{<<1} \oplus f(y_2) = x_2 \oplus y_1 \mid E_1] = 1$

2. 3rd *and* 4th *Query*: (3) $\mathsf{LR}_\xi(y_2, \mathbf{0}) = (z_1, z_2)$ and (4) $\mathsf{LR}_\xi(y_2', \Delta) = (z_1', z_2')$. Call this event by E_2.

   Lemma 2. $\Pr_{f \xleftarrow{\$} \mathcal{F}}[z_2 \oplus z_2' = y_2 \oplus y_2' \mid E_1, E_2] = 1$.

The above two lemmas are easy to verify from the Fig 5 and hence we skip the proofs. When distinguisher is interacting with $2n$-bit random permutations the probability $\Pr[z_2 \oplus z_2' = y_2 \oplus y_2' \mid E_1, E_2] \approx 1/2^{2n}$. So, we can use this event to make a distinguishing attack. Hence $\mathbf{Adv}^{\text{prp}}_{\mathsf{LR}_\xi}(4) \geq 1 - 1/2^{2n}$.

Remark 1. Similar attack can be carried out for the other simple variants, e.g. when $\xi(x) = \alpha \cdot x$ (the Galois field multiplication by the primitive element α) or any other linear function ξ (note that both rotation, or multiplication by a primitive element are linear over $GF(2)$ and $GF(2^n)$, respectively). In fact, with a closer look on the attack one can see that attack works for any function ξ such that $\Pr_{v \xleftarrow{\$} \{0,1\}^n}[\xi(\xi(v \oplus c_1)) \oplus \xi(\xi(v \oplus c_1)) = \Delta]$ is significantly high for some fixed constant Δ (depending on c_1 and c_2). Here the probability is computed over random choice of v. Note that this measurement is completely different from the spreading number considered in [8].

4 Security Analysis of LR with Non-palindrome Key-Assigning Function

Here we characterize all secure LR encryptions. Informally we prove that LR is secure if and only if the key-assigning is not palindrome. We have already seen the "only if" part which is more intuitive. However it is not obvious why non-palindrome key-assigning give a secure LR encryption. To understand the intuition let us first assume that $\sigma_1 \neq \sigma_r$, i.e. the first and last round keys are independent. Then the input to the first round function can be collided due to choices of plaintext (the first n-bit same as that of a previous plaintext). But we cannot control anymore collisions after that. It does not matter if we choose first n bits of plaintext same as the last n bits of the ciphertext as they are fed to independent random functions. The similar argument works when an attacker chooses a ciphertext. In general, for a non-palindrome σ, there must exist $r' < r/2 - 1$ such that $\sigma_i = \sigma_{r+1-i}$ for $i = 1, .., r'$ but $\sigma_i \neq \sigma_{r+1-i}$ for $i = r' + 1$. The similar argument would be applied to the random functions at rounds $r' + 1$ and $r - r'$. The random functions $\Gamma_{\sigma_{r'+1}}$ and $\Gamma_{\sigma_{r-r'}}$ at round $r' + 1$ and $r - r'$ are protecting a plaintext and ciphertext query respectively.

Theorem 1. *Let σ be an r-sequence. Then $\mathsf{LR}_{\Gamma\otimes\sigma}$ is (S)PRP if and only if σ is not palindrome and $r \geq 3$ (or $r \geq 4$ respectively). In this case, $\mathbf{Adv}^{\mathrm{prp}}_{\mathsf{LR}_{\Gamma\otimes\sigma}}(q, t)$ (or $\mathbf{Adv}^{\pm\mathrm{prp}}_{\mathsf{LR}_{\Gamma\otimes\sigma}}(q, t)$ for SPRP) is at most $\frac{(1+r^2)q^2}{2^n-1} + \frac{q^2}{2^{2n}}$.*

Corollary 1. *Let $r \geq 4$. Then $\mathsf{LR}_{\Gamma\otimes\sigma}$ is SPRP if it is PRP.*

The corollary is interesting as it says that any PRP LR for more than 4 round has to be SPRP. Note that it is not true for three rounds as we already know three round independent-keyed is PRP but not SPRP. This is a straightforward application of the theorem. We prove the theorem by using Patarin H-coefficient-technique as describe in Sec 2.1

Construction of the set of bad views S and computation of ϵ_1
For $1 \leq i \leq q$, we denote $M_i = (P_i, Q_i), C_i = (X_i, Y_i) \in \{0,1\}^n \times \{0,1\}^n$. We first define a set of bad views S (as we discuss in Patarin H coefficient technique). Given a view $\psi = ((M_1, C_1, \delta_1), \ldots, (M_q, C_q, \delta_q))$ we call P_j fresh if $P_j \neq P_i, Y_i, Y_j$ for all $i < j$. Similarly Y_j is fresh if $Y_j \neq P_i, Y_i, P_j$ for all $i < j$.

Definition 1. *A view $\psi = ((M_1, C_1, \delta_1), \ldots, (M_q, C_q, \delta_q)) \in (\{0,1\}^{2n} \times \{0,1\}^{2n} \times \{+1, -1\})^q$ is called* **bad** *if there is a j such that either P_j is not fresh and $\delta_j = -1$ or Y_j is not fresh and $\delta_j = 1$.*

Let S be the set of all bad views. Now we provide an upper bound of the probability that a view is bad when an adversary is interacting with a random permutation RP and its inverse RP^{-1}. We show that

$$\Pr[\mathrm{view}(A^{\mathsf{RP},\mathsf{RP}^{-1}}) \in S] \leq \epsilon_1 := \frac{q^2}{2^n - 1} \quad (1)$$

If the ith query is encryption (i.e. $\delta_i = 1$) then $C_i = (X_i, Y_i)$ is uniformly distributed over a set of size at least $2^{2n} - i + 1$. Thus $\Pr[Y_i = c] \leq 1/(2^n - 1)$ for any constant $c \in \{0,1\}^n$ provided $q \leq 2^n$ (o.w. the equation is obviously true). Thus Y_i is one of P_j or Y_j or P_i has probability at most $2i - 1$. Similar result is true when $\delta_i = -1$. Thus a view is bad view, has probability at most $q^2/(2^n - 1)$. So we have proved the Eq. 1.

Some Notations and Properties of Good Views. We say that $M_i = (P_i, Q_i)$ is fresh if $M_i \neq \mathsf{sw}(C_j)$, $j < i$. Similarly we define a fresh C_i. Let $r \geq 4$ and σ be a non-palindrome sequence such that r' is the size of the common prefix of σ and σ^{rev}. Note that $r' \leq r/2 - 1$. Given a good (non-bad) view $\psi = ((M_1, C_1, \delta_1), \ldots, (M_q, C_q, \delta_q))$ we define the following sets of query-indices

$$N_{\psi,P} = \{i : P_i \text{ is fresh}\},\ N_{\psi,Y} = \{i : Y_i \text{ is fresh}\},$$

$$N_{\psi,M} = \{i : M_i \text{ is fresh}\},\ N_{\psi,C} = \{i : C_i \text{ is fresh}\}.$$

In the following, we state some lemmas whose proofs are straightforward and easy to verify.

Lemma 3. *For all i, $P_i \neq Y_i$. We also have $i \in N_{\psi,P}$ or $i \in N_{\psi,Y}$ if $\delta_i = -1$ or $+1$ respectively. Let $M_i = \mathsf{sw}(C_j)$, then $j \in N_{\psi,C}$ (i.e. C_j is fresh) or $j \in N_{\psi,M}$ (i.e. M_j is fresh) if $j < i$ or $i < j$ respectively.*

For each query number i, we define two sets of round numbers $I'_i \subseteq I_i \subseteq [1..r]$ as follows:

1. $\delta_i = 1$: We define $I_i := [1..r]$ or $[2..r]$ or $[r'+1..r]$ if P_i is fresh or M_i is fresh or M_i is not fresh, respectively. $I'_i = I_i \setminus \{r-1, r\}$ (note that $r \geq 3$).
2. $\delta_i = -1$: We define $I_i := [1..r]$ or $[1..r-1]$ or $[1..r-r']$ if Y_i is fresh or C_i is fresh or C_i is not fresh, respectively. We define $I'_i = I_i \setminus \{1, 2\}$.

Some Observations on LR. Now we state some useful and easy to verify properties of the r-round LR computations $\mathsf{LR}(P, Q) = (X, Y)$. Let $u[\ell], v[\ell]$ denote the input and output of the internal function at the ℓth round.

Lemma 4. *The ℓth intermediate input $u[\ell] = (v[\ell-1] \oplus v[\ell-3] \oplus \ldots v[\ell\%2+1]) \oplus R$ where $R = P$ or Q if ℓ is odd or even respectively.*

If $\ell \notin I_i \setminus \{1, r\}$ then either $\ell \leq r'$ or $r - \ell \leq r'$. Moreover there is a $j < i$ such $r - \ell \in I_j$. In that case $u_i[\ell] = u_j[r-\ell]$ and $v_i[\ell] = v_j[r-\ell]$ (similar proof can be made as we did for distinguishing attack on palindrome key-assigning).[3] Thus $u_i[\ell]$ (or $v_i[\ell]$) for $\ell \in I_i, i \in [1..q]$ (we denote it by u_I) together determine all intermediate inputs (or outputs respectively). This can be further extended to the following result.

[3] If r' or $r - r' \notin I_i$ (i.e either M_i or C_i is not fresh) then $u_i[\ell] = u_j[r-\ell]$ where $\ell = r'+1$ or $r - r'$. However $\sigma_\ell \neq \sigma_{r-\ell}$ (by definition of r'). Hence independent random functions are applied to these same intermediate input.

Lemma 5. *$v_{I'} := \{v_i[\ell] : \ell \in I'_i, i \in [1..q]\}$ and ψ together determine all intermediate inputs and outputs.*

We denote the relation by the function $\mathcal{I}$, i.e. $\mathcal{I}(v_{I'}, \psi)$ is the tuple of all intermediate inputs. Now we see that if we choose $v_{I'}$ at random then the probability that all intermediate inputs in u_I are distinct is at least $(1 - \epsilon_2)$ where $\epsilon_2 := r^2 q^2/2^n$.

Proposition 1. *For any ℓ, ℓ' with $\sigma_\ell = \sigma_{\ell'}$, $\Pr_{v_{I'} \xleftarrow{\$}}[u_i[\ell] = u_j[\ell']] \leq 1/2^n$ where $u_i[\ell]$ and $u_j[\ell']$ are determined from $\mathcal{I}(v_{I'})$ and $v_{I'}$ is chosen at uniform distribution.*

Proof. We prove it in different cases. If $\ell, \ell' \in \{1, r\}$ then the probability is zero because the view ψ is good. In fact, if one of these is either 1 or r then the probability is $1/2^n$ as that one is constant (determined by ψ not by $v_{I'}$) and the other one is non-trivial linear function of $v_{I'}$. For any $\ell \notin \{1, r\}$, $u_i[\ell]$ is indeed non-trivial linear function of $v_{I'}$. Now if we show that the linear functions are different for $u_i[\ell]$ and $u_j[\ell]$ then by randomness of $v_{I'}$ the above probability is $1/2^n$. Let $j < i$ and $\ell = r' + 1$ (if $M_i = \mathsf{sw}(C_{i'-1})$) or $r - r'$ (if $C_i = \mathsf{sw}(M_{i'-1})$). Then $u_i[\ell] = u_{i'}[r - \ell]$ and $r - \ell - 1 \in I'_i$. Hence $v_j[r - \ell - 1]$ contributes to $u_i[\ell]$. If $i' \neq j$ then we are done. Otherwise note that if $\ell' = r - \ell$ then $\sigma_\ell \neq \sigma_{\ell'}$. So $\ell' \neq r - \ell$ and hence by above lemma the two linear functions are indeed different. ∎

Proof of Theorem 1. We apply Patarin's coefficient H-technique. We already have defined the set of bad views S and we know $\epsilon_1 := q^2/(2^n - 1)$. Let E be the event that for all ℓ, ℓ' with $\sigma_\ell = \sigma_{\ell'}$, $u_i[\ell] \neq u_j[\ell']$ where $\ell \in I_i$ and $\ell' \in I_j$. By the Proposition 1, we know that $\Pr[E] \leq \epsilon_2 := r^2 q^2/2^n$ since there are at most $r^2 q^2$ possible values of i, ℓ, j, ℓ'. So the number of possible $v_{I'}$ values such that $u_{I'}$ are all distinct is at least $2^{n|I'|}(1 - \epsilon_2)$. Given any such $v_{I'}$ (which determines the rest of the intermediate outputs) the probability that these are indeed the intermediate outputs is exactly $2^{-n|I|} = 2^{-n(|I'|+2q)}$. This is true since $|I| = |I'| + 2q$ and there are $|I|$'s distinct inputs for the internal random functions which takes some specific given values v_I. Thus, for any fixed good view ψ, $\Pr[\text{view} = \psi] \geq (1 - \epsilon_2)/2^{2nq}$ where $\epsilon_2 := r^2 q^2/2^n$. Hence we have proved our theorem by applying the Patarin's H-technique as described in Sec 2.1.

5 General Feistel Round Permutation

The LR round permutation can be expressed as $\mathsf{LR}_f(x_1, x_2) = \rho(f(x_1), x_1, x_2)$ where $\rho(v, x_1, x_2) = (v \oplus x_2, x_1)$, $v, x_1, x_2 \in \{0,1\}^n = \mathbb{F}$. So $\rho : \mathbb{F}^3 \to \mathbb{F}^2$ is a linear function which can be characterized by the matrix $M = \binom{L_1}{L_2}$ where $L_1 = (1, 0, 1) \in \mathbb{F}^3$ and $L_2 = (0, 1, 0) \in \mathbb{F}^3$. A *general Feistel* function $\mathsf{F}_{f,\rho} : \{0,1\}^{2n} \to \{0,1\}^{2n}$ with the internal function f and mix function ρ is defined by $\mathsf{F}_{f,\rho}(X) = \rho(f(X[1..n]), X)$ (see Fig 6). In practice, the internal function f is a strong cryptographic object and the mix function is a simple (mostly using xor or rotation or at most modular addition) efficiently computable function. We do

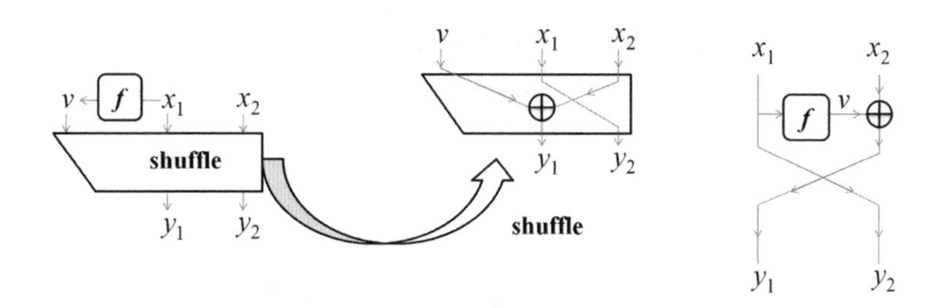

Fig. 6. A general LR or Feistel round permutation

not use or assume any cryptographic property on the mix function. However we require the Feistel function to be invertible (independent of the internal function) and hence we need some types of the mix function, called *shuffle.*

Definition 2 (Shuffle). *A function* $\rho : \{0,1\}^{3n} \to \{0,1\}^{2n}$ *is called* **shuffle** *if for any* $v \in \{0,1\}^n$, $\rho(v,\cdot,\cdot) := \rho_v$ *is a permutation over* $\{0,1\}^{2n}$ *and there exists a function* $\tau : \{0,1\}^{2n} \to \{0,1\}^n$ *such that* $\tau(\rho(v,x_1,x_2)) = x_1$, $\forall v, x_1, x_2 \in \{0,1\}^n$, *i.e. the function* $\rho_{v,1}^{-1}$ *is independent of* v *where* $\rho_v^{-1} = (\rho_{v,1}^{-1}, \rho_{v,2}^{-1})$. *Moreover it is called* **smooth** *if* ρ, ρ_v^{-1} *and* τ *are efficiently computable.*

Now we prove that (smooth) shuffle is the necessary and sufficient to have (efficiently computable) invertibility of the Feistel function.

Lemma 6. *The Feistel function* $\mathsf{F}_{f,\rho}$ *is permutation for all functions* f *if and only if the mix function* ρ *is shuffle. In this case,* $\mathsf{F}_{f,\rho}$ *and* $\mathsf{F}_{f,\rho}^{-1}$ *are efficiently computable if the shuffle* ρ *is smooth and the function* f *is efficiently computable.*

Proof. When ρ is a shuffle the inverse of the Feistel function can be shown to be $\mathsf{F}_{f,\rho}^{-1}(y_1, y_2) = \rho_v^{-1}(y_1, y_2)$ where $v = f(\tau(y_1, y_2))$. Hence $\mathsf{F}_{f,\rho}^{-1}$ is efficiently computable if f, τ, ρ_v^{-1} are efficiently computable. Clearly $\mathsf{F}_{f,\rho}$ is efficient if f and ρ are so. To prove the "only if" part it is easy to see that for all v, ρ_v must be invertible by choosing the constant internal function $f(x_1) = v$, $\forall x_1$. We show that the first n bit of the inverse does not depend on v. If not then for some v, v', $x_1 \neq x_1'$ and x_2, x_2' we have $\rho(v, x_1, x_2) = \rho(v', x_1', x_2')$. If we define a function f such that $f(x_1) = v$ and $f(x_1') = v'$ then the Feistel mapping is not injective as $\mathsf{F}_{f,\rho}(x_1, x_2) = \mathsf{F}_{f,\rho}(x_1', x_2')$. ∎

The r-round Feistel function (or permutation when we have shuffle) $\mathsf{F}_{(f_r,\rho_r)} \circ \ldots \circ \mathsf{F}_{(f_1,\rho_1)}$ is similarly denoted by $\mathsf{F}_{\mathbf{f},\boldsymbol{\rho}}$ where $\mathbf{f} = (f_1, \ldots, f_r)$ and $\boldsymbol{\rho} = (\rho_1, \ldots, \rho_r)$.

In case of linear shuffle functions ρ_i's are characterized by 2×3 matrix over $GF(2^n)$. The Lemma 7 characterizes all linear shuffles. More generally when we have linear shuffle over $GF(2)$ we have $2n \times 3n$ matrix over $GF(2)$. However we can have non-linear shuffle functions too. One such example is $\rho(v, x_1, x_2) = (v \boxplus (v \oplus x_2), x_1)$ where $\boxplus$ is modulo 2^n integer addition. A more complicated shuffle function may look like $\rho(v, x_1, x_2) = (y_1 := \pi_{v,x_1}(x_2), \pi'_{y_1}(x_1))$ where π_{v,x_1} and π'_{y_1} are any permutations. In fact, it is a general form of a shuffle if we assume that $\rho(v, x_1, \cdot)$ (equivalently $\tau(y_1, \cdot)$) is a permutation over $\{0,1\}^n$. This assumption is reasonable, otherwise we do not have complete diffusion in two rounds. In this paper we are only interested in linear shuffles. Now we state the version of Lemma 6 in case of linear shuffle. The proof is an immediate application of Lemma 6.

Lemma 7. *A linear function $M_{2\times 3}$ is shuffle if and only if*
(1) $c_1 \cdot M_{1*} \oplus c_2 \cdot M_{2*} = (0,1,0)$ for some pair of constants $(c_1, c_2) \in \mathbb{F}^2$ and
(2) $\mathsf{rank}(M_{*2}\ M_{*3}) = 2$, i.e. the 2×2 matrix $(M_{*2}\ M_{*3})$ is invertible.

5.1 PRP Attack on Three Round Linear-Mix Single-Key Feistel Function

Now we provide a PRP distinguishing attack on three-round single key Feistel function with any linear shuffles (may be different for each round). Let $\mathbf{f} = (f, f, f)$ and $\boldsymbol{\rho} = (\rho_1, \rho_2, \rho_3)$ be tuple of three linear shuffles. A very similar distinguishing attack as in Sec. 3 also works for $\mathsf{F} := \mathsf{F}_{\mathbf{f},\boldsymbol{\rho}}$. Our attack requires four encryption queries.

1. 1st *and* 2nd *Query*: (1) $\mathsf{F}(x_1, x_2) = (y_1, y_2)$ and (2) $\mathsf{F}(x_1, x'_2) = (y'_1, y'_2)$. Call this event by E_1. Let $c_1 \cdot Y_1 \oplus c_2 \cdot Y_2 = \rho_3^{-1}(Y_1, Y_2)[1..n]$ since ρ_3 is linear shuffle.

   Lemma 8. *There is a constant Δ, a linear function of $x_1, x_2, x'_2, y_1, y_2, y'_1$ and y'_2, such that*

   $$\Pr_{f \stackrel{\$}{\leftarrow} \mathcal{F}}[\rho_1(f(\tau), \tau, \mathbf{0})[1..n] = \rho_1(f(\tau'), \tau', \Delta)[1..n] \mid E_1] = 1$$

   where $\tau = c_1 \cdot y_1 \oplus c_2 \cdot y_2$, and $\tau' = c_1 \cdot y'_1 \oplus c_2 \cdot y'_2$.

2. 3rd *and* 4th *Query*: (3) $\mathsf{F}(\tau, \mathbf{0}) = (z_1, z_2)$ and (4) $\mathsf{F}(\tau', \Delta) = (z'_1, z'_2)$. Call this event by E_2.

   Lemma 9. *There is a constant Δ', a linear function of $x_1, x_2, x'_2, y_1, y_2, y'_1$ and y'_2, such that*

   $$\Pr_{f \stackrel{\$}{\leftarrow} \mathcal{F}}[c_1 \cdot (z_1 \oplus z'_1) \oplus c_2 \cdot (z_2 \oplus z'_2) = \Delta' \mid E_1, E_2] = 1$$

The above two lemmas are straightforward and tedious (the main thing is to compute Δ and Δ'). The idea of the proof is provided in the Figure 7. When distinguisher is interacting with $2n$-bit random permutations the above probability approximately $1/2^{2n}$. So we can use this event to make a distinguishing attack. So, at least four invocations of the underlying PRF are required to obtain a secure Feistel encryption.

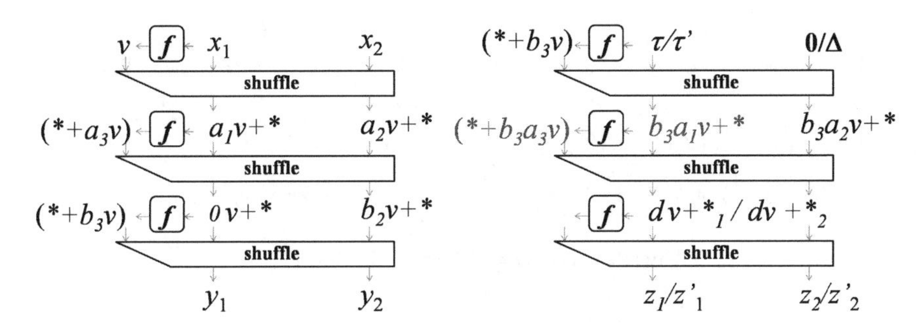

Fig. 7. It gives an idea how our attack works for 3-round Feistel with linear shuffle. The * denotes a linear function without involving v. We can express all internal variable by linear functions of v. The $\Delta' = *_1 - *_2$ and Δ is defined in a way such that the second internal inputs (for the queries $(\tau, \mathbf{0})$ and $\tau', \Delta)$ in the right half of the figure match. This can be done as the second input has non-zero coefficient in the first output and the coefficient of v (only unknown variable) are same for both queries.

6 SPRP Security Analysis of Single-Key 4-Round Feistel Function

In this section we first study the following simple variant of 4-round single-key LR. Let ρ be the LR shuffle and $\rho'(v, x_1, x_2) = (\xi(v) \oplus x_2, x_1)$ where $\xi(v) := v^{<<1}$ is one-bit left rotation. Let $\mathsf{LRv} = \mathsf{F}_{\mathcal{F}\otimes 1^4, \boldsymbol{\rho}}$ where $\boldsymbol{\rho} = (\rho', \rho, \rho, \rho)$. A similar security analysis would work for any function ξ with low spreading number, i.e. $\Pr_{v \xleftarrow{\$}}[\xi(v) \oplus v = c]$ is small for all c. We illustrated our design in Fig 3. In this section, we prove the SPRP security of this. In other words, we would that the Patarin's single-key SPRP LR-variant is not optimum and one invocation of PRF is completely of redundant. We follow the similar notation as in the proof of Theorem 1. In fact, the main idea of the proof remain same.

Let $f_K(P_i) = V_i$ and $f_K(Y_i) = W_i$. The four intermediate inputs of f_K during the computation $\mathsf{LR}'_{f^4}(P_i, Q_i) = (X_i, Y_i)$ are

$$P_i, \; a_i := V_i^{<<1} \oplus Q_i, \; b_i := X_i \oplus W_i, \; \text{ and } Y_i.$$

We want to prove that except the forced collisions (due to the choice of plaintexts or ciphertexts) all intermediate inputs are distinct with high probability given that a view is good or non-bad (the same definition of bad views as we have for Theorem 1). We have defined $N_{\psi,P}$ and $N_{\psi,Y}$. Given a good view ψ, let the pair $(\mathbf{v}, \mathbf{w})$ be called ψ-compatible if $v_i = v_j$ (or w_j) whenever $P_i = P_j$ or Y_j and $w_i = w_j$ whenever $Y_i = Y_j$ where $\mathbf{v} = (v_1, \ldots, v_q)$ and $\mathbf{w} = (w_1, \ldots, w_q)$. Let N denote the number of distinct P_i's and Y_i's (which are actually intermediate inputs).

Lemma 10. *Given a good view the number of ψ-compatible pairs is 2^{nN}. Among which there are at least $2^{nN}(1 - 13q^2/2^n)$ compatible elements give distinct $a_i :=$*

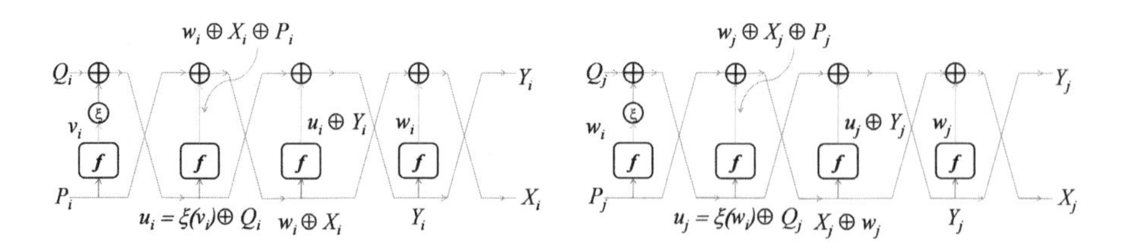

Fig. 8. Case: $\delta_j = +1$, $P_j = Y_i$ and $i < j$

$v_i^{<<1} \oplus Q_i, b_i = X_i \oplus w_i$, *for all* $1 \leq i \leq q$ *and they are different from* P_j*'s and* Y_j*'s for all* $1 \leq j \leq q$. *We call these* DI-compatible *(distinct-input compatible).*

Proof. We compute the probability of the complement event when we choose compatible pairs at random. We consider the case when $\delta_j = +1$ (i.e. an encryption query) and $P_j = Y_i$, $i < j$ (illustrated in Figure 8). In this case, the four intermediate inputs for i^{th} query are $P_i, Y_i, a_i = \xi(v_i) \oplus Q_i$ and $b_i = w_i \oplus X_i$. Since $P_j = Y_i$, the four intermediate inputs for j^{th} query are $P_j, Y_j, a_j = \xi(w_i) \oplus Q_j$ and $b_i = w_j \oplus X_j$. Note that v_i, w_i and w_j are chosen at random. Hence $(\mathbf{v}, \mathbf{w})$ is not DI-compatible due to the ith and j^{th} query has probability at most $13/2^n$. In particular, except the case for $b_i = a_j$, the probability is $1/2^n$ and there are 11 such possible collisions. The $\Pr[b_i = a_j] = \Pr[w_i \oplus \xi(w_i) = c] = 1/2^{n-1}$ (it can be easily checked and was shown in [8]) where $c = X_i \oplus Q_j$. The other cases can be proved similarly. Since there are $\binom{q}{2}$ pair of queries and for each pair the probability is bounded by $13/2^n$, we have proved that probability that a random compatible pair is DI-compatible is at least $(1 - 13/2^n)$. ∎

If a DI-compatible pair becomes all intermediate outputs then the all intermediate inputs are determined by these. Moreover these intermediate inputs are distinct. There are $N + 2q$ distinct intermediate inputs (including P_i's and Y_i's). Hence probability that the intermediate outputs are given by a specific DI-compatible pairs is exactly $2^{-n(N+2q)}$. So we have proved that

$$\Pr[\text{view} = \psi] \geq 2^{nN}(1 - 13q^2/2^n) \times 2^{-n(N+2q)} = \frac{1 - 13q^2/2^n}{2^{2nq}}.$$

Hence we have proved the SPRP-security of our proposal LRv.

Theorem 2. $\mathbf{Adv}_{\mathsf{LRv}}^{\pm\text{prp}}(q,t) \leq \frac{14q^2}{2^n - 1} + \frac{q^2}{2^{2n}}$.

7 Conclusion

This paper characterizes all secure LR constructions. So we know which LR are secure and which are not. If we make simple tweak in the LR-round then we can have secure single-key construction. Previously proposed tweak due to Patarin costs an extra invocation. In this paper we show that this extra invocation is redundant in case of SPRP design (4-round) but completely necessary in case of PRP design (3-round). We also provide a distinguishing attack on a wide class of single-key LR variants which invoke the underlying internal function 3 times. So 4-invocations is necessary for single-key LR type designs. Hence our proposed design is optimum. However we do not know yet whether there are any non-linear shuffles such that single-key Feistel with three rounds is SPRP.

Acknowledgement. This work was supported in part by the National Science Foundation, Grant CNS-0937267. Author would like to thank Donghoon Chang for his comments.

References

1. Halevi, S., Rogaway, P.: A tweakable enciphering mode. In: Boneh, D. (ed.) CRYPTO 2003. LNCS, vol. 2729, pp. 482–499. Springer, Heidelberg (2003)
2. Iwata, T., Kurosawa, K.: How to Re-use Round Function in Super-Pseudorandom Permutation. Information Security and Privacy, 224–235 (2004)
3. Koren, T.: On the construction of pseudorandom block ciphers, M.Sc. Thesis (in Hebrew), CS Dept., Technion, Israel (May 1989)
4. Luby, M., Rackoff, C.: How to construct pseudorandom permutations and pseudorandom functions. 2nd SIAM J. Comput. 17, 373–386 (1988)
5. Naor, M., Reingold, O.: On the Construction of Pseudorandom Permutations: Luby-Rackoff Revisited. J. Cryptology 12(1), 29–66 (1999)
6. National Bureau of Standards, Data encryption standard, Federal Information Processing Standard, PT U.S. Department of Commerce, FIPS PUB 46, Washington, DC (1977)
7. Patarin, J.: Pseudorandom permutations based on the DES scheme. In: Damgård, I.B. (ed.) EUROCRYPT 1990. LNCS, vol. 473, Springer, Heidelberg (1991)
8. Patarin, J.: How to construct pseudorandom and super pseudorandom permutations from one single pseudorandom pseudorandom function. In: Rueppel, R.A. (ed.) EUROCRYPT 1992. LNCS, vol. 658, pp. 256–266. Springer, Heidelberg (1993)
9. Patarin, J.: The "Coefficients H" Technique. Selected Areas in Cryptography 2008, 328–345 (2008)
10. Pieprzyk, J.: How to construct pseudorandom permutations from single pseudorandom functions. In: Damgård, I.B. (ed.) EUROCRYPT 1990. LNCS, vol. 473, pp. 140–150. Springer, Heidelberg (1991)
11. Sadeghiyan, B., Pieprzyk, J.: On necessary and sufficient conditions for the construction of super pseudorandom permutations. In: Matsumoto, T., Imai, H., Rivest, R.L. (eds.) ASIACRYPT 1991. LNCS, vol. 739, pp. 194–209. Springer, Heidelberg (1993)
12. Sadeghiyan, B., Pieprzyk, J.: A construction for super pseudorandom permutations from a single pseudorandom function. In: Rueppel, R.A. (ed.) EUROCRYPT 1992. LNCS, vol. 658, Springer, Heidelberg (1992)
13. Vaudenay, S.: Decorrelation: A Theory for Block Cipher Security. J. Cryptology 16(4), 249–286 (2003)
14. Zheng, Y., Matsumoto, T., Imai, H.: Impossibility and optimally results on constructing pseudorandom permutations. In: Quisquater, J.-J., Vandewalle, J. (eds.) EUROCRYPT 1989. LNCS, vol. 434, pp. 412–422. Springer, Heidelberg (1990)

Versatile Prêt à Voter: Handling Multiple Election Methods with a Unified Interface

Zhe Xia[1], Chris Culnane[1], James Heather[1], Hugo Jonker[2], Peter Y.A. Ryan[2], Steve Schneider[1], and Sriramkrishnan Srinivasan[1]

[1] Department of Computing, University of Surrey, Guildford GU2 7XH, U.K.
{z.xia,c.culnane,j.heather,s.schneider,s.srinivasan}@surrey.ac.uk
[2] Faculté des Sciences, de la Technologie et de la Communication, University of Luxembourg, L-1359 Luxembourg
{hugo.jonker,peter.ryan}@uni.lu

Abstract. A number of end-to-end verifiable voting schemes have been introduced recently. These schemes aim to allow voters to verify that their votes have contributed in the way they intended to the tally and in addition allow anyone to verify that the tally has been generated correctly. These goals must be achieved while maintaining voter privacy and providing receipt-freeness. However, most of these end-to-end voting schemes are only designed to handle a single election method and the voter interface varies greatly between different schemes. In this paper, we introduce a scheme which handles many of the popular election methods that are currently used around the world. Our scheme not only ensures privacy, receipt-freeness and end-to-end verifiability, but also keeps the voter interface simple and consistent between various election methods.

Keywords: Prêt à Voter, voting scheme, end-to-end verifiability, receipt-freeness, simple and consistent voter interface.

1 Introduction

In a traditional secret ballot election, voters mark their choice on a piece of paper and drop it into a box. The ballots are mixed together to break the link between the voter and her choice. Then these ballots are tallied under scrutiny. While the secret ballot meets its desired goals of ensuring voter privacy, lack of transparency and verifiability is considered a problem. There is no way for the voter to verify that her vote has contributed correctly to the tally and significant trust must be placed in the election officials to have carried out the election procedures and tally correctly.

End-to-end verifiable voting schemes aim to address these issues. These schemes allow voters to verify that their votes have contributed in the way they intended to the tally (individual verifiability) and in addition allow anyone to verify that the tally has been generated correctly (universal verifiability). These goals must be achieved while maintaining voter privacy. In addition, the voter

G. Gong and K.C. Gupta (Eds.): INDOCRYPT 2010, LNCS 6498, pp. 98–114, 2010.

should not be able to prove to others how she voted as adversaries can then coerce or bribe the voter to vote in a certain way (receipt-freeness).

1.1 Motivation

The most common election method is the First-Past-The-Post (FPTP) method, where a voter simply puts a mark next to the candidate of her choice. The majority of the existing end-to-end voting schemes are designed to handle FPTP elections. Some examples are Prêt à Voter (PaV) [9], [27], Scratch & Vote [2], Punchscan [24], Scantegrity [8], [7], Bingo Voting [6] and MarkPledge [18], [1].

However, many other election methods exist and are used widely. Typically, these allow the voter to indicate multiple preferences or allot a full or partial ranking of the candidates. A plethora of tallying methods also exist and are used around the world. In the Appendix, we provide a summary of the election methods we will discuss in this paper.

It is argued in the decision theory community that these ranked elections are superior as less votes are wasted and that they offer resistance to strategic voting. However, they introduce potential coercion problems. For example, if the election consists of a large number of candidates, a very large number of possible candidate rankings exist. Adversaries can force voters to cast their votes using specific orderings, and check whether ballots with these unique orderings appear among the cast ballots. This has been referred to as the Italian attack in the media and literature. We discuss Italian attacks further in Section 5.

Ranked election methods are typically less discussed in the end-to-end voting literature. There are a few notable exceptions. Several schemes [15], [31], [21] have been designed for Instant Runoff Voting (IRV) elections. Shuffle-Sum [5] handles both IRV and the Single Transferable Vote (STV), but it does not directly handle partial rankings. Condorcet elections can be handled in the scheme introduced in [10]. However, in order to foil the Italian attack, its user interface is quite different: instead of ranking the candidates, every voter is required to cast a number of ballots, where each ballot is a pairwise comparison of two candidates.

No generic end-to-end verifiable voting scheme exists that can handle a wide variety of election methods. This is an important consideration as several different elections, employing different election methods are often held at the same time. For example, on election day, a voter in the polling station may need to cast several ballots, each for a separate election using a different election method. The average voter is unable to understand complicated instructions and procedures to cast their vote. Therefore, the voting interface needs to be kept simple and consistent to avoid confusion. Two attempts, one with cryptography and one without cryptography, in this direction have been made in [32] and [25], where a generic solution to handle various election methods is introduced. Unfortunately, neither of them is receipt-free in ranked elections.

1.2 Our Contribution

In this paper, we gather together many diverse concepts and building blocks in the literature, unifying them into one generic scheme which handles a number of popular election methods. In addition to achieving the goals of voter privacy, receipt-freeness and end-to-end verifiability, our scheme has a simple and consistent voter interface across various election methods. In order to enable our scheme to be used in high-profile political elections, we also aim to achieve robustness so that the election can be run even in the presence of a minority of dishonest election officials and is able to recover from cheating when it is detected. We summarize our contributions in more specific detail in Section 6.

1.3 Structure of the Paper

In Section 2 we summarize the various cryptographic building blocks we will employ in our scheme. This is followed by an overview of the proposed scheme in Section 3. As discussed, the user interface is kept consistent for different election methods. The exact details of how the votes are processed for the different election methods is abstracted from the voter and we describe these details in Section 4. We then provide an informal analysis of the security properties of the scheme in Section 5 before concluding in Section 6.

2 Building Blocks

2.1 Paillier Cipher

The Paillier cipher [20] is an efficient, semantically secure public key cryptosystem which provides the additive homomorphic property. It is a fundamental building block of our scheme and works as follows: let n be an RSA modulus $n = pq$, where p and q are large primes. Let g be an integer of order a multiple of n modulo n^2, e.g. $g = n + 1$. The public key is (g, n), and the secret key is $\lambda = lcm((p-1),(q-1))$. To encrypt $m \in Z_n$, we randomly choose $r \in Z_n^*$ and compute the ciphertext $c = E_{pk}(m, r) = g^m r^n \pmod{n^2}$. To decrypt c, we compute $m = L(c^\lambda \bmod n^2)/L(g^\lambda \bmod n^2) \bmod n$, where the L-function takes input values from the set $S_n = \{u < n^2 | u = 1 \bmod n\}$ and $L(u) = (u-1)/n$.

- *Homomorphic property:* For two ciphertexts under the same public key $c_1 = E_{pk}(m_1, r_1)$ and $c_2 = E_{pk}(m_2, r_2)$, we have $E_{pk}(m_1, r_1) \cdot E_{pk}(m_2, r_2) = E_{pk}(m_1 + m_2, r_1 \cdot r_2)$. Moreover, for a value $k \in Z_n$, we have $E_{pk}(m, r)^k = E_{pk}(k \cdot m, r^k)$. This property is fundamental to the construction of our scheme and will be used extensively along with the Baudron homomorphic counter [3] detailed in the following section.
- *Paillier re-encryption:* Given a Paillier ciphertext $c = g^m r^n \pmod{n^2}$, a re-encryption of this ciphertext can be generated without knowledge of the secret key λ. A value $t \in Z_n^*$ is randomly selected and the re-encryption c' of c is calculated as $c' = c \cdot t^n = g^m (t \cdot r)^n \pmod{n^2}$.
- *Threshold Paillier:* The key pair for the Paillier cipher can be jointly generated by threshold parties, so that each party has a share of the secret key, but no single party knows the entire secret key. This technique can be found in [14], [12]. Moreover, ciphertexts can be decrypted by these threshold parties in a verifiable manner [13].

– *Verifiable shuffle for Paillier:* In a verifiable shuffle, a mix server receives a batch of ciphertexts, re-encrypts each ciphertext and then outputs the results in a random order. It also publishes a proof so that the correctness of the shuffle can be publicly verified, but the proof should not reveal the relationship between the inputs and the outputs. Verifiable shuffle for Paillier can be designed using existing techniques from [16], [4], [19], [22].

2.2 Baudron's Homomorphic Counter

Suppose there are $k+1$ candidates and the total number of eligible voters is M, where $M < 2^L$. We can define a set of counters $\{2^0, 2^L, 2^{2L}, \ldots, 2^{kL}\}$ as the election parameters, one for each candidate. Encryptions corresponding to each counter represent votes for the candidate who has been assigned the counter. Multiplying these encrypted votes together results in an encrypted counter where the received votes for each candidate is kept in a separate area of the counter without overflow between adjacent areas of the counter. For more technical details, please see [3].

2.3 Plaintext Equivalence Test (PET)

Suppose (g, n) is the Paillier public key and the secret key λ is shared among threshold parties. For two ciphertexts $c_1 = E_{pk}(m_1, r_1) = g^{m_1} {r_1}^n \pmod{n^2}$ and $c_2 = E_{pk}(m_2, r_2) = g^{m_2} {r_2}^n \pmod{n^2}$, the Plaintext Equivalent Test (PET) [30] can be used to check whether these two ciphertexts contain the same plaintext, without revealing either plaintext. Denote $c = c_1/c_2 = g^{m_1 - m_2}(r_1/r_2)^n \pmod{n^2}$. If $m_1 = m_2$, for some random value $r \in Z_n$, both c and c^r will contain the 0 plaintext. But if $m_1 \neq m_2$, c^r will contain a random plaintext. Therefore, if c^r is threshold decrypted, the result can tell whether the two ciphertexts contain the same plaintext without revealing either plaintext.

2.4 Binary Conversion and Plaintext Inequivalence Test (PIT)

For a Paillier ciphertext $c = E_{pk}(m, r)$, where the binary representation of its L-bit long plaintext is $m = m_1 m_2 \cdots m_L$, the Binary Conversion technique [28] can be used to convert the ciphertext c into separate encryptions of every individual bit of the plaintext m. This process can be illustrated as

$$E_{pk}(m_1 m_2 \cdots m_L, r) \to E_{pk}(m_1, r_1), E_{pk}(m_2, r_2), \ldots, E_{pk}(m_L, r_L)$$

Binary Conversion needs to be carried out by the threshold parties, and it is publicly verifiable. The above process can be reversed by anyone as:

$$E_{pk}(m, r') = \prod_{i=1}^{L} E_{pk}(m_{L+1-i}, r_{L+1-i})^{2^{i-1}}$$

Moreover, for any two Paillier ciphertexts which have been converted into the encryption of individual bits of the plaintext, the threshold parties can apply techniques in [11] to check whether these two ciphertexts contain the same plaintext, or which ciphertext contains larger plaintext. This phase is also publicly verifiable, and does not reveal either plaintext.

3 System Overview

We first describe the ballot generation phase in our proposed scheme. We then describe the vote capture phase, which is where the individual voter interacts with the system. This is the phase that is consistent from the voter's point of view, no matter what election method is being employed. The techniques employed in the vote processing phase are different depending on which election method is being used and the details are abstracted from the voter. We will describe the various vote processing techniques in Section 4.

- **Ballot generation:** All ballot forms are generated by a trusted party before the election. We trust this party for privacy, but the integrity of the election result does not rely on this party. Suppose there are 5 candidates in a sample election: Alice, Bravo, Charlie, Delta and Echo. They are assigned counters 2^0, 2^L, 2^{2L}, 2^{3L} and 2^{4L} respectively, where 2^L is larger than the total number of eligible voters in the election. These are published prior to the election as system parameters. A ballot, as shown in Figure 1, consists of two columns with a vertical perforation down the middle. The left hand column lists the candidate names in a random order. The barcode in the right hand column contains an encrypted value for each candidate and a proof. The encrypted values are called "onions" for historical reasons, and their ordering should match the candidate list. The proof proves that each onion encrypts a unique counter. To generate the proof, the party first generates a list of pseudo ciphertexts:

$$\{E_{pk}(2^0, 1), E_{pk}(2^L, 1), E_{pk}(2^{2L}, 1), E_{pk}(2^{3L}, 1), E_{pk}(2^{4L}, 1)\}$$

  Anyone can check that this list is well-formed because all the randomisations are 1. For each ballot, the party generates a proof that the onion list is a shuffle of the pseudo ciphertexts.

Fig. 1. A ballot form example

– **Vote capture:** To prevent the local officials from learning the candidate ordering, the ballots are delivered to the polling stations in sealed envelopes and these sealed ballots are handed to the voters during the voting phase. The voter opens the envelope to obtain the ballot and can then randomly decide whether to audit the ballot or use it to cast a vote. If the voter chooses to audit her ballot, she submits it to the local officials without marking her choice. After the onions have been threshold decrypted, anyone can check whether the candidate list can be reconstructed from the onions. Once a ballot is audited, it cannot be used to cast a vote, and the voter will be provided with another ballot. The voter can audit several ballots until she is satisfied. Then, the voter marks the ballot as instructed: selecting a single candidate, multiple candidates, or by providing a ranking of candidates. This is pictured in Figure 2.

Bravo	
Echo	X
Delta	
Charlie	
Alice	
	[barcode]

Bravo	4
Echo	1
Delta	5
Charlie	3
Alice	2
	[barcode]

Fig. 2. Completing the ballot form. (a) Single cross. (b) Preference list.

As follows, the voter separates the ballot along the perforation and shreds the left hand column. It is important to ensure via some process that the left hand column is destroyed before the voter is allowed to submit the vote. Otherwise, the voter can prove to an adversary how she voted. The remaining right hand column, as shown in Figure 3, contains the vote to be cast. The voter now brings it to the election officials and it is scanned into the election system and digitally signed. The voter retains the signed right hand side as her receipt. All the received votes are published on the bulletin board and the voter can check whether her vote has been recorded correctly by the election system. If not, the signed receipt can be used to initiate a complaint. The use of the Prêt à Voter style ballot form provides two advantages. Firstly, it is simple and very similar to the current paper based ballots that voters are already familiar with. Secondly, if the voter mistakenly casts an invalid vote, e.g. over-vote or under-vote, it can be discovered by the local officials before the vote is scanned and the voter can be assisted in casting a valid vote by being instructed appropriately and being provided with new ballots. Note that the election official cannot violate the vote secrecy by simply looking at the filled in right hand side, as the corresponding left hand side has been destroyed.

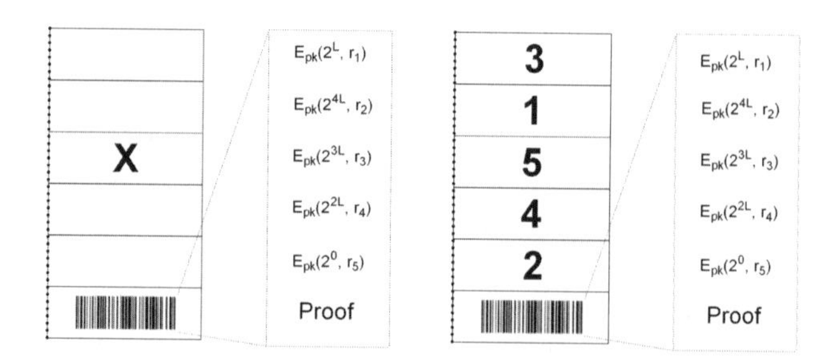

Fig. 3. The vote. (a) Single cross. (b) Preference list.

4 Vote Processing

As discussed earlier, the vote processing phase is transparent to voters. At a high level, this phase can be regarded as an oracle. If it was provided with the received votes and was told which election method is used, it will generate the election result based on that election method. Moreover, the oracle will output some data onto the bulletin board. The data is enough to publicly verify the tally, but it provides no information that can be used by an adversary to coerce voters.

4.1 Vote Processing for FPTP

The vote processing for FPTP elections is the same as in the Scratch & Vote scheme [2]. When the votes, as shown in Figure 3(a) are received, the proofs for each vote are checked and votes with invalid proofs are discarded. The proof is important to prevent malicious parties from casting invalid vote, e.g. negative vote(s) or multiple votes. For the remaining votes, the onion corresponding to the selected candidate is aggregated into a counter as described previously. Finally, this counter is threshold decrypted and the election result is announced along with the tally for each candidate.

In [31], a strategy is introduced to announce only the election winner without revealing any other information. Firstly, onions are aggregated into the counter. However, instead of decrypting the counter, the threshold parties can apply the Binary Conversion technique to convert it into separate encryptions of each bit of the plaintext. For our sample election, the plaintext is $5L$ bits long, and every L-bit block represents the received vote for a different candidate. The candidate with the most votes can be identified using the *plaintext inequivalence test* (PIT) and this can be publicly verified. The exact number of votes received by each candidate is kept secret in this method.

4.2 Vote Processing for Approval Voting

In Approval Voting, a voter can mark one or several crosses, and all the crosses are equally weighted. The vote processing for Approval Voting is similar to

FPTP. First, all proofs are checked and votes with invalid proof are removed. All onions corresponding to selected candidates in a ballot are then aggregated into a single ciphertext. From this point on the techniques and options are similar to those described above.

4.3 Vote Processing for Supplementary Vote

In Supplementary Vote elections, the votes, as shown in Figure 4, contain either one or two preferences. To tally the received votes, the proof of every vote is checked and all votes with invalid proofs are removed. The onions corresponding to the valid ballots are ordered according to the indicated preference. For example, the votes in Figure 4 will be ordered as $\{\theta_D\}$ and $\{\theta_B, \theta_A\}$ respectively (note that these θ values are ciphertexts and the subscripts are used for notational convenience). Now, the first onion in every vote is selected and aggregated into a counter. The threshold parties then apply the Binary Conversion technique to convert this ciphertext into a number of ciphertexts, where each encrypts the received votes for a particular candidate. After generating a pseudo ciphertext[1] $E_{pk}(Q, 1)$, where Q is the winning quota, the threshold parties apply the PIT to check whether some candidate has received more votes than the quota. If yes, the election ends and this candidate wins. Otherwise, the threshold parties identify the two candidates with the most votes using the PIT, and all other candidates are eliminated.

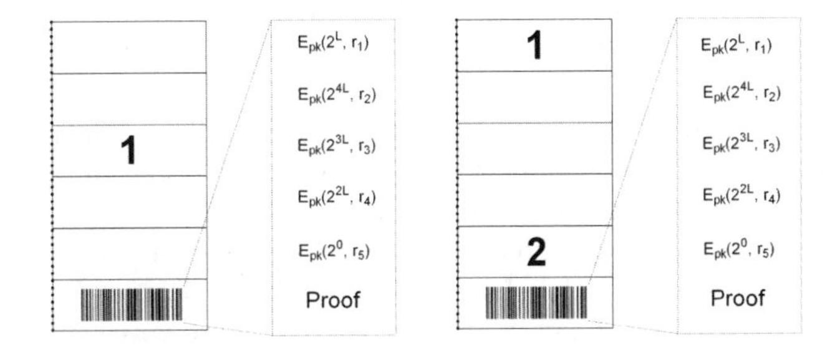

Fig. 4. Supplementary Vote. (a) One preference. (b) Two preferences.

Suppose *Bravo* and *Delta* are the two remaining candidates. We first generate two pseudo ciphertexts $\bar{\theta}_B = E_{pk}(2^L, 1)$ and $\bar{\theta}_D = E_{pk}(2^{3L}, 1)$ for them respectively. Then we shuffle all the votes using the verifiable shuffle. In the next step, for every vote in the outputs, the threshold parties apply the PET to compare its first onion with $\bar{\theta}_B$ and $\bar{\theta}_D$. If it matches with one of the pseudo ciphertexts, this vote will be sorted into the pile for that candidate. For these votes, the second preference will never be used. So if any vote has a second onion, it will

[1] Note that in the rest of this paper, if we mention pseudo ciphertext, we mean that the ciphertext is generated using the randomisation value 1. Therefore, anyone can verify that the pseudo ciphertext is well-formed.

be removed from the vote, e.g. θ_A will be removed from the vote $\{\theta_B, \theta_A\}$. For the other votes where the first onion does not match with any of the pseudo ciphertexts, we check whether it contains a second preference. If no, the vote will be removed from the tally. Otherwise, its first onion will be removed, leaving only the second onion, and we leave these votes in an unsorted pile. We now check the two piles of votes for the remaining candidates, if their difference is larger than the number of votes in the unsorted pile, the election ends and the remaining candidate with more votes wins. But if their difference is smaller than the number of votes in the unsorted pile, we aggregate all votes in the three piles into one ciphertext[2]. Then this ciphertext is threshold decrypted and one of the two remaining candidates with the most votes wins.

4.4 Vote Processing for Instance Runoff Voting (IRV)

To tally the received votes in IRV elections, proofs are checked and any vote with an invalid proof is discarded. Once again, the onions corresponding to the valid ballots are ordered according to the indicated preference e.g. the vote in Figure 3(b) will be re-written as $\{\theta_E, \theta_A, \theta_B, \theta_C, \theta_D\}$. The first onion of every vote is selected and aggregated into a counter. The threshold parties then apply the Binary Conversion technique to transfer this encrypted counter into a number of ciphertexts, where each encrypts the received votes for a candidate. After generating a pseudo ciphertext of the quota as $E_{pk}(Q, 1)$, the threshold parties can use the PIT to check whether some candidate has received more votes than the quota. If yes, the election ends, and this candidate wins. Otherwise, the threshold parties use the PIT to identify the candidate with the least votes and this candidate is eliminated. Suppose Alice is eliminated in the first round, a pseudo ciphertext for Alice is generated as $\bar{\theta}_A = E_{pk}(2^0, 1)$. Then all votes are inserted into the verifiable shuffle. In the outputs, the threshold parties apply the PET to locate the onion θ_A in every vote, and this value will be removed from the vote. Again, the first onion of every vote is selected and aggregated into a counter and then the threshold parties apply the Binary Conversion and PIT to check whether some candidate has received more votes than the quota. The above process is repeated until some candidate receives more votes than the quota or only one candidate remains. For a vote with partial rankings, if all preferences are removed, it is removed from the tally.

4.5 Vote Processing for Single Transferable Vote (STV)

In STV elections with fraction transfer, if some candidate receives more votes than the quota but not all seats are filled, the votes that exceed the quota need to be transferred. However, the transfer value is not an integer but a fraction. For example, if the quota is q and if a candidate receives m votes, where $m > q$, the transfer value will be $(m - q)/q$. Although this value can be treated as an integer if we multiply all the vote values by q, the Baudron counter can no more

[2] Note that at this moment, all votes in the three piles contain only one onion.

be used because there might be overflow between adjacent locations within the counter.

Here, the Shuffle-Sum techniques [5] to tally STV elections can be employed, but with modifications to allow partial ranking. Before the election, two pseudo ciphertexts $\bar{\theta}_0 = E_{pk}(0,1)$ and $\bar{\theta}_1 = E_{pk}(1,1)$ are generated. These two values are used to record which are the voter's genuine choices and which are appended choices. Every received vote will first be transferred into the so called *Preference-order table*, e.g. the vote in Figure 4(b) will be transferred as:

Candidate	θ_B	θ_A	θ_E	θ_D	θ_C
Preference	1	2	3	4	5
True/Fake	$\bar{\theta}_1$	$\bar{\theta}_1$	$\bar{\theta}_0$	$\bar{\theta}_0$	$\bar{\theta}_0$
Weight			$E_{pk}(w_v)$		

In the above table, the order of the voter's genuine preferences need to match with the vote, but the appended preferences can be in any order. Anyone can verify that the above table is correctly generated. The following procedures are similar to the Shuffle-Sum protocol: the above table can be transferred between the *First-preference table* and the *Candidate-elimination table* to elect some candidate or to eliminate some candidate from the vote. Note that in any case if all the genuine choices are eliminated, the vote will be discarded. This can be checked that in the third row of the *Preference-order table*, the encrypted value under the first preference encodes plaintext 0.

We will not go into the details of the Shuffle-Sum protocol. For more information, please see [5]. But superior than the Shuffle-Sum protocol, our scheme always enable us to check whether all seats are filled in the first round using the homomorphic tallying. The process is the same as in Supplementary Vote and IRV elections. Therefore, we only need to apply the mixnets tallying if all seats are not filled in the first round.

4.6 Vote Processing for Condorcet Voting

In Condorcet Voting, the Binary Conversion technique or PET/PIT are not used, and the proof in the vote does not need to be checked. Instead, each received vote is interpreted as follows: e.g. the ciphertexts in the vote in Figure 3(b) will be arranged as per the preferences as $\{\theta_E, \theta_A, \theta_B, \theta_C, \theta_D\}$. A group of ciphertext triples are then generated.

$$\{\theta_E, \theta_A, \bar{\theta}_1\}\ \{\theta_E, \theta_B, \bar{\theta}_1\}\ \{\theta_E, \theta_C, \bar{\theta}_1\}\ \{\theta_E, \theta_D, \bar{\theta}_1\}\ \{\theta_A, \theta_B, \bar{\theta}_1\}$$
$$\{\theta_A, \theta_C, \bar{\theta}_1\}\ \{\theta_A, \theta_D, \bar{\theta}_1\}\ \{\theta_B, \theta_C, \bar{\theta}_1\}\ \{\theta_B, \theta_D, \bar{\theta}_1\}\ \{\theta_C, \theta_D, \bar{\theta}_1\}$$

In the above group, for each ciphertext triple, the first two ciphertexts are taken from the vote with the same order, and the third one is a pseudo encryption of plaintext 1, $\bar{\theta}_1 = E_{pk}(1,1) = g^1 \cdot 1^n \pmod{n^2}$.

Similarly, another group of ciphertext triples are generated, in which the first two ciphertexts are taken from the vote in the reverse order, and the third value[3]

[3] Note that in Paillier cipher, the plaintext space is Z_n so we cannot directly encrypt the plaintext -1, but we can encrypt $n-1$ instead.

is a pseudo encryption of -1, $\bar{\theta}_{-1} = E_{pk}(-1, 1) = g^{-1} \cdot 1^n \pmod{n^2}$.

$$\{\theta_D, \theta_C, \bar{\theta}_{-1}\}\ \{\theta_D, \theta_B, \bar{\theta}_{-1}\}\ \{\theta_D, \theta_A, \bar{\theta}_{-1}\}\ \{\theta_D, \theta_E, \bar{\theta}_{-1}\}\ \{\theta_C, \theta_B, \bar{\theta}_{-1}\}$$
$$\{\theta_C, \theta_A, \bar{\theta}_{-1}\}\ \{\theta_C, \theta_E, \bar{\theta}_{-1}\}\ \{\theta_B, \theta_A, \bar{\theta}_{-1}\}\ \{\theta_B, \theta_E, \bar{\theta}_{-1}\}\ \{\theta_A, \theta_E, \bar{\theta}_{-1}\}$$

Now, we treat all the ciphertext triples in the above two groups as inputs and insert them into the verifiable shuffle. Note that in the outputs, the last value of the ciphertext triple is no longer a pseudo ciphertext i.e. its randomisation is no longer 1. Then, for each of the output ciphertext triples, the threshold parties decrypt the first two values. Now, in exactly half of the result triples, the two candidates are in the alphabetic order, and in the other half, they are in the reverse alphabetic order. All triples where the two candidates are in the reversed alphabetic order are now removed. The remaining triples are as follows:

$$\{Alice, Bravo, \theta_1\}\ \{Alice, Charlie, \theta_1\}\ \{Alice, Delta, \theta_1\}$$
$$\{Alice, Echo, \theta_{-1}\}\ \{Bravo, Charlie, \theta_1\}\ \{Bravo, Delta, \theta_1\}$$
$$\{Bravo, Echo, \theta_{-1}\}\ \{Charlie, Delta, \theta_1\}\ \{Charlie, Echo, \theta_{-1}\}$$
$$\{Delta, Echo, \theta_{-1}\}$$

After all the received votes are interpreted in the above format, a pairwise comparison of every two candidates is done to check which candidate is more preferred by the voters. For example, to compare *Alice* and *Bravo*, the triple in every vote which contains these two candidates is selected. The third values of these triples is aggregated into one ciphertext using the additive homomorphic property. If this ciphertext is decrypted, a positive plaintext indicates that Alice is more preferred than Bravo, and a negative plaintext indicates the opposite. If there exists a candidate who wins every pairwise comparison, that candidate is elected. In case the tally does not result a winner, some other additional methods must be used to determine the winner.

5 System Analysis

In this section, we present an informal analysis of our proposed scheme.

– *Privacy and receipt-freeness:* The random candidate ordering provides voter privacy. If the left hand column as depicted in Figure 2 is detached and destroyed, the right hand column does not reveal how the voter voted. In addition, to see if our scheme provides receipt-freeness we must consider the Italian attack. As discussed earlier, if a ranked election contains a large number of candidates, adversaries can coerce the voter to cast the vote with a unique ordering of candidates, and then they can check whether this pattern has appeared in the list of cast ballots. Our scheme never reveals the entire plaintext vote and is therefore robust against this attack. A variant of the Italian attack can be found in [29]. For example, the adversary can coerce the voter to put a very unpopular candidate as the first preference and a very popular candidate as the second preference. In any round, if the unpopular candidate is eliminated but there is no transfer from this eliminated candidate to that popular candidate, the adversaries will know that the voter did not follow the instructions. In our scheme, the transfer history is kept secret during the vote processing phase and therefore, our scheme is robust against this variant of the Italian attack as well.

- *End-to-end verifiability:* In our scheme, voters can use the "cut-and-choose" method to check whether the ballots are correctly generated. This ensures that the voter's intention will be correctly encoded. Each voter is also provided with a receipt, and the voter can check that the vote has been recorded by the system. These two actions provide individual verifiability. As the entire vote processing phase can be publicly verified, our scheme also achieves universal verifiability.
- *Robustness:* We have ensured that the various steps in the vote processing phase can either be done by any party without requiring the knowledge of the secret key or by a threshold set of parties. Therefore, so long as there exists a quorum of honest threshold parties, the correct election result will always be generated.
- *Complexity:* Our scheme has been designed so that it handles both homomorphic tallying and mixnet based tallying. Hence we can tailor the design of the vote processing phase based on different election methods, so that the election result can always be output in the most efficient manner. In FPTP and Approval Voting elections, all received votes are tallied using the homomorphic property. Hence the election result can be generated very efficiently. In Supplementary Vote, IRV and STV elections, the received votes also can be tallied using the homomorphic property if the election winner(s) can be identified in the first round. Otherwise, we need to use mixnets in the vote processing phase, and this phase may contain several rounds. In Condorcet voting, we interpret each of the received vote into a number of data triples, where each data triple pairwise compares two candidates. This process is done by mixnets and its complexity is quadratic in the number of candidates.
- *Usability:* Voters only need to be involved in the vote capture phase, and their experience is simple and consistent for various election methods. Also, the ballot form in our proposed scheme is very similar to those used in current paper based elections around the world.

5.1 Open Problems

A number of avenues to improve our scheme exist and we list a few. Note that some of the identified problems are common to existing voting schemes.

- *Authority knowledge attack:* All ballots are generated by a single party. Hence we need to trust this party for privacy and receipt-freeness. Generating the ballots in a distributed fashion is desirable, because it ensures no one but the voter ever learns the candidate ordering. However, achieving distributed ballot generation is not easy in our scheme. There are three major obstacles: Firstly, how to prove the ballot is well-formed in the distributed fashion. Secondly, how to print the ballot without the printer(s) learning the candidate

ordering. And thirdly, how to ensure robustness so that the scheme can be run even in the presence of some dishonest election officials. Solving one or two of the above obstacles might be possible, but it is still an open problem whether all these three obstacles can be solved at the same time.
- *Chain voting attack:* If adversaries can successfully smuggle a blank ballot form out of the polling station, they can use this ballot to coerce voters. They mark an initial ballot with the candidate of their choice and give it to a voter entering the polling station. If the voter emerges with another blank ballot, she is rewarded. The adversaries can repeat this attack with the next voter using the new ballot. Ryan and Peacock have discussed this attack in [26], and some of their countermeasures also work for our scheme.
- *Randomisation attack:* Adversaries can coerce voters to bring out their receipts with some unique pattern, e.g. the cross always at the top or the ranking is in the ascend order. Although they do not know how these voters have cast their votes, they effectively force the voter to vote randomly. In FPTP elections, there exists a countermeasure to foil the randomisation attack. If the voter was coerced to bring out her receipt with the cross at the top, she can keep auditing ballots until she receives one with her favorite candidate at the top. But in ranked elections, such a countermeasure is not so effective and the voters may still be coerced to cast their votes randomly.
- *Usability:* Voters with certain specific disabilities may not be able to tear the paper ballot apart along the perforation. In future works we hope to investigate how to improve the accessibility of our schemes for these voters.
- *Scalability:* Our scheme is not suitable for elections with a large number of candidates. Since the candidate list is in random order, voters may have difficulty to find candidates in a long candidate list, especially in ranked elections. Moreover, the size of the homomorphic counter will become very large if there are many candidates, and some of the building blocks will become inefficient.

6 Conclusion

We have introduced a generic end-to-end verifiable voting scheme that handles many of the currently used election methods. We believe our work is an important step forward from the voter's point of view, keeping the voting experience simple and consistent no matter what election method is employed. Our work is based on the success of many existing concepts and building blocks, and we also contribute several improvements to these previous works:

- Lundin [17] and Popoveniuc et. al. [23] have earlier discussed how end-to-end verifiable voting schemes can be designed using the modular approach. For example, they discuss how the front-end and back-end of Prêt à Voter [9], [27] and Punchscan [24] can be designed in a mix-and-match fashion. However, they only focus on a single election method. In this paper, we extend the concept to multiple election methods and illustrate that multiple back-ends, each for a different election method, can be designed in the modular fashion, with a unified front-end.

- We introduce a very simple and straightforward solution to the Shuffle-Sum protocol [5] to handle partial ranking, and we also show that the Shuffle-Sum protocol can be designed much more efficiently if the election result can be tallied in the first round.
- We introduce a novel tallying method for Condorcet elections. Compared to the scheme in [10], we have simplified the voter interface without introducing extra complexity in the tallying phase.

In the future, we hope to introduce further enhancements to mitigate the open problems.

Acknowledgement

This work was supported by the UK Engineering and Physical Sciences Research Council (EPSRC) under grant EP/G025797 and the FNR (National Research Fund) Luxembourg under project SeRVTS–C09/IS/06. We would also like to thank the anonymous reviewers for their comments on the paper.

References

1. Adida, B., Neff, A.: Efficient receipt-free ballot casting resistant to covert channel. In: EVT 2009 (2009)
2. Adida, B., Rivest, R.L.: Scratch & Vote: self-contained paper-based cryptographic voting. In: WPES 2006, pp. 29–40 (2006)
3. Baudron, O., Fouque, P.-A., Pointcheval, D., Stern, J., Poupard, G.: Practical multi-candidate election system. In: PODC 2001, pp. 274–283 (2001)
4. Benaloh, J.: Towards simple verifiable elections. In: WOTE 2006, pp. 61–68 (2006)
5. Benaloh, J., Moran, T., Ramchen, K., Naish, L., Teague, V.: Shuffle-Sum: coercion resistant verifiable tallying for STV voting. In: IEEE Transactions on Information Forensics and Security (December 2009)
6. Bohli, J.-M., Müller-Quade, J., Röhrich, S.: Bingo Voting: secure and coercion-free voting using a trusted random number generator. In: Alkassar, A., Volkamer, M. (eds.) VOTE-ID 2007. LNCS, vol. 4896, pp. 111–124. Springer, Heidelberg (2007)
7. Chaum, D., Carback, R., Clark, J., Essex, A., Popoveniuc, S., Rivest, R.L., Ryan, P.Y.A., Shen, E., Sherman, A.T.: Scantegrity II: end-to-end verifiable for optical scan election systems using invisible ink confirmation codes. In: EVT 2008 (2008)
8. Chaum, D., Essex, A., Carback, R., Clark, J., Popoveniuc, S., Sherman, A.T., Vora, P.: Scantegrity: end-to-end voter verifiable optical-scan voting. Journal of IEEE Security & Privacy 6(3), 40–46 (2008)
9. Chaum, D., Ryan, P.Y.A., Schneider, S.: A practical voter-verifiable election scheme. In: di Vimercati, S.d.C., Syverson, P.F., Gollmann, D. (eds.) ESORICS 2005. LNCS, vol. 3679, pp. 118–139. Springer, Heidelberg (2005)
10. Clarkson, M., Myers, A.: Coercion-resistant remote voting using decryption mixes. In: FEE 2005 (2005)
11. Damgård, I., Fitzi, M., Kiltz, E., Nielsen, J.B., Toft, T.: Unconditional secure constant-rounds multi party computation for equality, comparison, bits and exponentiation. In: Halevi, S., Rabin, T. (eds.) TCC 2006. LNCS, vol. 3876, pp. 285–304. Springer, Heidelberg (2006)

12. Damgård, I., Koprowski, M.: Practical threshold RSA signatures without a trusted dealer. In: Pfitzmann, B. (ed.) EUROCRYPT 2001. LNCS, vol. 2045, pp. 152–165. Springer, Heidelberg (2001)
13. Fouque, P.-A., Poupard, G., Stern, J.: Sharing decryption in the context of voting or lotteries. In: Frankel, Y. (ed.) FC 2000. LNCS, vol. 1962, Springer, Heidelberg (2001)
14. Fouque, P.-A., Stern, J.: Fully distributed threshold RSA under standard assumptions. In: Boyd, C. (ed.) ASIACRYPT 2001. LNCS, vol. 2248, Springer, Heidelberg (2001)
15. Heather, J.: Implementing STV securely in Prêt à Voter. In: CSF 2007, pp. 157–169 (2007)
16. Jakobsson, M., Juels, A., Rivest, R.L.: Making mix nets robust for electronic voting by randomized partial checking. In: USENIX Security Symposium, pp. 339–353 (2002)
17. Lundin, D.: Component based electronic voting systems. In: Proceedings of IAVoSS Workshop on Trustworthy Elections (WOTE 2007), Ottawa, Canada, pp. 11–16 (2007)
18. Neff, A.: Practical high certainly intent verification for encrypted votes. VoteHere document (2004)
19. Nguyen, L., Safavi-Naini, R., Kurosawa, K.: Verifiable shuffles: a formal model and a Paillier-based efficient construction with provable security. In: Jakobsson, M., Yung, M., Zhou, J. (eds.) ACNS 2004. LNCS, vol. 3089, pp. 61–75. Springer, Heidelberg (2004)
20. Paillier, P.: Public-key cryptosystems based on discrete logarithms residues. In: Stern, J. (ed.) EUROCRYPT 1999. LNCS, vol. 1592, pp. 223–238. Springer, Heidelberg (1999)
21. Peng, K., Bao, F.: A design of secure preferential e-voting. In: Ryan, P.Y.A., Schoenmakers, B. (eds.) VOTE-ID 2009. LNCS, vol. 5767, pp. 141–156. Springer, Heidelberg (2009)
22. Peng, K., Boyd, C., Dawson, E.: Simple and efficient shuffling with provable correctness and ZK privacy. In: Shoup, V. (ed.) CRYPTO 2005. LNCS, vol. 3621, pp. 188–204. Springer, Heidelberg (2005)
23. Popoveniuc, S., Vora, P.: A framework for secure electronic voting. In: Proceedings of IAVoSS Workshop On Trustworthy Elections (WOTE 2008), Leuven, Belgium (2008)
24. Punchscan, http://www.punchscan.org
25. Rivest, R.L., Smith, W.D.: Three voting protocols: ThreeBallot, VAV, and Twin. In: Proceedings of the 2nd USENIX/ACCURATE Electronic Voting Technology Workshop (EVT 2007), Boston, MA (2007)
26. Ryan, P.Y.A., Peacock, T.: Threat analysis of cryptographic election schemes. Technical Report of University of Newcastle, CS-TR:971 (2006)
27. Ryan, P.Y.A., Schneider, S.: Prêt à Voter with re-encryption mixes. In: Gollmann, D., Meier, J., Sabelfeld, A. (eds.) ESORICS 2006. LNCS, vol. 4189, pp. 313–326. Springer, Heidelberg (2006)
28. Schoenmakers, B., Tuyls, P.: Efficient binary conversion for paillier encrypted values. In: Vaudenay, S. (ed.) EUROCRYPT 2006. LNCS, vol. 4004, pp. 522–537. Springer, Heidelberg (2006)
29. Teague, V., Ramchen, K., Naish, L.: Coercion-resistant tallying for STV voting. In: EVT 2008 (2008)
30. Ting, P.-Y., Huang, X.-W.: Distributed paillier plaintext equivalence test. International Journal of Network Security 6(3), 258–264 (2008)

31. Wen, R., Buckland, R.: Minimum disclosure counting for the alternative vote. In: Ryan, P.Y.A., Schoenmakers, B. (eds.) VOTE-ID 2009. LNCS, vol. 5767, pp. 122–140. Springer, Heidelberg (2009)
32. Xia, Z., Schneider, S., Heather, J., Ryan, P.Y.A., Lundin, D., Peel, R., Howard, P.: Prêt à Voter: All-In-One. In: WOTE 2007, pp. 47–56 (2007)

Appendix

We briefly summarise the vote casting procedure and tallying details of the various election methods considered in this paper for reference.

- **First-Past-The-Post (FPTP):** In FPTP elections, the voter simply puts a mark next to her preferred candidate and the candidate with the most votes wins. FPTP is used in various elections in Canada, India, the UK and some elections in the US.
- **Approval Voting:** In Approval Voting, the voter can indicate multiple preferences up to a set maximum. All votes carry the same weight and the candidates with the most votes wins. Approval Voting is currently used in some local council elections in the UK.
- **Supplementary Vote:** In Supplementary Vote elections, the voter marks her ballot as follows: she first marks 1 next to her most favourite candidate. If she also has a second preference, she marks 2 next to this candidate. Tallying takes place in at most two rounds. In the first round, only the first preference on every ballot is taken into account. If some candidate receives more than half of the votes, the election ends and this candidate wins. If no candidate wins in the first round, all candidates apart from those in the first two places are eliminated. Now, ballots with the first preference for one of the eliminated candidates are checked for their second preferences. If a ballot does not have a second preference or its second preference is also for one of the eliminated candidates, it is discarded. Otherwise, its second preference is treated as its first preference. Now, one of the two non eliminated candidate with the most votes wins. Supplementary Voting is used to elect mayors in the UK (e.g. the London Mayor Elections) and a variant is used in the Sri Lankan presidential elections.
- **Instance Runoff Voting (IRV):** IRV is also sometimes called alternative vote or preferential voting. In IRV elections, voters rank candidates based on their preference and depending on the election, ranking all candidates may be mandatory or optional. The winner is determined by a quota, which is normally half of the received votes. In the first round of tallying, only the first preference of every ballot is taken into account. If some candidate receives more votes than the quota, the candidate wins and the election ends. Otherwise, the candidate with the least votes will be eliminated and all ballots with the highest preference for this voter will be redistributed among the remaining candidates, based on the next preference. In case the next preference is empty or all the remaining preferences are for eliminated candidates, the ballot is discarded. The process is repeated until a winner is

found. Currently, IRV is used in some local government elections in the US, New Zealand and Malta. Moreover, there are proposals to replace the FPTP system used for parliamentary elections in the UK with the IRV.

- **Single Transferable Vote (STV):** Unlike the other methods described so far, in STV, more than one candidate is elected. To cast a vote, the voter ranks as few or as many candidates as she likes. To be elected, a candidate must receive more votes than a set quota. In the first phase of tallying, only the first preference on every ballot is taken into account. If no one receives more votes than the quota, the candidate with the least votes will be eliminated, and all ballots for this candidate will be redistributed among the remaining candidates, based on the next preference. However, if some candidates receive more votes than the quota, they will be elected and it will be checked if all seats are filled. If yes, the election ends. Otherwise, the tally continues so as to fill the remaining seats. Elected candidates do not need votes they receive over the quota and the surplus votes are transferred to the remaining candidates based on the next preference[4]. The above process is repeated until all seats are filled. STV is a popular election method and is currently used to elect the lower house of parliament in some territories in Australia, in local government elections in Scotland as well as some elections in Northern Ireland.
- **Condorcet voting:** In Condorcet elections, the voter provides a full ranking of the candidates. This allows every candidate to be compared to every other candidate. For each pairwise combination, it is checked which candidate is more preferred by the voters and the election winner is the candidate who wins every pairwise combination. Note that this process does not always result in a winner and special methods are required to determine the winner, but these techniques are out of the scope of this paper. The Condorcet method is not used in political elections, but it can be found in popular media, e.g. elections organised by MTV.

[4] Note that there are several different methods to transfer the surplus votes. For example, fraction transfer is used in Australia and Scotland, and random transfer is used in Northern Ireland.

Cryptographic Hash Functions: Theory and Practice

Bart Preneel

Katholieke Universiteit Leuven and IBBT
Dept. Electrical Engineering-ESAT/COSIC,
Kasteelpark Arenberg 10 Bus 2446, B-3001 Leuven, Belgium
bart.preneel@esat.kuleuven.be

Abstract. Cryptographic hash functions are an essential building block for security applications. Until 2005, the amount of theoretical research and cryptanalysis invested in this topic was rather limited. From the hundred designs published before 2005, about 80% was cryptanalyzed; this includes widely used hash functions such as MD4 and MD5. Moreover, serious shortcomings have been identified in the theoretical foundations of existing designs. In response to this hash function crisis, a large number of papers has been published with theoretical results and novel designs. In November 2007, NIST announced the start of the SHA-3 competition, with as goal to select a new hash function family by 2012. About half of the 64 submissions were broken within months. We present a brief outline of the state of the art of hash functions half-way the competition and attempt to identify open research issues.

Cryptographic hash functions map input strings of arbitrary length to short fixed length output strings. They were introduced in cryptology in the 1976 seminal paper of Diffie and Hellman on public-key cryptography [4]. Hash functions can be used in a broad range of applications: to compute a short unique identifier of a string (e.g. for a digital signature), as one-way function to hide a string (e.g. for password protection), to commit to a string in a protocol, for key derivation and for entropy extraction.

Until the late 1980s, there were few hash function designs and most proposals were broken very quickly after their introduction. The first theoretical result is the construction of a collision-resistance hash function based on a collision-resistant compression function, proven independently by Damgård [3] and Merkle [10] in 1989. Around the same time, the first cryptographic algorithms were proposed that are intended to be fast in software; the hash functions MD4 [14] and MD5 [15] fall in this category. Both were picked up quickly by application developers as they were ten times faster than DES; in addition they were not patent-encumbered and they posed less export problems than an encryption algorithm. As a consequence, hash functions were also used to construct MAC algorithms (e.g., HMAC as analyzed by Bellare et al. [2,1]) and even block ciphers and stream ciphers.

G. Gong and K.C. Gupta (Eds.): INDOCRYPT 2010, LNCS 6498, pp. 115–117, 2010.

During the 1990s, a growing number of hash functions were proposed [13], but unfortunately very few of these designs have withstood cryptanalysis. Notable results were obtained by Dobbertin, who found collisions for MD4 in 1995 [5]. Very few theoretical results were available in the area. At the same time however, MD5 and SHA-1, the latter introduced in 1995 by NIST (National Institute for Standards and Technology, US) [7], were deployed in an ever growing number of applications, resulting in the name "Swiss army knifes" of cryptography.

Wang et al. made substantial progress in the differential cryptanalysis of hash functions of the MD4 type: in 2004 they found collisions for MD4 by hand and for MD5 in a few minutes [17]. They managed to reduce the cost of collisions for SHA-1 by three orders of magnitude [16]. Suddenly hash functions moved to the center stage in cryptology: many new theoretical results were obtained, new designs were proposed and the cryptanalytic techniques of Wang et al. were further developed. Today RIPEMD-160 [6] seems to be one of the few older 160-bit hash functions for which no shortcut attacks are known. In 2002, NIST introduced the SHA-2 family of hash functions [8] with as goal to match the security levels provided by 3-DES and AES (output results of 224 to 512 bits). Even if attempts to cryptanalyzed SHA-2 have failed so far, there is a concern that the attacks of Wang et al. would also apply to these functions, which have design principles that are quite similar to those of SHA-1.

In November 2007, NIST announced that it would organize an open competition to select the SHA-3 algorithm [11]. In October 2008, 64 candidates were submitted; 51 of these were admitted to the first round and in July 2009, 14 were selected for the second round. In December 2010, NIST will announce 4 to 6 finalists; the final winner will be announced in the second Quarter of 2012.

Our talk presents an overview of the state of hash functions. We discuss the main theoretical results, describe some of the most important attacks, including the rebound attack [9]. Next we give an update on the status of the SHA-3 competition and explain why SHA-3 will be a hash function that is very different from SHA-2. One can expect that the SHA-3 competition will result in a robust hash function with a good performance, that will co-exist with SHA-2. One can also expect that NIST will standardize a tree mode for hash functions to obtain improved performance on multi-core processors (see [3,12] and several SHA-3 submissions). For the long term, we face the challenging problem to design an efficient hash function for which the security can be reduced to a mathematical problem that is elegant and for which we have a convincing security reduction.

References

1. Bellare, M.: New proofs for NMAC and HMAC: security without collision-resistance. In: Dwork, C. (ed.) CRYPTO 2006. LNCS, vol. 4117, pp. 602–619. Springer, Heidelberg (2006)
2. Bellare, M., Canetti, R., Krawczyk, H.: Keying hash functions for message authentication. In: Koblitz, N. (ed.) CRYPTO 1996. LNCS, vol. 1109, pp. 1–15. Springer, Heidelberg (1996)

3. Damgård, I.B.: A design principle for hash functions. In: Brassard, G. (ed.) CRYPTO 1989. LNCS, vol. 435, pp. 416–427. Springer, Heidelberg (1990)
4. Diffie, W., Hellman, M.E.: New directions in cryptography. IEEE Trans. on Information Theory IT-22(6), 644–654 (1976)
5. Dobbertin, H.: Cryptanalysis of MD4. Journal of Cryptology 11(4), 253–271 (1998); See also Gollmann, D. (ed.): FSE 1996. LNCS, vol. 1039, pp. 53–69. Springer, Heidelberg (1996)
6. Dobbertin, H., Bosselaers, A., Preneel, B.: RIPEMD-160: a strengthened version of RIPEMD. In: Gollmann, D. (ed.) FSE 1996. LNCS, vol. 1039, pp. 71–82. Springer, Heidelberg (1996)
7. FIPS 180-1, Secure Hash Standard, Federal Information Processing Standard (FIPS), Publication 180-1, National Institute of Standards and Technology, US Department of Commerce, Washington D.C. (April 17, 1995)
8. FIPS 180-2, Secure Hash Standard, Federal Information Processing Standard (FIPS), Publication 180-2, National Institute of Standards and Technology, US Department of Commerce, Washington D.C. (August 26, 2002) (Change notice 1 published on December 1, 2003)
9. Lamberger, M., Mendel, F., Rechberger, C., Rijmen, V., Schläffer, M.: Rebound distinguishers: Results on the full Whirlpool compression function. In: Matsui, M. (ed.) ASIACRYPT 2009. LNCS, vol. 5912, pp. 126–143. Springer, Heidelberg (2009)
10. Merkle, R.C.: One way hash functions and DES. In: Brassard, G. (ed.) CRYPTO 1989. LNCS, vol. 435, pp. 428–446. Springer, Heidelberg (1990)
11. NIST SHA-3 Competition, `http://csrc.nist.gov/groups/ST/hash/`
12. Pal, P., Sarkar, P.: PARSHA-256 A new parallelizable hash function and a multithreaded implementation. In: Johansson, T. (ed.) FSE 2003. LNCS, vol. 2887, pp. 347–361. Springer, Heidelberg (2003)
13. Preneel, B.: Analysis and design of cryptographic hash functions, Doctoral Dissertation, Katholieke Universiteit Leuven (1993)
14. Rivest, R.L.: The MD4 message digest algorithm. In: Menezes, A., Vanstone, S.A. (eds.) CRYPTO 1990. LNCS, vol. 537, pp. 303–311. Springer, Heidelberg (1991)
15. Rivest, R.L.: The MD5 message-digest algorithm. Request for Comments (RFC) 1321, Internet Activities Board, Internet Privacy Task Force (April 1992)
16. Wang, X., Yin, Y.L., Yu, H.: Finding collisions in the full SHA-1. In: Shoup, V. (ed.) CRYPTO 2005. LNCS, vol. 3621, pp. 1–16. Springer, Heidelberg (2005)
17. Wang, X., Yu, H.: How to break MD5 and other hash functions. In: Cramer, R. (ed.) EUROCRYPT 2005. LNCS, vol. 3494, pp. 19–35. Springer, Heidelberg (2005)

Cryptanalysis of Tav-128 Hash Function

Ashish Kumar[1], Somitra Kumar Sanadhya[2], Praveen Gauravaram[3,⋆], Masoumeh Safkhani[4], and Majid Naderi[4]

[1] Indian Institute of Technology, Kharagpur, WB, India
[2] Indraprastha Institute of Information Technology-Delhi, New Delhi, India
[3] Technical University of Denmark, Denmark
[4] Iran University of Science and Technology, Tehran, Iran
kumarashish.iitkgp@gmail.com, somitra@iiitd.ac.in, p.gauravaram@mat.dtu.dk, m_safkhani@iust.ac.ir, m_naderi@iust.ac.ir

Abstract. Many RFID protocols use cryptographic hash functions for their security. The resource constrained nature of RFID systems forces the use of light weight cryptographic algorithms. Tav-128 is one such 128-bit light weight hash function proposed by Peris-Lopez *et al.* for a low-cost RFID tag authentication protocol. Apart from some statistical tests for randomness by the designers themselves, Tav-128 has not undergone any other thorough security analysis. Based on these tests, the designers claimed that Tav-128 does not posses any trivial weaknesses. In this article, we carry out the first third party security analysis of Tav-128 and show that this hash function is neither collision resistant nor second preimage resistant. Firstly, we show a practical collision attack on Tav-128 having a complexity of 2^{37} calls to the compression function and produce message pairs of arbitrary length which produce the same hash value under this hash function. We then show a second preimage attack on Tav-128 which succeeds with a complexity of 2^{62} calls to the compression function. Finally, we study the constituent functions of Tav-128 and show that the concatenation of nonlinear functions A and B produces a 64-bit permutation from 32-bit messages. This could be a useful light weight primitive for future RFID protocols.

Keywords: Hash function, Tav-128, Cryptanalysis, RFID, Compression function.

1 Introduction

RFID technology has gained wide acceptance in the market-place in the last decade. Due to the security vulnerabilities found in various RFID protocols and implementations, researchers have been focusing on designing secure RFID protocols. RFID tags have hard constraints on their size, the chip area and power

⋆ Author is supported by the Danish Council for Independent Research - Technology and Production Sciences (FTP) and Natural Sciences (FNU) grant number 274-09-0096.

G. Gong and K.C. Gupta (Eds.): INDOCRYPT 2010, LNCS 6498, pp. 118–130, 2010.

consumption. Due to these constraints public key cryptography becomes generally impractical for use in the RFID protocols. Consequently, researchers have relied on cryptographic hash functions [13,22,21,9,7] or block ciphers [10] in the design of RFID protocols. Unlike block ciphers, hash functions do not require exchange of a key before operation of the protocol and hence RFID community has thought them to be simpler to implement. However, Feldhofer and Rechberger [11] have shown that the number of gates required to implement most common hash functions on an RFID chip is much higher than that for some block ciphers.

It is interesting to note that in NIST's ongoing SHA-3 hash function competition [17], among the hash functions selected for the second round [19] of the competition, only CubeHash with hash values upto 512 bits has been reported with a hardware implementation of less than 10,000 Gate Equivalents (GE) [2]. In a recent study on the FPGA implementation of fourteen second round 256-bit SHA-3 candidates [12], it was reported that only Keccak, CubeHash and Luffa hash functions have throughput to area ratio better than the current NIST standard SHA-2 [18]. On a similar note, challenges involved in the design of compact dedicated hash functions with large hash values and some hash function proposals based on the compact block cipher PRESENT [4] were discussed in [5]. Very recently, another lightweight hash function has been proposed by Aumasson *et al.* [1].

Tav-128 is a 128-bit lightweight cryptographic hash function developed by Peris-Lopez *et al.* [20] which is utilized in an RFID authentication protocol proposed by the designers. Peris-Lopez *et al.* remark that Tav-128 "can be fitted in low-cost RFID tags and provides a suitable security level for most applications". Apart from some statistical randomness tests on Tav-128 by the designers themselves, the hash function has not undergone any other thorough security analysis. Based on these tests, the designers claimed that Tav-128 does not posses any trivial weaknesses. In this work, we carry out the first third party analysis of Tav-128 and show that it is not a strong hash function. In particular, we show that practical collisions can be found for messages of any arbitrary length. Using ideas from our collision attack, we also show a second preimage attack on Tav-128.

Results: In this work we show the following.

1. A collision attack on Tav-128 with a complexity $\approx 2^{37}$.
2. A second preimage attack on Tav-128 with a complexity $\approx 2^{62}$.
3. Suggestion for a low cost 64-bit permutation primitive based on Tav-128.

The organization of the paper is as follows. In § 2, we present the notation used and describe the security requirements of a cryptographic hash function. In § 3 the structure of Tav-128 is explained. In § 4, we describe our collision attack and provide colliding message pairs for Tav-128. We show a second preimage attack on Tav-128 in § 5. We conclude with some open problems on the constructions of light weight primitives for secure protocols in § 6.

2 Notation and Preliminaries

2.1 Notation

In this work, the following notation are used.
$+$: Addition modulo 2^{32}.
$-$: Subtraction modulo 2^{32}.
$\|$: Concatenation of two quantities.
$\lll$: Left bit-shift operator (on 32-bit quantities).
$\ggg$: Right bit-shift operator (on 32-bit quantities).

2.2 Security Requirements of Cryptographic Hash Functions

A cryptographic hash function produces a fixed length digest, called the hash value, for an arbitrary sized message. It must satisfy some or more of the following properties, depending on the intended use of that hash function [16].

Preimage Resistance: A hash function $h(.)$ is preimage resistant if it is computationally infeasible to find an x for *any* given y such that $h(x) = y$.

Second Preimage Resistance: A hash function $h(.)$ is second preimage resistant if it is computationally infeasible to find an x_2 given *any* x_1 such that $h(x_1) = h(x_2)$ and $x_1 \neq x_2$.

Collision Resistance: A hash function $h(.)$ is collision resistant if it is computationally infeasible to find a pair (x_1, x_2), $x_1 \neq x_2$ such that $h(x_1) = h(x_2)$.

By "computational infeasibility", we mean that the complexity of an algorithm to break any one of the security properties is not less than the generic attack for breaking that property. For a hash function producing an ℓ-bit hash value, the complexity of birthday attack is $2^{\ell/2}$, and that for the preimage and the second preimage attack is 2^{ℓ}. Note that for some iterated hash function modes such as those based on Merkle-Damgård construction [8,15], the complexity to find a second preimage for a target message of 2^t blocks is $2^{\ell-t}$ [14]. If an attack can be described against any one of these properties and that attack has better complexity than these generic attacks then it is known as a "breaking" of the hash function.

3 The Tav-128 Hash Function

As previously mentioned, Tav-128 has been designed as a lightweight hash function to be used in an RFID authentication protocol [20] proposed by the designers. Tav-128 is intended to be a cryptographically secure hash function. The designers of Tav-128 note that choosing an output size of 128-bit is governed by the fact that a hash size of 64-bit would imply that "finding collisions is a relatively easy task due to the birthday paradox (around 2^{32} operations)".

The hash function outputs a digest of 128-bit for a message of any length. The structure of Tav-128 is as shown in Figure 1. A message of length $32 \times k$

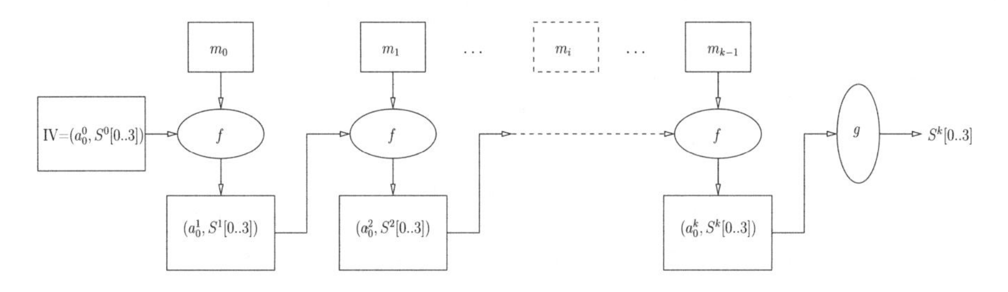

Fig. 1. The Merkle-Damgård structure of Tav-128. Compression function f is $\{0,1\}^{32} \times \{0,1\}^{160} \rightarrow \{0,1\}^{160}$. Final transformation g simply outputs the final state S.

is hashed by following the Merkle-Damgård [8,15] mode of operation where the compression function $f : \{0,1\}^{32} \times \{0,1\}^{160} \rightarrow \{0,1\}^{160}$ is iterated k times. The final truncation transformation g outputs last four 32-bit words of the 160-bit internal state. The reference implementation of Tav-128 from [20] is provided in Appendix B.

3.1 The Compression Function *f*

Each call to the f function with a 32-bit message m updates 5 variables, each of which is of size 32 bits. These 5 variables are the register a_0 and the states $S[0], S[1], S[2]$ and $S[3]$. The f function utilizes 4 functions A, B, C and D to update these variables and also uses two internal variables h_0 and h_1 in this process. The functions C and D together is called $C\&D$ function and its definition can be seen in the reference implementation of Tav-128 provided in Appendix B. The two functions A and B are shown in Table 1. The schema of the compression function is described in Figure 2. The symbol $\oplus$ in Figure 2 denotes bit wise XOR of two 32-bit integers.

Table 1. Nonlinear functions A and B in Tav-128

$A(h_0, m)$	$B(h_1, m)$
`for(i=0; i<32; i++)`	`for(i=0; i<32; i++)`
$h_0 = (h_0 \ll 1) + (h_0 + m) \gg 1$	$h_1 = (h_1 \gg 1) + (h_1 \ll 1) + h_1 + m$

For Tav-128 to be considered as a collision resistant hash function, it should take 2^{64} compression function calls to find a collision and 2^{128} compression function calls to mount a (second) preimage attack. Note that Tav-128 has an internal state of 160 bits but outputs a hash value of 128 bits. Hence, long message second preimage attack [14] is cheaper than the brute force second preimage attack for a target message of size more than 2^{32} message blocks. For a target message of 2^{32} blocks, the cost of both second preimage attacks is the same.

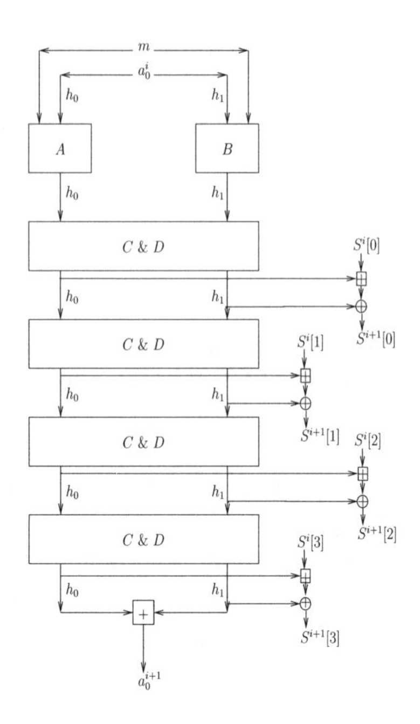

Fig. 2. Schema of the compression function f of Tav-128

4 Collision Attack on Tav-128

The compression function of Tav-128 first initializes the variables h_0 (resp. h_1) with a constant and then a non-linear function A (resp. B) updates this value depending on the 32-bit message m.

The authors of the hash function justify the inclusion of these two functions by stating that [20] "*We have also tried to include a filter phase (corresponding to algorithms A and B) in the input of the Tav-128 function, in order to avoid the attacker to have direct access to any bit of the internal state. Not having this possibility, some attacks that have been found on other cryptographic primitives in the past are precluded.*".

In § 4.1, we investigate if the claim above is true and whether the application of these two functions weaken or strengthen the hash design. As described in § 4.1, we find that the inclusion of these two functions, at least for the IV chosen, indeed improves the security.

4.1 Non-existence of Collisions at the Level of Functions A and B

We note that the functions $A(h_0, \cdot)$ and $B(h_1, \cdot)$ are not permutations. It is easy to find collisions on m in either of the two functions A or B. Some such examples are presented in Table 2. Some more analysis of functions A and B is presented in Appendix A.

Table 2. Examples of collision in A and B functions

S. No.	Function A			Function B		
	h_0	m	$A(h_0, m)$	h_1	m	$B(h_1, m)$
1	0x768c7e74	0x74093e01 0x09057d79	0x6dabc1e3	0x768c7e74	0x4505b289 0x62ee7bbd	0x3d8fd817
2	0x0	0x1bd18de3 0x554216d3	0x099587a6	0x0	0x51b70ece 0x17cac654	0xd3502587

Despite the fact that collisions in A and B are easily found, it does not appear easy to find collisions in *both* A and B *simultaneously*. Note that the only place where the message m is used in the compression function f is in functions A and B. Further computations in functions C and D only operate on the intermediate values h_0 and h_1 and state variables $S[0], S[1], S[2]$ and $S[3]$. Therefore, if a message pair m_1 and m_2 could be found such that $A(h_0, m_1) = A(h_0, m_2)$ and $B(h_1, m_1) = B(h_1, m_2)$ for the IV specified values $h_0 = h_1 =$ 0x768c7e74, then the pair (m_1, m_2) will constitute a collision for the full hash function Tav-128. In order to find collisions in Tav-128, we therefore investigate collisions in the concatenation of outputs of functions A and B.

To search for collisions in $A(h_0, m) \| B(h_1, m)$ we used the following strategy.

1. Create a table of size 2^{32} corresponding to all 32-bit messages m containing the triplet $(A(h_0, m), B(h_1, m), m)$.
2. Sort this table.
3. Look for a pair of adjacent rows in the sorted table where the first two entries are equal.

Since the size of the file in the strategy above would be $\approx$ 128 Gigabytes, sorting it would become computationally expensive on a standard PC. The standard sorting methods having complexity $\mathcal{O}(n \ \log \ n)$ will result in an effort of the order of $2^{45.3}$ for n =128 GB $= 2^{40}$ bits. Therefore, we resorted to the use of a disk sorting algorithm. Disk sorting algorithms typically sort a file in multiple *runs* or passes of sorting, sorting a part of the file in each pass. The sorted parts of the file are merged in subsequent passes. These algorithms never require more memory than that which is physically available on the system. Due to hardware limitations (we used a standard PC with 4 GB RAM and about 500 GB hard disk), we could not efficiently used disk sorting on the complete file in a single trial. The temporary files created in various runs of the disk sorting routine would exhaust the entire free space on the hard disk. Therefore, we modified the strategy slightly by dividing the file into 16 chunks of roughly equal size and then using disk sorting on each file of approximately 8 GB. Finally we combined these 16 files by using the merge sort algorithm. For sorting individual files, we used disk based sorting algorithm *psort* [3]. For merging individual files, we wrote our own merge sort routine.

Our search reveals that there is *no* pair of messages on which both A and B functions collide starting from the IV. Thus there does not exist any collision on 32-bit messages at the level of A and B function.

4.2 On the Map $(m\|m) \to A(h_0, m)\|B(h_1, m)$

As remarked by the designers of Tav-128 [20], individual functions A and B are quite efficient, requiring very few gates to implement them. Therefore they seem to have potential applications in light weight protocols. However, our discussion in § 4.1 shows that it is very easy to find collisions in these individual functions. Therefore the use of these functions in any application where collisions in these functions could cause loss of security is immediately ruled out. Contrary to what one would expect, however, we have found that the 64-bit map $(m\|m) \to A(h_0, m)\|B(h_1, m)$ *is* a permutation for the IV value $h_0 = h_1 =$ `0x768c7e74`. Since this is a light weight permutation, constructed from two light weight primitives, it may be a useful tool for future protocols requiring low cost constructs.

Note: The map $(m\|m) \to A(h_0, m)\|B(h_1, m)$ is a permutation for the given IV. However, this does not rule out the possibility that it may not be a permutation for a random IV. We did not find such an IV in our (small) experiments, but since there are 2^{32} possible IV's one cannot be certain. If there does exist an IV $h_0 = h_1 = \alpha$ such that $A(\alpha, m_1) = A(\alpha, m_2)$ and $B(\alpha, m_1) = B(\alpha, m_2)$ for 32-bit messages m_1 and m_2, then it gives another collision attack on Tav-128 as follows. Randomly attempt to find a message m_0 such that the variable a_0 at the end of one round of Tav-128 comes out equal to α. If we succeed to find such a message then there exists a 2-block message pair $(m_0\|m_1)$ and $(m_0\|m_2)$ which collides on Tav-128. Intuitively, the success probability of this attack is expected to be worse than the collision attack on Tav-128 we propose in the following section.

4.3 Finding Collisions at the Level of $C\&D$ Function

From Figure 2, we note that if the intermediate values h_0 and h_1 collide for two different messages immediately after one application of the $C\&D$ function, then $S[0]$ will have the same value for both the messages. Since the message does not get used in subsequent computations, all the register values (h_0, h_1, $S[0]$, ..., $S[3]$) and hence the final hash output will also be same for these messages.

We used a strategy similar to the one described in § 4.1 to generate such message pairs. We restate it here for clarity.

1. Create a table of size 2^{32} corresponding to all 32-bit messages m containing the triplet (h_0, h_1, m), where the values of the variables h_0 and h_1 are after one application of the $C\&D$ function.
2. Sort this table.
3. Look for a pair of adjacent rows in the sorted table where the first two entries are equal.

The pair of messages thus obtained (third entries in the two adjacent rows above) are colliding for the full 160-bit internal state of Tav-128 after one application of the $C\&D$ function. Therefore these are the message pairs which collide for the hash function. We obtained 11 message pairs satisfying these requirements. For two of these pairs, the colliding hash value and the intermediate values of h_0 and h_1 after the first application of $C\&D$ function are presented in Table 3.

The cost of collision attack. Recall that our attack finds collisions after one application of the $C\&D$ function and there are 4 applications of this function in the Tav-128. Thus each random search of suitable message pair in our attack costs about $\frac{1}{4}^{\text{th}}$ of the cost of evaluating the hash value for that message pair. The estimate of effort of our attack is about 2^{32} calls to about $\frac{1}{4}^{\text{th}}$ of the hash function, followed by sorting the lists of h_0 and h_1 and subsequent search for colliding pair. In fact, we experimentally found that the cost of finding a message pair which satisfies our requirements is a little less than 2^{29}, rather than the expected 2^{32}. The whole process of creating the files, sorting them, merging to create one file and finally searching for colliding pairs on this merged file took less than 1 working day on a standard PC. We estimate this effort to be about 2^{37} calls of Tav-128. This is significantly below the birthday bound of 2^{64} for a hash function producing 128-bit digests.

4.4 Colliding Message Pairs

Two pairs of colliding messages are presented in Table 3. Both the message pairs are 32 bits and the hash output $H(M)$ is 128 bits. h_0 and h_1 are the intermediate values of the variables after one iteration of the $C\&D$ function. Since the chaining variables $h_0, h_1, S[0], S[1], S[2]$ and $S[3]$, and hence the full 160-bit state $(a_0, S[0...3])$, are all equal for each message pair, appending any arbitrary message at the end of the colliding pair will still collide for Tav-128. Given the results in Table 3, it is trivial to construct messages of any length $\geq$ 32-bit which will collide for Tav-128.

Table 3. Colliding message pairs for Tav-128

S.No.	M	h_0	h_1	$H(M)$
1	0x80e19efb 0x8e474d73	0x9feaad6c	0x58b49a48	11a208c1 822c7b31 c41dd0a4 10a9c8c0
2	0x1f399d6d 0x90adacf0	0xa148201c	0x97094b03	dd7d4e3a b426513a 6631c011 9241384f

5 Second Preimage Attack

In § 4, we exploited the non randomness in the functions A, B and $C\&D$ of the compression function of Tav-128 to mount a cheap collision attack on the hash function. We use a similar approach to find a second preimage for Tav-128 for a given target message $M = m_1^{(1)} \| m_1^{(2)} \| m_1^{(3)} \| \ldots \| m_1^{(N)}$, where each $m_1^{(i)} \in \{0,1\}^{32}$.

To find a second preimage, the adversary starts from the first block of M and looks for another message block $m_2^{(1)}$ such that $m_1^{(1)} \neq m_2^{(1)}$ but h_0 and h_1 values after the first application of the $C\&D$ function collide. If the adversary can find such a message $m_2^{(1)}$ then the second preimage of M would be $m_2^{(1)} \| m_1^{(2)} \| m_1^{(3)} \| \ldots \| m_1^{(N)}$. Otherwise, the adversary keeps $m_1^{(1)}$ unchanged and does the same exhaustive search on the second message block $m_1^{(2)}$. The adversary repeats this process until a message pair $m_1^{(i)}, m_2^{(i)}$ can be found which is different and yet collides for Tav-128 after one application of the $C\&D$ function. Whenever the adversary finds such a pair, it produces the second preimage of M under Tav-128 as $M' = m_1^{(1)} \| m_1^{(2)} \| \ldots \| m_2^{(i)} \| m_1^{(i+1)} \| \ldots \| m_1^{(N)}$. We can see that M and M' are unequal in the ith block and equal otherwise.

The cost of second preimage attack. We denote the 64-bit $h_0 \| h_1$ part of the internal chaining value after one application of the $C\&D$ function in the ith iteration of the compression function by $C\&D(m_j^{(i)})$, where $m_j^{(i)}$ denotes the ith block of jth message. In the ith iteration, the adversary searches for a 32-bit message $m_2^{(i)}$ such that $C\&D(m_1^{(i)}) = C\&D(m_2^{(i)})$, where $m_1^{(i)}$ is the ith block of the given target message M. Hence, in the ith iteration the adversary looks for a second preimage for $C\&D(m_1^{(i)})$ which has the length 64 bits. The complexity to find a second preimage would be about 2^{64} calls to the compression function up to the end of the first call to the $C\&D$ which is about $\frac{1}{4}^{th}$ of the Tav-128 compression function. This roughly equals 2^{62} calls to the compression function. However, for each iteration of Tav-128, we have at most 2^{32} choices for $m_2^{(i)}$. Hence, the expected length of the given target message is $\frac{2^{64}}{2^{32}} = 2^{32}$ message blocks. In other words, the message for which second preimage can be found using our attack must have a length of the order of 2^{32} blocks of 32-bits each, i.e. 2^{37} bits. Although it is unrealistic to consider such a long message for an RFID application, our attack nonetheless shows an undesirable property of Tav-128. The cost of our attack is 2^{62} calls to the compression function of Tav-128 which is significantly below the theoretical bound of 2^{128} for a hash function producing 128-bit hash values.

6 Conclusions and Open Problems

In this paper we presented a practical collision attack and a second preimage attack on the hash function Tav-128. Both these attacks demonstrate that the security of Tav-128 hash function reduces to only $1/4^{\text{th}}$ of it. Following our attacks, we state that Tav-128 is not a cryptographically secure hash function.

The construction of a library of light weight primitives having collision resistance and difficult inversion property is an open problem. Such a library will certainly be of use to designers of light weight secure protocols, not limited to just RFID applications. The cost comparison of such primitives and conditions on the optimal cost of an individual primitive also remain interesting open problems.

References

1. Aumasson, J.-P., Henzen, L., Meier, W., Naya-Plasencia, M.: Quark: A lightweight hash. In: Mangard, S., Standaert, F.-X. (eds.) CHES 2010. LNCS, vol. 6225, pp. 1–15. Springer, Heidelberg (2010)
2. Baldwin, B., Byrne, A., Hamilton, M., Hanley, N., McEvoy, R.P., Pan, W., Marnane, W.P.: FPGA implementations of SHA-3 candidates: Cubehash, grostl, LANE, shabal and spectral hash. In: Núñez, A., Carballo, P.P. (eds.) 12th Euromicro Conference on Digital System Design, Architectures, Methods and Tools, DSD, pp. 783–790. IEEE Computer Society, Los Alamitos (2009)
3. Bertasi, P., Bressan, M., Peserico, E.: Yet Another Fast Stable Sorting Software. In: Experimental Algorithms. LNCS, vol. 5526, pp. 76–78. Springer, Heidelberg (2009)
4. Bogdanov, A., Knudsen, L.R., Leander, G., Paar, C., Poschmann, A., Robshaw, M.J.B., Seurin, Y., Vikkelsoe, C.: PRESENT: An Ultra-Lightweight Block Cipher. In: Paillier, P., Verbauwhede, I. (eds.) CHES 2007. LNCS, vol. 4727, pp. 450–466. Springer, Heidelberg (2007)
5. Bogdanov, A., Leander, G., Paar, C., Poschmann, A., Robshaw, M.J.B., Seurin, Y.: Hash Functions and RFID Tags: Mind the Gap. In: Oswald, E., Rohatgi, P. (eds.) CHES 2008. LNCS, vol. 5154, pp. 283–299. Springer, Heidelberg (2008)
6. Brassard, G.: CRYPTO 1989. LNCS, vol. 435. Springer, Heidelberg (1990)
7. Choi, E.Y., Lee, S.-M., Lee, D.H.: Efficient RFID Authentication Protocol for Ubiquitous Computing Environment. In: Enokido, T., Yan, L., Xiao, B., Kim, D.Y., Dai, Y.-S., Yang, L.T. (eds.) EUC-WS 2005. LNCS, vol. 3823, pp. 945–954. Springer, Heidelberg (2005)
8. Damgård, I.: A Design Principle for Hash Functions. In: Brassard [6], pp. 416–427
9. Dimitriou, T.: A Lightweight RFID Protocol to protect against Traceability and Cloning attacks. In: First International Conference on Security and Privacy for Emerging Areas in Communications Networks (SecureComm 2005), Athens, Greece, pp. 56–66. IEEE Computer Society Press, Los Alamitos (September 2005)
10. Feldhofer, M., Dominikus, S., Wolkerstorfer, J.: Strong Authentication for RFID Systems Using the AES Algorithm. In: Joye, M., Quisquater, J.-J. (eds.) CHES 2004. LNCS, vol. 3156, pp. 357–370. Springer, Heidelberg (2004)
11. Feldhofer, M., Rechberger, C.: A Case Against Currently Used Hash Functions in RFID Protocols. In: Meersman, R., Tari, Z., Herrero, P. (eds.) OTM 2006 Workshops. LNCS, vol. 4277, pp. 372–381. Springer, Heidelberg (2006)
12. Gaj, K., Homsirikamol, E., Rogawski, M.: Fair and comprehensive methodology for comparing hardware performance of fourteen round two SHA-3 candidates using FPGAs. In: Mangard, S., Standaert, F.-X. (eds.) CHES 2010. LNCS, vol. 6225, pp. 264–278. Springer, Heidelberg (2010)
13. Henrici, D., Müller, P.: Hash-based Enhancement of Location Privacy for Radio-Frequency Identification Devices using Varying Identifiers. In: PerCom Workshops, pp. 149–153. IEEE Computer Society, Los Alamitos (2004)
14. Kelsey, J., Schneier, B.: Second Preimages on n-bit Hash Functions for Much Less than 2^n Work. In: Cramer, R. (ed.) EUROCRYPT 2005. LNCS, vol. 3494, pp. 474–490. Springer, Heidelberg (2005)
15. Markle, R.: One way Hash Functions and DES. In: Brassard [6], pp. 428–446
16. Menezes, A.J., van Oorschot, P.C., Vanstone, S.A.: Handbook of Applied Cryptography. CRC Press, Boca Raton (1997), `http://www.cacr.math.waterloo.ca/hac/`

17. National Institute of Standards and Technology. Announcing Request for Candidate Algorithm Nominations for a New Cryptographic Hash Algorithm (SHA-3) Family (November 2007), http://csrc.nist.gov/groups/ST/hash/documents/FR_Notice_Nov07.pdf with the Docket No: 070911510751201 (Accessed on 22/09/2010)
18. NIST. FIPS PUB 180-2-Secure Hash Standard (August 2002), http://csrc.nist.gov/publications/fips/fips180-2/fips180-2.pdf (accessed on 23/09/2010)
19. NIST. Second Round Candidates. Official notification from NIST (2009), http://csrc.nist.gov/groups/ST/hash/sha-3/Round2/submissions_rnd2.html (accessed on 22/09/2010)
20. Peris-Lopez, P., Castro, J.C.H., Estévez-Tapiador, J.M., Ribagorda, A.: An Efficient Authentication Protocol for RFID Systems Resistant to Active Attacks. In: Denko, M.K., Shih, C.-s., Li, K.-C., Tsao, S.-L., Zeng, Q.-A., Park, S.H., Ko, Y.-B., Hung, S.-H., Park, J.-H. (eds.) EUC-WS 2007. LNCS, vol. 4809, pp. 781–794. Springer, Heidelberg (2007)
21. Rhee, K., Kwak, J., Kim, S., Won, D.: Challenge-Response Based RFID Authentication Protocol for Distributed Database Environment. In: Hutter, D., Ullmann, M. (eds.) SPC 2005. LNCS, vol. 3450, pp. 70–84. Springer, Heidelberg (2005)
22. Weis, S.A., Sarma, S.E., Rivest, R.L., Engels, D.W.: Security and Privacy Aspects of Low-Cost Radio Frequency Identification Systems. In: Hutter, D., Müller, G., Stephan, W., Ullmann, M. (eds.) Security in Pervasive Computing. LNCS, vol. 2802, pp. 201–212. Springer, Heidelberg (2004)

A Some Properties of Functions *A* and *B*

In this section we comment on some combinatorial properties of the functions A and B. As already mentioned, these functions are not permutations over input messages.

A.1 Analyzing Function *A*

The `for` loop runs 32 times and updates h_0. Let the initial value of h_0 before the loop starts be $h_{0,0}$ and the updated value of h_0 after the ith iteration be $h_{0,i}$. In step i, the loop will perform the following operation.

$$h_{0,i} = (h_{0,i-1} \ll 1) + (h_{0,i-1} + m) \gg 1.$$

We therefore analyze the following equation.

$$x = (y \ll 1) + (y + m) \gg 1 \qquad (1)$$

In the equation above, there are 3 variables: x, y and m. Consider the problem of solving for y given x and m. We make the following observations on this problem.

1. For all pairs of x and m, there exist two distinct values of y satisfying Equation 1.

2. Let these two values of y be y_1 and y_2. The difference of y_1 and y_2 is always `0x55555555` (or its additive inverse modulo 2^{32}, *i.e.*`0xaaaaaaab` since the addition in Equation 1 is modulo 2^{32}). Note that the 32-bit constant `0x55555555` represents alternating sequence of 0 and 1 bits.
3. Note that Equation 1 can be written as

$$(x - y \ll 1) = (\text{something}) \gg 1.$$

The most significant bit (msb) of the rhs is always zero, hence the msb of $(x - y \ll 1)$ must also be always 0.

Despite the inversion of a single step of Equation 1 being trivial for a message m (given x and y), the problem of inverting the `for` loop (i.e. finding an m given $h_{0,0}$ and $h_{0,32}$) is difficult.

A.2 Analyzing Function B

Similar to the analysis above, let us try to study the following equation.

$$(y \ll 1) + (y \gg 1) + y + m = x \qquad (2)$$

We would like to solve Equation 2 for y, given the values of x and m. It is interesting that unlike in the case of Equation 1, this time we may or may not be able to solve the equation. The following cases can occur:

1. For some values of (x, m), there is no solution for y. For example: $x =$ `0x7409c642` and $m =$ `0x3d303017`.
2. For some values of (x, m), there is exactly one value of y satisfying Equation 2. E.g. $x =$ `0x152e1fdb` and m=`0x3d77b373`, for which y=`0xcfeafa67`.
3. For some values of (x, m), there are two values of y satisfying Equation 2. E.g. $x =$ `0x4a6cd8f5` and $m =$ `0x10dff39b`, for which $y =$ `0x5995f863` and `0xa2ba8aac`.

The probability of occurrence of the three cases above are roughly $\frac{1}{3}$.

Similar to the case of function A, the inversion of the `for` loop (computing m from $h_{1,0}$ and $h_{1,32}$ where the symbols have similar meaning) corresponding to 32 calls to Equation 2 is a difficult problem.

B Tav-128 Reference Code from [20]

```
/*************************************************************************/
Process the input a1 modifying the accumulated hash a0 and the state
/*************************************************************************/
void tav(unsigned long *state, unsigned long *a0, unsigned long *a1)
{
unsigned long h0,h1;
int i,j,r1,r2,nstate;
```

```
/* Initialization */
r1=32; r2=8; nstate=4;
h0=*a0; h1=*a0;

/* A - Function */
for(i=0;i<r1;i++){h0=(h0<<1)+((h0+(*a1))>>1);}
/* B - Function */
for(i=0;i<r1;i++){h1=(h1>>1)+(h1<<1)+h1+(*a1);}

/ * C & D - Function */
for(j=0;j<nstate;j++) {
   for(i=0;i<r2;i++) {
         /* C - Function */
        h0^=(h1+h0)>>3;
            h0=((((h0>>2)+h0)>>2)+(h0<<3) +(h0<<1))^0x736B83DC;
     /* D - Function */
         h1^=(h1^h0)>>1;
         h1=(h1>>4)+(h1>>3)+(h1<<3)+h1;
   } // round-r2
   state[j]+=h0;
   state[j]^=h1;
} // state

/* a0 updating */
 *a0=h1+h0;
}

/*******************************************************************************/
Initialization of the state and a0 with random values obtained from www.random.org
/*******************************************************************************/

void init state(unsigned long *state, unsigned long *a0)
{
   state[0]=0xa92be51d;
   state[1]=0xba9b1ef0;
   state[2]=0xc234d75a;
   state[3]=0x845c2e03;
   a0[0]=0x768c7e74;
}
```

Near-Collisions for the Reduced Round Versions of Some Second Round SHA-3 Compression Functions Using Hill Climbing

Meltem Sönmez Turan[1] and Erdener Uyan[2]

[1] Computer Security Division, National Institute of Standards and Technology, USA
[2] Institute of Applied Mathematics, Middle East Technical University, Turkey
meltem.turan@nist.gov, uerdener@metu.edu.tr

Abstract. A hash function is near-collision resistant, if it is hard to find two messages with hash values that differ in only a small number of bits. In this study, we use hill climbing methods to evaluate the near-collision resistance of some of the second round SHA-3 candidates. We practically obtained *(i)* 184/256-bit near-collision for the 2-round compression function of Blake-32; *(ii)* 192/256-bit near-collision for the 2-round compression function of Hamsi-256; *(iii)* 820/1024-bit near-collisions for 10-round compression function of JH. Among the 130 possible reduced variants of Fugue-256, we practically observed collisions for 7 variants (e.g. $(k, r, t) = (1, 2, 5)$) and near-collisions for 26 variants (e.g. 234/256 bit near-collision for $(k, r, t) = (2, 1, 8)$).

Keywords: Hash functions, Near-collisions, SHA-3 Competition.

1 Introduction

Hill climbing methods are simple heuristic algorithms that aim to provide "good" solutions to "hard" optimization problems in short running times. These algorithms start with a random point and iteratively improve it by making small changes, then they terminate after converging to a local optimum. They are successful for problems for which the value of the problem at a specific point gives some information about "close" points [24]. For the well known traveling salesman problem, these methods get within approximately 10-15% of optimal solution in relatively short time [9].

There are many hard search problems in the field of cryptography, such as factorization of RSA numbers, finding secret key in symmetric cryptosystems or building efficient components with good cryptographic properties. However, the success of the simple optimization techniques have been very limited in most of these problems (e.g. [4]). One of the reasons of the failure is that most of the cryptographic problems (such as searching for the secret key) have only one single solution and no other "good" solutions. Another reason is due to the discontinuity of the most cryptographic functions, i.e. small changes in the input usually result in random looking changes in the outputs. Clark in his PhD thesis

G. Gong and K.C. Gupta (Eds.): INDOCRYPT 2010, LNCS 6498, pp. 131–143, 2010.

[5] claims that these techniques might give significant and surprising results if used in the right way. Searching for cryptographically strong Boolean functions is one of the cryptographic problems that benefit from these methods [15,16,8,17].

After the announcement of the SHA-3 hash competition by National Institute of Standards and Technology (NIST) [18], the submitted hash functions have been a prolific source of new cryptographic problems. A secure cryptographic hash function is expected to resist collision, second preimage and preimage attacks [18]. Moreover, resisting other attacks such as partial preimage and near-collision attacks increases the confidence in the algorithm.

Truncating some of the output hash bits might be necessary for compatibility of systems or desired for the efficiency purposes. In such cases, near-collision results have significant importance, since the output differences may diminish after a truncation operation and collisions may be obtained [6,10].

Hill climbing methods seems to be more promising for searching near-collisions compared to other type of attacks, since the problem does not have a singular solution, but rather many optimal solutions. We select a subset of the 14 second round SHA-3 candidates, namely Blake [2], Fugue [7], Hamsi [20] and JH [27], and analyze the security of their compression functions using a simple hill climbing method. In our approach, we give differences to the input message blocks, and we select the subset of algorithms such that each algorithm has a different type of message loading. In Blake, 512-bit message blocks are injected using 10 different permutations, whereas in Fugue, 32-bit message blocks are directly assigned to the state. In Hamsi, 32-bit message block is expanded by a linear mapping and then the expansion is loaded to the state. In JH, the message block is XORed to different parts of input and output chaining value. Evaluation of the remaining candidates is left as a future study.

We observed that for some of the reduced versions of these candidates, the hill climbing method produced better results compared to the generic random search. We practically present near-collision examples for reduced compression functions of Blake-32, Fugue-256, Hamsi-256 and JH-256 that were obtained in short running times.

Organization of the paper is as follows. In Section 2, generic methods to find near collisions are discussed. Then, in Section 3, the proposed hill climbing method is described. Section 4, the results we obtained for reduced versions of Blake, Fugue, Hamsi and JH are presented. Finally, the results are summarized in Section 5.

2 Near-Collisions

A hash function H is near-collision resistant, if it is "hard" to find two messages with hash values that differ in only a small number of bits [14]. An *l/n-bit near-collision* is obtained, whenever two messages m_1, m_2 satisfying

$$weight(H(m_1) \oplus H(m_2)) = n - l \tag{1}$$

are found, where *weight* represents the Hamming weight. If the attacker initializes the hash functions with a value different from its original initial value, the result is a *pseudo near-collision*. If the same pseudo initial value is used for both messages m_1 and m_2, the result is a *semi-free-start near-collision*, whereas if a difference is introduced in the initial values, the result is a *free-start near-collision* [22].

Let h be a compression function which takes a m-bit message block and an n-bit chaining value as inputs and generates the n-bit chaining value as the output. An l/n-bit near-collision on h is obtained whenever two message blocks M_1, M_2 and two chaining values CV_1, CV_2 satisfying

$$weight(h(M_1, CV_1) \oplus h(M_2, CV_2)) = n - l \tag{2}$$

are found. Clearly, $l = n$ corresponds to a collision on the compression function.

A generic method to find near-collisions for a compression function is to generate (M^i, C_1^i, C_2^i) values $(i > 0)$, such that $H(M^i, C_1^i) = C_2^i$ and then compare the C_2^i's to find the closest pair. The method approximately requires $\sqrt{2^n/\binom{n}{l}}$ compression function calls with approximately same amount of memory, to find an l/k-bit near collision [11].

3 Hill Climbing Method

If the compression function h has strong diffusion properties, for a randomly chosen message M and input chaining value CV, the Hamming weight of

$$h(M, CV) \oplus h(M, CV \oplus \delta) \tag{3}$$

is approximately $\frac{n}{2}$, where δ is an n-bit vector with small Hamming weight. However if the diffusion of δ is not satisfied, $h(M, CV)$ and $h(M, CV \oplus \delta)$ might be correlated, i.e., the value of $h(M, CV)$ might provide some exploitable information about the value of $h(M, CV \oplus \delta)$ (e.g. some bits positions may always be equal). In such cases, the hill climbing algorithms to find near-collisions may work better than the generic approaches.

The aim of our hill climbing method is to minimize the function

$$f_{M_1,M_2}(x) = weight(h(M_1, x) \oplus h(M_2, x)) \tag{4}$$

where $x \in \{0,1\}^n$, for given message blocks M_1 and M_2. Let CV be a randomly chosen chaining value. We define the set of k-bit neighbors of an n-bit CV as

$$S_{CV}^k = \{x \in \{0,1\}^n | weight(CV \oplus x) \leq k\}. \tag{5}$$

Clearly, the size of S_{CV}^k is equal to $\sum_{i=0}^k \binom{n}{i}$.

For message blocks M_1 and M_2, we define a chaining value CV to be *k-opt*, if

$$f_{M_1,M_2}(CV) = \min_{x \in S_{CV}^k} f_{M_1,M_2}(x). \tag{6}$$

The hill climbing method proposed in this section works as follows. Given a pair of message blocks M_1 and M_2, we randomly select a candidate chaining value

CV and calculate $f_{M_1,M_2}(CV)$. Then, we search the set S_{CV}^k to find a better chaining value. If found, our candidate is updated. Then, a new search is started in the k-bit neighbor of the new candidate. The algorithm terminates whenever a *k-opt* chaining value is obtained. The pseudocode of the method is presented in Algorithm 1.

Algorithm 1. HILLCLIMBING(M_1, M_2, k)

```
Randomly select CV;
f_best = f_{M1,M2}(CV);
while (CV is not k-opt)
    CV = x such that x ∈ S^k_CV with f(x) < f_best;
    f_best = f_{M1,M2}(CV);
return (CV, f_best)
```

Given current CV, the next candidate can be selected in two ways. In the first way, the first chaining value that has lower f value is chosen and this approach is known as the *greedy gradient ascent*. In the second way, the best chaining value in S_{CV}^k is chosen and this approach is known as the *steepest ascent*. We made preliminary experiments to compare the efficiency of both approaches, using same parameter k and we observe that the greedy approach results in better near-collisions in same running times. For example, the greedy approach results are on the average 2% better than the steepest ascent results, when we run the algorithm for the 2-round compression function Blake.

4 Experimental Results

Searching S_{CV}^ks with larger k (> 3) values might result in better near-collisions, but the method is no longer efficient. Moreover, when k is large, it is harder to find correlated $h(M, CV)$ and $h(M, CV \oplus \delta)$, where weight of δ is k. For our experiments, we use k values less than or equal to 2.

To evaluate the success of the hill climbing method, it is compared to generic search approaches. However, the approach presented in Section 2 requires huge amount of memory that makes it harder to compare, since the hill climbing method requires no significant memory. Hence, we propose a modified generic search with no memory requirement, however with a larger time complexity.

In the proposed approach, we randomly try input chaining values CV to minimize

$$weight(h(M_1, CV) \oplus h(M_2, CV)) \tag{7}$$

for a given M_1 and M_2 pair. For a secure compression function, the distribution of the expression in (7) is Binomial with parameters n and $1/2$ and observing a value with weight l, i.e. obtaining an l/n-bit near-collision, requires approximately $2^n/\binom{n}{l}$ compression function calls with almost no memory requirement. Table 1 shows the expected time complexity to obtain l/n-bit near-collisions for a compression function with 256, 512 or 1024-bit output, using this approach.

Table 1. Approximate time complexity to obtain l/n-bit near-collisions

l/n	Complexity ($\approx$)
128/256, 256/512, 512/1024	2^4
151/256, 287/512, 553/1024	2^{10}
166/256, 308/512, 585/1024	2^{20}
176/256, 323/512, 606/1024	2^{30}
184/256, 335/512, 623/1024	2^{40}
191/256, 345/512, 638/1024	2^{50}
197/256, 354/512, 651/1024	2^{60}

We repeat our experiments approximately 2^{25} times and consider our method successful, whenever we obtained an l/n-near collision with $l \geq 184$ for $n = 256$, $l \geq 335$ for $n = 512$ and $l \geq 623$ for $n = 1024$. These bounds are achievable by the generic random search with complexity of 2^{40} as given in Table 1.

4.1 Blake-32

Blake [2] is based on the HAIFA iteration mode with a compression function that uses a modified version of the stream cipher ChaCha. The compression function of Blake-32 inputs 256-bit CV, 512-bit message block, 128-bit salt and 64-bit counter and outputs 256-bit CV. The function is composed of 10 rounds and in each round, the nonlinear function G that operates on four words is applied to columns and diagonals of the state.

In our experiments, 1-bit difference to the input message blocks are given and the counter and the salt are fixed to zero. For 1-round compression function of Blake-32, we easily obtained 252/256-bit near-collisions. These near-collisions are obtained whenever we give a 1-bit difference to the $9th$, $11th$, $13th$ or $15th$ word of the message blocks (due to the fixed permutations). Then, we consider 1.5-round compression function in which the half round corresponds to the applications of G to the columns of the state. The best result we obtained for 1.5-round and 2-round Blake is 209/256-bit and 184/256-bit near collisions, respectively (See Table 2). For larger rounds, the hill climbing method did not provide significantly better results compared to the generic random search.

The results presented in this paper are obtained by giving input difference to only the message bits. Giving additional differences to input chaining value, salt and counter as in [1] increases the flexibility of the attacker, however decreases the practicability of the attack. Another flexibility for the attackers is to start the attack on a middle round, instead of the first round of the compression function as in [1,25]. To compare the available results, we run our algorithm for 4-round compression function for a couple of days. Comparison of near-collision attacks on Blake-32 is given in Table 3.

The result on the compression function is conjectured to be expandable to a semi-free start near-collision attack on reduced round Blake-32, by choosing

Table 2. Example Near Collisions for the Compression Function of Blake

1-Round Compression Function: 252/256-bit Near Collision	
M_1	8f4a6174 719e5909 41112fdc e5fa805a 1bdea684 b491ec4a 4deb8a83 5f31cf20 6a111277 4b6ff9f9 3f210a47 67388c82 a54cbe2a 3ac0d8e6 8042a2a5 c0549b9e
M_2	8f4a6174 719e5909 41112fdc e5fa805a 1bdea684 b491ec4a 4deb8a83 5f31cf20 6a111277 4b6ff9f9 3f210a47 67388c82 a54cbe2a 3ac0d8e6 8042a2a5 c0549b1e
CV	c34a1a90 c6955a4e c0c7e9ab cbf5b76c fbab3691 3368498b a8801cd7 20267316
$h(M_1, CV) \oplus h(M_2, CV)$	00000000 80000000 00000000 00000080 01000000 00000000 80000000 00000000

1.5-Round Compression Function: 209/256-bit Near Collision	
M_1	4ffcdfb9 5429ec40 18f9d1d6 c2b5b039 09c31d11 18d1bc19 532edb9c 58e3664a f757e1bf 6b0acf84 6d01bd05 0ec90891 a439a1bf c8de2b0e be5a524a ae843e5a
M_2	4ffcdfb9 5429ec40 18f9d1d6 c2b5b039 09c31d11 18d1bc19 532edb9c 58e3664a f757e1bf 6b0acf84 6d01bd05 0ec90891 a439a1bf c8de2b0e be5a524a ae843eda
CV	67134117 63e4044d 1a0bbd2b b99824e3 cb638884 8b8d284f 13977bba ad75b3a0
$h(M_1, CV) \oplus h(M_2, CV)$	00006020 80080801 88008008 80808898 412300a1 03003810 99100081 b1008118

2-Round Compression Function: 184/256-bit Near Collision	
M_1	3bd4eee9 035c9cd7 d35de9f7 cd3ab897 6f4fc516 e117aa80 ff72acc8 05c22424 87aa2e99 cec2210d 2fd0974b 652e8e26 37acc0e7 5a7a7157 c5bb6f9b 7853cda1
M_2	3bd4eee9 035c9cd7 d35de9f7 cd3ab897 6f4fc516 e117aa80 ff72acc8 05c22424 87aa2e99 cec2210d 2fd0974b 652e8e26 37acc0e7 5a7a7157 c5b96f9b 7853cda1
CV	c25dd2cd 2030a7b6 0fc043e8 5a0b5096 f084c81f 1f90d7d6 af48e019 34cd3554
$h(M_1, CV) \oplus h(M_2, CV)$	01c40003 180ac188 20818018 31442186 13309080 0858600b 143a4041 7f3144d0

short messages such that the padding and the message fits one message block, i.e. the length of the padded message is 512-bits. This is possible since padding operation does not require to process an extra block.

4.2 Fugue

Fugue, designed by Halevi et al. [7], is a sponge-like design inspired by the hash function Grindahl. Fugue-256 is based on the F-256 function that uses a large internal state of thirty 32-bit words. F-256 operates 32-bit message blocks using a round transformation that consists of the following operations; *(i)* **TIX**(I) that loads the 32-bit message blocks to the state, *(ii)* **ROR3** that rotates the state by three columns, *(iii)* **CMIX** that mixes columns and *(iv)* **SMIX** that applies a nonlinear substitution to the first four columns of the state. The pseudocode of F-256 is given in Algorithm 2.

Table 3. Comparison of results on reduced-round compression function of Blake-32

Paper	Rounds	Complexity	Type	Difference
✓	1	2^1	252/256-bit near-collision	Message
✓	1.5	$< 2^{26}$	209/256-bit near-collision	Message
✓	2	$< 2^{26}$	184/256-bit near-collision	Message
[25]	4 (4-7)	2^{21}	152/256-bit near-collision	Message, CV
✓	4	$2^{37.39}$	182/256-bit near-collision	Message
[1]	4 (3-6)	2^{56}	232/256-bit near-collision	Message, CV, Salt, Counter

The third party analysis of Fugue is very limited compared to other candidates. First external analysis of Fugue is done by Khovratovich [12], in which structures are used to find collisions. However the attack requires 2^{352} time and memory complexity which is significantly higher than the generic attack complexities. Aumasson and Phan [3] show an efficient distinguisher for a slightly modified version of the finalization of Fugue.

Algorithm 2. F-256($M_1, \ldots, M_m, CV_0, \ldots, CV_7, k, r, t$)

```
for i ← 0 to 21
   S_i = 0;
for i ← 22 to 29
   S_i = CV_{i−22};
for i ← 1 to m
   TIX(M_i);
   for j ← 1 to k
      ROR3; CMIX; SMIX;
for i ← 1 to r × k
   ROR3; CMIX; SMIX;
for i ← 1 to t
   S_4+ = S_0; S_15+ = S_0; ROR15; SMIX;
   S_4+ = S_0; S_16+ = S_0; ROR14; SMIX;
return (S_1, S_2, S_3, S_4, S_15, S_16, S_17, S_18.)
```

Table 4. Summary of best results for different reduced versions of F-256

(k, r, t)	Best Near-collision result
(1,1,1),(1,1,2),(1,2,1),(1,2,2), (1,2,3),(1,2,4),(1,2,5)	Collision
(1,1,3),(1,2,6),(1,3,1),(1,3,2), (1,3,3),(1,3,4),(1,3,5),(1,3,6), (1,3,7),(1,3,8),(2,1,1),(2,1,2), (2,1,3),(2,1,4),(2,1,5),(2,1,6), (2,1,7),(2,1,8)	$\geq$ 231/256-bit near-collision
(1,1,4),(1,1,5),(1,2,7),(1,2,8), (1,3,9),(1,3,10),(2,1,9),(2,1,10)	$\geq$ 184/256-bit near-collision

Designers also proposed parameterized Fugue, $F[n, k, s, r, t]$ where n is the output size, k is the number of sub-rounds per round transformation, s is the state size, r is the number of rounds in the first phase of finalization and t is the number of rounds in the second phase of the final transformation. For the output size 256, the default values are $F[8, 2, 30, 5, 13]$. In our experiments, 130 ($= 2 \times 5 \times 13$) reduced versions based on the selection of k, r and t are evaluated with the default values of n and s. Experiments are repeated using 32-bit random messages without considering the padding scheme. For each version, we repeat the experiment 2^{25} times and the results better than 184/256-bit near-collisions are summarized in Table 4. Table 5 gives examples for three of these cases.

Table 5. Example near-collisions for Fugue

(k,r,t)=(1,2,5): Collision	
M_1	490f3725
M_2	0c021472
CV	54091f45 2b019af6 6950d523 2542deba 7ec4fc2a e5672d97 13f9d54c 51a838f9
$h(M_1, CV) \oplus h(M_2, CV)$	00000000 00000000 00000000 00000000 00000000 00000000 00000000 00000000

(k,r,t)=(2,1,8): 234/256-bit near-collision	
M_1	02442aec
M_2	94b7d6b8
CV	d8a49539 55b27d25 8b51ff28 90b8aeab c6921ca9 40de0de3 b83e522c 93df1165
$h(M_1, CV) \oplus h(M_2, CV)$	00000000 00000000 50e18100 216b3040 00000000 00000000 00000000 02030400

(k,r,t)=(2,1,10):185/256-bit Near collision	
M_1	c6699e14
M_2	3679710d
CV	d2c5af72 6d17c1ff e1341948 49df91d7 3be47b3c bb4bf4a9 16f631d7 9f6282b4
$h(M_1, CV) \oplus h(M_2, CV)$	20029e40 3d911e80 92200ce0 a4015824 ac94b07c 82b4a500 00000000 03d24000

4.3 Hamsi

Hamsi, designed by Küçük [20], is based on the concatenate-permute-truncate design strategy. The compression function of Hamsi-256 inputs a 32-bit message block and a 256-bit chaining value and outputs a 256-bit chaining value. The compression function acts on a state of 512 bits, which can be considered as a 4x4 matrix of 32-bit words.

First, 32-bit message block is expanded to 256 bits using a linear code (128,16, 70) over $\mathbb{F}_4$. Then, the expanded message and the chaining value, each of being eight 32-bit words is loaded to the state of Hamsi-256. Then, the state is XORed with the predefined constants and a round counter and each of the 128 columns of the state goes through a 4x4 s-box. Finally, a linear transformation L, is applied to the four independent diagonals of the state. The compression function has 3 rounds, and a round transformation contains addition of constants, substitution and diffusion operations.

Nikolic [19] found 231/256-bit near-collisions for the compression function of Hamsi-256 for fixed message blocks. Wang et al. [26] improved the attack and practically showed 233/256-bit near-collisions for the compression function of Hamsi-256. Another improvement is proposed by Yun-qiang and Ai-lan using a genetic algorithm [28].

In our experiments, no input difference is given to the chaining values and two random 32-bits message blocks are chosen as input. In all previous attacks

Table 6. Comparison of near-collision results on Hamsi-256

Paper by	Rounds	$\lvert\delta_{CV}\rvert$	$\lvert\delta_M\rvert$	Result
Nikolic	3	14	0	(256-25)/256-bit NC
Wang et al.	3	16	0	(256-23)/256-bit NC
Aumasson et al.	3	6	0	(256-25)/256-bit NC
Yun-qiang et al.	3	4	0	(256-20)/256-bit NC
✓	1	0	$\geq$ 70 bits	(256-24)/256-bit NC
✓	2	0	$\geq$ 70 bits	(256-64)/256-bit NC

Table 7. Example Near-collisions for the Compression Function of Hamsi

1- Round Compression Function: 232/256-bit Near-collision	
M_1	22e20185
M_2	dd1dfe7a
CV	f6bf6de4 13429c65 b149b61a af8ed58d e3068bc8 e0397375 22866132 a8c5d4d3
$h(M_1, CV) \oplus h(M_2, CV)$	00042000 80040000 28040100 10000000 40080802 c8080000 00040000 0801004b

2- Round Compression Function: 192/256-bit Near-collision	
M_1	cf15a470
M_2	2287860c
CV	5b0ef41a f6933669 9d50a0b1 f3a0d239 63d65d26 fdca6f81 1509bfea f6e73e66
$h(M_1, CV) \oplus h(M_2, CV)$	8810058e 00021462 c330a008 7224440b 02008812 31040d80 8a9c0060 0c028448

[19,26,28], weight of input difference given to the chaining values is smaller that the weight of the output difference. So, the attackers inherently assume that they have already obtained better near-collisions [21]. Moreover, in all attacks, the message expansion is avoided by not giving any difference to the message blocks. As also suggested by the designer [21], near collisions by giving difference to the message blocks are harder to obtain. Comparison of the results and the near-collision examples obtained for the 1- and 2-round compression function are provided in Table 6 and 7, respectively.

4.4 JH

JH, defined by Wu [27], is an iterated hash function with a novel mode of operation. In the compression function of JH, the 1024-bit chaining value and the 512-bit message block are compressed into the 1024-bit chaining value. Initially, the lower half of the state is XORed with the input message block and then the bijection function E is applied. Then, the upper half of the state is XORed with

Table 8. Near-collisions for the compression function of JH

Rounds	Near-collision	Complexity
1	1023/1024	$2^{20.31}$
2	1020/1024	$2^{18.57}$
3	1019/1024	$2^{19.20}$
4	1013/1024	$2^{19.80}$
5	1005/1024	$2^{25.01}$
6	991/1024	$2^{27.57}$
7	942/1024	$2^{20.71}$
8	907/1024	$2^{24.24}$
9	816/1024	$2^{19.77}$
10	820/1024	$2^{23.24}$

Table 9. Example near-collisions for 9-round and 10-round compression function of JH

9- Round Compression Function: 816/1024-bit Near Collision								
M_1	7b6d6a9e	464d09e1	86410000	35aeff35	db02a693	1da2914e	0e340511	4bb9b2df
	9847eb69	ab7422cd	efa4d5ed	eb7c248f	c09f84f4	8e71652f	c8af1bed	911a8de6
M_2	9b6d6a9e	464d09e1	86410000	35aeff35	db02a693	1da2914e	0e340511	4bb9b2df
	9847eb69	ab7422cd	efa4d5ed	eb7c248f	c09f84f4	8e71652f	c8af1bed	911a8de6
CV	64cdd586	e453fbab	60c0a125	a596b15e	22735167	8d69b439	b8039dd3	327bacbb
	55685b28	5a717a0b	e1cc05c8	fc607792	fc31f4cb	49ff1ca2	be3aba98	1618e6a3
	da5021d9	895c668b	ab40f1c5	6526e807	4074d5b1	e8141140	63bc2df1	8f738ba6
	5def4921	0385997f	da7b308d	30f64dd7	56a7301e	64bc927a	da94cded	3ede8236
$h(M_1, CV) \oplus h(M_2, CV)$	54504100	45114010	50045455	40400101	41444001	15450001	00554501	11041044
	44004114	10004501	10455441	04115401	40551514	14105014	01500441	01501004
	b0010405	04010514	44511000	54001541	05100545	04144510	10040144	00514404
	11445500	45005400	01000400	01100014	44040455	44440000	05000405	45441440
10- Round Compression Function: 820/1024-bit Near Collision								
M_1	2dcdeb76	ed262d2f	16c56a55	90cb76fa	59e71f06	765a5e59	6aa1ba10	24fe14b1
	aaa28629	918fea7f	da88deba	87110630	ca28d5ed	83465471	be02a361	2df6564f
M_2	2bcdeb76	ed262d2f	16c56a55	90cb76fa	59e71f06	765a5e59	6aa1ba10	24fe14b1
	aaa28629	918fea7f	da88deba	87110630	ca28d5ed	83465471	be02a361	2df6564f
CV	faa3c300	af6a90ae	b49356e2	6994afd8	ef1a1119	5a43864d	d2a9b5f1	bcc08129
	468a89c5	df2c42eb	8abe5884	f3688af1	98978ec7	b63c05a3	5af13a34	43c52bc2
	2313f9b7	e8013174	2a3389ff	439c0432	ad4ab2e8	23934359	33a12345	52a427f7
	bbae8074	2bf65083	ec04ee67	21e2e376	20760866	ad6f586e	97837de8	22c7c119
$h(M_1, CV) \oplus h(M_2, CV)$	d0848cda	80b0560d	00000000	00000000	00000000	00000000	24981865	56b25240
	4a83359e	400c1b6b	00000000	00000000	00000000	00000000	13709c6e	db64dc89
	06e12007	4490779e	00000000	00000000	00000000	00000000	e417dc75	f465014e
	44496142	3105c9a0	00000000	00000000	00000000	00000000	5404400f	a8013ca8

the input message block The bijective function includes a grouping function, the round function (run 35 times), an additional substitution layer together with a de-grouping function. Basic building blocks of the compression function are two 4×4 s-boxes and a $(4, 2, 3)$ Maximum Distance Separable (MDS) code over $GF(2^4)$.

In [23], Rijmen et al. found 1008/1024-bit semi-free-start near-collision for 19 rounds of JH for all hash sizes with $2^{156.77}$ compression function calls and $2^{143.70}$ byte memory complexity, and 768/1024-bit semi-free-start near-collision for 22 rounds with $2^{156.56}$ compression function calls and the same memory complexity, employing the rebound attack [13].

In our experiments, we choose two 512-bit random messages with 1-byte difference, and without considering the padding block, the attack is successful up to 10 rounds of the compression function of JH (out of 35) and the best results are summarized in Table 8.

Table 9 provides example near-collisions for 9 and 10 round compression function of JH. An interesting observation is that for the 10-round example, four 32-bit parts of equal bits are evenly spaced showing that they are not scattered around as in other examples. This is due to the permutation function used in JH. Linear transformation of JH causes prorogation of differences on two 4-bit words into a single word, this property might further be exploited by choosing specific message blocks to find better near-collisions.

5 Conclusion

In this study, we propose a simple hill-climbing method to find near-collisions for the reduced round compression functions of some of the round two SHA-3 candidates. The method produced better results compared to the generic random search, when the diffusion of chaining value bits is not fully satisfied.

We run the algorithms approximately 2^{25} times and compared the best obtained near-collision to the one obtained with 2^{40} complexity with generic random search.

We practically obtained *(i)* 184/256-bit near-collision for the 2-round compression function of Blake-32; *(ii)* 192/256-bit near-collision for the 2-round compression function of Hamsi-256; *(iii)* 820/1024-bit near-collisions for 10-round compression function of JH. For Fugue, it is possible to define 130 different reduced versions by the selection of the parameter (k, r, t). We obtained collisions for 7 reduced cases, near-collisions with distance less than 25 for 18 cases and near collisions with distance less than or equal to 72 for 8 cases.

The results obtained in this study do not affect the security of the hash functions against preimage, second preimage and collision attacks, but rather give a security margin of the compression functions against near-collision attacks. Since Fugue, Hamsi and JH process an additional message block including the padding, the results cannot be directly extended to the hash function. For 2-round hash function Blake-32, by selecting message blocks that include the padding, the results are conjectured to be expandable to a semi-free start near-collision attack.

As a future study, we plan to incorporate some hash function specific information to the search algorithm to achieve better near-collisions. Evaluating other SHA-3 candidates is left as another future study.

Acknowledgment

The authors would like to thank the anonymous reviewers for their comments and Çağdaş Çalık for his suggestions and help in coding.

References

1. Aumasson, J.-P., Guo, J., Knellwolf, S., Matusiewicz, K., Meier, W.: Differential and Invertibility Properties of BLAKE. In: Hong, S., Iwata, T. (eds.) FSE 2010. LNCS, vol. 6147, pp. 318–332. Springer, Heidelberg (2010)
2. Aumasson, J.-P., Henzen, L., Meier, W., Phan, R.C.-W.: SHA-3 Proposal BLAKE. Submission to NIST (2008)
3. Aumasson, J.-P., Phan, R.C.-W.: Distinguisher for Full Final Round of Fugue-256. Second Round SHA-3 Conference, Santa Barbara (2010)
4. Borghoff, J., Knudsen, L.R., Matusiewicz, K.: Hill Climbing Algorithms and Trivium. In: Selected Areas in Cryptography, 17th Annual International Workshop, SAC 2010, Ontario, Canada, August 12-13. LNCS, Springer, Heidelberg (to appear 2010)
5. Clark, J.A.: Metaheuristic Search as a Cryptological Tool. PhD thesis, Department of Computer Science, University of York (2001)
6. Gauravaram, P.: Cryptographic Hash Functions: Cryptanalysis, Design and Applications. PhD thesis, Information Security Institute, Queensland University of Technogy, Australia (June 2007)
7. Halevi, S., Hall, W.E., Jutla, C.S.: The Hash Function Fugue. Submission to NIST (updated) (2009)
8. Izbenko, Y., Kovtun, V., Kuznetsov, A.: The Design of Boolean Functions by Modified Hill Climbing Method. Information Technology: New Generations, 356–361 (2009)
9. Johnson, D.S., Mcgeoch, L.A.: The Traveling Salesman Problem: A Case Study in Local Optimization. In: Aarts, E.H.L., Lenstra, J.K. (eds.) Local Search in Combinatorial Optimization, pp. 215–310 (1997)
10. Kelsey, J.: SHA-160: A Truncation Mode for SHA256 (and most other hashes). Halloween Hash Bash Workshop (2005), http://csrc.nist.gov/groups/ST/hash/documents/Kelsey_Truncation.pdf
11. Kelsey, J., Lucks, S.: Collisions and Near-Collisions for Reduced-Round Tiger. In: Robshaw, M.J.B. (ed.) FSE 2006. LNCS, vol. 4047, pp. 111–125. Springer, Heidelberg (2006)
12. Khovratovich, D.: Cryptanalysis of Hash Functions with Structures. In: Jacobson Jr., M.J., Rijmen, V., Safavi-Naini, R. (eds.) SAC 2009. LNCS, vol. 5867, pp. 108–125. Springer, Heidelberg (2009)
13. Mendel, F., Rechberger, C., Schläffer, M., Thomsen, S.S.: The Rebound Attack: Cryptanalysis of Reduced Whirlpool and Grøstl. In: Dunkelman, O. (ed.) Fast Software Encryption. LNCS, vol. 5665, pp. 260–276. Springer, Heidelberg (2009)

14. Menezes, A.J., van Oorschot, P.C., Vanstone, S.A.: Handbook of Applied Cryptography. CRC Press, Boca Raton (2001)
15. Millan, W., Clark, A.: Smart Hill Climbing Finds Better Boolean Functions. In: Workshop on Selected Areas in Cryptology 1997, Workshop Record, pp. 50–63 (1997)
16. Millan, W., Clark, A.: Boolean Function Design Using Hill Climbing Methods. In: Pieprzyk, J.P., Safavi-Naini, R., Seberry, J. (eds.) ACISP 1999. LNCS, vol. 1587, pp. 1–11. Springer, Heidelberg (1999)
17. Millan, W., Clark, A., Dawson, E.: Heuristic Design of Cryptographically Strong Balanced Boolean Functions. In: Nyberg, K. (ed.) EUROCRYPT 1998. LNCS, vol. 1403, pp. 489–499. Springer, Heidelberg (1998)
18. National Institute of Standards and Technology. Announcing Request for Candidate Algorithm Nominations for a New Cryptographic Hash Algorithm (SHA-3) Family. Federal Register 27(212),62212–62220 (2007), `http://csrc.nist.gov/groups/ST/hash/documents/FR_Notice_Nov07.pdf`
19. Nikolic, I.: Near Collisions for the Compression Function of Hamsi-256. CRYPTO rump session (2009)
20. Küçük, Ö.: The Hash Function Hamsi. Submission to NIST (2008)
21. Küçük, Ö.: The Hash Function Hamsi (presentation). In: Second SHA-3 Conference, Santa Barbara (2010)
22. Preneel, B.: Analysis and Design of Cryptographic Hash Functions. PhD thesis, Katholieke Universiteit Leuven (1993)
23. Rijmen, V., Toz, D., Varici, K.: Rebound Attack on Reduced-Round Versions of JH. In: Hong, S., Iwata, T. (eds.) Fast Software Encryption, FSE 2010, Seoul,Korea. LNCS, p. 18. Springer, Heidelberg (2010)
24. Russell, S.J., Norvig, P., Candy, J.F., Malik, J.M., Edwards, D.D.: Artificial Intelligence: A Modern Approach. Prentice-Hall, Inc., Upper Saddle River (1996)
25. Su, B., Wu, W., Wu, S., Dong, L.: Near-Collisions on the Reduced-Round Compression Functions of Skein and BLAKE. Cryptology ePrint Archive, Report 2010/355 (2010), `http://eprint.iacr.org/2010/355.pdf`
26. Wang, M., Wang, X., Jia, K., Wang, W.: New Pseudo-Near-Collision Attack on Reduced-Round of Hamsi-256. Cryptology ePrint Archive, Report 2009/484 (2009), `http://eprint.iacr.org/2009/484.pdf`
27. Wu, H.: The Hash Function JH. Submission to NIST, (updated) (2009)
28. Yun-qiang, L.I., Ai-lan, W.: Near Collisions for the Compress Function of Hamsi-256 Found by Genetic Algorithm. Cryptology ePrint Archive, Report 2010/423 (2010), `http://eprint.iacr.org/2010/423.pdf`

Speeding Up the Wide-Pipe: Secure and Fast Hashing

Mridul Nandi[1] and Souradyuti Paul[2,3]

[1] C.R. Rao AIMSCS Institute, Hyderabad*
[2] National Institute of Standards and Technology
Security Technology Group
100 Bureau Dr., MS 8931
Gaithersburg, MD 20899, United States
[3] Katholieke Universiteit Leuven, Dept. ESAT/COSIC,
Kasteelpark Arenberg 10,
B–3001, Leuven-Heverlee, Belgium
mridul.nandi@gmail.com, souradyuti.paul@nist.gov

Abstract. In this paper we propose a new sequential mode of operation – the *Fast wide pipe* or FWP for short – to hash messages of arbitrary length. The mode is shown to be (1) *preimage-resistance preserving*, (2) *collision-resistance-preserving* and, most importantly, (3) *indifferentiable* from a random oracle up to $\mathcal{O}(2^{n/2})$ compression function invocations. In addition, our rigorous investigation suggests that any variants of Joux's multi-collision, Kelsey-Schneier 2nd preimage and Herding attack are also ineffective on this mode. This fact leads us to conjecture that the indifferentiability security bound of FWP can be extended beyond the birthday barrier. From the point of view of efficiency, this new mode, for example, is *always* faster than the Wide-pipe mode when both modes use an identical compression function. In particular, it is nearly twice as fast as the Wide-pipe for a reasonable selection of the input and output size of the compression function. We also compare the FWP with several other modes of operation.

1 Introduction

A hash function $H : \{0,1\}^* \longrightarrow \{0,1\}^n$ is a mathematical function which takes as input a binary string of arbitrary length and outputs a binary string of finite length. A secure hash function can be applied in many applications such as data authentication, digital signature, commitment protocols and password protection. A very popular trend of designing a hash function is executing a *fixed-input-length* (FIL) compression function in a sequential mode as many times as to take the whole message as input. Many practical hash functions,

* The large part of the work has been done while working in The George Washington University.

G. Gong and K.C. Gupta (Eds.): INDOCRYPT 2010, LNCS 6498, pp. 144–162, 2010.

Table 1. Comparison among several hash modes of operation with respect to *indifferentiability* attacks. All numbers are in bits. By Input and Output in the table, we mean bits into and bits out of the compression function.

Mode	Hash-length	Input	Output	Rate	Lower Bound	Upper Bound	Condition
MD[17]	n	a	$b = n$	$a - b$	0	0	$a > b$
MDP[10]	n	a	$b = n$	$a - b$	$n/2$	$n/2$	$a > b$
Wide-pipe[15]	n	a	$b = 2n$	$a - b$	$\approx n$	$\approx n$	$a > b$
Sponge[2]	n	a	$b = a$	$a - n$	$n/2$	$n/2$	$a > n$
JH[23,4]	n	a	$b = a$	$a/2$	$n/3$	n	$a > 2n$
FWP	$\boldsymbol{n}$	$\boldsymbol{a}$	$\boldsymbol{b = 2n}$	$\boldsymbol{a - \frac{b}{2}}$	$\boldsymbol{n/2}$	$\boldsymbol{n}$	$\boldsymbol{a > b/2}$

such as MD4 [20], MD5 [21], SHA-0 [18], SHA-1 [19] follow the aforementioned design paradigm. These hash functions precisely have two components: (1) a compression function and (2) a mode of operation.

This paper is all about design and analysis of a new hash mode of operation, which is named the *Fast wide pipe* or FWP for short.

Related work. The classical Merkle-Damgärd mode is the most widely used and most studied hash mode of operation. [17,8]. The mode is simple and collision-resistance-preserving.[1] All the practical hash functions mentioned before are based on the Merkle-Damgärd mode. The landscape is no longer the same. A telltale proof of declining interest of the designers in this mode is that none of the 51 hash functions competing in the ongoing NIST hash function competition uses the classical Merkle-Damgärd mode. The main reasons for discarding this mode by the designers are a few influential attacks: Length extension attack, multi-collision attack [11], Kelsey-Schneier 2nd preimage attack [14] and Herding attack [12]. On the positive side, the slow and gradual departure of the classical Merkle-Damgärd hash mode has motivated two new lines of research which go nearly hand in hand: (1) design of new modes of operation and (2) development of new security frameworks to analyze hash functions. The first line of research has indeed resulted in a number of new modes of operation – Wide-pipe [15], HAIFA [5], Sponge [2], EMD [1], JH[23] are some of them. One of the major results of the second line of research is the *indifferentiability framework* developed by Maurer *et al.* [16]. Against this framework, we measure the extent to which a hash function is behaving as a random oracle under a suitable assumption on the underlying compression function. Informally speaking if a hash function is *indifferentiable* from a random oracle then, for example, it does not come under length extension attack (assuming the underlying compression function is a FIL random oracle). It is, therefore, important that a new mode of operation is both *collision-resistance-preserving* and *indifferentiable* from a random oracle. Another crucial issue is to recognize that a hash function *indifferentiable*

[1] In a *collision-resistance-preserving* hash function collision resistance of a compression function implies collision resistance of the entire hash function.

from a random oracle does not guarantee that it is *collision-resistance-preserving* (e.g. modes of operation designed in [7] are not *collision-resistance-preserving*, although they are *indifferentiable* from random oracles[1]). These two properties should be analyzed separately [1].

Our contribution. To make a hash function resistant against Joux's multi-collision-type attacks, Lucks has proposed to make the intermediate chaining values of the Merkle-Damgärd mode twice as long as the final hash value; this mode is known as the Wide-pipe mode [15]. Suppose the compression function in a Merkle-Damgärd based hash function is defined as $C : \{0,1\}^{m+n} \to \{0,1\}^n$. Lucks has, very rightly, advocated to use a compression function $C : \{0,1\}^{m+2n} \to \{0,1\}^{2n}$ to avoid Joux's multi-collision-type attacks [11,13]. We call this compression function Lucks' compression function. The message and chaining input to the Lucks' compression function are m and $2n$ bits. Using any Lucks' compression function $C : \{0,1\}^{m+2n} \to \{0,1\}^{2n}$ we design a hash function FWP, where the message and the chaining input to the compression function are $m+n$ and n bits; we, thus, speed up the hashing operation by allowing $m+n$ bits of message instead of just m bits per compression function invocation . At the same time we prove that the FWP mode is *collision-resistance preserving* and *indifferentiable* from a random oracle up to $\mathcal{O}(2^{n/2})$ compression function invocations. The fact that the FWP does not come under Joux's multi-collision-type attacks, such as Kelsey-Schneier 2nd preimage attack, leaves open the possibility to extend the *indifferentiability* bound beyond the birthday barrier.

In Table 1, we compare our results with several other competing hash modes with respect to indifferentiability attacks. Against other attacks such as collision all the modes perform almost identically. It is readily observable that the FWP outperforms all other modes in at least one of the three properties, namely Rate, Lower Bound and Upper Bound in Table 1. The important features of the FWP are pointed out below.

1. FWP performs better than the Wide-pipe with respect to the rate of the hash function. For example, when the input size of the compression function is three times the output size – which is a reasonable choice – FWP is twice faster than the Wide-pipe.
2. Efficiency-wise, FWP has similar performance as Sponge and JH. However, there is a strong evidence that the indifferentiability security bound of FWP can be extended beyond $n/2$ bits, while there already exists an attack on the Sponge with work factor $n/2$ bits.

2 Notation and Convention

In addition to the above notation, we shall use another set of notation in the context of indifferentiability results of the hash function FWP. They are described in Sect. 5.1.

Table 2. Notation

$\{0,1\}^{\leq l}$	$\{\varepsilon\} \cup \{0,1\} \cup \{0,1\}^2 \cup \{0,1\}^3 \cup \ldots \cup \{0,1\}^l$
$[x, y]$	The set of integers $x, x+1, \ldots, y$
$a\|\|b$	concatenation of a and b
$\|X\|$	Size of set X; Bit-length if X is a string
$\mathsf{pad}(M)$	The sequence of bits after padding M
fixed-input-length	Fixed input length
variable-input-length	Variable input length
FWP	Fast wide-pipe

3 The New Mode *Fast Wide Pipe* or FWP

In this section we define a new sequential mode of operation *Fast Wide Pipe* (or FWP for short) for hashing messages of length up to 2^{64} bits.

Diagrammatic representation of the mode FWP is given in Fig. 1. An algorithmic description is in Algorithm 1. The padding rule $\mathsf{pad}(M)$ is the execution of the following operation: append t zero bits and a 64-bit encoding of $|M|$ to the message M. Select the least integer $t \geq 0$ such that $|M|+t+n+64 = 0 \bmod l$ (see

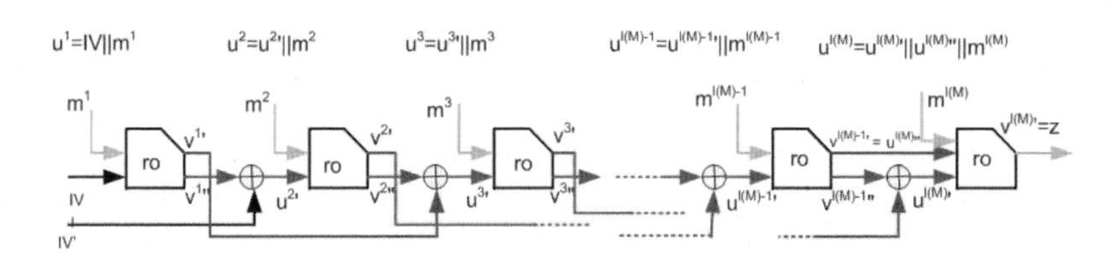

Fig. 1. The new mode FWP. Message $M = m^1 m^2 \ldots m^{\ell(M)}$ is hashed by FWP^{ro}. The symbols are described in Table 3.

Algorithm 1. The FWP mode of operation with the compression function C (*i.e.*, FWP^C)

Input: Message M
Output: Hash output h of size n bits
Initialize: $h_{-1} = h'_{-1} = 0^n$
1: $M_0||M_1||\ldots||M_{k-1} = \mathsf{pad}(M)$ where $|M_i| = l$ for all $i < k-1$ and $|M_{k-1}| = l - n$;
2: $(h_{k-2}, h'_{k-2}) = \text{FWP}_t^C(h_{-1}, h'_{-1}, M_0, M_1, \ldots, M_{k-2})$; /* See subfunction below */
3: $C(h_{k-2}||h'_{k-2}||M_{k-1}) = h''_{k-1}||h'_{k-1}$;
4: return hash output $h = h'_{k-1}$;
Subfunction $\text{FWP}_t^C(h_{-1}, h'_{-1}, M_0, M_1, \ldots, M_{k-2})$
5: **for** $i = 0$ to $k-2$ **do**
6: $C(h_{i-1}||M_i) = h''_i||h'_i$;
7: $h_i = h''_i \oplus h'_{i-1}$;
8: **end for**
9: **return** (h_{k-2}, h'_{k-2});

Algorithm 1 for the notation). We now make attempts to analyze the security of the FWP. For the sake of simplicity, we assume $l - n \geq 64$ which ensures that the length-encoding is completely included in the last block. The entire analysis can be modified easily to include the case $l - n < 64$.

4 Security of the FWP: Resistance against Collision and Preimage Attacks

The main results of this section are two theorems which prove that the collision and the preimage attacks on the FWP mode can be reduced to similar attacks on the underlying compression function (see Algorithm 1 for the definition of the FWP mode). In other words, the theorems show that finding collision and preimage on the FWP are at least as hard as finding collision and preimage on the compression function.

Before establishing the security results, we first define the following functions. The functions $C_T, C_B : \{0,1\}^{l+n} \rightarrow \{0,1\}^n$ are defined as $C_T(x) = h'$ and $C_B(x) = h''$ where $C(x) = h''||h'$ (the compression function C of the FWP is defined in Algorithm 1).

Theorem 1. *If the compression function C_T is preimage resistant so is the FWP^C.*

Proof. The theorem can be verified easily by observing the last block of FWP^C.

Glancing at the XOR operations, one may be tempted to conclude that the FWP may be vulnerable against the generalized birthday attack [22]. The following theorem drives away such fears.

Theorem 2. *If the compression function C_T is collision resistant so is the FWP^C.*

Proof. To prove the theorem we need to prove that, if there exists an adversary who finds a pair of messages (M, M') such that FWP$^C(M)$=FWP$^C(M')$ and $M \neq M'$ then there exists an adversary who can find $X \neq X'$ such that $C_T(X) = C_T(X')$.

Suppose an adversary finds a pair (M, M') such that FWP$^C(M)$=FWP$^C(M')$ and $M \neq M'$. Now there are two possible cases.

Case 1: $|M| \neq |M'|$. Suppose that the number of message-blocks in $\mathsf{pad}(M)$ and $\mathsf{pad}(M')$ are a and b where $a \neq b$. Note, as per our definition of C and FWP^C, $M_{a-1} \neq M'_{b-1}$ due to the length padding. Now, FWP$^C(M)$=FWP$^C(M')$ implies $C_T(h_{a-2}||h'_{a-2}||M_{a-1}) = C_T(g_{b-2}||g'_{b-2}||M'_{b-1})$. Therefore, we get a collision on C_T.

Case 2: $|M| = |M'|$. Suppose that the number of message-blocks in $\mathsf{pad}(M)$ is a. Now there are two cases.

Case 2(a): $C_T(h_{a-2}||h'_{a-2}||M_{a-1}) = C_T(g_{a-2}||g'_{a-2}||M'_{a-1})$ where $h_{a-2}||h'_{a-2}||M_{a-1} \neq g_{a-2}||g'_{a-2}||M'_{a-1}$. Therefore, we obtain a collision

on C_T.
CASE 2(B): $C_T(h_{a-2}||h'_{a-2}||M_{a-1}) = C_T(g_{a-2}||g'_{a-2}||M'_{a-1})$
where $h_{a-2}||h'_{a-2}||M_{a-1} = g_{a-2}||g'_{a-2}||M'_{a-1}$.

The above equation implies that $\mathrm{FWP}_t^C(0^n||0^n||M_0||\ldots||M_{a-2})= \mathrm{FWP}_t^C(0^n||0^n||M'_0||\ldots||M'_{a-2})$ which in turn implies collision on C_T by Lemma 1 (the definition of FWP_t^C is provided in Algorithm 1). Now the only remaining part needed to complete the proof is the proof of Lemma 1 which is provided below.

The following lemma has been used in Theorem 2. It will be further used to obtain some indifferentiability results of the FWP^C in Sect. 5.

Lemma 1. *If the compression function C_T is collision resistant then the FWP_t^C is free-start collision resistant for fixed length messages. In other words, if there exists an adversary who finds two triples $(h_{-1}, h'_{-1}, M) \neq (g_{-1}, g'_{-1}, M')$ such that $|M| = |M'|$ ($|M|$ is a multiple of l) and $FWP_t^C(h_{-1}, h'_{-1}, M) = FWP_t^C(g_{-1}, g'_{-1}, M')$, then there exists an adversary who finds $X \neq X'$ such that $C_T(X) = C_T(X')$.*

Proof. Suppose there exists an adversary who finds two triples $(h_1, h'_1, M) \neq (g_1, g'_1, M')$ such that $|M| = |M'|$ (the number of message-blocks in M is a) and $\mathrm{FWP}_t^C(h_{-1}, h'_{-1}, M) =\mathrm{FWP}_t^C(g_{-1}, g'_{-1}, M')$. In order to obtain a pair $X \neq X'$ such that $C_T(X) = C_T(X')$ we need to check at most a equations whether they are satisfied:

$$C(h_{i-1}, M_i) \stackrel{?}{=} C(g_{i-1}, M'_i) \text{ where } i = a-1, \ldots, 0.$$

We claim that the above verification will produce some m with $0 \leq m \leq a-1$ such that $C_T(h_{m-1}, M_m) = C_T(g_{m-1}, M'_m)$ and $(h_{m-1}, M_m) \neq (g_{m-1}, M'_m)$ and thus, the lemma is proved. This claim can be proved by the following crucial observation on FWP_t^C.

OBSERVATION: For all $i \in [0, a-1]$, $(h_i, h'_i) = (g_i, g'_i)$ implies one of the following two statements: (1) $(h_{i-1}, M_{i-1}) \neq (g_{i-1}, M'_{i-1})$ which implies collision on C_T, (2) $(h_{i-1}, M_{i-1}) = (g_{i-1}, M'_{i-1})$ which implies $(h_{i-1}, h'_{i-1}) = (g_{i-1}, g'_{i-1})$.

Next, we move on to analyze the FWP in a different security framework known as the indifferentiability framework.

5 Security of the FWP Mode: Indifferentiable from a Random Oracle

In this section we discuss the indifferentiability property of the FWP mode. In the context of hash function, an important use of the indifferentiability framework developed by Maurer *et al.* [16] is the determination of whether a *variable-input-length* hash function behaves reasonably randomly when the underlying compression function is a *fixed-input-length* random oracle. There is a considerable chance for the reader to be lured into believing that the collision resistance preservation (described in Sect. 4) and the indifferentiability property of a hash function may be related. In particular, one may be inclined to intuiting that one property implies the other. Such intuition is not true [1]. These two properties are orthogonal and need to be analyzed separately.

5.1 Preliminaries: Introduction to Indifferentiability Framework

We begin with the definition of a random oracle; this useful object will be used frequently in the subsequent discussion.

Definition 1 (Random oracle). *A random oracle is a function* $\mathsf{RO}: X \to Y$ *chosen uniformly at random from the set of all* $|Y|^{|X|}$ *functions that map* $X \to Y$*. In other words, a function* $\mathsf{RO}: X \to Y$ *is a random oracle if and only if, for each* $x \in X$*, the* $\mathsf{RO}(x)$ *is chosen uniformly at random from* Y*.*

Corollary 1. *If a function* $\mathsf{RO}: X \to Y$ *is a random oracle then*

$$\Pr[\mathsf{RO}(x) = y | \mathsf{RO}(x_1) = y_1,\ \mathsf{RO}(x_2) = y_2,\ \ldots,\ \mathsf{RO}(x_q) = y_q] = \frac{1}{|Y|}$$

where $x \notin \{x_1, x_2, \ldots, x_q\}$*,* $y \in Y$ *and* $q \in \mathbb{Z}$*.*

Now we introduce the indifferentiability framework and briefly discuss its significance. The following definition is a slightly modified version of the original definition provided in [16,7].

Definition 2 (Indifferentiability framework). *[16] A Turing machine* T *with oracle access to an ideal primitive* $\mathcal{F}$ *is said to be* $(t_{\mathcal{A}}, t_S, q, \sigma, \varepsilon)$*-indifferentiable from an ideal primitive* $\mathcal{G}$ *if there exists a simulator* S *such that for any distinguisher* $\mathcal{A}$ *the following equation is satisfied:*

$$\mathbf{Adv}_{\mathcal{A}}((T, \mathcal{F}), (\mathcal{G}, S)) = |\Pr[\mathcal{A}^{T,\mathcal{F}} = 1] - \Pr[\mathcal{A}^{\mathcal{G},S} = 1]| < \varepsilon$$

The simulator S *is an interactive algorithm which has oracle access to* $\mathcal{G}$ *and runs in time at most* t_S*. The distinguisher* $\mathcal{A}$ *runs in time at most* $t_{\mathcal{A}}$ *and makes at most* q *queries. The total message blocks queried by* $\mathcal{A}$ *is at most* σ*.*

Briefly, the significance of *indifferentiability* property is described as follows: Suppose, an ideal primitive $\mathcal{G}$ (e.g. a *variable-input-length* random oracle) is indifferentiable from an algorithm T based on another ideal primitive $\mathcal{F}$ (e.g. a *fixed-input-length* random oracle). In such case, any cryptographic system $\mathcal{P}$ based on $\mathcal{G}$ is as secure as the $\mathcal{P}$ based on $T^{\mathcal{F}}$ (i.e., $T^{\mathcal{F}}$ replaces $\mathcal{G}$ in $\mathcal{P}$). See [16] for more on that.

Pictorial Description of Def. 2(Fig. 2). In the figure, five entities involved in Def. 2 are shown with an example. Suppose, the oracle Turing machine T, the ideal primitives $\mathcal{F}$, $\mathcal{G}$ are, respectively, a hash function H, random oracles ro and RO. The exchange of queries and responses is also shown in the figure. Note that it is forbidden to issue queries in the opposite directions. For example, the hash function H can send a query to ro and receive response, but never the other way round. In this setting, Def. 2 addresses the degree to which any computationally bounded adversary is unable to distinguish between Option 1 and Option 2. □

5.2 Indifferentiability Framework for FWP: Designing a Simulator S

In this section we describe the entities of Fig. 2 with respect to the hash function FWP: $\{0,1\}^{\leq 2^{64}} \rightarrow \{0,1\}^n$. The mode FWP is defined in Fig. 1. In the rest of the paper the H is understood to be the FWP hash function. The *fixed-input-length* random oracle $\mathsf{ro} : \{0,1\}^{r+n} \rightarrow \{0,1\}^{2n}$ is the compression function invoked by the FWP mode. The *variable-input-length* random oracle RO is defined as $\mathsf{RO} : \{0,1\}^{\leq 2^{64}} \rightarrow \{0,1\}^n$. The only remaining part to complete the indifferentiability framework is designing a simulator S. This section is devoted to that. The fifth entity of Fig. 2, which is an arbitrary distinguisher $\mathcal{A}$, is discussed in Sect. 5.3. We kick off with the notation.

Notation. Table 3 provides the notation useful to follow our *indifferentiability* results on the new hash function FWP. Note that the notation can be very easily adapted to any hash function based on a sequential mode of operation. □

Now we define a few terms – in relation to Fig. 1 and 2 – which will be used to arrive at our main indifferentiability results of Sect. 5.3.

Queries and lists. We now define various types of queries and lists (or arrays) that can potentially be used by a distinguisher to separate a hash function from a random oracle. The first assumption is that a distinguisher does not resubmit to an oracle a query whose response is already known. This is a valid assumption because, in our case, an identical oracle – any of FWP hash function, ro, RO and S of Fig. 2 – gives identical response to an identical query (it would be further clear when we shall concretely define the simulator S). Our next assumption is that, unless otherwise specified, a query is known to be submitted by the distinguisher. In the present case, we are not interested in queries submitted by the simulator S or by the hash function FWP. Now we define two special types of queries.

Definition 3 (*Short* and *long* query). *A query submitted to S or ro is defined as a* short *query. Similarly, a query submitted to FWP or RO is defined as the* long *query (see Fig. 2).*

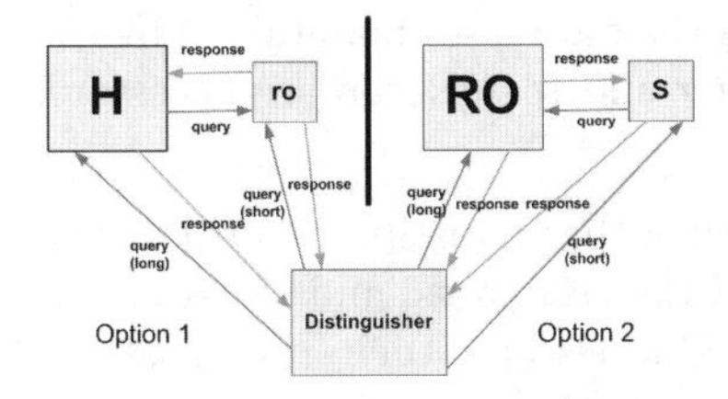

Fig. 2. The entities and their behavior involved in the *indifferentiability* framework of Def. 2; $T \equiv H$, $\mathcal{F} \equiv \mathsf{ro}$, $\mathcal{G} \equiv \mathsf{RO}$, $S \equiv$ simulator (see description above). In Sect. 5.2, H is the FWP hash function.

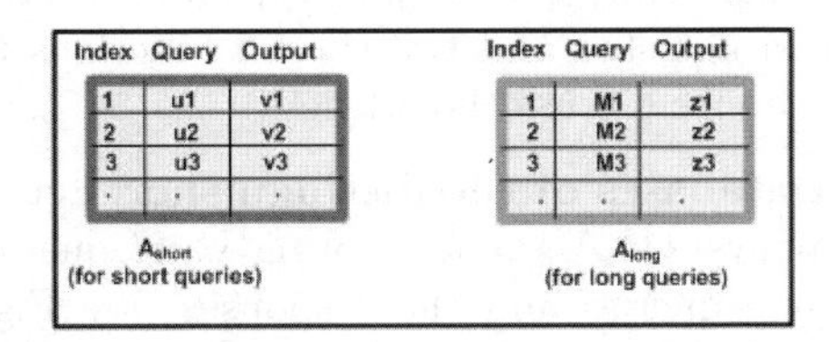

Fig. 3. Several databases maintained by the distinguisher

Table 3. The notation used in the *indifferentiability* framework for FWP (see Fig. 1)

Symbol	bit-length	Description
$\mathsf{A}_{\text{short}}$, A_{long}	-	Current query-response arrays
$\mathsf{A}_{\text{inter}}$	-	Array for intermediate query-responses
$A[i, i-1, \ldots j]$	-	Array (or bit-string) A truncated between indices i and j
$\mathcal{A}$	-	A distinguisher
$\mathcal{A}'$	-	Modification of the distinguisher $\mathcal{A}$
$\ell(M)$	-	Number of compression function calls to hash M
λ	0	Empty String
M	$\leq 2^{64}$	Message $M = m^1 m^2 \ldots m^{\ell(M)}$
m^k, $m^{\ell(M)}$	$r, r-n$	Messages of kth and $\ell(M)$th compression functions ($k < m^{\ell(M)}$)
MesgVer	-	Message verification algorithm
MesgRecon	-	Message reconstruction algorithm
q, σ	-	Maximum number of queries and blocks used by distinguisher
ro, RO	-	Random oracles
$\mathcal{S}$	-	Set of reconstructed messages given a *short query*
S	-	The simulator
$t_{\mathcal{A}}$, t_S	-	Time of $\mathcal{A}$ and S
$u^{k\prime}$	n	Chaining input to kth compression function ($k < m^{\ell(M)}$)
$u^{\ell(M)\prime\prime} \| u^{\ell(M)\prime}$	$2n$	Chaining input to $\ell(M)$th compression function
u^k, $u^{\ell(M)}$	$r+n, r+n$	Total input to kth and $\ell(M)$th compression functions
$v^{k\prime}$, $v^{k\prime\prime}$	n, n	Two halves of output from kth compression function
v^k	$2n$	Total output from kth compression function
z	n	Final hash value

At this time it is important to discuss a subclass of *short* and *long* queries known as *trivial* queries. For easy understanding, we try to introduce the notion without the rigors of mathematical notation as much as possible; however, our treatment is logically sound and foolproof. The motivation behind the determination of *trivial* queries is that their outputs are implied by the previous queries and their responses, no matter whether the distinguisher is interacting with Option 1 or Option 2 of Fig. 2. Therefore, *trivial* queries cannot be used to distinguish between two systems, even if they satisfy specific 'bad' conditions. Before we formally define *trivial* queries, some discussion on the databases maintained by the distinguisher and two special functions MesgVer and MesgRecon are necessary. We first discuss them briefly.

Databases of the distinguisher. Let us assume that a distinguisher uses two arrays: (1) $\mathsf{A}_{\text{short}}$ for storing *short* queries and the responses, and (2) A_{long} for *long* queries and the responses (see Fig. 3). Queries and their responses are indexed by the time they are submitted. Note that the simulator S can access $\mathsf{A}_{\text{short}}$ but not A_{long}.

Discussion on algorithms MesgVer and MesgRecon. Informally speaking, MesgVer is a function which takes two inputs – the current list $\mathsf{A}_{\text{short}}$, a *long query* M – to verify whether the *long query* M is a valid message for the hash mode FWP. What it essentially does is compute all compression function inputs

Algorithm 2. Message verification algorithm $\mathsf{MesgVer}(\cdot,\cdot)$

Input: Array $\mathsf{A}_{\text{short}}$, bit-string M ($|M| \leq 2^{64}$)
Output: A bit b
(See Table 3 and Fig. 1 for the notation.)
1: Set $b = 1$;
2: **for** $i = 1$ to $\ell(M)$ **do**
3: Compute u^i from m^i, v^{i-1}, v^{i-2};
4: **if** $\nexists v$ such that $(u^i, v) \in \mathsf{A}_{\text{short}}$ **then**
5: **return** $b = 0$;
6: **else**
7: Compute v^i using u^i and $\mathsf{A}_{\text{short}}$;
8: **end if**
9: **end for**
10: **return** b;

– u^1, u^2, ..., $u^{\ell(M)}$ – sequentially and checks whether they exist in $\mathsf{A}_{\text{short}}$. The MesgVer algorithm has been described in Algorithm 2.

The MesgRecon algorithm, in some sense, works in the opposite direction. It takes the current list $\mathsf{A}_{\text{short}}$ and a *short query* x as inputs and reconstructs a set of messages $\mathcal{S}$ such that each message $M \in \mathcal{S}$ is a valid message for FWP mode and, moreover, the input to the last compression function is x. The algorithm is described in Algorithm 3. □

Now we are ready to define the *trivial queries.*

Definition 4 (Trivial short query). *A* short query x *is a* trivial short query *if the following conditions hold:*

- *MesgRecon*$(\mathsf{A}_{short}, x) = \{M\}$.
- *The* M *has been queried previously as a* long *query (i.e.* $\exists v$ *such that* $(M, v) \in \mathsf{A}_{long}$*).*

Definition 5 (Trivial long query). *A* long query M *is a* trivial long query *if the following conditions hold:*

- *MesgVer*(A_{short}, M)*=1. Suppose the final input* $u^{\ell(M)}$ *computed in MesgVer*(A_{short}, M) *is the* ith *query in* A_{short}.
- *MesgRecon*$(\mathsf{A}_{short}[i-1, \ldots, 2, 1], u^{\ell(M)}) = \{M\}$.

The *nontrivial short* and *long* queries are obvious from the above definitions.

Definition 6 (Nontrivial queries). *A* short query x *is a* nontrivial short query *if it is not a* trivial short *query. Similarly, a* long query M *is a* nontrivial long query *if it is not a* trivial long *query.*

At this point it is useful to, once more, remember the motivation behind separating the trivial queries from all queries. The distinguisher may communicate with $(\mathsf{FWP}, \mathsf{ro})$ or (RO, S). Irrespective of whether it is communicating with $(\mathsf{FWP}, \mathsf{ro})$ or (RO, S), the responses of the trivial queries should be implied by the previous

Algorithm 3. Message reconstruction algorithm $\mathsf{MesgRecon}(\cdot,\cdot)$

Input: Array $\mathsf{A}_{\mathrm{short}}$, bit-string x ($|x| = r + n$)
Output: A set of reconstructed messages $\mathcal{S}$
Assumption: For simplicity we assume $r - n \geq 64$. This makes 64-bit length-encoding in the last
message block. If $r - n < 64$ then we need more than one block to determine the length. (See Table 3 and Fig. 1 for the notation.)

1: Compute $\ell(M)$ from $x[64, \ldots, 2, 1]$;
2: Break $x = u^{\ell(M)'} || v^{\ell(M)-1'} || m^{\ell(M)}$ such that $v^{\ell(M)-1'} = x[r, r-1, \ldots, r-n+1]$;
3: Construct $G = \{(u, v) \in \mathsf{A}_{\mathrm{short}} \mid v[n, \ldots, 2, 1] = v^{\ell(M)-1'}\}$;
4: **if** $|G| \neq 1$ **then**
5:     **return** $\mathcal{S} = \emptyset$;
6: **end if**
7: **for** $i = \ell(M) - 1$ to 1 **do**
8:     $m^i = u[r, \ldots, 2, 1]$;
9:     Compute $v^{i-1'} = u^{i+1'} \bigoplus v^{i''}$;
10:     **if** $i \neq 1$ **then**
11:         Construct $G = \{(u, v) \in \mathsf{A}_{\mathrm{short}} \mid v[n, \ldots, 2, 1] = v^{i-1'}\}$;
12:         **if** $|G| \neq 1$ **then**
13:             **return** $\mathcal{S} = \emptyset$;
14:         **end if**
15:     **else**
16:         **if** $u[r+n, \ldots, r+1] = IV$ and $v^{i-1'} = IV'$ **then**
17:             Compute $M = m^1 m^2 \ldots m^{\ell(M)}$;
18:             **return** $\mathcal{S} = \{M\}$;
19:         **else**
20:             **return** $\mathcal{S} = \emptyset$;
21:         **end if**
22:     **end if**
23: **end for**

query-responses. Therefore, the trivial queries do not help a distinguisher to differentiate between $(\mathsf{FWP}, \mathsf{ro})$ and (RO, S) (see Fig. 2). We have just concretely defined the trivial queries in Def. 4and 5. However, we still cannot say whether the *trivial* queries indeed fulfil the motivation until we prove the existence of a compatible simulator. Such a simulator S is described below.

Our design of indifferentiability framework is now complete, except establishing a property that shows, under trivial queries, both (FWP, ro) and (RO, S) behave identically, if they are supplied with identical $\mathsf{A}_{\mathrm{short}}$ and $\mathsf{A}_{\mathrm{long}}$. We capture this property in the following lemma.

Lemma 2. *Suppose, for a distinguisher $\mathcal{A}$, the lists A_{short} and A_{long} are identical for both (FWP, ro) and (RO, S) after the ith query. Then the following statements are true.*

1. *If M is the $(i+1)th$* trivial long query *then the probability distributions of $FWP^{\mathsf{ro}}(M)$ and $\mathsf{RO}(M)$ are identical.*

Algorithm 4. The simulator $S(\cdot)$

Input: *short query* x
Output: $2n$-bit string v
1: $\mathcal{S}$=$\mathsf{MesgRecon}(\mathsf{A}_{\text{short}}, x)$;
2: **if** $|S| = 1$ **then**
3: **return** $v = \mathsf{RO}(M)$; /* $\mathcal{S} = \{M\}$ */
4: **end if**
5: **return** $v = \mathsf{ro}(x)$;

2. *If x is the $(i+1)th$* trivial short query *then the probability distributions of $S(x)$ and $\mathsf{ro}(x)$ are identical.*

Proof. The proof is immediate from the construction of the simulator S which is described in Algorithm 4.

5.3 Bounding the Advantage of an Arbitrary Distinguisher

After designing the simulator S in the previous section, now we are left with the most important part of the paper: to compute an ε as a function of $(t_{\mathcal{A}}, t_S, q, \sigma)$ (see Def. 2). To that end, we first design an arbitrary oracle algorithm $\mathcal{A}$ (see Algorithm 5 in Appendix B) that separates $(\text{FWP}, \mathsf{ro})$ from (RO, S).

Algorithm 5 is characterized by two functions: (1) the $f_{query}(\cdot,\cdot)$ which computes the next query, and (2) the $f_{cond}(\cdot,\cdot)$ which decides whether the system is (FWP, ro) or (RO, S). Both the functions take the arrays $\mathsf{A}_{\text{short}}$, A_{long} as inputs. To bound the advantage of $\mathcal{A}$, we slightly modify $\mathcal{A}$ to design $\mathcal{A}'$ which is described in Algorithm 6 of Appendix B. We now discuss the algorithms briefly.

Discussion on Algorithm 5 and 6. Both $\mathcal{A}$ and $\mathcal{A}'$ have identical query function f_{query}. We only modify f_{cond} of $\mathcal{A}$ to design f'_{cond} of $\mathcal{A}'$. The additional parts of $\mathcal{A}'$ are placed within boxes in Algorithm 6. The algorithm $\mathcal{A}'$, in addition to $\mathsf{A}_{\text{short}}$ and A_{long}, uses an extra array $\mathsf{A}_{\text{inter}}$ which, using a function $\mathsf{MesgDecom}(M_i)$, stores all intermediate inputs and outputs for any *long query* M_i applied to FWP. Our main task is to define a suitable f'_{cond} such that the following inequality holds:

$$\max_{\mathcal{A}} |\Pr[\mathcal{A}(\text{FWP}, \mathsf{ro}) = 1] - \Pr[\mathcal{A}(\mathsf{RO}, S) = 1]| \leq \max_{\mathcal{A}'} \Pr[\mathcal{A}'(\text{FWP}, \mathsf{ro}) = 1] \quad (1)$$

where the maximum values of the right hand side and the left hand side are based on the suitable choices of (1) f_{query} and f_{cond}, and (2) f'_{cond} respectively. It is easy to show that he above inequality implies $\mathbf{Adv}_{\mathcal{A}}((\text{FWP}, \mathsf{ro}), (\mathsf{RO}, S)) \leq \max_{\mathcal{A}} \Pr[\mathcal{A}'(\text{FWP}, \mathsf{ro}) = 1]$. We now define a suitable f'_{cond} recursively.

Definition 7 (f'_{cond} of Algorithm 6). *The definition is divided into two complementary parts.*
(1) Let the ith query computed by f_{query} of $\mathcal{A}'$ be a nontrivial long query *denoted by M_i. Then $f'_{cond} = 1$ if one or more following conditions are satisfied.*

- *Collision between the final input for the current* long query M_i *and the final input for some previous* long query M_j. *That is,* $u_i^{\ell(M_i)} = u_j^{\ell(M_j)}$ *for some* $j < i$.
- *Collision between the final input for the current* long query M_i *and some intermediate input for some previous* long query M_j. *That is,* $u_i^{\ell(M_i)} = u_j^k$ *for some* $j \leq i$ *and* $k < \ell(M_j)$.
- *Collision between some intermediate input for the current* long query M_i *and the final input for some previous* long query M_j. *That is,* $u_i^k = u_j^{\ell(M_j)}$ *for some* $j < i$ *and* $k < \ell(M_i)$.
- *Collision between the final input for the current* long query M_i *and some previous* short query x_j. *That is,* $u_i^{\ell(M_i)} = x_j$ *for some* $j < i$.

Otherwise $f'_{cond} = 0$.
(2) Let the ith query computed by f_{query} *of* $\mathcal{A}'$ *be a* nontrivial short query *denoted by* x_i. *Then* $f'_{cond} = 1$ *if the following condition is satisfied.*

- *Collision between the current* short query x_i *and the final input for some previous* long query M_j. *That is,* $x_i = u_j^{\ell(M_j)}$ *for some* $j < i$.

Otherwise $f'_{cond} = 0$.

Now we state the following theorem.

Theorem 3. *Under Def. 7 of* f'_{cond} *the following inequality holds.*

$$\mathbf{Adv}_{\mathcal{A}}((FWP, \mathsf{ro}), (\mathsf{RO}, S)) \leq \max_{\mathcal{A}'} \Pr[\mathcal{A}'(FWP, \mathsf{ro}) = 1].$$

Proof. The theorem has been proved for a general domain extension in [3]. Note that, in the present case, the event $\mathcal{A}'(\text{FWP}, \mathsf{ro}) = 1$ is also an event invoked by $\mathcal{A}(\text{FWP}, \mathsf{ro})$ according to Def. 7 – exactly this event has been termed a Bad event for a GDE in [3]. So by using Theorem 1 of [3] we have our result.

In the remainder of the section we strive to obtain an upper bound ε on $\max_{\mathcal{A}'} \Pr[\mathcal{A}'(\text{FWP}, \mathsf{ro}) = 1]$. According to Theorem 3, ε is an upper bound on $\mathbf{Adv}_{\mathcal{A}}((\text{FWP}, \mathsf{ro}), (\mathsf{RO}, S))$ too.

We have two databases $\mathsf{A}_{\text{short}}$ and $\mathsf{A}_{\text{inter}}$ which essentially store all invocations to ro. Each element of $\mathsf{A}_{\text{short}}$ and $\mathsf{A}_{\text{inter}}$ is of the form (u, v) where $u \in \{0,1\}^{r+n}$ and $v \in \{0,1\}^{2n}$. We denote the ith pair by $\mathsf{A}_{\text{short}}(i) = (\mathsf{A}_{\text{short}}(i,1), \mathsf{A}_{\text{short}}(i,2))$ and $\mathsf{A}_{\text{inter}}(i) = (\mathsf{A}_{\text{inter}}(i,1), \mathsf{A}_{\text{inter}}(i,2))$.

Whenever we add a pair (u, v) to $\mathsf{A}_{\text{inter}}$ it corresponds to a pair (M, i) such that when we compute $FWP^{\mathsf{ro}}(M)$, the ith intermediate input, output are u and v respectively. Note, when $i = \ell(M)$ $FWP^{\mathsf{ro}}(M) = v[2n, 2n-1, \ldots n+1]$.

We define the following bad events. It mainly considers one of the following cases: (1) the unexpected collisions in the first or last half of the outputs of ro which are stored in one of the two databases $\mathsf{A}_{\text{short}}$ and $\mathsf{A}_{\text{inter}}$ during query-responses of $\mathcal{A}'$ and (2) collision on the least significant n bits of inputs of ro stored in $\mathsf{A}_{\text{inter}}$ with least significant n bits of inputs of ro stored in one of the two lists.

1. Type-1 bad. $\mathsf{A}_{\text{short}}$ vs. $\mathsf{A}_{\text{short}}$ for output collision: If $\mathsf{A}_{\text{short}}(i,2)[n,n-1,\ldots 1] = \mathsf{A}_{\text{short}}(i',2)[n,n-1,\ldots 1]$ or $\mathsf{A}_{\text{short}}(i,2)[2n,2n-1,\ldots n+1] = \mathsf{A}_{\text{short}}(i',2)[2n,2n-1,\ldots n+1]$ for some $i \neq i'$.
2. Type-2 bad. $\mathsf{A}_{\text{short}}$ vs. $\mathsf{A}_{\text{inter}}$ for output collision: If $\mathsf{A}_{\text{short}}(i,2)[n,n-1,\ldots 1] = \mathsf{A}_{\text{inter}}(i',2)[n,n-1,\ldots 1]$ or $\mathsf{A}_{\text{short}}(i,2)[2n,2n-1,\ldots n+1] = \mathsf{A}_{\text{inter}}(i',2)[2n,2n-1,\ldots n+1]$ for some i,i' such that the following is not true:

   $\mathsf{A}_{\text{inter}}(i',2)$ corresponds to the pair (M,j) and the computation of $FWP^{\mathsf{ro}}(M)$ up to $j-1$ intermediate input is already in the list $\{\mathsf{A}_{\text{short}}(r) : r \leq j-1\}$ and the jth intermediate input is $\mathsf{A}_{\text{inter}}(i',2)$.
3. Type-3 bad. $\mathsf{A}_{\text{inter}}$ vs. $\mathsf{A}_{\text{inter}}$ for output collision: If $\mathsf{A}_{\text{short}}(i,2)[n,n-1,\ldots 1] = \mathsf{A}_{\text{inter}}(i',2)[n,n-1,\ldots 1]$ or $\mathsf{A}_{\text{short}}(i,2)[2n,2n-1,\ldots n+1] = \mathsf{A}_{\text{inter}}(i',2)[2n,2n-1,\ldots n+1]$ for some i,i' such that the pairs corresponding to $\mathsf{A}_{\text{short}}(i,2)$ and $\mathsf{A}_{\text{inter}}(i',2)$ are not identical.
4. Type-4 bad. $\mathsf{A}_{\text{inter}}$ vs. both list for input collision: $\mathsf{A}_{\text{inter}}(i,1)[n,n-1,\ldots 1] = \mathsf{A}_{\text{inter}}(i',1)[n,n-1,\ldots 1]$ or $\mathsf{A}_{\text{inter}}(i,1)[n,n-1,\ldots 1] = \mathsf{A}_{\text{short}}(j,1)[n,n-1,\ldots 1]$ for some $i \neq i'$.

Lemma 3. *If f'_{cond} (see definition 7) returns 1 then at least one of the above four types of bad events occurs.*

Proof. The proof is immediate.

Note that for a short query we add one element to $\mathsf{A}_{\text{short}}$ and for a long query we add $\ell = \ell(M)$ elements to A_{int}. In total we update σ elements in two databases after q queries, where σ is the total number of blocks in all q queries (both short and long). We define bad^i to be one of the Bad events when we add ith element, $1 \leq i \leq \sigma$. The complement of the event is denoted by good^i. We estimate the following probability for different possible cases:

$$\Pr[\mathsf{bad}^i \mid \wedge_{j=1}^{i-1} \mathsf{good}^j].$$

We divide Bad events into two cases based on whether the ith update (u,v) is on $\mathsf{A}_{\text{short}}$ or on $\mathsf{A}_{\text{inter}}$.

- Case 1. Bad event on the update of $\mathsf{A}_{\text{short}}$: It can happen in two ways. Either the adversary correctly guesses u which already exists in $\mathsf{A}_{\text{inter}}$ or the outputs collide accidentally with one of the previous outputs stored in $\mathsf{A}_{\text{short}}$ or $\mathsf{A}_{\text{inter}}$ given that the guess is not correct. Note that if the guess is not correct then the input u is fresh and its output is uniformly distributed. The collision occurs in one of the n-bits with probability at most $2(i-1)/2^n$. Moreover, if u appears as jth intermediate input of $FWP^{\mathsf{ro}}(M)$ for some M such that (M,j) corresponding to an element of $\mathsf{A}_{\text{inter}}$ then the type-4 bad event occurs with probability $(i-1)/2^n$.

  Now, given that good event, all information to $\mathcal{A}$ so far, is independent of the internal computation. So the guess is correct with some internal input having the probability bounded by $(i-1)/2^n$. So

$$\Pr[\mathsf{bad}^i \mid \wedge_{j=1}^{i-1} \mathsf{good}^j] \leq 4(i-1)/2^n.$$

– Case 2. Bad event on the update of $\mathsf{A}_{\text{inter}}$: This probability can be bounded by random oracle collision probability as the input u freshly appears due to the good event. The following can be shown easily:
$\Pr[\text{type-4 } \mathsf{bad}^i | \wedge_{j=1}^{i-1} \mathsf{good}^j] \leq (i-1)/2^n, \Pr[\text{type-2 or 3 } \mathsf{bad}^i | \wedge_{j=1}^{i-1} \mathsf{good}^j] \leq 2(i-1)/2^n$ and hence $\Pr[\mathsf{bad}^i | \wedge_{j=1}^{i-1} \mathsf{good}^j] \leq 3(i-1)/2^n$.

Combining all these cases we obtain that the probability of bad event is at most $\sigma(\sigma-1)/2^{n-1}$.

Now we state our indifferentiability results.

Theorem 4. *The FWP hash is $(t_{\mathcal{A}}, t_S, q, \sigma, \varepsilon^*)$-indifferentiable in the random oracle model for the compression function, for any $t_{\mathcal{A}}$, with $t_S = \ell \cdot \mathrm{O}(q^2)$ and $\varepsilon^* = \sigma^2/2^{n-1}$ where the simulator S is described in Algorithm 4.*

6 Resistance of FWP against Some Recent Attacks

One of the most significant works in hash function cryptanalysis in recent times is the discovery of the multi-collision attack on the Merkle-Damgärd mode [11]. Using similar technique as multi-collision attack, Kelsey and Schneier devised another very influential attack that recovered 2nd preimage with work lower than the brute-force when long messages were used in the Merkle-Damgärd mode. These two attacks do not work on the FWP mode. Any variants of these types of attacks do not seem to work too on the FWP transform. The above two attacks crucially rely on the intermediate collisions on n-bit chaining values which cannot be adjusted by message modification. The FWP mode has $2n$-bit chaining value which also cannot be adjusted by message modification. Therefore, the complexity of such attacks on the FWP mode appears to be no less than the brute-force. The same argument applies to the FWP's resistance to Herding attack [12] too. In the full version of the paper we shall provide further evidence why the FWP should be able to resist all variants of the above attacks.

6.1 Comparison of the FWP with Other Modes

The highlight of the FWP mode is that the compression function takes n bits of previous chaining value while produces $2n$ bits of ouput. With the emergence of new types of attacks on the Merkle-Damgärd mode (see Sect. 6), it has been found necessary that the compression function output should be at least $2n$ bits to generate n bits of hash output. This type of constructions is known as the Wide-pipe mode propounded by Lucks [15] (see Fig. 4 (c)). Many modern hash functions use this type of mode [9] to defend against multi-collision type attacks. The main problem with that mode is that the $2n$ bits of chaining value, which are fed into the next compression function, reduce the bandwidth of the message-block and, thereby, impede the speed of the hash function. To skirt this difficulty the Sponge construction with $2n$ bits of compression function output has been proposed [2] (see Fig. 4(d)). Unfortunately this construction collapses as easily as Merkle-Damgärd mode against all the attacks of Sect. 6. Another competing proposal is the HAIFA [5] mode. The HAIFA mode can be viewed as a special Merkle-Damgärd mode with an additional counter injected into each compression function call. This extra counter is very useful to thwart the attacks described in [13,12]. However, the price to pay is the reduction of bandwidth for message in each compression function call, resulting in slower performance. In addition, the HAIFA mode is still as weak against Joux's multi-collision attack as the old Merkle-Damgärd mode.

7 Conclusion and Open Problems

This paper proposes a new sequential mode of operation, known as FWP, to hash messages of arbitrary length. The mode is *collision-resistance-preserving*, *preimage-resistance-preserving* and *indifferentiable* from a random oracle up to $\mathcal{O}(2^{n/2})$ compression function invocations. The mode is also shown to be more efficient than the *Wide-pipe* mode. Comparison of the FWP with other proposals has been outlined. No known attacks have so far been found in this mode, indicating that it may be possible to stretch the *indifferentiable* security bound of the mode beyond the birthday barrier of $2^{n/2}$. We leave this as an open problem.

Acknowledgements

The first author was supported in part by the National Science Foundation, Grant CNS-0937267.

References

1. Bellare, M., Ristenpart, T.: Multi-Property-Preserving Hash Domain Extension and the EMD Transform. In: Lai, X., Chen, K. (eds.) ASIACRYPT 2006. LNCS, vol. 4284, pp. 299–314. Springer, Heidelberg (2006)
2. Bertoni, G., Daemen, J., Peeters, M., Assche, G.V.: On the Indifferentiability of the Sponge Construction. In: Smart, N.P. (ed.) EUROCRYPT 2008. LNCS, vol. 4965, pp. 181–197. Springer, Heidelberg (2008)
3. Bhattacharya, R., Mandal, A., Nandi, M.: Indifferentiability Characterization of Hash Functions and Optimal Bounds of Popular Domain Extensions. In: Roy, B., Sendrier, N. (eds.) INDOCRYPT 2009. LNCS, vol. 5922, pp. 199–218. Springer, Heidelberg (2009)
4. Bhattacharyya, R., Mandal, A., Nandi, M.: NandiSecurity Analysis of the Mode of JH Hash Function. In: Hong, S., Iwata, T. (eds.) FSE 2010. LNCS, vol. 6147, pp. 168–191. Springer, Heidelberg (2010)
5. Biham, E., Dunkelman, O.: A framework for iterative hash functions – HAIFA. In: Second NIST Cryptographic Hash Workshop 2006 (2006)
6. Brassard, G. (ed.): CRYPTO 1989. LNCS, vol. 435. Springer, Heidelberg (1990)
7. Coron, J.-S., Dodis, Y., Malinaud, C., Puniya, P.: Merkle-Damgård Revisited: How to Construct a Hash Function. In: Shoup, V. (ed.) CRYPTO 2005. LNCS, vol. 3621, pp. 430–448. Springer, Heidelberg (2005)
8. Damgård, I.: A Design Principle for Hash Functions. In: Brassard [6], pp. 416–427

9. Rivest, R. et al.: The MD6 Hash Function 16
10. Hirose, S., Park, J.H., Yun, A.: A Simple Variant of the Merkle-Damgård Scheme with a Permutation. In: Kurosawa, K. (ed.) ASIACRYPT 2007. LNCS, vol. 4833, pp. 113–129. Springer, Heidelberg (2007)
11. Joux, A.: Multicollisions in Iterated Hash Functions: Application to Cascaded Constructions. In: Franklin, M. (ed.) CRYPTO 2004. LNCS, vol. 3152, pp. 306–316. Springer, Heidelberg (2004)
12. Kelsey, J., Kohno, T.: Herding Hash Functions and the Nostradamus Attack. In: Vaudenay, S. (ed.) EUROCRYPT 2006. LNCS, vol. 4004, pp. 183–200. Springer, Heidelberg (2006)
13. Kelsey, J., Schneier, B.: Second Preimages on n-Bit Hash Functions for Much Less than 2^n Work. In: Cramer, R. (ed.) EUROCRYPT 2005. LNCS, vol. 3494, pp. 474–490. Springer, Heidelberg (2005)
14. Klimov, A., Shamir, A.: New Cryptographic Primitives Based on Multiword T-Functions. In: Roy, B., Meier, W. (eds.) FSE 2004. LNCS, vol. 3017, pp. 1–15. Springer, Heidelberg (2004)
15. Lucks, S.: A failure-friendly design principle for hash functions. In: Roy, B. (ed.) ASIACRYPT 2005. LNCS, vol. 3788, pp. 474–494. Springer, Heidelberg (2005)
16. Maurer, U.M., Renner, R., Holenstein, C.: Indifferentiability, impossibility results on reductions, and applications to the random oracle methodology. In: Naor, M. (ed.) TCC 2004. LNCS, vol. 2951, pp. 21–39. Springer, Heidelberg (2004)
17. Merkle, R.C.: One Way Hash Functions and DES. In: Brassard [6], pp. 428–446
18. NIST. Secure hash standard. In: Federal Information Processing Standard, FIPS-180 (1993)
19. NIST. Secure hash standard. In: Federal Information Processing Standard, FIPS 180-1 (April 1995)
20. Rivest, R.: The MD4 message-digest algorithm. In: Menezes, A., Vanstone, S.A. (eds.) CRYPTO 1990. LNCS, vol. 537, pp. 303–311. Springer, Heidelberg (1991)
21. Rivest, R.: The MD5 message-digest algorithm. IETF RFC 1321 (1992)
22. Wagner, D.: A Generalized Birthday Problem. In: Yung, M. (ed.) CRYPTO 2002. LNCS, vol. 2442, pp. 288–303. Springer, Heidelberg (2002)
23. Wu, H.: The JH Hash Function. In: The 1st SHA-3 Candidate Conference

A Comparison of Modes

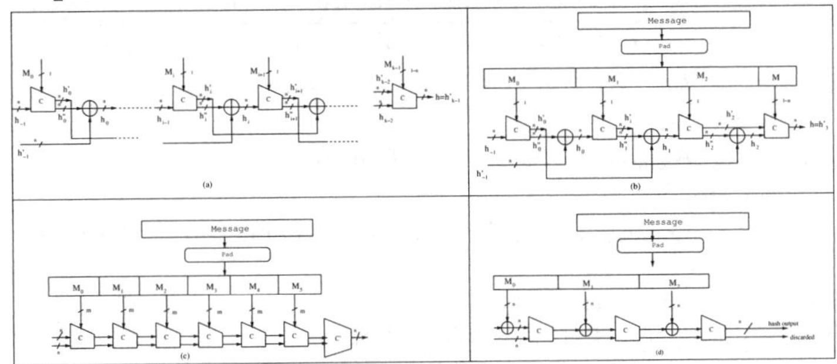

Fig. 4. (a) The FWP mode. (b) A four-block example of the FWP mode. (c) The Widepipe mode. (d) The Sponge mode.

B Arbitrary Distinguisher

Algorithm 5. An arbitrary distinguisher $\mathcal{A}(\cdot,\cdot)$ telling apart (FWP, ro) and (RO, S)

Input: An oracle $\mathcal{O}_{small} : \{0,1\}^{r+n} \rightarrow \{0,1\}^{2n}$ /* $\mathcal{O}_{small}$ is either ro or S */
An oracle $\mathcal{O}_{big} : \{0,1\}^{\leq 2^{64}} \rightarrow \{0,1\}^{n}$ /* $\mathcal{O}_{big}$ is either FWP or RO */
Output: A bit b

1: Initialize: $\mathsf{A}_{\text{short}}, \mathsf{A}_{\text{long}} = \emptyset$;
2: **for** $i = 1$ to q **do**
3: $(X_i, tag) = f_{query}(\mathsf{A}_{\text{short}}, \mathsf{A}_{\text{long}})$; /* $tag = 0, 1$ implies *long*, *short* queries */
4: **if** $tag = 0$ **then**
5: $M_i = X_i$, $z_i \longleftarrow \mathcal{O}_{big}(M_i)$;
6: $\mathsf{A}_{\text{long}} = \mathsf{A}_{\text{long}} \cup \{(M_i, z_i)\}$; /* Updating A_{long} */
7: $b = f_{cond}(\mathsf{A}_{\text{short}}, \mathsf{A}_{\text{long}})$;
8: **if** $b = 1$ **then**
9: **return** b; /* The system is (FWP, ro) */
10: **end if**
11: **end if**
12: **if** $tag = 1$ **then**
13: $x_i = X_i$, $y_i \longleftarrow \mathcal{O}_{small}(x_i)$;
14: $\mathsf{A}_{\text{short}} = \mathsf{A}_{\text{short}} \cup \{(x_i, y_i)\}$; /* Updating $\mathsf{A}_{\text{short}}$ */
15: $b = f_{cond}(\mathsf{A}_{\text{short}}, \mathsf{A}_{\text{long}})$;
16: **if** $b = 1$ **then**
17: **return** b; /* The system is (FWP, ro) */
18: **end if**
19: **end if**
20: **end for**
21: **return** $b = 0$; /* The system is (RO, S) */

Algorithm 6. Algorithm $\mathcal{A}'(\cdot,\cdot)$ computing Bad events

Input: An oracle $\mathcal{O}_{small} : \{0,1\}^{r+n} \to \{0,1\}^{2n}$, /* $\mathcal{O}_{small}$ is ro */
An oracle $\mathcal{O}_{big} : \{0,1\}^{\leq 2^{64}} \to \{0,1\}^{n}$ /* $\mathcal{O}_{big}$ is FWP */
Output: A bit b

1: Initialize: $\mathsf{A}_{\text{short}}$, $\mathsf{A}_{\text{long}} = \emptyset$, $\boxed{\mathsf{A}_{\text{inter}} = \emptyset,}$ Bad=0;
2: **for** $i = 1$ to q **do**
3: $(X_i, tag) = f_{query}(\mathsf{A}_{\text{short}}, \mathsf{A}_{\text{long}})$; /* $tag = 0, 1$ implies *long*, *short* queries */
4: **if** $tag = 0$ **then**
5: $M_i = X_i$, $z_i \longleftarrow \mathcal{O}_{big}(M_i)$;
6: $\mathsf{A}_{\text{long}} = \mathsf{A}_{\text{long}} \cup \{(M_i, z_i)\}$; /* Updating A_{long} */
7: $\boxed{\mathsf{A}_{\text{inter}} = \mathsf{A}_{\text{inter}} \cup \mathsf{MesgDecom}(M_i);}$ /* Updating A_{inter} */
8: $\boxed{b = f'_{cond}(\mathsf{A}_{\text{short}}, \mathsf{A}_{\text{long}}, A_{inter});}$ /* Checking condition for Bad event */
9: **if** $b = 1$ **then**
10: **return** b; /* Bad event */
11: **end if**
12: **end if**
13: **if** $tag = 1$ **then**
14: $x_i = X_i$, $y_i \longleftarrow \mathcal{O}_{small}(x_i)$;
15: $\mathsf{A}_{\text{short}} = \mathsf{A}_{\text{short}} \cup \{(x_i, y_i)\}$; /* Updating $\mathsf{A}_{\text{short}}$ */
16: $\boxed{b = f'_{cond}(\mathsf{A}_{\text{short}}, \mathsf{A}_{\text{long}}, A_{inter});}$ /* Checking condition for Bad event */
17: **if** $b = 1$ **then**
18: **return** b; /* Bad event */
19: **end if**
20: **end if**
21: **end for**
22: **return** $b = 0$; /* Good event */

New Boomerang Attacks on ARIA

Ewan Fleischmann[1], Christian Forler[2], Michael Gorski[1], and Stefan Lucks[1]

[1] Bauhaus-University Weimar, Germany
{Ewan.Fleischmann,Michael.Gorski,Stefan.Lucks}@uni-weimar.de
[2] Sirrix AG, Germany
c.forler@sirrix.com

Abstract. ARIA [5] is a block cipher proposed at ICISC'03. Its design is very similar to the Advanced Encryption Standard (AES). The authors propose that on 32-bit processors, the encryption speed is at least 70% of that of the AES. It is claimed to offer a higher security level than AES. In this paper we present three new attacks of reduced round ARIA which shows some weaknesses of the cipher. Moreover, our attacks have the lowest memory complexity compared to existing attacks on ARIA.

Keywords: block ciphers, differential cryptanalysis, boomerang attack, ARIA.

1 Introduction

The ARIA block cipher [5] was presented at ICISC'03. Its design is very similar to the advanced encryption standard (AES/Rijndael) [4]. ARIA employs two kinds of S-Boxes and two types of substitution layers which are different between even and odd rounds. They skip using a MixColumns operation and use an 16×16 binary matrix with branch number 8 in their diffusion layer. The authors propose that ARIA can increase the efficiency in 8-bit and 32-bit software implementations in comparison to AES. Moreover, they claim to have better security against all existing attacks on block ciphers.

Wu et al. [9] showed that there exist good impossible differentials to break up to 6 rounds of ARIA. Later Li et al. [7] presented also some impossible differential attacks of up to 6 rounds of ARIA. In this paper we apply another technique on ARIA which is called the boomerang attack [8].

The boomerang attack is a strong extension to differential cryptanalysis [2] in order to break more rounds than plain differential attacks can, since the cipher is treated as a cascade of two sub-ciphers, using short differentials in each sub-cipher. These differentials are combined in an adaptive chosen plaintext and ciphertext attack to exploit properties of the cipher that have a high probability. Biryukov [3] proposed a similar boomerang attack on the AES-128 which can break up to 5 and 6 out of 10 rounds. Table 1 summarizes the existing results on ARIA and our new attacks.

The paper is organized as follows: In Section 2 we give a brief description of the ARIA block cipher. In Section 3 we describe the boomerang attack.

G. Gong and K.C. Gupta (Eds.): INDOCRYPT 2010, LNCS 6498, pp. 163–175, 2010.

Table 1. Comparison of attacks on ARIA

Attack	# Rounds	Data	Memory	Time	Source
Impossible Differential	5	$2^{71.3}$ CP	2^{72} mem*	$2^{71.6}$	[7]
Meet-in-the-Middle Attack	5	2^{25} CP	$2^{122.5}$ mem*	$2^{65.4}$	[10]
Boomerang Attack	5	2^{109} ACPC	2^{57} mem	2^{110}	Sec. 4
Integral Attack	5	$2^{27.5}$ CP	$2^{27.5}$ mem*	$2^{76.7}$	[6]
Impossible Differential	6	2^{121} CP	2^{121} mem*	2^{112}	[9]
Impossible Differential	6	$2^{120.5}$ CP	2^{121} mem*	$2^{104.5}$	[7]
Impossible Differential	6	2^{113} CP	2^{113} mem*	$2^{121.6}$	[7]
Integral Attack	6	$2^{124.4}$ CP	$2^{124.4}$ mem*	$2^{172.4}$	[6]
Meet-in-the-Middle Attack	6	2^{56} CP	$2^{122.5}$ mem*	$2^{121.5}$	[10]
Boomerang Attack	6	2^{128} KP	2^{56} mem	2^{108}	Sec. 5
Meet-in-the-Middle Attack	7	2^{120} CP	2^{187} mem*	$2^{185.3}$	[10]
Boomerang Attack	7	2^{128} KP	2^{184} mem	2^{236}	Sec. 5
Meet-in-the-Middle Attack	8	2^{56} CP	2^{252} mem*	$2^{251.6}$	[10]

CP: Chosen Plaintexts, KP: Known Plaintexts, ACPC: Adaptive Chosen Plaintexts and Ciphertexts, mem: memory usage in blocks.
* We estimated the memory usage of this attack since it was not mentioned in the paper.

In Section 4, we present a boomerang attack on 5-round ARIA. In Section 5 we propase an attack on a 6-rounds as well as on 7-rounds of ARIA. We conclude the paper in Section 6.

2 Description of ARIA

ARIA is a substitution and permutation network whose structure is based on the advanced encryption standard (AES) [4]. ARIA uses data blocks of 128 bits with an 128, 192 or 256-bit key. A different number of rounds is used depending on the length of the key, 10, 12, or 14 rounds when a 128, 192 or 256-bit key is used, respectively. ARIA contains two kinds of S-boxes and two types of substitution layers which are different between even and odd rounds. The diffusion layer of ARIA uses a 16×16 binary matrix with a branch number 8.[1] ARIA is claimed to be more efficient in 8-bit and 32-bit software implementations than the AES. The plaintexts are treated as a 4 x 4 byte matrix, which is called the state.

[1] A differential *branch number* of a linear transformation M is given by

$$\min_{a \neq 0}\{w_b(a) + w_b(M(a))\}$$

where $w_b(a)$ is the number of non-zero elements of the vector a. We always mean differential branch number whenever we write branch number [4].

A round applies three operations to the state: **Substitution Layer (SL).** ARIA uses two S-Boxes S_1 and S_2 and also their inverse S_1^{-1}, S_2^{-1}, where S_1 is the same S-Box used for the AES. Each S-Box is defined to be an affine transformation of the inversion function over $\mathrm{GF}(2^8)$.

$$S_1, S_2 : GF(2^8) \rightarrow GF(2^8),$$

$$S_1 : x \mapsto A \cdot x^{-1} \oplus a,$$

where

$$A = \begin{pmatrix} 1\,0\,0\,0\,1\,1\,1\,1 \\ 1\,1\,0\,0\,0\,1\,1\,1 \\ 1\,1\,1\,0\,0\,0\,1\,1 \\ 1\,1\,1\,1\,0\,0\,0\,1 \\ 1\,1\,1\,1\,1\,0\,0\,0 \\ 0\,1\,1\,1\,1\,1\,0\,0 \\ 0\,0\,1\,1\,1\,1\,1\,0 \\ 0\,0\,0\,1\,1\,1\,1\,1 \end{pmatrix} \quad \text{and} \quad a = \begin{pmatrix} 1 \\ 1 \\ 0 \\ 0 \\ 0 \\ 1 \\ 1 \\ 0 \end{pmatrix}.$$

$$S_2 : x \mapsto B \cdot x^{247} \oplus b,$$

where

$$B = \begin{pmatrix} 0\,1\,0\,1\,1\,1\,1\,0 \\ 0\,0\,1\,1\,1\,1\,0\,1 \\ 1\,1\,0\,1\,0\,1\,1\,1 \\ 1\,0\,0\,1\,1\,1\,0\,1 \\ 0\,0\,1\,0\,1\,1\,0\,0 \\ 1\,0\,0\,0\,0\,0\,0\,1 \\ 0\,1\,0\,1\,1\,1\,0\,1 \\ 1\,1\,0\,1\,0\,0\,1\,1 \end{pmatrix} \quad \text{and} \quad b = \begin{pmatrix} 0 \\ 1 \\ 0 \\ 0 \\ 0 \\ 1 \\ 1 \\ 1 \end{pmatrix}.$$

ARIA has two types of S-box layers for even and odd rounds as shown in Figures 2 and 3. Type 1 is used in the odd rounds and type 2 is used in the even rounds.

Before the first round, an initial ARK operation is applied and the DL operation is omitted in the last round. The bytes coordinates of a 4 x 4 state matrix are labeled as in Figure 1.

0	4	8	12
1	5	9	13
2	6	10	14
3	7	11	15

Fig. 1. Byte coordinates of a 4 x 4 state matrix of ARIA

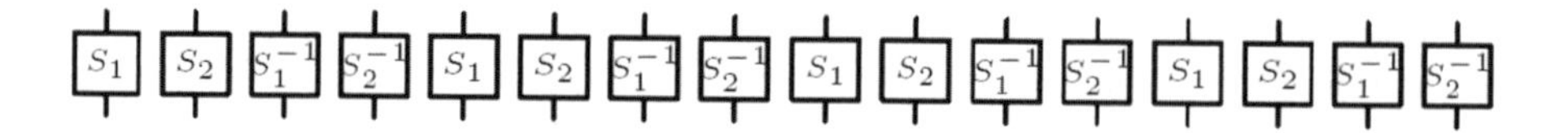

Fig. 2. S-box layer SL_1

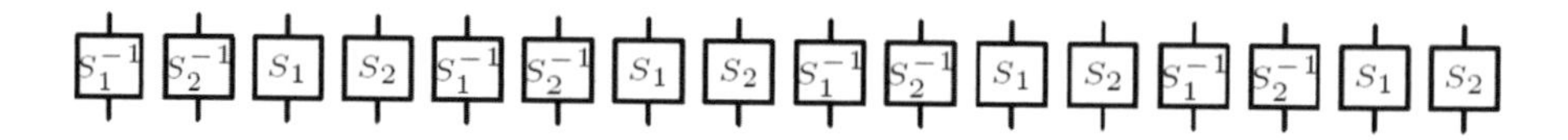

Fig. 3. S-box layer SL_2

Diffusion Layer (DL). A mapping $\mathrm{GF}(2^8)^{16} \rightarrow \mathrm{GF}(2^8)^{16}$ is performed which is given by

$$(x_0, x_1, \ldots, x_{15}) \mapsto (y_0, y_1, \ldots, y_{15}),$$

where

$$\begin{aligned}
y_0 &= x_3 \oplus x_4 \oplus x_6 \oplus x_8 \oplus x_9 \oplus x_{13} \oplus x_{14}, \\
y_1 &= x_2 \oplus x_5 \oplus x_7 \oplus x_8 \oplus x_9 \oplus x_{12} \oplus x_{15}, \\
y_2 &= x_1 \oplus x_4 \oplus x_6 \oplus x_{10} \oplus x_{11} \oplus x_{12} \oplus x_{15}, \\
y_3 &= x_0 \oplus x_5 \oplus x_7 \oplus x_{10} \oplus x_{11} \oplus x_{13} \oplus x_{14}, \\
y_4 &= x_0 \oplus x_2 \oplus x_5 \oplus x_8 \oplus x_{11} \oplus x_{14} \oplus x_{15}, \\
y_5 &= x_1 \oplus x_3 \oplus x_4 \oplus x_9 \oplus x_{10} \oplus x_{14} \oplus x_{15}, \\
y_6 &= x_0 \oplus x_2 \oplus x_7 \oplus x_9 \oplus x_{10} \oplus x_{12} \oplus x_{13}, \\
y_7 &= x_1 \oplus x_3 \oplus x_6 \oplus x_8 \oplus x_{11} \oplus x_{12} \oplus x_{13}, \\
y_8 &= x_0 \oplus x_1 \oplus x_4 \oplus x_7 \oplus x_{10} \oplus x_{13} \oplus x_{15}, \\
y_9 &= x_0 \oplus x_1 \oplus x_5 \oplus x_6 \oplus x_{11} \oplus x_{12} \oplus x_{14}, \\
y_{10} &= x_2 \oplus x_3 \oplus x_5 \oplus x_6 \oplus x_8 \oplus x_{13} \oplus x_{15}, \\
y_{11} &= x_2 \oplus x_3 \oplus x_4 \oplus x_7 \oplus x_9 \oplus x_{12} \oplus x_{14}, \\
y_{12} &= x_1 \oplus x_2 \oplus x_6 \oplus x_7 \oplus x_9 \oplus x_{11} \oplus x_{12}, \\
y_{13} &= x_0 \oplus x_3 \oplus x_6 \oplus x_7 \oplus x_8 \oplus x_{10} \oplus x_{13}, \\
y_{14} &= x_0 \oplus x_3 \oplus x_4 \oplus x_5 \oplus x_9 \oplus x_{11} \oplus x_{14}, \\
y_{15} &= x_1 \oplus x_2 \oplus x_4 \oplus x_5 \oplus x_8 \oplus x_{10} \oplus x_{15}.
\end{aligned}$$

Round Key Addition (ARK). The round keys are derived from the key using the key schedule which uses a 3-round 256-bit Feistel cipher. We skip its description since we do not use it in our attack. We refer the reader to [5] for more details.

3 The Boomerang Attack

The boomerang attack was introduced by Wagner [8]. It is a strong extension of differential cryptanalysis which uses two differentials that cover half of the cipher each instead of one differential for the whole cipher. In general the less rounds a differential covers, the higher its probability will be. Using two highly probable short differential which have a higher probability in combination than a long differential will decrease the complexity of an attack based on these differential cryptanalysis in some cases. This is true for the AES where boomerang attacks appear to be much stronger than the differential cryptanalysis.

Two plaintexts (P, P') are called a *pair*, while two pairs (P, P', O, O') are called a *quartet*. We split the boomerang attack into two steps: The *boomerang distinguisher step* and the *key recovery step*. The boomerang distinguisher is used to find all plaintexts sharing a desired difference that depends on the choice of the differential. These plaintexts are used in the key recovery step afterwards to recover subkey bits for the initial round key.

Distinguisher Step

During the *distinguisher step* we treat the cipher as a cascade of two sub-ciphers $E_K(P) = E1_K(E0_K(P))$, where K is the key used for encryption and decryption. Since we always use the same key we omit the key K and write $E(P) = E1(E0(P))$ instead. We assume that the differential $\alpha \rightarrow \beta$ for $E0$ occurs with probability p, while the differential $\gamma \rightarrow \delta$ for $E1$ occurs with probability q, where α, β, γ and δ are differences of intermediate encryption values. The backward direction $E0^{-1}$ and $E1^{-1}$ of the differential for $E0$ and $E1$ are denoted by $\alpha \leftarrow \beta$ and $\gamma \leftarrow \delta$ and occur with probability p and q respectively. The attack works as follows:

1. Choose a pool of s plaintexts P_i, $i \in \{1, \ldots, s\}$ uniformly at random and compute a pool $P'_i = P_i \oplus \alpha$.
2. Ask for the encryption of P_i, i.e., $C_i = E(P_i)$ and ask for the encryption of P'_i, i.e., $C'_i = E(P'_i)$.
3. Compute the new ciphertexts $D_i = C_i \oplus \delta$ and $D'_i = C'_i \oplus \delta$.
4. Ask for the decryption of D_i, i.e., $O_i = E^{-1}(D_i)$ and ask for the decryption of D'_i, i.e., $O'_i = E^{-1}(D'_i)$.
5. For each pair (O_i, O'_i), $i \in \{1, \ldots, s\}$ check if $O_i \oplus O'_i$ is equal to α and store the quartet (P_i, P'_i, O_i, O'_i) into a set Θ if true.

A pair (P_i, P'_i), $i \in \{1, \ldots, s\}$ with the difference α satisfies the differential $\alpha \rightarrow \beta$ with probability p. The output of $E0$ is A_i and A'_i, i.e., $E0(P_i) = A_i$ and $E0(P'_i) = A'_i$ have a certain difference $\beta = A_i \oplus A'_i$ with probability p. Using the ciphertexts C_i and C'_i we can compute the new ciphertexts $D_i = C_i \oplus \delta$ and $D'_i = C'_i \oplus \delta$. Let $B_i = E1^{-1}(D_i)$ and $B'_i = E1^{-1}(D'_i)$ are the decryption of D_i and D'_j with $E1^{-1}$ $i \in \{1, \ldots s\}$. A difference δ leads to a difference γ after passing $E1^{-1}$ with probability q. Since $\delta = C_i \oplus D_i$ and $\delta = C'_i \oplus D'_i$ we know that $\gamma = A_i \oplus B_i$ and $\gamma = A'_i \oplus B'_i$ with probability q^2. Since we also know, that $A_i \oplus A'_i = \beta$ with probability p, it follows that $(A_i \oplus B_i) \oplus (A_i \oplus A'_i) \oplus (A'_i \oplus B'_i) = \gamma \oplus \beta \oplus \gamma = \beta = (B_i \oplus B'_i)$ holds with probability $p \cdot q^2$. A β difference leads to an α difference after passing the differential $E0^{-1}$ with probability p. Thus, a pair of plaintexts (P_i, P'_i) with $P_i \oplus P'_i = \alpha$ generates a new pair of plaintexts

(O_i, O'_i) where $O_i \oplus O'_i = \alpha$ with probability $p^2 \cdot q^2$. A quartet containing these two pairs is defined as:

Definition 1. *A quartet (P_i, P'_i, O_i, O'_i) which satisfies*

$$\begin{aligned}P_i \oplus P'_i &= \alpha = O_i \oplus O'_i,\\ A_i \oplus A'_i &= \beta = B_i \oplus B'_i,\\ A_i \oplus B_i &= \gamma = A'_i \oplus B'_i,\\ C_i \oplus D_i &= \delta = C'_i \oplus D'_i,\end{aligned}$$

*is called a **right quartet** which occurs with probability $Pr_c = p^2 \cdot q^2$. A quartet (P_i, P'_i, O_i, O'_i) which only satisfies the condition $P \oplus P'_i = \alpha = O_i \oplus O'_i$ is called a **wrong quartet**.*

Figure 4 displays the structure of the boomerang step. Any attacker who applies a boomerang distinguisher does not know the internal states A_i, A'_j, B_i, B'_j, since he can only apply a chosen plaintext and ciphertext attack on the cipher. The set Θ which is the output of the boomerang distinguisher, therefore contains right and wrong boomerang quartets. It is impossible to form another distinguisher which separates the right and the wrong boomerang quartets, since the interior differences β and γ cannot be computed.

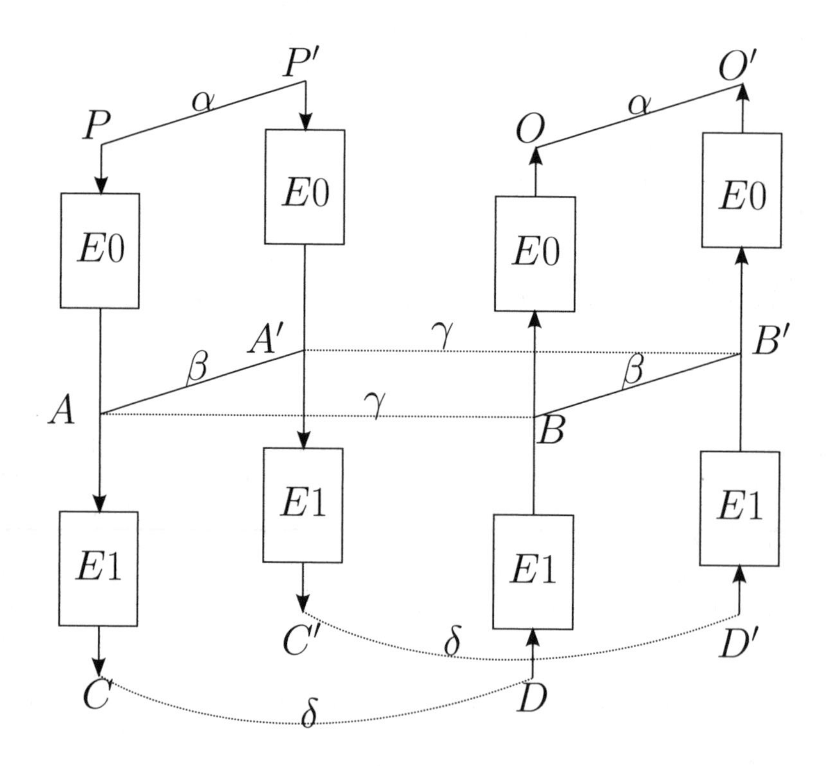

Fig. 4. The boomerang attack

Key Recovery Step

The second step of the boomerang attack is the *key recovery step*. From now on, an adversary operates on the set Θ that was stored by the boomerang distinguisher. Let k be some key bits of the last round keys derived from the cipher keys K. Let $d_k(C)$ be the one round partial decryption of C under the key k (the size of k is usually much smaller than the size of K). The key recovery step works as follows:

- For each key-bit combination of k
  1. Initialize a counter for each key-bit combination with zero.
  - For all quartets (P_i, P'_j, O_i, O'_j) stored in Θ
    2. Ask for the encryption of P_i, P'_i, O_i, O'_i and obtain the ciphertext quartet (C_i, C'_j, D_i, D'_j) respectively. Decrypt the ciphertexts C_i, C'_j, D_i, D'_j, i.e., $\bar{C}_i = d_k(C_i), \bar{C}'_j = d_k(C'_j), \bar{D}_i = d_k(D_i)$ and $\bar{D}'_j = d_k(D'_j)$.
    3. Test whether the differences $\bar{C}_i \oplus \bar{D}_i$ and $\bar{C}'_j \oplus \bar{D}'_j$ have a desired difference an attacker would expect depending on the differential being used. Increase a counter for the used key-bits if the difference is fulfilled in both pairs.
4. Output the key-bits k with the highest counter as the right one and perform an exhaustive key search on the remaining key bits.

Four cases can be differentiated in Step 3, since Θ contains right and wrong quartets and the key-bit combination k can either be right or wrong. A right quartet encrypted with the right key bits will have the desired difference needed to pass the test in Step 3 with probability 1. Hence, the counter for the right key bits is increased. The three other cases are: a right quartet is used with false key bits (Pr_{cK_f}), a wrong quartet is used with the correct key-bits (Pr_{fK_c}) or a wrong quartet is used with a false key-bit combination (Pr_{fK_f}). The probabilities in the three later cases are very small and for our analysis only the biggest one counts. For simplicity we set

$$Pr_{cK_f} = Pr_{fK_c} = Pr_{fK_f} =: Pr_{\mathit{filter}}.$$

The probability that a quartet in one of the three undesirable cases is counted for a certain key bit combination is Pr_{filter}. The differentials have to be chosen such that the counter of the correct key bits is significantly higher than the counter of each false key bit combination. If the differentials have a high probability the key recovery step outputs the correct key-bits in Step 4 with a high probability much faster than exhaustive search.

4 A Boomerang Attack on 5-Round ARIA

In this section we mount a boomerang attack on 5-round ARIA-128. Note that this attack works on the other versions as well. The cipher is treated as $E(P) =$

$E1(E0(P))$, where a differential for $E0$ containing rounds 1 to 3 and a differential for $E1$ is covering rounds 4 to 5. We apply a key recovery attack to retrieve 56 key-bits of the first round. The notations used in our attack are defined as:

- P_i, O_i are plaintexts.
- C_i, D_i are ciphertexts.
- a is a known non-zero byte difference.
- $*$ is a non-zero byte differences.
- ? is an unknown byte differences.

4.1 The Differential for $E0$

Considering the S-box being used there are 126 values which occur with probability 2^{-7}, one with probability 2^{-6} and 129 with probability 0. We choose the b difference such that it transforms into an a difference with probability 2^{-6}. Thus, the non-zero differences in bytes 3, 4, 6, 8, 9, 13 and 14 in the input difference α of the differential for $E0$ transforms into an a difference in bytes 3, 4, 6, 8, 9, 13 and 14 with probability 2^{-42}. DL_1 then leaves an a difference in byte 0, while the remaining bytes become zero. Since ARK is linear it does not alter this difference. SL_2 produces a non-zero difference in byte 0 and DL_2 spreads this difference in bytes 3, 4, 6, 8, 9, 13 and 14. At the end of the differential we obtain a difference called β_{out} where all the 16 bytes of the state difference are unknown. We have to guarantee that the differential for $E0^{-1}$ starts with a difference which is equal to β_{out} in order to get the correct α difference after $E0^{-1}$. We discuss this below in more details. The probability of the differential for $E0$, i.e., the transformation of an α difference into a β_{out} difference, is given by

$$Pr(\alpha \rightarrow \beta_{out}) = 2^{-42}.$$

The differential $E0$ is shown in Figure 5.

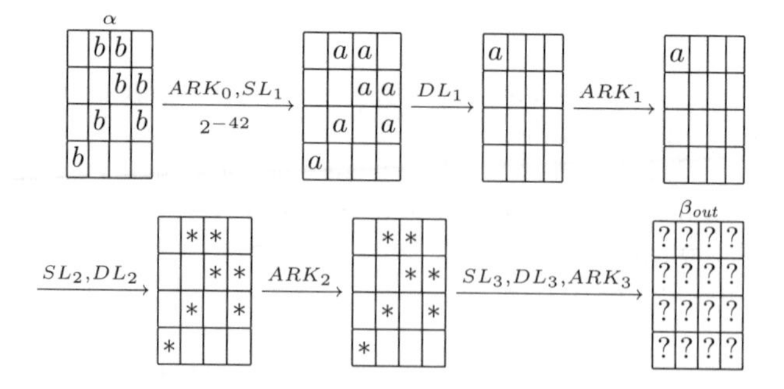

Fig. 5. The differential for $E0$

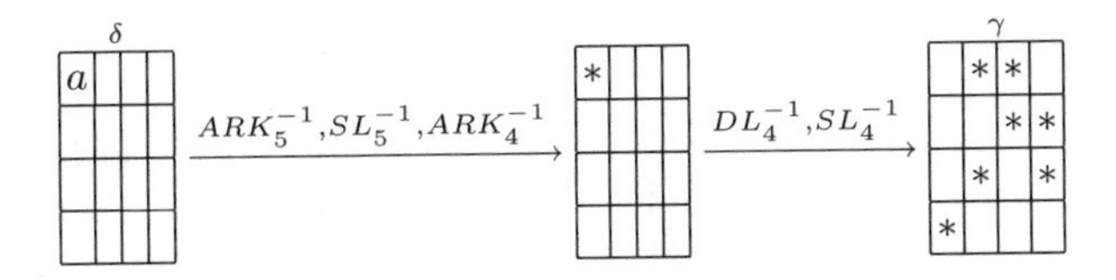

Fig. 6. The differential for $E1^{-1}$

4.2 The Differential for $E1^{-1}$

The ciphertext difference δ consists of one a difference in byte 0 and a zero difference in the remaining bytes. The non-zero difference remains after the inverse of round 5. The DL_4^{-1} operation spreads this non-zero difference to bytes 3, 4, 6, 8, 9, 13 and 14. Remember that there is no DL_5 operation in the last round of ARIA. We call the γ the output difference of the differential. The probability of $E1^{-1}$ is $\Pr(\gamma \leftarrow \delta) = 1$. The differential $E1^{-1}$ is shown in Figure 6.

4.3 The Differential for $E0^{-1}$

For the following steps we need that the output difference β_{out} of the differential for $E0$ is equal to the input difference β_{in} for the differential for $E0^{-1}$. Note that β_{in} and β_{out} are not only equal in the same positions of non-zero differences but are also equal in each byte. We compute the probability that this actually happens. From the boomerang condition inside the cipher for two differences γ_1 and γ_2 we know that

$$\beta_{out} \oplus \gamma_1 \oplus \gamma_2 = \beta_{in}$$

holds with some probability. When γ_1 and γ_2 are equal in all the bytes, we simply write γ. We compute the probability for that to occur below. Thus, the above condition reduces to:

$$\beta_{out} \oplus \gamma \oplus \gamma = \beta_{out} = \beta_{in} \tag{1}$$

Using the differentials above, the differences β_{in} and β_{out} are equal with probability 2^{-56}. This is the probability that the 7 non-zero bytes in γ_1 are equal to the 7 non-zero bytes in γ_2.

Let A, A', B, B' be the internal state after SL_3 in the forward direction when encrypting P, P', O, O', respectively. The notation from Figure 4 is used. Since DL is linear γ can be expressed as

$$\gamma = K_3 \oplus DL_3(A) \oplus K_3 \oplus DL_3(B) = DL_3(A \oplus B) \tag{2}$$

and as

$$\gamma = K_3 \oplus DL_3(A') \oplus K_3 \oplus DL_3(B') = DL_3(A' \oplus B'). \tag{3}$$

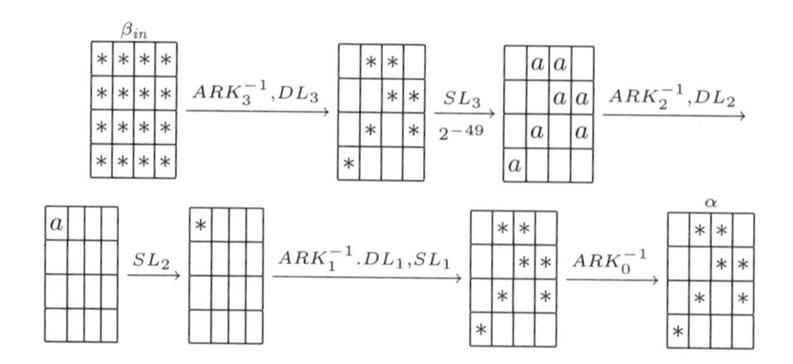

Fig. 7. The differential for $E0^{-1}$

Equations (2) and (3) can be combined, which leaves $A \oplus A' = B \oplus B'$. In other words, DL_3 can be undone with probability 1 due to the boomerang condition (1). This means that we know exactly that after DL_3 in the backward direction bytes 3, 4, 6, 8, 9, 13, and 14 are non-zero while the remaining bytes are zero. There are several cases for which an a difference in bytes 3, 4, 6, 8, 9, 13, and 14 occurs after SL_3. There are $7 \cdot 127$ cases, each with probability 2^{-43}, there is one case with probability 2^{-42} and there are 126^7 cases, each with probability 2^{-49}. Thus on average, after SL_3 an a difference in bytes 3, 4, 6, 8, 9, 13, and 14 occurs with probability $(2^{-6.93})^7 = 2^{-48.79}$. DL_2 outputs an a difference in byte 0 and a zero difference in the remaining bytes. SL_2 then transforms the a difference in byte 0 into a non-zero difference, which is spread into the bytes 3, 4, 6, 8, 9, 13, and 14 after DL_1. The output difference α of the differential for $E0^{-1}$ contains these non-zero and zero differences. The differential for $E0^{-1}$ has the probability $\Pr(\alpha \leftarrow \beta_{in}) \approx 2^{-49}$ to occur. It is shown in Figure 7.

4.4 The Attack

The adversary first collects data and stores the filtered data in the set ϕ. A key-search is then applied to the remaining quartets in ϕ in order to find 56 bits of K_0. Let k_0 be a 56-bit subkey in the position of bytes 3, 4, 6, 8, 9, 13, and 14. Let $e_{0,k}(X)$ be the partial encryption of X under the subkey k before DL_1 is applied. The attack is as follows:

1. Choose 2^{53} structures S_j, $j \in \{1, 2, \ldots, 2^{53}\}$ each consists of 2^{56} plaintexts $P_{i,j}$, $i \in \{1, 2, \ldots, 2^{56}\}$ which have all possible values in seven bytes (3, 4, 6, 8, 9, 13, and 14). Ask for the encryption of the $P_{i,j}$ to obtain the ciphertexts $C_{i,j}$, i.e., $C_{i,j} = E(P_{i,j})$. Compute $P'_{i,j} = P_{i,j} \oplus \alpha$ and ask for the encryption of $P'_{i,j}$ to obtain $C'_{i,j}$, i.e., $C'_{i,j} = E(P'_{i,j})$.
2. For each ciphertext $C_{i,j}$ compute a new ciphertext $D_{i,j} = C_{i,j} \oplus \delta$, where δ is a fixed 128-bit value with a non-zero value a in byte 0 and zero in the remaining bytes. For each ciphertext $C'_{i,j}$ compute a new ciphertext $D'_{i,j} = C'_{i,j} \oplus \delta$, respectively.

3. Ask for the decryption of the $D_{i,j}$ and $D'_{i,j}$ to obtain the new ciphertexts $O_{i,j}$, i.e., $O_{i,j} = E^{-1}(D_{i,j})$ and $O'_{i,j}$, i.e., $O'_{i,j} = E^{-1}(D'_{i,j})$, respectivly.
4. Store only those quartets $(P_{i,j}, P'_{l,j}, O_{i,j}, O'_{l,j})$ in the set ϕ where $O_{i,j} \oplus O'_{l,j}$ have a non-zero difference in bytes 3, 4, 6, 8, 9, 13, and 14 and a zero difference in the remaining bytes.
5. For each 56-bit candidate key k
   - Set a counter to zero.
   For each quartet passing the test in Step 4:
   5.1. Partially encrypt the plaintext quartet $(P_{i,j}, P'_{l,j}, O_{i,j}, O'_{l,j})$, i.e., $\bar{P_{i,j}} = e_{0,k}(P_{i,j})$, $\bar{P'_{l,j}} = e_{0,k}(P'_{l,j})$, $\bar{O_{i,j}} = e_{0,k}(O_{i,j})$ and $\bar{O'_{l,j}} = e_{0,k}(O'_{l,j})$.
   5.2. Increase the counter for the used 56-bit subkey k by one if $\bar{P_{i,j}} \oplus \bar{P'_{l,j}}$ and $\bar{O_{i,j}} \oplus \bar{O'_{l,j}}$ have a difference of a in all the bytes 3, 4, 6, 8, 9, 13, and 14.
6. Output the 56-bit subkey k with the highest counter.

4.5 Analysis of the Attack

We have 2^{53} structures which contain 2^{55} plaintext pairs of the desired difference each. Thus, we expect about $\#PP = 2^{53} \cdot 2^{55} = 2^{108}$ quartets in total. Since we use structures of all possible values in seven bytes, we get 2^{55} pairs with the desired difference needed in the differential $E0$. Thus, the transition in this differential happens with probability one in this case. A right quartet occurs with probability

$$\begin{aligned} Pr_c &= \Pr(\alpha \rightarrow \beta_{out}) \cdot (\Pr(\gamma \leftarrow \delta))^2 \cdot \Pr(\gamma_1 = \gamma_2) \cdot \Pr(\alpha \leftarrow \beta_{in}) \\ &= 1 \cdot 1 \cdot 2^{-56} \cdot 2^{-49} = 2^{-105}, \end{aligned}$$

since all the differential conditions are fulfilled. A random difference of two plaintexts has 9 zero byte difference with probability $Pr_f = 2^{-72}$. Thus, after Step 4 we have about $\#C = 2^{108} \cdot 2^{-105} = 2^3$ right and approximately $\#F = 2^{164} \cdot 2^{-72} = 2^{92}$ wrong quartets. A quartet passes the test in Step 5.2 with probability $Pr_{filter} = 2^{-112}$, since we have a 56-bit filtering condition on both pairs of a quartet. Thus, $\#CK_c = 2^3$ right boomerang quartets for the right key and $\#FK_c = \#F \cdot Pr_{filter} = 2^{92} \cdot 2^{-112} = 2^{-20}$ wrong quartets for each wrong subkey guess remain after this step.

Using the Poisson distribution[2] we can compute the success rate of our attack. The probability that the counter of a wrong key is at least 3 assuming $Y_i \sim Poisson(\mu = 2^{-20})$ is

$$\Pr(Y \geq 3) = e^{-2^{-20}} \cdot \frac{(2^{-20})^3}{3!} \approx 2^{-62}.$$

[2] Normally, we would use a binomial distribution but we use the Poisson distribution as an approximation. $X \sim Poisson(\mu)$ means that the random variable X follows the Poisson distribution with mean μ.

For all the $2^{56}-1$ wrong keys used in our analysis we expect about $2^{56} \cdot 2^{-62} = 2^{-6}$ wrong keys which have a count of at least 3 quartets. The probability that the right key has a count of at least 3 quartets using $Z \sim Poisson(\mu = 2^3)$ is

$$\Pr(Z \geq 3) \approx 0.98.$$

We can increase the success probability by increasing the number of quartets which also increases the data and time complexity of the attack.

Each structure of data can be analyzed sequentially. Thus, the total memory complexity is determined by Step 1 to 3, which is about $2 \cdot 2^{56} = 2^{57}$ blocks and additionally 2^{56} counters. The memory complexity of Step 4, 5.1 and 5.2 (is negligible compared to the memory complexity of the first two steps). The time complexity of Step 1 to 3 is $2 \cdot 2^{56} = 2^{57}$ encryptions. Since we have to run these steps for each structure of data the time complexity of the attack is about $2^{53} \cdot 2^{57} = 2^{110}$ 5-round encryptions. The data complexity is of size about $2^{53} \cdot 2^{56} = 2^{109}$ adaptive chosen plaintexts.

5 Boomerang Attack on 6 and 7-Round ARIA

The attack of the previous section can be easily extended to a 6-round attack on ARIA-192 and ARIA-256. Therefore, we need the following property of ARIA.

Property 1. The round key addition (ARK) and the diffusion layer (DL) can be interchanged, due to its linearity.

Using this property we can change the order of DL_5 and ARK_5. Thus, we can use our 5-round boomerang distinguisher to apply a 6-round attack in the following way. We add one round after the boomerang distinguisher as shown in Figure 8. We can guess 7 byte of K_6 at bytes 3, 4, 6, 8, 9, 13 and 14. This allows us to choose the desired difference ϕ such that after SL_6 in backward direction a known difference a occurs in each of these bytes. The DL_5 operation then outputs an a difference in byte 0 while the remaining bytes become zero. From this point the 5-round boomerang distinguisher works as explained above. Using the attack in the previous sections and the improvements on the boomerang attack presented by Biham et al. [1] we obtain the following results. Since we have to guess 56 bits after the distinguisher, using the notation from [1], we have $m_b = r_b = 0$, $m_f = r_f = 56$, $t_f = 38.08$. For our attack to work we need about $2^3 \cdot 2^{105} \cdot 2^{-1} = 2^{107}$ structures. The data complexity of the attack is 2^{128} known plaintexts using the data reduction technique from [1]. The memory complexity of the attack is $2^{m_b+m_f} + 2^{r_b+t_f} \approx 2^{56}$. The expected time complexity of the attack is $N(2+2^{2r_b+t_f-n-1}+2^{m_b+t_b+2t_f-n-1}+2^{m_f+2t_b+t_f-n-1}) \approx 2^{108}$ memory accesses. This attack can be applied to all instances of ARIA.

We can easily extend our 6-round attack guessing an entire 128 bit subkey from either the top or the bottom of the cipher. Thus, the memory complexity increases to $2^{56} \cdot 2^{128} = 2^{184}$ and the expected time complexity increases to about $2^{108} \cdot 2^{128} = 2^{236}$ memory accesses. Thus, our attack on 7-rounds of ARIA is only applicable to ARIA-192 and ARIA-256.

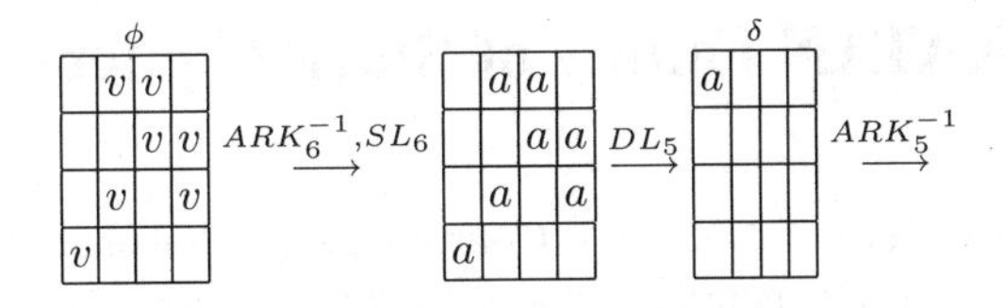

Fig. 8. The round after the distinguisher

6 Conclusion

In this paper we have shown some new attacks on the ARIA block cipher. Our attacks on 6 and 7-rounds of ARIA have the lowest memory complexity compared with existing attacks. This paper shows some weaknesses of reduced versions of ARIA, but the full round ARIA remains still secure.

References

[1] Biham, E., Dunkelman, O., Keller, N.: New Results on Boomerang and Rectangle Attack. Cryptology ePrint Archive, Report 2002/041 (2002), `http://eprint.iacr.org/`

[2] Biham, E., Shamir, A.: Differential Cryptanalysis of DES-like Cryptosystems. In: Menezes, A., Vanstone, S.A. (eds.) CRYPTO 1990. LNCS, vol. 537, pp. 2–21. Springer, Heidelberg (1991)

[3] Biryukov, A.: The Boomerang Attack on 5 and 6-Round Reduced AES. In: Dobbertin, H., Rijmen, V., Sowa, A. (eds.) AES 2005. LNCS, vol. 3373, pp. 11–15. Springer, Heidelberg (2005)

[4] Daemen, J., Rijmen, V.: The Design of Rijndael: AES - The Advanced Encryption Standard. Springer, Heidelberg (2002)

[5] Kwon, D., Kim, J., Park, S., Sung, S.H., Sohn, Y., Song, J.H., Yeom, Y., Yoon, E.-J., Lee, S., Lee, J., Chee, S., Han, D., Hong, J.: New Block Cipher: ARIA. In: Lim, J.-I., Lee, D.-H. (eds.) ICISC 2003. LNCS, vol. 2971, pp. 432–445. Springer, Heidelberg (2004)

[6] Li, P., Sun, B., Li, C.: Integral Cryptanalysis of ARIA. In: Pre-proceeding of Inscrypt 2009 (2009)

[7] Li, P.Z.R., Sun, B., Li, C.: New Impossible Differential Cryptanalysis of ARIA. Cryptology ePrint Archive, Report 2008/227 (2008), `http://eprint.iacr.org/`

[8] Wagner, D.: The Boomerang Attack. In: Knudsen, L.R. (ed.) FSE 1999. LNCS, vol. 1636, pp. 156–170. Springer, Heidelberg (1999)

[9] Wu, W., Zhang, W., Feng, D.: Impossible Differential Cryptanalysis of Reduced-Round ARIA and Camellia. J. Comput. Sci. Technol. 22(3), 449–456 (2007)

[10] Tang, X., Sun, B., Li, R., Li, C.: A meet-in-the-middle attack on aria. Cryptology ePrint Archive, Report 2010/168 (2010), `http://eprint.iacr.org/`

Algebraic, AIDA/Cube and Side Channel Analysis of KATAN Family of Block Ciphers

Gregory V. Bard[1,⋆], Nicolas T. Courtois[2], Jorge Nakahara Jr.[3,⋆⋆], Pouyan Sepehrdad[3], and Bingsheng Zhang[4,⋆⋆⋆]

[1] Fordham University, NY, USA
[2] University College London, Gower Street, London, WC1E 6BT, UK
[3] EPFL, Lausanne, Switzerland
[4] Cybernetica AS, Estonia and University of Tartu, Estonia
bard@fordham.edu, n.courtois@ucl.ac.uk,
{jorge.nakahara,pouyan.sepehrdad}@epfl.ch, zhang@ut.ee

Abstract. This paper presents the first results on AIDA/cube, algebraic and side-channel attacks on variable number of rounds of all members of the KATAN family of block ciphers. Our cube attacks reach 60, 40 and 30 rounds of KATAN32, KATAN48 and KATAN64, respectively. In our algebraic attacks, we use SAT solvers as a tool to solve the quadratic equations representation of all KATAN ciphers. We introduced a novel pre-processing stage on the equations system before feeding it to the SAT solver. This way, we could break 79, 64 and 60 rounds of KATAN32, KATAN48, KATAN64, respectively. We show how to perform side channel attacks on the **full 254-round** KATAN32 with one-bit information leakage from the internal state by cube attacks. Finally, we show how to reduce the attack complexity by combining the cube attack with the algebraic attack to recover the full 80-bit key. Further contributions include new phenomena observed in cube, algebraic and side-channel attacks on the KATAN ciphers. For the cube attacks, we observed that the same maxterms suggested more than one cube equation, thus reducing the overall data and time complexities. For the algebraic attacks, a novel pre-processing step led to a speed up of the SAT solver program. For the side-channel attacks, 29 linearly independent cube equations were recovered after 40-round KATAN32. Finally, the combined algebraic and cube attack, a leakage of key bits after 71 rounds led to a speed up of the algebraic attack.

Keywords: algebraic, cube, side-channel attacks, cryptanalysis, lightweight block ciphers for RFID tags.

⋆ This work was supported by the US National Science Foundation Grant No. DMS-0821725 to Prof. William A. Stein and the SAGE community. The computing power provided yielded the results of Section 3.

⋆⋆ This work was supported by the National Competence Center in Research on Mobile Information and Communication Systems (NCCR-MICS), a center of the SNF under grant number 5005-67322.

⋆⋆⋆ This work is supported by Estonian Science Foundation, grant #8058 and #8124, the European Regional Development Fund through the Estonian Center of Excellence in Computer Science (EXCS) and ICT doctoral school.

G. Gong and K.C. Gupta (Eds.): INDOCRYPT 2010, LNCS 6498, pp. 176–196, 2010.

1 Introduction

This paper describes our findings of cube attacks [16], also known as AIDA [30], algebraic [11] and side-channel (cube) attacks [17] applied to a variable number of rounds of all members of the KATAN family of block ciphers [13]. As far as we are aware of, this is the first paper detailing these attacks on the KATAN ciphers.

The cube attack is a kind of algebraic technique that exploits the existence of low-degree polynomial equations in the output of cryptographic algorithms. An attractive feature of the cube attack is that it requires only black-box access to the cryptographic function, that is, the knowledge of internal details of the target function is not required. Informally, if the decomposition of algebraic multivariate polynomial equations at the output of a target cipher has degree at most $d+1$, then linear equations on unknown key bits can potentially be extracted, provided that at most 2^d computations (encryptions) are feasible. Thus, the basic setting is key-recovery, but distinguish-from-random variants have been demonstrated in [2]. In [16] the cube attack has been applied to reduced-round variants of the Trivium [14] stream cipher.

Algebraic cryptanalysis exploit the multivariate polynomial system of equations in ANF format representing a given cipher [11]. The aim is to solve such systems for the unknown key (usually in a known-plaintext setting, but often we use chosen-plaintext attack to reduce the running time). Typically, quadratic equations are the main target representation. We convert the ANF equations to CNF format and feed it to a SAT solver, in our case MiniSat [18] and CryptoMiniSat [26]. Furthermore, we introduce a novel pre-processing step on the system of equations before giving it to a SAT solver. This allows us to break a larger number of rounds. Ultimately, we combine the cube and algebraic attacks on reduced-round KATAN ciphers.

We also combined the cube and side-channel techniques to attack the full-round KATAN32, following the model in [17]. Table 4 summarizes all the attack complexities in this paper. Up to the moment, we know of no other independent attacks on the KATAN ciphers, even for reduced-round versions.

This paper is organized as follows: Section 2 briefly describes the KATAN family of block ciphers; Section 3 provides some theoretical background on algebraic attacks; Section 4 gives theoretical framework and our experimental findings on AIDA/cube attacks; Section 5 combines both attacks; Section 6 describes side-channel cube attack; Section 7 concludes the paper.

2 The KATAN Family of Block Ciphers

KATAN is a family of lightweight, hardware-oriented block ciphers consisting of three variants with 32, 48 and 64-bit blocks. For all KATAN ciphers, key size is of 80 bits ($n=80$), and they all iterate 254 rounds [13]. The block size is used as suffix to designate each cipher member, as KATAN32, KATAN48 and KATAN64. The design of these ciphers was inspired by the stream cipher Trivium [14]. The structure of KATAN32 cipher consists of two LFSR's, called L_1 and L_2, loaded with the plaintext and then transformed by two nonlinear Boolean functions, f_a and f_b as follows (Table 1 lists the bit sizes and the indices x_i and y_j of L_1 and L_2).

$$f_a(L_1) = L_1[x_1] + L_1[x_2] + (L_1[x_3] \cdot L_1[x_4]) + L_1[x_5] \cdot \mathrm{IR}) + k_a$$
$$f_b(L_2) = L_2[y_1] + L_2[y_2] + (L_2[y_3] \cdot L_2[y_4]) + L_2[y_5] \cdot L_2[y_6]) + k_b$$

where IR is the output of an LFSR i.e. $L_1[x_5]$ is used whenever IR $= 1$. The values of IR for each round are specified in [13]. For the i-th round, $k_a = k_{2i}$ and $k_b = k_{2i+1}$ that is, only two key bits are used per round. The output of each of these functions is loaded to the least significant bits (LSB) of the other LFSR, after they are left-shifted. This operation is performed in an invertible manner.

For KATAN48, f_a and f_b are each applied twice per round, so that the LFSR's are clocked twice (but the same pair of key bits are reused). For KATAN64, each Boolean function is applied three times per round, again with the same pair of key bits reused three times.

The selection of bits x_i and y_i in f_a and f_b are listed in Table 1. In this report, plaintext and ciphertext bits are numbered in **right-to-left order** starting from 0. Thus, for instance, a plaintext block for KATAN32 will be numbered as $p = (p_{31}, \ldots, p_0)$. The key schedule algorithm of all KATAN ciphers is a linear mapping that expands an 80-bit key K to 508 subkey bits according to

$$k_i = \begin{cases} K_i, & \text{for } 0 \le i \le 79 \\ k_{i-80} + k_{i-61} + k_{i-50} + k_{i-13}, & \text{otherwise} \end{cases}$$

Thus, the subkey of the i-th round is $k_a||k_b = k_{2i}||k_{2i+1}$.

Table 1. Parameters for the f_a and f_b functions

Cipher	$\lvert L_1\rvert$	$\lvert L_2\rvert$	x_1	x_2	x_3	x_4	x_5	y_1	y_2	y_3	y_4	y_5	y_6
KATAN32	13	19	12	7	8	5	3	18	7	12	10	8	3
KATAN48	19	29	18	12	15	7	6	28	19	21	13	15	6
KATAN64	25	39	24	15	20	11	9	38	25	33	21	14	9

After r rounds, at most $2 * r$ key bits are mixed with the internal state since two key bits are xored per round. Thus, at least 40 rounds are needed before complete key diffusion for any KATAN cipher is achieved. Further details about these ciphers can be found in [13].

For analyses purposes, the numbering of the key bits in the user key in our attacks is $K = (K_{79}, \ldots, K_0)$.

3 Algebraic Attacks Using SAT Solvers

Algebraic cryptanalysis is a type of cryptographic attack that relies on solving a multivariate polynomial representation of a given cipher or hash function. It was initially formulated as early as 1949 by Shannon [29]. Algebraic attacks, since the controversial paper of [11], have been applied to several stream ciphers [1,9,10] and is able to break some of them but it has not been successful in breaking real-life block ciphers, except Keeloq [7,21]. Compared to statistical analysis, such as linear and differential cryptanalysis, algebraic attacks require a comparatively small number of text pairs. The adversary formulates the cipher as a polynomial multivariate system of equations. This representation is usually over small finite fields like $\mathbf{GF}(2)$. This system of equations is often sparse, since efficient implementations of real-world systems require a low gate-count. In the subsequent stage, the adversary solves the system. The problem of solving such system is NP hard in general and is recognized as the MQ problem. An instance of an MQ problem is a set of functions

$$f_1(x_1,\dots,x_n) = y_1, \quad f_2(x_1,\dots,x_n) = y_2 \quad ,\dots, \quad f_m(x_1,\dots,x_n) = y_m$$

where f_i can always be converted to a quadratic polynomial by introducing new variables. Notice that n is the number of variables and m is the number of equations. Let $c = \frac{m}{n}$ denote the degree of "overdefinition" of a system [4]. Hence, $c = 1$ denotes exactly a defined system, $c > 1$ an overdefined system and $c < 1$ an underdefined system. The polynomial representation of most block ciphers are overdefined or if not then $c \approx 1$. It turned out that the more the system is overdefined and sparse the easier it is to be solved [11]. For one instance of KATAN cipher, $c < 1$. This because in all versions, the key size is larger than the block size. Thus, more than one pair is required to find the correct key.

There are multiple methods for solving such systems. The traditional method uses the Gröbner basis approach such as Buchberger [6] or F4 [20] and F5 [19] algorithms. The drawback of such methods is memory, implying that after a while the algorithm outputs the result or it crashes due to running out of memory. This is true particularly for large systems, but they are usually faster than other methods for small systems and when the characteristic of field q is not 2. When $q = 2$, more efficient methods were proposed, such as converting these equations to Boolean expressions in Conjunctive Normal Form (CNF) [4] and deploying various SAT-solver programs. Other strategies include the XL family [12,11], the recent MutantXL [15,25], ElimLin [8] and the Raddum-Semaev [27] algorithms. We focus on SAT-solver based methods in this paper. From now on we will work only with the field $\mathbf{GF}(2)$. To solve such polynomial system by SAT solvers, the attacker initially converts the system from Algebraic Normal Form (ANF) to CNF. There is an efficient conversion method due to Bard-Courtois-Jefferson [4]. We also use a direct method which we call "local interpolation". Let assume that the total degree of the equations is at most 6. We proceed as follows:

- If there are equations which contain more than 6 variables, split these long XORs into several shorter XORs with at most 6 variables, by adding extra variables, for example $abc + def + gh = 1$ becomes $abc + def = x$ and $x + gh = 1$.
- for each equation, convert the Boolean function to CNF and write it explicitly.

The concatenation of these CNFs gives a file with extension .cnf on which we can apply any SAT solver. The magic number 6 originates from [4], and is called the cutting number—sometimes 4, 5, or rarely 7 is optimal instead.

The area of SAT Solving has seen tremendous progress over the last years. Many problems (e.g. in hardware and software verification) and in our application in cryptanalysis that seemed to be completely out of reach a decade ago can now be handled routinely. Besides, new algorithms and better heuristics, refined implementation techniques turned out to be vital for this success. New SAT solvers can now solve large systems in reasonable time. Since 2002, almost each year a SAT Race competition [28] was established. In 2007 and 2010 respectively, MiniSat [18] and CryptoMiniSat [26] won the Gold prizes. We used these two SAT solvers in our analysis but since the timings of MiniSat were faster, we do not report CryptoMinisat results in this paper.

3.1 Straightforward Algebraic Attack on KATAN Using SAT Solvers

One instance of KATAN32 can be represented as $8,620$ very sparse quadratic equations with $8,668$ variables, KATAN48 as $24,908$ equations and $24,940$ variables and KATAN64 as $49,324$ equations and $49,340$ variables. As can be observed, the system is underdefined. That is because the key size is larger than block size for all versions. To have a defined or an overdefined system, we need multiple samples.

A summary of our results is in Table 4. We used the "guess and determine" algebraic attack initially proposed in [4]. This implies that we fix t bits of the key and then we show that recovering the other $80-t$ bits is faster than exhaustive search. This is represented in the column titled "Fixed" in Table 4. In fact, we fix t LSB of the key, since heuristically we obtained better results than fixing the t MSB of the key. We used the graph partitioning method by Wong and Bard [31] to derive the best state variables to fix, but it did not bring about anything better than using the heuristic of fixing the least t significant bits of the key. Note, if we fix t bits, the algebraic attack is solving a system of equations to recover the $80-t$ remaining bits.

We represent the time complexity of the SAT Solver (MiniSat) in seconds using a 3 Ghz CPU. Note that our algebraic attacks are in the chosen-plaintext scenario, except in some rare cases as noted. We noticed that chosen-plaintext attack is much stronger against KATAN family than known-plaintext (KP) attack. In our attacks, we followed the following structure for the chosen plaintexts for KATAN32: $p_{i+1} = ((p_i \gg 19) + 1) \ll 19$ and $p_{i+1} = ((p_i \gg 29) + 1) \ll 29$ for KATAN48 and $p_{i+1} = ((p_i \gg 39) + 1) \ll 39$ for KATAN64 for $i \geq 1$, where p_i is the i-th plaintext we pick and p_1 can be arbitrary and $\gg$ and $\ll$ are shift right and shift left respectively and $+$ is xor. Note that bits $19, 29, 39$ are exactly bit 0 of L_1 register for KATAN32, KATAN48 and KATAN64 respectively. This choice of the bits makes the SAT solver run faster. Moreover, we believe it is fair to assume each round encryption of KATAN takes at least 3 CPU cycles. This yields a comparison between the complexity of our attacks and exhaustive key search.

Deploying the straightforward method of converting ANF to CNF and then feeding it to a SAT solver, we could break up to 79 rounds of KATAN32 and 64 rounds of KATAN48 and 60 rounds of KATAN64. But, we can do better by performing a pre-processing on the system of equations before applying it to a SAT solver. Using this pre-processing (see next section), we could break 79 rounds of KATAN32. We only tried this method on KATAN32 equations. Further research would apply this technique to other members of the family.

3.2 The Pre-processing SAT-Solver Attack

In this attack, we use the equations generated as described earlier and solve them with the SAT solver MiniSat [18]. It is simpler to formulate KATAN as a sparse system. But, this may not be the best representation for a SAT solver. One characteristic of these equations is that there are many of them with the form $x = y$, as well as $x = 0$, $y = 1$ and more rarely $x + y = 1$. Also, in a typical example (78 rounds, 45 key bits fixed and 20 CPs of KATAN32) there are $51,321$ total equations. Naturally one wants to take advantage of these special equations, during pre-processing, to create a smaller system which has fewer variables and equations.

More precisely, the four heuristics of a CNF problem are (1) the number of variables, (2) the number of clauses, (3) the average number of symbols per clause, and (4) the total number of symbols in the system. The pre-processing algorithm that we describe in the next section is designed on the principle of primarily reducing (1) and (2) while causing the minimum possible increase in (3) and (4). To be specific, at each iteration, a substitution will be made and this substitution reduces (1) and (2) by one, and the substitution is selected in the style of the "greedy algorithm" using (4) as the criterion.

The following pre-processing algorithm, due to Bard, is a refinement of the "massaging" algorithm of [4] and so we call it "turbo-massage". Starting with the equations that were generated, we ran the pre-processing algorithm; after that, we converted the polynomials into a CNF problem, according to [4] and ran MiniSat on that CNF problem to get a solution. We will explain the pre-processors here and refer the reader to [4] or [3] for the process of converting a polynomial system into a CNF problem.

3.3 The Turbo-Massage Pre-processing Algorithm

As described before, the equations can be thought of as a series of polynomials $f_1(x) = 0$, $f_2(x) = 0, \ldots$. We define the operation "fully-substitute" as follows: Let $f(x)$ be a polynomial with some monomial μ. To fully-substitute $f(x)$ into $g(x)$ on μ means to

- Write $g(x)$ in the form $g(x) = \mu h_1(x) + h_2(x)$.
- Write $f(x)$ in the form $f(x) = \mu + h_3(x)$.
- Replace $g(x)$ with $h_1(x)h_3(x) + h_2(x)$, which is mathematically equivalent, because in any satisfying solution x, we would have $\mu = h_3(x)$.
- By clever use of data structures, this can be made highly efficient.

Observe that for the four common forms: $x = y$, as well as $x = 0$, $y = 1$, and more rarely $x + y = 1$, the "fully-substitute" definition does what one would do if solving a system of equations with a pencil and paper. For more higher weight $f(x)$, understanding what it does to $g(x)$ is more complex.

We must also use a non-standard definition of the weight of a polynomial $f(x)$. We define it to be the number of monomials in $f(x)$, but excluding the constant $+1$ from the tabulation. The reason for this is that if $f(x)$ has weight w according to this modified definition, then 2^{w-1} conjunctive normal form clauses of length w will be required to represent the polynomial, assuming all the monomials are already defined. The total number of symbols is then $w2^{w-1}$ and so minimizing w is crucial in keeping the CNF problem small and thus solvable. We now perform the following algorithm:

- **INPUT:** A system of polynomial equations over $\mathbf{GF}(2)$, and a weight-limit w_{max}.
- Mark all polynomials "unused."
- While the set of unused polynomials is not empty do:

- Locate the lowest weight unused polynomial $f(x)$.
- If $f(x)$ exceeds w_{max}, terminate.
- Mark $f(x)$ as "used."
- If $f(x)$ has weight 1, then select μ to be the only monomial in $f(x)$.
- If $f(x)$ has weight 3 or higher, select μ to be the monomial which appears least frequently in the entire system of equations.
- If $f(x)$ has weight exactly 2, select μ to be the monomial which appears most frequently in the entire system of equations.
- For any polynomial $g(x)$ containing the monomial μ, simply "fully-substitute" $f(x)$ into $g(x)$ on μ.

The "turbo-massaging" algorithm will always terminate, because eventually every polynomial has been marked used. In practice, it will terminate early, where all unused polynomials are of weight w_{max} or higher. If there are n "used" polynomials, then there will be n monomials which appear nowhere in the entire system except in exactly one polynomial. This is, of course, the monomial μ which was chosen when that polynomial was getting used. In our system of equations, it was almost always the case (by an overwhelming margin) that μ was degree one. And so, each used polynomial effectively amputates one variable from the polynomial system of equations.

The special case of weight 2 deserves explanation. When f has weight 1, there is no decision to be made, but it is noteworthy that the weight of g will decrease. When f has weight 2, then the weight of g will not change during the "fully-substitute" operation, except in some odd cases like substituting $x = y$ into $zx + zy + w + x + y = 0$, where the weight goes from 5 to 1 instantly. Since the weight is not likely to change and we are eliminating a monomial, it makes sense to eliminate a common monomial. When the weight of f is 3 or more, then the weight of g will increase. If we choose μ to be very popular, appearing k times, then the total weight of the system will increase by $k(w-2)$. Thus, it makes sense to keep the weight growth bounded and choose μ to be rare. This heuristic was found after an enormous number of iterations of "trial-and-error."

For example, in the 78-round, 20 CP, 45-bit-key case of KATAN32, the weight went from $110,726$ to $101,516$ after $47,032$ polynomials got used. Furthermore, the "fully substitute" function was called $330,587$ times. There were $50,033$ equations at this point, down from $51,321$, representing $1,288$ equations that became $0 = 0$. In other words, the original system was not full rank. The average weight of a polynomial, using the modified definition of weight, was roughly 2.02898. The system had $53,993$ distinct monomials, plus $2,005$ variables which appeared only in degree 1 monomials, and ended with a CNF problem of $55,398$ variables, and $156,010$ clauses. The conversion process, which must be run only once and not 2^{45} times, takes between 20 and 29 minutes in all the cases explored here.

It should be noted that in other polynomial systems it might be the case that μ is often quadratic or higher degree. It remains open if one should force μ to be linear when possible. This is a question that the authors hope to investigate shortly. As it comes to pass, $w_{max} = 2$ turned out to be slightly better than $w_{max} = 3$ for this problem, but in other cases up to $w_{max} = 5$ has been used.

A minor note for algebraic geometers familiar with the concept of a Macaulay matrix [23] in the Lazard [22] family of algorithms (including F4 [20], XL and their variants

[12,11]) is that this algorithm is like a Gaussian Elimination on that matrix, but stopping early. The pivoting strategy used is reducible to the Markowitz pivoting algorithm [24]. However, the "fully-substitute" is not the same in this case, as adding $x = y$ to $zx + zy + w + x + y = 0$ would result in $zx + zy + w = 0$. On the other hand, fully-substituting $x = y$ into $zx + zy + w + x + y = 0$ would result in $zx + zx + w + x + x = 0$ which turns into $w = 0$. As you can see, full-substitution is distinct from adding, and is very similar to what a mathematician would do if solving a system of polynomial equations with a pencil and paper.

3.4 Results

The first result was 76 rounds, 20 CP and 45 (fixed) key bits of KATAN32, broken faster than by brute force. To extend this result, we explored using fewer key bits, and more rounds. First we conducted the above process for 20 CPs, and for 76, 77, 78, 79 and 80 rounds. Every case was run 50 times.

Because we fixed 45 bits of the key, and so assuming one nano-second per round for a brute force attacker, our attack against r rounds is faster than brute force if and only if it runs in t seconds with $2^{45}t < r2^{80}10^{-9}$ or more plainly $t < r2^{35}10^{-9} \approx r(34.3597\cdots)$. We also ran trials with 43 bits of the key fixed for 76 rounds and there the threshold would be 4 times greater or $137.439r$ seconds and for 41 bits of the key $549.755r$ seconds.

The running times are given in Table 2. Observe the enormous variance in each trial. In some cases, the fastest run is $1000\times$ faster than the slowest. This is very typical in SAT-solver-based cryptanalysis. We excluded the three fastest and slowest trials and took the mean and standard deviation of the remaining 44 trials.

The running time of 2^{45} executions, all added together, is the sum of 2^{45} samples from independent random variables. Therefore, the central limit theorem applies and regardless of the actual distribution of running times, if the mean is m_1 and stdev is σ_1, the sum of 2^{45} of them will be normally distributed and have a mean of $2^{45}m_1$ and a standard deviation of $2^{22.5}\sigma_1$. Since σ/m is an important instrument in gauging the reliability of a normal sample, it is interesting to note here that σ/m (for the sum of 2^{45} execution times) would be $2^{-22.5}(\sigma_1/m_1)$ which is phenomenally tiny. Thus, the running time of the real-world attacker would be essentially constant.

Notice, that we claim that the 2^{45} running times are independent, but we do not claim that they are identically distributed. On the other hand, one could conceive of a cipher where one key bit was ignored by the cipher, in which case the running times for two keys which differ only in that bit would be highly dependent. These cases are of pedagogical interest only, because no cipher designer would ever do that.

As can be seen in Table 2, we are between 80.75 and 2.39 times faster than brute force search for up to and including 79 rounds. In the case of 80 rounds, out of 50 trials, 29 of them timed-out after 1 hour. Since this is majority, it is not possible that the mean is less than the required 2748.77 seconds, and so we are not faster than brute-force for 80 rounds. For 43 key bits and 41 key bits, the attack becomes vastly more efficient. But, we cannot test 39 key bits, as the time-out value would have to be set to 167,125 seconds or roughly 46 hours, for each of 50 processes.

Table 2. Running time and some statistical results for different number of rounds of the preprocessed equations for KATAN32. The running times are in second.

# of rounds	76	77	77	78	79	80	76	76
fixed	45	45 first batch	45 second batch	45	45	45	43	41
1	2.89	1.00	2.43	11.04	17.05	59.62	1.50	1.75
2	3.15	2.16	3.69	11.54	24.97	64.61	5.48	1.91
3	3.39	2.25	4.01	14.51	26.86	100.28	15.75	3.36
4	3.39	3.39	4.12	15.83	28.82	135.34	25.88	3.77
5	4.61	3.93	4.40	19.17	54.27	157.10	34.81	5.17
6	6.73	4.16	4.44	24.99	57.02	166.41	39.92	5.65
7	8.29	4.22	4.65	51.46	60.72	230.60	39.97	8.64
8	8.46	4.58	4.72	63.04	64.08	277.04	45.06	11.35
9	11.54	4.81	5.07	86.06	70.34	353.45	50.19	21.71
10	13.15	4.84	6.41	89.89	89.17	354.07	50.79	35.31
11	17.19	4.96	6.81	109.21	109.86	402.56	52.09	41.71
12	17.62	5.44	10.08	115.86	130.28	423.76	60.94	53.7
13	23.64	5.62	14.54	141.19	137.77	433.73	75.35	55.77
14	26.60	5.74	15.03	148.91	145.05	463.78	102.91	61.6
15	27.69	5.83	18.16	161.49	210.29	516.65	116.01	78.29
16	37.32	6.80	18.51	163.23	217.28	687.88	121.89	84.18
17	38.04	7.64	19.51	206.66	269.08	1163.48	123.25	87.51
18	39.67	8.38	21.31	218.43	326.69	1591.56	123.36	104.76
19	48.68	9.54	21.35	230.86	402.61	2180.93	124.39	108.29
20	50.63	10.08	21.57	236.17	408.39	3261.20	131.54	128.62
21	56.51	11.32	22.06	241.45	537.16	3274.25	132.67	138.37
22	62.53	13.81	22.41	248.64	547.32	29 timeouts	134.03	166.93
23	66.03	15.72	22.63	256.66	718.58		207.34	170.14
24	81.25	16.69	27.15	293.66	780.44		208.48	182.83
25	88.88	17.47	28.45	319.31	873.25		233.40	183.9
26	101.43	17.86	32.39	377.06	893.29		258.52	185.41
27	115.13	19.19	45.27	455.50	949.06		300.38	200.08
28	127.09	19.63	49.92	504.97	1007.55		326.94	223.6
29	176.33	22.76	54.80	593.65	1223.91		374.62	246
30	200.26	24.29	54.82	822.36	1244.11		387.17	248.05
31	224.75	29.68	73.71	854.80	1388.40		444.42	254.58
32	243.36	30.09	82.72	880.31	1436.00		449.31	256.05
33	258.53	33.27	85.42	1111.59	1632.59		542.73	263.13
34	278.53	34.02	85.56	1118.54	1838.31		829.13	275.75
35	294.99	35.62	97.22	1197.05	1864.98		905.35	304.75
36	353.49	35.94	97.76	1388.38	1875.87		954.94	305.1
37	407.02	43.33	103.34	1449.29	2031.08		1217.79	305.18
38	423.38	43.65	111.18	1514.89	2038.93		1367.94	328.86
39	475.98	48.18	118.48	1517.73	2167.55		1390.52	352.89
40	506.67	48.22	119.15	1533.10	2262.50		1618.79	356.23
41	687.95	49.96	184.91	1538.97	2369.57		2234.32	403.7
42	842.95	73.62	222.26	1689.96	2413.38		2455.77	407.63
43	942.88	106.69	226.48	1894.40	2495.42		2668.97	418.7
44	2387.95	133.21	335.07	2031.93	2641.90		3246.26	427.04
45	2400.12	186.39	456.45	2375.14	2960.11		3326.73	429.21
46	3722.62	201.89	662.92	2682.71	3460.90		3530.63	555.35
47	4471.28	302.66	815.38	2837.97	4023.81		7157.16	577.3
48	> 6000	344.63	976.94	3731.61	4129.64		9378.05	6248.59
49	> 6000	433.70	2378.61	> 6000	4212.65		> 10,000	6763.91
50	> 6000	524.56	> 6000	> 6000	> 6000		> 10,000	9655.8
Threshold-time	2611.34	2645.70	2645.70	2680.06	2714.42	2748.78	10445.35	41781.40
# faster	47	50	49	48	49	21	48	50
Median	95.16	17.67	30.42	348.19	883.27	n/a	245.96	184.66
Mean of all but 6	463.21	38.98	100.88	768.47	1146.77	*n/a*	868.70	205.97
Stdev of all but 6	957.09	60.29	168.93	786.56	1054.09	*n/a*	1381.91	154.29
Kurtosis of all but 6	9.51	9.19	9.30	0.17	-0.12	n/a	9.33	-0.46
Times faster than brute force	5.64	67.87	26.23	3.49	2.37	n/a	12.02	202.85
Mean of log	4.648	2.914	3.650	5.938	6.369	*n/a*	5.706	4.810
Stdev of log	1.820	1.185	1.421	1.380	1.396	*n/a*	1.513	1.319
Kurtosis of log	-0.582	-0.470	-0.620	-0.514	-0.985	*n/a*	-0.985	0.867

In addition to MiniSat, we ran all 50 instances with CryptoMiniSat [26], a SAT-Solver constructed specifically for cryptography by Mate Soos. However, it was consistently slower than MiniSat. We suspect that this is the case because CryptoMiniSat was intended to minimize the impact of long-XORs, which are normally very damaging to the running time of SAT-solver methods; however, we have no long-XORs in our equations, in fact, no sum was longer than 5 symbols after pre-processing, excluding the constant monomial.

3.5 The Gibrat Hypothesis

In [4], [8] as well as [3], Bard hypothesized that the true distribution of the running times of a CNF-problem in a polynomial-system-based SAT problem follows the Gibrat distribution. That is to say, that the logarithm of the running time is normal. The running times here were such that their standard deviations exceeded the mean. If the distribution of the running time were normal, having $\sigma > \mu$ would imply a very significant fraction of the running times would be negative. Therefore, it is not possible that the running time is normally distributed. On the other hand, we also tabulated the mean and standard deviation of the logarithm.

The ratio of the mean and standard deviation of the logarithm of running times is much more reasonable. The kurtosis is the typical measurement of the "normalness" of a distribution and the kurtosis of the logarithms of the running times are far closer to 1 (and are in fact within ± 1) than the kurtosis of the running times themselves (which had kurtoses over 9). So the hypothesis that the running times are Gibrat, from [4], seems well-justified for these examples.

3.6 A Strange Phenomena

We were perplexed to discover that solving 77 rounds was far easier than solving 76 rounds or 78 rounds. Therefore, we ran the experiments again, with both sets of results listed in the Table 2 as first batch and second batch. As you can see, in both cases, 77 rounds is much easier than 76 or 78—and with a very large margin. Moreover, this remained true as well in our experiments with CryptoMiniSat. As random variables, the ith iteration of the 76 round attack and the ith iteration of the 77 round attack had absolute correlation of $0.060419\cdots$ and likewise between 77 and 78 it was $-0.09699\cdots$. These extremely low correlations make it safe to hypothesize that the running times are independent and this removes the possibility that the effect is an artifact of some methodology error. Note, the formula for correlation that we used is

$$\mathrm{Cor}(X,Y) = \frac{\mathrm{E}(X-\mu_x)\mathrm{E}(Y-\mu_y)}{\sigma_x\sigma_y}$$

as is standard. Moreover, we observed the same behaviour when dealing with the size of the vertex separator in the variable-sharing graph representation of polynomial system of equations of KATAN32 using the strategy described in [31]. For KATAN32, the size of vertex separator is not increasing with the number of rounds and as a matter of fact it fluctuates. We offer no explanation as to the cause of the weakness of the 77-round version of KATAN32.

4 AIDA/Cube Attacks

AIDA/cube attacks [16] are generic key-recovery attacks that can be applied to cryptosystems in a black-box setting, that is, the internal structure of the target cipher is unknown. An important requirement is that the output from the cryptosystem can be represented as a low-degree decomposition multivariate polynomial in Algebraic Normal Form (ANF), called master polynomial, in the key and the plaintext. This attack does not depend on the knowledge of the master polynomial, which may be dense, or whose representation is so large that it cannot even be stored.

Let $p(x_1,\dots,x_n,v_1,\dots,v_m)$ denote a master polynomial over $\mathbf{GF}(2)$ in ANF, with x_i, $1 \leq i \leq n$ and $v_j, 1 \leq j \leq m$, the public variables (plaintext, IV bits) and v_j the secret key variables. We assume the adversary is allowed to query the master polynomial at values x_i (that is, a chosen-plaintext, chosen-IV setting) of its choice (these are also called tweakable parameters) and obtain the resulting bit from the master polynomial. This way, the adversary obtains a system of polynomial equations in terms of secret variables only. The ultimate goal of the attack is to solve this system of equations, which reveals the key variables v_j. For this attack, the master polynomial is decomposed as follows:

$$p(x_1,\dots,x_n,v_1,\dots,v_m) = t_I \cdot p_{S(I)} + q(x_1,\dots,x_n,v_1,\dots,v_m)$$

where t_I is a monomial containing only public variables from an index set $I \subset \{1,2,\dots,n\}$ called cube or hypercube; '+' stands for bitwise xor; $p_{S(i)}$ is called the superpoly of I in p. The superpoly of I in p does not contain any common variable with t_I and each monomial in q does not contain at least one variable from I, since they have all been factored out in $p_{S(I)}$. The $p_{S(I)}$ of interest are linear mappings in terms of v_j's. Any t_I that leads to a linear $p_{S(I)}$ in key bits is called maxterm. The output of the offline phase of the attack consists of linear equations in the user key bits directly. Further, Gaussian elimination allows one to reconstruct the user key (independent of the key schedule algorithm). For instance, let

$$p(x_1,x_2,x_3,v_1,v_2,v_3,v_4) = x_2.x_3.v_3 + x_1.x_2.v_1 + x_2.v_4 + x_1.x_3.v_2.v_3 + x_1.x_2.v_2 + 1$$

Let $I = \{1,2\}$, so that $t_I = x_1.x_2$ and we have the following decomposition

$$p(x_1,x_2,x_3,v_1,v_2,v_3,v_4) = x_1.x_2.p_{S(I)} + q$$

where $p_{S(I)} = v_1 + v_2$ and $q = x_2.x_3.v_3 + x_2.v_4 + x_1.x_3.v_2.v_3 + 1$.

The main motivation for this decomposition of the master polynomial is that the symbolic sum over $\mathbf{GF(2)}$ of all evaluations of p by assigning all possible binary values to the variables in I (and a fixed value, usually 0, to all the public variables not in I) is exactly $p_{S(I)}$, the superpoly of t_I in p. This is the fundamental theorem in [16]. In the example,

$$\begin{aligned}\bigoplus_{x_i, i\in I} p(x_1,x_2,x_3,v_1,v_2,v_3,v_4) = {} & p(0,0,x_3,v_1,v_2,v_3,v_4)+ \\ & p(0,1,x_3,v_1,v_2,v_3,v_4)+ \\ & p(1,0,x_3,v_1,v_2,v_3,v_4)+ \\ & p(1,1,x_3,v_1,v_2,v_3,v_4) = v_1 + v_2 = p_{S(I)}\end{aligned}$$

since $t_I = 0$ whenever either of x_1, x_2 is zero. In q, since each monomial does not contain at least one of the variables in t_I, each monomial will appear an even number of times in the summation of p and the xor sum will be zero.

The cube attack has a pre-processing (offline) and an online phase. In the former, the aim is to find monomials t_I's that lead to linear superpolys. The maxterms are not key dependent, so they need to be computed only once per master polynomial, for a fixed number of rounds. For each maxterm, the adversary computes the coefficients of the v_j's, effectively reconstructing the ANF of the superpoly of each t_I. This step is performed by linearity tests [5]. The main issue in the pre-processing is to find the correct combination of $|I|$ public variables (out of n) x_i that result in **linear superpolys**. Since the exact form of the master polynomial is unknown, this step is heuristic and consists in randomly choosing the cube variables and using linearity tests to check the superpolys. This phase is performed only once for a given cipher and a fixed number of rounds.

Besides the linearity tests there are also 'constant' tests that are used to determine the constant terms 0 or 1 in the superpoly's. The public variables not in the maxterms should be set to the same fixed value in both phases. After a sufficient[1] number of linearly independent (LI) superpolys have been found, the online phase starts by evaluating the superpolys, that is, summing up p over all the values of the corresponding maxterm, $\bigoplus_{x_i, i \in I} p$ and deriving the value of the linear combination of secret v_j bits. If the degree of t_I is d, each xor sum requires 2^d evaluations of p (which implies a chosen-plaintext setting). Thus, the time and data complexities are proportional to the maximum degree d among all maxterms.

The online complexity is proportional to 2^{d_i} encryptions, for a superpoly whose maxterm has d_i variables, since the ciphertexts have to be collected (and xored) for this same amount of chosen plaintexts. If t LI superpolys are available, then $\sum_{i=1}^{t} 2^{d_i}$ encryptions will be needed to recover each superpoly. On the other hand, if the key size is k bits, then 2^{k-t} encryptions shall be enough to recover the remaining unknown part of the key. In total, the time complexity becomes $2^{k-t} + \sum_{i=1}^{t} 2^{d_i}$.

4.1 Cube Attack on KATAN32

Table 5 shows cubes and maxterms for 50-round KATAN32. The maxterm is shown in hexadecimal (the bits set to '1' are the selected bits) for a compact description in the tables in the appendix. We used Gaussian elimination to select LI equations. Experimentally, not all ciphertext bits leak information on the key bits (cube equations). We found that the same maxterm can be used for different key equations, for distinct cipher bits. This means that we can save data and computational complexity during the online phase. Out of the 46 maxterms obtained in total, we observed that the maxterm $\texttt{1B8EE77B}_x$ gives equations $k_2 + k_{12}$ and $k_8 + k_{26} + 1$ and a similar phenomenon

[1] An ideal quantity is a trade-off between the number of linearly independent superpolys and the effort to recover the remaining key bits.

happened for the maxterm EB3AEAE6$_x$ and 9CF75766$_x$. Thus, the data complexity becomes $43 \cdot 2^{20} = 2^{25.42}$ and the time complexity is $2^{25.42} + 2^{80-46} \approx 2^{34}$ 50-round KATAN32 computations.

Table 6 shows cubes and maxterms for 60-round KATAN32. Out of the 41 maxterms obtained in total, we observed that the maxterm EF2FF9EF$_x$ gives equations $k_{26} + 1$ and $k_{22} + k_{32} + 1$ and a similar phenomenon happened for the maxterm B7F2DFDF$_x$. Thus, the data complexity becomes $39 \cdot 2^{25} \approx 2^{30.28}$ CP and time complexity is $2^{30.28} + 2^{80-41} = 2^{39}$ 60-round KATAN32 encryptions.

In all our cube attacks, we ran 10,000 linearity tests and then we tested the equations for 50 distinct random keys to be sure they are correct.

4.2 Cube Attack on KATAN48

Table 8 shows cubes and maxterms for 40-round KATAN48. All 31 obtained maxterms have degree 20. The data complexity is $31 \cdot 2^{20} \approx= 2^{24.95}$ CP. The memory cost is negligible. The computational complexity is $2^{24.95} + 2^{49} \approx 2^{49}$ 40-round KATAN48 computations, which is dominated by the exhaustive search for the remaining 49 key bits.

4.3 Cube Attack on KATAN64

Table 9 shows cubes and maxterms for 30-round KATAN64. All 25 maxterms found have degree 16. The data complexity of the attack is $25 \cdot 2^{16} \approx 2^{20.64}$ CP. The memory cost is negligible. Since only two subkey bits are used per round, there are at most 60 key bits involved in 30 rounds. The time complexity is $2^{20.64} + 2^{60-25} \approx 2^{35}$ 30-round KATAN64 computations.

5 Combining Cube and Algebraic Attacks

The bottleneck in cube attacks is that after some rounds, the degree of maxterms becomes large. Therefore, it takes a long time to find a linear superpolynomial. But still, if we even get a few linear superpolynomials, it would help to reduce the complexity of the classical algebraic attack. In fact, the overall complexity would be the sum of those two complexities. For a small number of rounds, algebraic attacks are successful, but for larger number of rounds it becomes slower. In such cases, the result of cube attacks and classical algebraic attacks can be combined. For instance, observing Table 4, we have obtained a 3-bit condition on the key bits for 71-round KATAN32 using cube attacks with time complexity $2^{29.58}$. The complexity of algebraic attack alone is $2^{66.60}$. Binding these two attacks reduces the complexity of algebraic attack by $1/8$ because it reduces the number of keys to be guessed from 35 to 32. In fact, we need to guess 3 bits less in order to get the same complexity. So, in Table 4, this reduces our complexity to $2^{63.60}$.

6 Side-Channel Attack for Full-Round KATAN32

In this section we consider side-channel attack models such as [17] in which internal cipher data leaks after r rounds, where $r < 254$, of some full-round KATAN cipher. On one hand, such data is supposed to have been independently captured by some side channels for instace, power or timing analysis or electromagnetic emanations (which is a strong assumption). On the other hand, for our attack setting, only one bit of the cipher state is needed.

The position of the internal cipher data that leaks is selected by the adversary such that its polynomial representation has low degree d and it can be regarded as ciphertext bit c_j after r rounds. Unlike [17], though, we consider c_j to be error free, that is, noise-free. Cube attacks are further employed to derive information on the key from c_j. In this setting, the same bit c_j is supposed to be accessible after each encryption of 2^d CP by the adversary. The adversary chooses different cubes in order to obtain new equations from c_j, all of which are mutually linearly independent.

Table 3. Maxterms, 29 cube equations from ciphertext bit c_{19} from 40-round KATAN32

Maxterm	Degree	Cube equation	Cipher bit
41356548_x	12	k_4	c_{19}
$2464E14C_x$	12	k_{15}	c_{19}
$1EA26848_x$	12	k_5+1	c_{19}
$E3516900_x$	12	k_1+k_{16}	c_{19}
$4A8E6888_x$	12	$k_0+k_{17}+1$	c_{19}
$EBD02900_x$	12	$k_3+k_{10}+1$	c_{19}
$A0867A0C_x$	12	$k_{14}+k_{17}+1$	c_{19}
$C0C34C43_x$	12	$k_4+k_{10}+k_{19}$	c_{19}
$E2A54302_x$	12	$k_{11}+k_{15}+k_{23}$	c_{19}
$9C045983_x$	12	$k_2+k_7+k_{11}+k_{16}+k_{24}+k_{26}$	c_{19}
$bd30cb11_x$	15	k_{13}	c_{19}
$7c366259_x$	16	k_{18}	c_{19}
$2cd5f264_x$	16	$k_6+k_{15}+1$	c_{19}
$b7351759_x$	18	$k_3+k_{18}+k_{23}$	c_{19}
$cf9df815_x$	19	k_3+1	c_{19}
$75e471ee_x$	19	$k_{24}+1$	c_{19}
$65765d7a_x$	19	$k_0+k_{10}+k_{16}+k_{18}+k_{19}+k_{26}+k_{30}+k_{43}$	c_{19}
$ab7f3a4b_x$	20	k_7	c_{19}
$b61d73f9_x$	20	k_8+1	c_{19}
$3d7f3476_x$	20	k_2+k_{19}	c_{19}
$e4f636be_x$	20	k_6+k_{16}	c_{19}
$acd1bbf6_x$	20	$k_{12}+k_{20}+k_{29}$	c_{19}
$bdcddcac_x$	20	$k_{16}+k_{21}+k_{26}+1$	c_{19}
$deff1456_x$	20	$k_7+k_9+k_{18}+k_{26}$	c_{19}
$37d7d2b3_x$	20	$k_{16}+k_{23}+k_{26}+k_{43}$	c_{19}
$d7035eef_x$	20	$k_4+k_8+k_{14}+k_{18}+1$	c_{19}
$ad754de7_x$	20	$k_2+k_{16}+k_{19}+k_{20}+k_{26}+k_{43}$	c_{19}
$17dfaa6d_x$	20	$k_{13}+k_{18}+k_{21}+k_{22}+k_{23}+k_{26}+k_{30}+1$	c_{19}
$6afeaf85_x$	20	$k_0+k_9+k_{18}+k_{24}+k_{25}+k_{26}+k_{27}+k_{30}$	c_{19}

Table 4. Attack complexities on KATAN family of block ciphers (memory complexity is negligible)

Cipher	# Rounds	$Time_1$	$Time_2$	Data	Fixed	Attack	Source
KATAN32	40		11 sec	3 KP	0	MiniSat, LI conv.	Sect. 3.1
	50	2^{34}		$2^{25.42}$ CP	0	AIDA/Cube	Sect. 4.1
	50		11 sec	3 KP	0	MiniSat, LI conv.	Sect. 3.1
	60	2^{39}		$2^{30.28}$ CP	0	AIDA/Cube	Sect. 4.1
	60		18 sec	3 KP	0	MiniSat, LI conv.	Sect. 3.1
	65		1.81 min	3 KP	0	MiniSat, LI conv.	Sect. 3.1
	66		8.85 min	3 KP	0	MiniSat, LI conv.	Sect. 3.1
	67		26 sec	3 KP	30	MiniSat, LI conv.	Sect. 3.1
	68		2.55 min	3 KP	30	MiniSat, LI conv.	Sect. 3.1
	69		47.76 min	3 KP	35	MiniSat, LI conv.	Sect. 3.1
	70		1.64 min	10 CP	35	MiniSat, LI conv.	Sect. 3.1
	71		3.58 min	10 CP	35	MiniSat, LI conv.	Sect. 3.1
	71		3.58 min	10 CP	35	MiniSat & Cube, LI conv.	Sect. 5
	75		12.50 h	3 KP	35	MiniSat, LI conv.	Sect. 3.1
	76		1.59 min	20 CP	45	MiniSat, BCJ conv./Pre-Proc	Sect. 3.2
	76		4.1 min	20 CP	43	MiniSat, BCJ conv.Pre-Proc	Sect. 3.2
	76		3.08 min	20 CP	41	MiniSat, BCJ conv./Pre-Proc	Sect. 3.2
	77		18 sec	20 CP	45	MiniSat, BCJ conv./Pre-Proc	Sect. 3.2
	78		5.80 min	20 CP	45	MiniSat, BCJ conv./Pre-Proc	Sect. 3.2
	79		14.72 min	20 CP	45	MiniSat, BCJ conv./Pre-Proc	Sect. 3.2
	254	2^{51}		$2^{23.80}$ CP	0	Side-Channel	Sect. 6
KATAN48	40	2^{49}		$2^{24.95}$ CP	0	AIDA/Cube	Sect. 4.2
	40		2 sec	5 CP	40	MiniSat, LI conv.	Sect. 3.1
	50		7 sec	5 CP	40	MiniSat, LI conv.	Sect. 3.1
	60		13.18 min	5 CP	40	MiniSat, LI conv.	Sect. 3.1
	61		7.12 min	5 CP	45	MiniSat, LI conv.	Sect. 3.1
	62		11.86 min	10 CP	40	MiniSat, LI conv.	Sect. 3.1
	63		17.47 min	10 CP	45	MiniSat, LI conv.	Sect. 3.1
	64		6.42 h	5 CP	40	MiniSat, LI conv.	Sect. 3.1
KATAN64	30	2^{35}		$2^{20.64}$ CP	0	AIDA/Cube	Sect. 4.3
	40		2 sec	5 CP	40	MiniSat, LI conv.	Sect. 3.1
	50		12 sec	5 CP	40	MiniSat, LI conv.	Sect. 3.1
	60		3.17 h	5 CP	40	MiniSat, LI conv.	Sect. 3.1

$Time_1$: time complexity unit for attacking r rounds is number of r-round KATAN computations.
$Time_2$: clock time for algebraic attacks; KP: known plaintext; CP: chosen plaintext;
LI converter: local interpolation converter; BCJ: Bard-Courtois-Jefferson converter
negl: negligible, Pre-Proc: preprocessed system of equations

In this model, only very few internal cipher bits are allowed to leak. In our case, only a single internal bit will be used. Assume one can get the value of internal bit c_{19} after 40 rounds (c.f. Table 3). We can recover 29 key bits via cube attack with data complexity $10 \cdot 2^{12} + 2^{15} + 2 \cdot 2^{16} + 2^{18} + 3 \cdot 2^{19} + 12 \cdot 2^{20} = 2^{23.80}$ CP. The remaining key bits are recovered by brute force. This brings about the time complexity of 2^{51} encryptions to attack the full 254-round KATAN32.

Table 5. Maxterms, LI cube equations and ciphertext bit for 50-round KATAN32

Maxterm	Degree	Cube equation	Cipher bit
$\texttt{1B8EE77B}_x$	20	k_2+k_{12}	c_{10}
$\texttt{1B8EE77B}_x$	20	$k_8+k_{26}+1$	c_{31}
$\texttt{B1FF633A}_x$	20	$k_{16}+1$	c_{10}
$\texttt{EB3AEAE6}_x$	20	k_{22}	c_9
$\texttt{EB3AEAE6}_x$	20	$k_{16}+k_{20}+k_{28}+k_{31}+k_{35}+k_{42}$	c_{28}
$\texttt{AEF689F9}_x$	20	$k_{16}+k_{32}+1$	c_{30}
$\texttt{56CE3DFA}_x$	20	k_{12}	c_{10}
$\texttt{7FDAA996}_x$	20	k_5	c_{10}
$\texttt{D77FE20E}_x$	20	k_{25}	c_{28}
$\texttt{AF19DFB4}_x$	20	k_6	c_{12}
$\texttt{DC3C97ED}_x$	20	$k_4+k_{11}+1$	c_{30}
$\texttt{61BC7B9F}_x$	20	k_5+k_{20}	c_9
$\texttt{23B35FD7}_x$	20	k_0	c_{12}
$\texttt{9CF75766}_x$	20	$k_{15}+1$	c_6
$\texttt{9CF75766}_x$	20	k_7+1	c_{26}
$\texttt{E58FB7CA}_x$	20	$k_{22}+k_{26}$	c_{26}
$\texttt{C7E6C7CB}_x$	20	$k_{11}+k_{21}+k_{31}+1$	c_{10}
$\texttt{3E3BE3EA}_x$	20	$k_{13}+1$	c_{10}
$\texttt{3FFCCD62}_x$	20	$k_8+k_{10}+k_{13}+k_{14}+k_{20}+k_{24}$	c_{30}
$\texttt{EF4FD985}_x$	20	$k_6+k_{11}+k_{22}+k_{32}+k_{33}+k_{37}$	c_7
$\texttt{2F6D66FA}_x$	20	k_3+1	c_{31}
$\texttt{BE5E19F3}_x$	20	$k_2+k_7+k_9+k_{12}+k_{16}+k_{17}+k_{21}+k_{26}+k_{27}+k_{30}+k_{34}+k_{39}+k_{43}$	c_{29}
$\texttt{ECDD58BD}_x$	20	$k_6+k_7+k_{16}+k_{17}+k_{20}+k_{23}+k_{25}+k_{26}+k_{27}+k_{34}+1$	c_{29}
$\texttt{DEBCFB22}_x$	20	$k_4+k_{17}+k_{28}+k_{35}+k_{37}+k_{45}$	c_8
$\texttt{FE3E09D7}_x$	20	$k_{11}+k_{22}$	c_9
$\texttt{F83B3AEB}_x$	20	k_{18}	c_{29}
$\texttt{BACCAF37}_x$	20	$k_2+k_{12}+k_{14}+k_{22}$	c_{12}
$\texttt{7FD07B66}_x$	20	$k_{15}+k_{31}+1$	c_8
$\texttt{BAFEA8D3}_x$	20	k_{27}	c_{10}
$\texttt{AF6AAE75}_x$	20	k_1+1	c_9
$\texttt{3ADC3DD7}_x$	20	$k_0+k_2+k_{12}+k_{20}+k_{24}+k_{31}$	c_{10}
$\texttt{A7D3F749}_x$	20	$k_2+k_{10}+k_{12}+k_{21}+k_{23}+k_{30}$	c_8
$\texttt{8FF7D615}_x$	20	$k_{16}+k_{18}+k_{28}+1$	c_{28}
$\texttt{AF88BDFA}_x$	20	k_{17}	c_9
$\texttt{FE1A11FF}_x$	20	$k_5+k_{14}+k_{19}+k_{22}+k_{31}+1$	c_{12}
$\texttt{DFF9C30D}_x$	20	k_{23}	c_{29}
$\texttt{95BF5D4D}_x$	20	$k_2+k_4+k_7+k_8+k_{10}+k_{14}+k_{16}+k_{18}+k_{22}+k_{24}+k_{26}+k_{35}+1$	c_6
$\texttt{9DF2EE93}_x$	20	$k_{21}+k_{38}+1$	c_{27}
$\texttt{6DD3973B}_x$	20	$k_{12}+k_{14}+k_{29}$	c_{11}
$\texttt{A271A7FF}_x$	20	$k_1+k_9+k_{10}+k_{16}+k_{20}+k_{21}+1$	c_{10}
$\texttt{F6CDFA15}_x$	20	$k_8+k_{14}+k_{15}+k_{18}+k_{24}+k_{29}+k_{33}+k_{40}$	c_{29}
$\texttt{5E5AB5EB}_x$	20	$k_{12}+k_{14}+k_{16}+k_{20}+k_{22}+k_{25}+k_{26}+k_{39}$	c_{31}
$\texttt{FD847AF6}_x$	20	$k_{24}+k_{33}+1$	c_9
$\texttt{F770ECEC}_x$	20	$k_5+k_{16}+k_{24}+k_{26}+k_{27}+k_{30}+k_{32}+k_{34}+k_{35}+k_{36}+k_{43}$	c_5
$\texttt{FBAE4E3A}_x$	20	$k_7+k_{28}+k_{29}+k_{33}+k_{44}+1$	c_{27}
$\texttt{EEB6A9A7}_x$	20	$k_6+k_9+k_{16}+k_{18}+k_{19}+k_{25}+k_{28}+k_{37}+k_{38}+k_{41}+k_{43}++k_{48}$	c_7

Table 6. Maxterms, cube degree and equations and ciphertext bit for 60-round KATAN32

Maxterm	Degree	Cube equation	Cipher bit
B6F7FAFD$_X$	25	$k_{30}+1$	c_{31}
EFE7F6FA$_X$	25	k_{38}	c_{11}
F7CDFFCE$_X$	25	$k_{28}+k_{30}+k_{32}+k_{36}+k_{40}+k_{42}+k_{49}+k_{50}+1$	c_{31}
FEBF7EAE$_X$	25	$k_{28}+k_{30}+k_{40}+k_{54}+1$	c_{14}
63D7FFF7$_X$	25	$k_{16}+k_{26}+k_{30}+k_{38}+k_{40}+k_{43}+k_{44}$	c_{27}
3F7BBF5F$_X$	25	$k_{8}+k_{18}+k_{20}+k_{26}+k_{32}$	c_{17}
EF2FF9EF$_X$	25	$k_{26}+1$	c_{12}
EF2FF9EF$_X$	25	$k_{22}+k_{32}+1$	c_{31}
FFF3F573$_X$	25	$k_{15}+k_{17}+k_{19}+k_{23}+k_{32}+k_{33}+k_{35}+k_{37}+k_{41}+k_{43}+k_{44}+1$	c_{11}
DA9EFFF7$_X$	25	$k_{4}+k_{10}+k_{14}+k_{22}+k_{24}+k_{29}+k_{30}+k_{49}+1$	c_{11}
FFD6BABF$_X$	25	k_{46}	c_{31}
FF5777F6$_X$	25	k_{47}	c_{11}
B7F2DFDF$_X$	25	$k_{16}+1$	c_{16}
B7F2DFDF$_X$	25	$k_{13}+k_{16}+1$	c_{17}
BDF7FD97$_X$	25	$k_{27}+k_{34}$	c_{11}
7FDEBFDC$_X$	25	$k_{14}+k_{21}+k_{22}+k_{41}$	c_{15}
EFF9B7ED$_X$	25	$k_{13}+k_{31}+1$	c_{16}
FFF5FDCC$_X$	25	$k_{19}+k_{23}+k_{27}+k_{35}+k_{38}+k_{43}+k_{48}+1$	c_{9}
EEBB7DF7$_X$	25	$k_{22}+1$	c_{16}
F7C6EDFF$_X$	25	$k_{8}+k_{18}+k_{35}+k_{43}+1$	c_{14}
FE3BF77E$_X$	25	$k_{17}+k_{33}+1$	c_{15}
9FFE7FAE$_X$	25	$k_{10}+k_{18}+k_{20}+k_{24}+1$	c_{17}
EFFFDD9A$_X$	25	$k_{38}+k_{39}$	c_{12}
FEF7779B$_X$	25	$k_{23}+k_{27}$	c_{14}
CFFF7BE6$_X$	25	k_{12}	c_{14}
EABFF73F$_X$	25	$k_{32}+k_{36}+1$	c_{13}
BC7FCF7F$_X$	25	$k_{2}+1$	c_{14}
FADFECFB$_X$	25	$k_{43}+1$	c_{14}
DDD3FF3F$_X$	25	$k_{28}+1$	c_{13}
EB67DDFF$_X$	25	$k_{16}+k_{26}+k_{28}+k_{35}+k_{40}+k_{44}+1$	c_{30}
FEEFB8FE$_X$	25	$k_{30}+k_{32}+k_{42}$	c_{31}
3CFFEF7E$_X$	25	$k_{14}+k_{16}+k_{20}+k_{22}+1$	c_{13}
DFEFF4DD$_X$	25	k_{15}	c_{31}
AFBEFDCF$_X$	25	k_{10}	c_{17}
DAF9FFED$_X$	25	$k_{28}+k_{32}+k_{36}+k_{45}+1$	c_{18}
DF733FEF$_X$	25	$k_{16}+k_{21}+1$	c_{17}
BF7BEE6F$_X$	25	$k_{1}+1$	c_{13}
FFEE57FC$_X$	25	$k_{11}+k_{29}+k_{33}+k_{35}+k_{37}+k_{39}+k_{41}+k_{44}+k_{45}+k_{49}+k_{50}+k_{51}+k_{59}+k_{60}+1$	c_{11}
FA7ECFFD$_X$	25	$k_{17}+k_{22}+k_{24}+k_{27}+k_{28}+k_{33}+k_{34}+k_{39}$	c_{16}
B9E77FFE$_X$	25	$k_{10}+k_{20}+k_{24}+k_{26}+k_{28}+k_{30}+k_{36}+k_{37}+k_{38}+k_{44}+k_{53}+1$	c_{11}
FF37EDEB$_X$	25	k_{50}	c_{29}

Table 7. Maxterms, cube degree and equations and ciphertext bit for 71-round KATAN32

Maxterm	Degree	Cube equation	Cipher bit
FFFFD5FC$_x$	27	$k_{52}+k_{54}+k_{60}$	c_{14}
FFFFF7C5$_x$	27	$k_{31}+k_{51}+k_{53}+k_{54}+k_{64}+k_{71}$	c_{31}
FFF3FA7F$_x$	28	$k_{46}+1$	c_{18}

Table 8. Maxterms, cube degree and equations and ciphertext bit for 40-round KATAN48

Maxterm	Degree	Cube equation	Cipher bit
66140B44FE81$_x$	20	k_3+k_8	c_{37}
2096B841C6F2$_x$	20	k_0+k_6	c_{18}
2004D819B69F$_x$	20	k_8	c_{46}
01E07456499B$_x$	20	$k_3+k_{12}+k_{18}+1$	c_{13}
874108B1E347$_x$	20	k_2+k_9	c_{43}
85DF1310A226$_x$	20	k_{11}	c_{43}
D9F00150D11E$_x$	20	$k_1+k_{12}+1$	c_{12}
204D49C8B56C$_x$	20	k_4	c_{16}
3000F607DC4E$_x$	20	k_6+1	c_{43}
75045046CC5E$_x$	20	k_9	c_{15}
8D705440E2CB$_x$	20	$k_5+k_{14}+k_{18}$	c_{42}
5024603E9A37$_x$	20	$k_3+k_6+k_8+k_{10}+k_{23}+1$	c_{16}
5024603E9A37$_x$	20	k_1+k_2+1	c_{44}
3034E083566D$_x$	20	k_7+1	c_{18}
81E48D04DB19$_x$	20	$k_1+k_3+k_5+k_{14}+k_{15}+1$	c_{41}
41482473ADB4$_x$	20	$k_4+k_{19}+1$	c_{42}
3D4635605382$_x$	20	k_{13}	c_{39}
51902406CABF$_x$	20	$k_3+k_8+k_{10}+1$	c_{44}
583088DB0C6E$_x$	20	$k_1+k_2+k_9+k_{16}$	c_{16}
5040C4CE9AF1$_x$	20	$k_0+k_2+k_4+k_{21}+1$	c_{41}
7749008CBAC1$_x$	20	$k_1+k_5+k_8+k_{10}+k_{11}+k_{13}+1$	c_{42}
96940C46139E$_x$	20	$k_0+k_1+k_9+k_{12}+k_{14}+k_{20}+1$	c_{41}
96800804FF5B$_x$	20	$k_0+k_3+k_8+k_9+k_{15}+k_{17}+1$	c_{46}
211013326F3D$_x$	20	$k_4+k_8+k_{10}+k_{16}+k_{25}$	c_{15}
18574012A577$_x$	20	$k_6+k_8+k_{12}+k_{14}+k_{29}+1$	c_{47}
81668801EE97$_x$	20	$k_0+k_1+k_5+k_6+k_8+k_9+k_{14}+k_{15}+k_{31}+1$	c_{46}
3050A044F5ED$_x$	20	$k_1+k_8+k_9+k_{13}+k_{15}+k_{24}+1$	c_{46}
42BF16A44AA0$_x$	20	$k_7+k_9+k_{22}$	c_{10}
50AD4122AA1F$_x$	20	$k_{10}+k_{27}$	c_{12}
0AAB9004F89D$_x$	20	$k_0+k_3+k_{10}+k_{11}+k_{26}$	c_{20}
C884A100FCF3$_x$	20	$k_3+k_{11}+k_{13}+k_{14}+k_{15}+k_{17}+k_{28}$	c_{18}

Table 9. Maxterms, cube degree and equations and ciphertext bit for 30-round KATAN64

Maxterm	Degree	Cube equation	Cipher bit
0CB0C29808C10001$_x$	16	k_5	c_{44}
2E2128800020305A$_x$	16	k_4	c_7
10E2002920014471$_x$	16	$k_1+k_5+k_{12}$	c_{47}
0A12042100446263$_x$	16	$k_8+k_{10}+k_{19}$	c_{12}
029290CC02C10140$_x$	16	k_2	c_5
AE0C032002100492$_x$	16	k_9	c_9
4241092108534C00$_x$	16	k_1	c_{44}
0E0864A20828A800$_x$	16	k_0	c_{56}
4104901087403083$_x$	16	k_7	c_8
44010B12812A0124$_x$	16	k_3	c_{49}
0200A0D00305E08A$_x$	16	k_3+k_{10}	c_{48}
041102168238A802$_x$	16	k_6	c_9
439C00A810940044$_x$	16	$k_3+k_8+k_{17}$	c_9
60910A0B93000802$_x$	16	k_1+k_8	c_{47}
018C084049C98003$_x$	16	$k_0+k_1+k_2+k_8+k_{11}$	c_8
3C1500040080C097$_x$	16	k_4+k_{15}	c_{48}
0800FD4900016180$_x$	16	$k_5+k_9+k_{18}$	c_{54}
002091443A501C40$_x$	16	k_2+k_{13}	c_{45}
1027118032506001$_x$	16	$k_1+k_5+k_{10}+k_{21}$	c_{10}
0080DC00814454A8$_x$	16	$k_5+k_7+k_{14}$	c_{49}
11320C0241095220$_x$	16	$k_4+k_5+k_7+k_9+k_{15}+k_{20}+k_{24}$	c_{50}
8E200808003A8D40$_x$	16	$k_3+k_6+k_{12}+k_{16}$	c_{51}
00458C3220521011$_x$	16	$k_0+k_2+k_5+k_{10}+k_{11}+k_{13}+k_{20}$	c_{11}
4024935C01018048$_x$	16	$k_0+k_5+k_9+k_{11}+k_{22}$	c_{49}
8004007882307052$_x$	16	$k_0+k_6+k_{12}+k_{23}$	c_6

7 Conclusions

This paper described algebraic, AIDA/cube and side-channel attacks on the KATAN family of block ciphers [13]. A new feature observed in cube attacks is that the same maxterm suggests more than one linear independent equation on the key bits. This phenomenon leads to a reduction in the data complexity of our attacks.

For algebraic attacks, deploying pre-processing step on the system of equations before feeding it to the SAT solvers decreases the complexity of the attack for KATAN32. As topic for further research, this method can be tried on other family members.

In the side-channel attack for KATAN32, we observed significant leakage from bit 19 after 40 rounds. More specifically, we could recover 29 linear independent equations on the key bits. Surprisingly enough, this bit position is exactly the LSB of register L_1. This finding is similar to the structure of chosen plaintexts picked in attacking various versions using SAT solvers (Sect. 3.1). We leave similar side-channel analysis of KATAN48 and KATAN64 as future work.

Table 4 summarizes the attack complexities on the KATAN family of block ciphers. In this table, we keep two different time complexities: Time_1 and Time_2, since there is no straightforward and unique way to convert one into the other. Recall that Time_1

measures the effort in number of encryptions, while Time_2 measures the effort in clock time. The former is used for attacks that explicitly perform partial encryption or decryption, while the latter is used for attacks related to internal operations in SAT solvers.

References

1. Ars, G., Faugère, J.-C.: An Algebraic Cryptanalysis of Nonlinear Filter Generators using Gröbner Bases. Technical report, INRIA research report (2003), `https://hal.ccsd.cnrs.fr/`
2. Aumasson, J.P., Dinur, I., Meier, W., Shamir, A.: Cube Testers and Key Recovery Attacks on Reduced-Round MD6 and Trivium. In: Dunkelman, O. (ed.) Fast Software Encryption. LNCS, vol. 5665, pp. 1–22. Springer, Heidelberg (2009)
3. Bard, G.: Algebraic Cryptanalysis. Springer, Heidelberg (2009)
4. Bard, G., Courtois, N., Jefferson, C.: Efficient Methods for Conversion and Solution of Sparse Systems of Low-Degree Multivariate Polynomials over GF(2) via SAT-Solvers. Presented at ECRYPT workshop Tools for Cryptanalysis eprint/2007/024 (2007)
5. Blum, M., Luby, M., Rubinfeld, R.: Self testing/correcting with applications to numerical problems. In: ACM STOC, pp. 73–83 (1990)
6. Buchberger, B.: An Algorithm for Finding the Basis Elements of the Residue Class Ring of a Zero Dimensional Polynomial Ideal. PhD thesis, Johannes Kepler University of Linz, JKU (1965)
7. Courtois, N., Bard, G., Wagner, D.: Algebraic and Slide Attacks on Keeloq. In: Nyberg, K. (ed.) FSE 2008. LNCS, vol. 5086, pp. 97–115. Springer, Heidelberg (2008)
8. Courtois, N., Bard, G.V.: Algebraic Cryptanalysis of the Data Encryption Standard. In: Galbraith, S.D. (ed.) IMA Int. Conf 2007. LNCS, vol. 4887, pp. 152–169. Springer, Heidelberg (2007)
9. Courtois, N., Meier, W.: Algebraic Attacks on Stream Ciphers with Linear Feedback. In: Biham, E. (ed.) EUROCRYPT 2003. LNCS, vol. 2656, pp. 345–359. Springer, Heidelberg (2003)
10. Courtois, N., O'Neil, S., Quisquater, J.: Practical Algebraic Attacks on the Hitag2 Stream Cipher. In: Samarati, P., Yung, M., Martinelli, F., Ardagna, C.A. (eds.) ISC 2009. LNCS, vol. 5735, pp. 167–176. Springer, Heidelberg (2009)
11. Courtois, N., Pieprzyk, J.: Cryptanalysis of Block Ciphers with Overdefined Systems of Equations. In: Zheng, Y. (ed.) ASIACRYPT 2002. LNCS, vol. 2501, pp. 267–287. Springer, Heidelberg (2002)
12. Courtois, N., Shamir, A., Patarin, J., Klimov, A.: Efficient Algorithms for Solving Overdefined Systems of Multivariate Polynomial Equations. In: Preneel, B. (ed.) EUROCRYPT 2000. LNCS, vol. 1807, pp. 392–407. Springer, Heidelberg (2000)
13. De Canniére, C., Dunkelman, O., Knezević, M.: Katan and ktantan - a family of small and efficient hardware-oriented block ciphers. In: Clavier, C., Gaj, K. (eds.) CHES 2009. LNCS, vol. 5747, pp. 272–288. Springer, Heidelberg (2009)
14. De Canniére, C., Preneel, B.: Trivium. In: Robshaw, M.J.B., Billet, O. (eds.) New Stream Cipher Designs. LNCS, vol. 4986, pp. 244–266. Springer, Heidelberg (2008)
15. Ding, J., Buchmann, J., Mohamed, M.S.E., Mohamed, W.S.A., Weinmann, R.-P.: MutantXL algorithm. In: Proceedings of the 1st International Conference in Symbolic Computation and Cryptography, pp. 16–22 (2008)
16. Dinur, I., Shamir, A.: Cube attacks on tweakable black box polynomials. In: Joux, A. (ed.) EUROCRYPT 2009. LNCS, vol. 5479, pp. 278–299. Springer, Heidelberg (2010)

17. Dinur, I., Shamir, A.: Side Channel Cube Attacks on Block Ciphers. IACR ePrint Archive, ePrint 127 (2009)
18. Een, N., Sorensson, N.: Minisat - A SAT Solver with Conflict-Clause Minimization. In: Giunchiglia, E., Tacchella, A. (eds.) SAT 2003. LNCS, vol. 2919, pp. 502–518. Springer, Heidelberg (2004)
19. Faugère, J.: A new efficient algorithm for computing Gröbner bases without reduction to zero (F5). In: Symbolic and Algebraic Computation - ISSAC, pp. 75–83 (2002)
20. Faugère, J.C.: A new effcient algorithm for computing Gröbner bases (F4). Journal of Pure and Applied Algebra 139(1), 61–88 (1999)
21. Indesteege, S., Keller, N., Dunkelman, O., Biham, E., Preneel, B.: A Practical Attack on Keeloq. In: Smart, N.P. (ed.) EUROCRYPT 2008. LNCS, vol. 4965, pp. 1–18. Springer, Heidelberg (2008)
22. Lazard, D.: Gröbner-bases, Gaussian elimination and resolution of systems of algebraic equations. In: van Hulzen, J.A. (ed.) ISSAC 1983 and EUROCAL 1983. LNCS, vol. 162, Springer, Heidelberg (1983)
23. Macaulay, F.S.: The algebraic theory of modular systems. Cambridge Mathematical Library (1916)
24. Markovitz, H.M.: The Elimination Form of the Inverse and Its Application to Linear Programming. Management Science, 225–269 (1957)
25. Mohamed, M.S.E., Mohamed, W.S.A.E., Ding, J., Buchmann, J.: MXL2: Solving Polynomial Equations over GF(2) using an Improved Mutant Strategy. In: Buchmann, J., Ding, J. (eds.) PQCrypto 2008. LNCS, vol. 5299, pp. 203–215. Springer, Heidelberg (2008)
26. Nohl, K., Soos, M.: Solving Low-Complexity Ciphers with Optimized SAT Solvers. In: EUROCRYPT (2009)
27. Raddum, H., Semaev, I.: New technique for solving sparse equation systems. In: Cryptology ePrint Archive (2006), `http://eprint.iacr.org/2006/475`
28. SAT. Sat Race Competition, `http://www.satcompetition.org/`
29. Shannon, C.E.: Claude Elwood Shannon Collected Papers. Wiley-IEEE Press, Piscataway (1993)
30. Vielhaber, M.: Breaking ONE.FIVIUM by AIDA an Algebraic IV Differential Attack. In: Cryptology ePrint Archive, report 413 (2007)
31. Wong, K.K.H., Bard, G.: Improved Algebraic Cryptanalysis of QUAD, Bivium and Trivium via Graph Partitioning on Equation Systems. In: ACISP (2010)

A Maxterms and Cube Indexes

The maxterms listed bellow represent the ones with the smallest maxterms. All cube equations listed are linearly independent (LI). We use hex format to describe maxterms such that those one bits are selected bits. e.g. $0000000f_x$ means the maxterm cube indices are 0, 1, 2 and 3. For KATAN32, plaintext/ciphertext bits are numbered as $p = (p_{31}, \ldots, p_0)$. For KATAN48, the bit numbering is $p = (p_{47}, \ldots, p_0)$. For KATAN64, the bit numbering is $p = (p_{63}, \ldots, p_0)$.

The Improbable Differential Attack: Cryptanalysis of Reduced Round CLEFIA*

Cihangir Tezcan

École Polytechnique Fédérale de Lausanne
EDOC-IC BC 350 Station 14 CH-1015 Lausanne, Switzerland
cihangir.tezcan@epfl.ch

Abstract. In this paper we present a new statistical cryptanalytic technique that we call improbable differential cryptanalysis which uses a differential that is less probable when the correct key is used. We provide data complexity estimates for this kind of attacks and we also show a method to expand impossible differentials to improbable differentials. By using this expansion method, we cryptanalyze 13, 14, and 15-round CLEFIA for the key sizes of length 128, 192, and 256 bits, respectively. These are the best cryptanalytic results on CLEFIA up to this date.

Keywords: Cryptanalysis, Improbable differential attack, CLEFIA.

1 Introduction

Statistical attacks on block ciphers make use of a property of the cipher so that an incident occurs with different probabilities depending on whether the correct key is used or not. For instance, differential cryptanalysis [1] considers characteristics or differentials which show that a particular output difference should be obtained with a relatively high probability when a particular input difference is used. Hence, when the correct key is used, the predicted differences occur more frequently. In a classical differential characteristic the differences are fully specified and in a truncated differential [2] only parts of the differences are specified.

On the other hand, impossible differential cryptanalysis [3] uses an impossible differential which shows that a particular difference cannot occur for the correct key (i.e. probability of this event is exactly zero). Therefore, if these differences are satisfied under a trial key, then it cannot be the correct one. Thus, the correct key can be obtained by eliminating all or most of the wrong keys.

In this paper we describe a new variant of differential cryptanalysis in which a given differential holds with a relatively small probability. Therefore, when the correct key is used, the predicted differences occur less frequently. In this respect, the attack can be seen as the exact opposite of (truncated) differential cryptanalysis. For this reason, we call this kind of differentials *improbable differentials*

* This work was done when the author was a research assistant at Institute of Applied Mathematics, Middle East Technical University, Ankara, Turkey.

G. Gong and K.C. Gupta (Eds.): INDOCRYPT 2010, LNCS 6498, pp. 197–209, 2010.

and we call the method *improbable differential cryptanalysis*. Early applications of improbable events in differential attacks were mentioned in [4] and [5].

Accurate estimates of the data complexity and success probability for many statistical attacks are provided by Blondeau et al. in [6,7] but these estimates work for the cases when an incident is more probable for the correct key. We make necessary changes on these estimates to be able to estimate data complexity and success probability of improbable differential attacks.

Moreover, we show that improbable differentials can be obtained when suitable differentials that can be put on the top or the bottom of an impossible differential exist. Expanding an impossible differential to an improbable differential in this way can be used to distinguish more rounds of the cipher from a random permutation; or it can be turned into an improbable differential attack that covers more rounds than the impossible differential attack.

CLEFIA [8] is a 128-bit block cipher developed by Sony Corporation that has a generalized Feistel structure of four data lines. Security evaluations done by the designers [8,9] show that impossible differential attack is one of the most powerful attacks against CLEFIA for they provided 10, 11, and 12-round impossible differential attacks on CLEFIA for 128, 192, and 256-bit key lengths, respectively. In [10,11], Tsunoo et al. provided new impossible differential attacks on 12, 13, and 14-round CLEFIA for 128, 192, and 256-bit key lengths, respectively. Moreover, in [12], Zhang and Han provided a 14-round impossible differential attack on 128-bit keyed CLEFIA but due to the arguments on the time complexity, it remains unknown whether this attack scenario is successful or not.

In this work, we expand the 9-round impossible differentials introduced in [10] to 10-round improbable differentials and use them to attack 13, 14, and 15-round CLEFIA for 128, 192, and 256-bit key lengths, respectively. To the best of our knowledge, these are the best cryptanalytic results on CLEFIA. The paper is organized as follows: The description of the improbable differential attack, estimates of the data complexity, and expansion of impossible differentials to improbable differentials are given in Sect. 2. The notation and the description of CLEFIA is given in Sect. 3. In Sect. 4, we expand 9-round impossible differentials on CLEFIA to 10-round improbable differentials and use them to attack 13, 14, and 15-round CLEFIA. We conclude our paper with Sect. 5.

2 Improbable Differential Cryptanalysis

Statistical attacks on block ciphers make use of a property of the cipher so that an incident occurs with different probabilities depending on whether the correct key is used or not. We denote the probability of observing the incident under a wrong key with p and p_0 denotes the probability of observing the incident under the correct key.

In previously defined statistical differential attacks on block ciphers, a differential is more probable for the correct key than a random key (i.e. $p_0 > p$). Moreover, an impossible differential attack uses an impossible differential that is not possible when tried with the correct key (i.e. $p_0 = 0$). We define the improbable differential attack as a statistical differential attack in which a given differential is less probable than a random key (i.e. $p_0 < p$). Hence, improbable differential attacks can be seen as the exact opposite of differential attacks.

We aim to find a differential with α input difference and β output difference so that these differences are observed with probability p_0 for the correct key and with probability p for a wrong key where $p_0 < p$. One way of obtaining such differences is by finding nontrivial differentials that have α input difference and an output difference other than β, or vice versa. Hence these differentials reduce the probability of observing the differences α and β under the correct key.

We define an improbable differential as a differential that does not have the output difference β with a probability p', when the input difference is α. Thus, p' denotes the total probability of nontrivial differentials having α input difference with an output difference other than β. Hence for the correct key, probability of observing the α and β differences (i.e. satisfying the improbable differential) becomes $p_0 = p \cdot (1 - p')$. Note that p_0 is larger than $p \cdot (1 - p')$ if there are nontrivial differentials having α input difference and β output difference. Hence the attacker should check the existance of such differentials.

An improbable differential can be obtained by using a miss in the middle [3] like technique which we call the almost miss in the middle technique. Let α difference becomes δ with probability p_1 after r_1 rounds of encryption and β difference becomes γ after r_2 rounds of decryption as shown in Fig. 1. With the assumption that these two events are independent, if δ is different than γ, then α difference does not become β with probability $p' = p_1 \cdot p_2$ after $r_1 + r_2$ rounds of encryption. Note that p_1 and p_2 equal to 1 in the miss in the middle technqiue. Furthermore, we define an expansion method for constructing an improbable differential from an impossible differential in Sect. 2.2.

Note that the impossible differential attacks can be seen as a special case of improbable differential attacks where the probability p' is taken as 1.

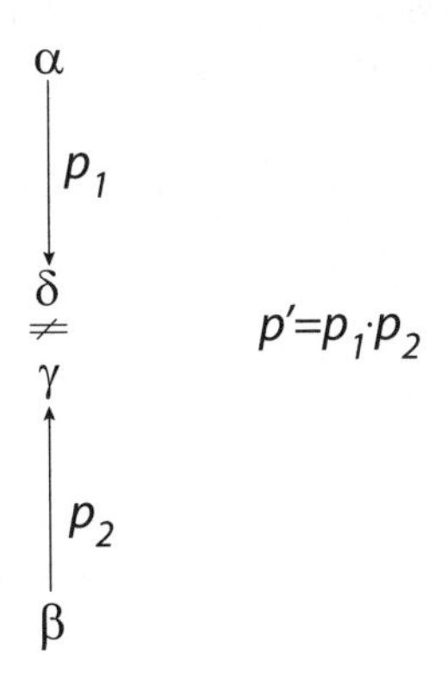

Fig. 1. Almost miss in the middle technique

2.1 Data Complexity and Success Probability

Since p_0 is less than p, our aim is to use N plaintext pairs and count the hits that every guessed subkey gets and expect that the counter for the correct subkey to be less than a threshold T. Number of hits a wrong subkey gets can be seen as a random variable of a binomial distribution with parameters N, p (and a random variable of a binomial distribution with parameters N, p_0 for the correct subkey). We denote the *non-detection* error probability with p_{nd} which is the probability of the counter for the correct subkey to be higher than T. And we denote the *false alarm* error probability with p_{fa} which is the probability of the counter for a wrong subkey to be less than or equal to T. Therefore, the success probability of an improbable differential attack is $1 - p_{nd}$.

Accurate estimates of the data complexity and success probability for many statistical attacks are provided by Blondeau et al. in [6,7] and these estimates can be used for improbable differential attacks with some modifications. Unlike improbable differential cryptanalysis, in most of the statistical attacks $p_0 > p$ and this assumption is made throughout [6]. Hence, we need to modify the approximations N', N'' and N_∞ of the number of required samples N that are given in [6] for the $p_0 < p$ case in order to use them for improbable differential attacks. We first define the Kullback-Leibler divergence which plays an important role in these estimates.

Definition 1 (Kullback-Leibler divergence [13]). *Let P and Q be two Bernoulli probability distributions of parameters p and q. The Kullback - Leibler divergence between P and Q is defined by*

$$D(p||q) = p \ \ln\left(\frac{p}{q}\right) + (1-p) \ \ln\left(\frac{1-p}{1-q}\right).$$

Secondly, we modify Algorithm 1 of [6] for the $p_0 < p$ case which computes the exact number of required samples N and corresponding relative threshold $\tau := \frac{T}{N}$ to reach error probabilities less than (p_{nd}, p_{fa}). The estimates for non-detection and false alarm error probabilities are denoted by $G_{nd}(N, \tau)$ and $G_{fa}(N, \tau)$.

Algorithm 1. from [6], modified for the $p_0 < p$ Case

Input: p_0, p, p_{nd}, p_{fa}
Output: N, τ
$\tau_{min} := p_0,\ \tau_{max} := p$
repeat
  $\tau := \frac{\tau_{min}+\tau_{max}}{2}$
  Compute N_{nd} such that $\forall N > N_{nd},\ G_{nd}(N, \tau) \leq p_{nd}$
  Compute N_{fa} such that $\forall N > N_{fa},\ G_{fa}(N, \tau) \leq p_{fa}$
  if $N_{nd} > N_{fa}$ **then** $\tau_{min} = \tau$
  else $\tau_{max} = \tau$
until $N_{nd} = N_{fa}$
$N := N_{nd}$
Return N, τ

N_{nd} and N_{fa} can be calculated by a dichotomic search and the following Equations 1 and 2 can be used for the estimates $G_{nd}(N,\tau)$ and $G_{fa}(N,\tau)$, respectively. The number of samples obtained from the algorithm with these estimates is denoted by N_∞.

Theorem 1 ([14]). *Let p_0 and p be two real numbers such that $0 < p_0 < p < 1$ and let τ such that $p_0 < \tau < p$. Let Σ_0 and Σ_k follow a binomial law of respective parameters (N, p_0) and (N, p). Then as $N \to \infty$,*

$$P(\Sigma_0 \geq \tau N) \sim \frac{(1-p_0)\sqrt{\tau}}{(\tau - p_0)\sqrt{2\pi N(1-\tau)}} e^{-ND(\tau||p_0)}, \tag{1}$$

and

$$P(\Sigma_k \leq \tau N) \sim \frac{p\sqrt{1-\tau}}{(p-\tau)\sqrt{2\pi N\tau}} e^{-ND(\tau||p)}. \tag{2}$$

A simple approximation N' of N is defined in [6] when the relative threshold is chosen as $\tau = p_0$ which makes non-detection error probability p_{nd} of order 1/2. We define N' for the $p_0 < p$ case as in [15]:

Proposition 1. *For a relative threshold $\tau = p_0$, a good approximation of the required number of pairs N to distinguish between the correctly keyed permutation and an incorrectly keyed permutation with false alarm probability less than or equal to p_{fa} is*

$$N' = -\frac{1}{D(p_0||p)}\left[\ln\left(\frac{\nu \cdot p_{fa}}{\sqrt{D(p_0||p)}}\right) + 0.5\ln(-\ln(\nu \cdot p_{fa}))\right] \tag{3}$$

where

$$\nu = \frac{(p-p_0)\sqrt{2\pi p_0}}{p\sqrt{(1-p_0)}}.$$

In [6] a good approximation of N' which is also valid for the $p_0 < p$ case is defined as follows

$$N'' = -\frac{\ln(2\sqrt{\pi}p_{fa})}{D(p_0||p)}. \tag{4}$$

2.2 Improbable Differentials from Impossible Differentials

An improbable differential can be obtained by combining a differential (or two) with an impossible differential in order to obtain improbable differentials covering more rounds. Let $\delta \nrightarrow \gamma$ be an impossible differential and $\alpha \to \delta$ and $\gamma \leftarrow \beta$ be two differentials with probabilities p_1 and p_2, respectively. Then we can construct improbable differentials $\alpha \nrightarrow \gamma$, $\delta \nrightarrow \beta$ and $\alpha \nrightarrow \beta$ with probabilities p' equal to p_1, p_2 and $p_1 \cdot p_2$ as shown in Fig. 2.

This expansion method can be used to construct improbable differentials to distinguish more rounds of the cipher from a random permutation; or an impossible differential attack can be turned into an improbable differential attack

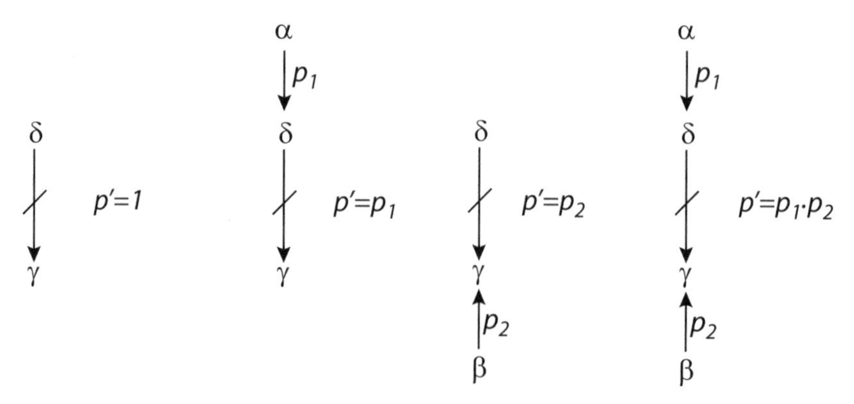

Fig. 2. Expansion of an impossible differential to improbable differentials

on more rounds of the cipher when suitable differentials $\alpha \rightarrow \delta$ or $\gamma \leftarrow \beta$ exist. However, such a conversion might require more data to obtain the correct key and hence result in higher data and time complexity. If the size of the guessed key decreases in the converted improbable differential attack, so does the memory complexity. The guessed subkeys can be represented by one bit of an array in impossible differential attacks. However, we need to keep counters for the subkeys in improbable differential attacks and hence the memory complexity is higher when the same number of subkeys are guessed.

3 Notation and the CLEFIA

3.1 Notation

We use the notations provided in Table 1 in the following sections.

3.2 CLEFIA

CLEFIA is a 128-bit block cipher having a generalized Feistel structure with four data lines. For the key lengths of 128, 192, and 256 bits, CLEFIA has 18,

Table 1. Notation

$a_{(b)}$	b denotes the bit length of a
$a\|b$	Concatenation of a and b
$[a, b]$	Vector representation of a and b
a^t	Transposition of a vector a
$a \oplus b$	Bitwise exclusive-OR (XOR) of a and b
$[x^{\{i,0\}}, x^{\{i,1\}}, x^{\{i,2\}}, x^{\{i,3\}}]$	i-th round output data
Δa	XOR difference for a

22, and 26 rounds. Each round contains two parallel F functions, F_0 and F_1 and their structures are shown in Fig. 3 where S_0 and S_1 are 8×8-bit S-boxes. The two matrices M_0 and M_1 that are used in the F-functions are defined as follows.

$$M_0 = \begin{pmatrix} 0x01 & 0x02 & 0x04 & 0x06 \\ 0x02 & 0x01 & 0x06 & 0x04 \\ 0x04 & 0x06 & 0x01 & 0x02 \\ 0x06 & 0x04 & 0x02 & 0x01 \end{pmatrix}, \qquad M_1 = \begin{pmatrix} 0x01 & 0x08 & 0x02 & 0x0a \\ 0x08 & 0x01 & 0x0a & 0x02 \\ 0x02 & 0x0a & 0x01 & 0x08 \\ 0x0a & 0x02 & 0x08 & 0x01 \end{pmatrix}.$$

The encryption function uses four 32-bit whitening keys (WK_0, WK_1, WK_2, WK_3) and $2r$ 32-bit round keys ($RK_0, \ldots, RK_{2r-1}$) where r is the number of rounds. We represent the bytes of a round key as $RK_i = RK_{i,0}|RK_{i,1}|RK_{i,2}|RK_{i,3}$. The encryption function ENC_r is shown in Fig. 4.

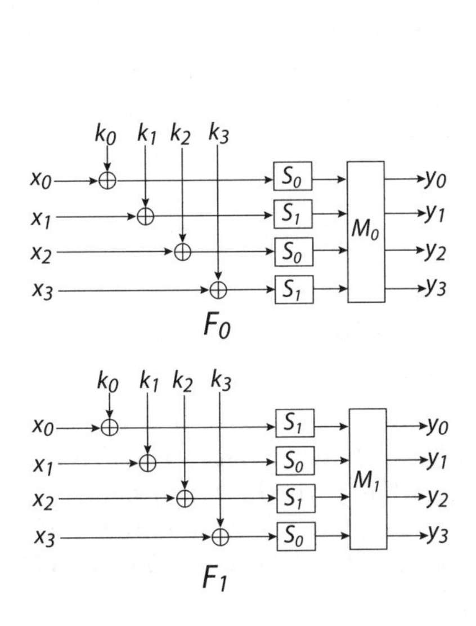

Fig. 3. F_0 and F_1 functions

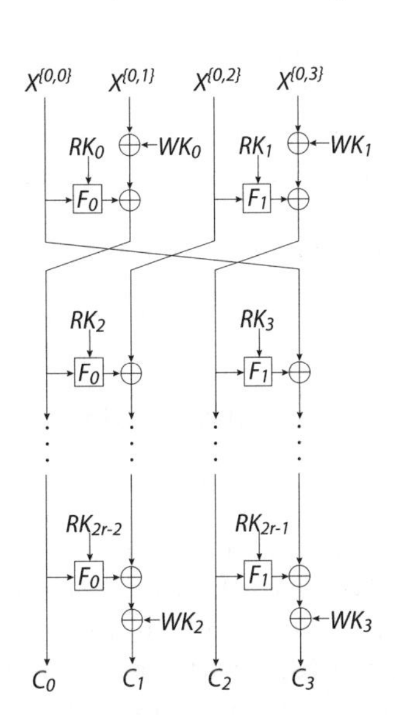

Fig. 4. Encryption function

4 Improbable Differential Attacks on CLEFIA

In this section, we present 10-round improbable differentials and introduce an improbable differential attack on 13-round CLEFIA with key length of 128 bits. We also introduce improbable differential attacks on 14 and 15-round CLEFIA for key lengths 196 and 256 bits in Appendix A and B. Moreover, we provide a practical improbable differential attack on 6-round CLEFIA in Appendix C. In these attacks our aim is to derive the round keys and we do not consider the key scheduling part as done in [9,10,11].

4.1 10-Round Improbable Differentials

We will use the following two 9-round impossible differentials that are introduced in [10],

$$[0_{(32)}, 0_{(32)}, 0_{(32)}, [X, 0, 0, 0]_{(32)}] \nrightarrow_{9r} [0_{(32)}, 0_{(32)}, 0_{(32)}, [0, Y, 0, 0]_{(32)}]$$
$$[0_{(32)}, 0_{(32)}, 0_{(32)}, [0, 0, X, 0]_{(32)}] \nrightarrow_{9r} [0_{(32)}, 0_{(32)}, 0_{(32)}, [0, Y, 0, 0]_{(32)}]$$

where $X_{(8)}$ and $Y_{(8)}$ are non-zero differences. We obtain 10-round improbable differentials by adding the following one-round differentials to the top of these 9-round impossible differentials,

$$[[\psi, 0, 0, 0]_{(32)}, \zeta_{(32)}, 0_{(32)}, 0_{(32)}] \rightarrow_{1r} [0_{(32)}, 0_{(32)}, 0_{(32)}, [\psi, 0, 0, 0]_{(32)}]$$
$$[[0, 0, \psi, 0]_{(32)}, \zeta'_{(32)}, 0_{(32)}, 0_{(32)}] \rightarrow_{1r} [0_{(32)}, 0_{(32)}, 0_{(32)}, [0, 0, \psi, 0]_{(32)}]$$

which hold when the output difference of the F_0 function is ζ (resp. ζ') when the input difference is $[\psi, 0, 0, 0]$ (resp. $[0, 0, \psi, 0]$). We choose ψ and corresponding ζ and ζ' depending on the difference distribution table (DDT) of S_0 in order to increase the probability of the differential. One can observe that the values 10, 8, 6 and 4 appear 9, 119, 848 and 5037 times in the DDT of S_0, respectively. When ψ, ζ and ζ' is chosen according to these differences, the average probability of the 10-round improbable differentials becomes

$$p' = ((9 \cdot 10 + 119 \cdot 8 + 848 \cdot 6 + 5037 \cdot 4)/256)/6013 \approx 2^{-5.87}.$$

4.2 Improbable Differential Attack on 13-Round CLEFIA

We put one additional round on the plaintext side and two additional rounds on the ciphertext side of the 10-round improbable differentials to attack first 13 rounds of CLEFIA that captures RK_1, $RK_{23,1} \oplus WK_{2,1}$, RK_{24}, and RK_{25}.

We place the whitening key WK_2 at the XOR with the 11th-round output word $x^{\{11,2\}}$ and XOR with RK_{23}. Moreover, we place the whitening key WK_1 at the XOR with the first round output word $x^{\{1,2\}}$, as shown in Fig. 5. These movements are equivalent transformations.

Data Collection. For a single choice of ψ and corresponding ζ values, we choose 2^K structures of plaintexts where the first word $x^{\{1,0\}}$ and the second, third and fourth bytes of the second word $x^{\{1,1\}}$ are fixed (similarly, we fix the first, second and fourth bytes of the second word $x^{\{1,1\}}$ for a choice of ψ and ζ'). We construct pairs where the first byte (resp. third byte) of the second word $x^{\{1,1\}}$ has the difference ψ, the third word $x^{\{1,2\}}$ has the difference ζ (resp. ζ') and the fourth word $x^{\{1,3\}}$ has the same difference with the output of F_1, which is obtained from the guessed round key RK_1, when the input difference of F_1 is ζ (resp. ζ'). Such a structure proposes $2 \cdot 6013 \cdot 2^{71}$ pairs.

We keep the ciphertext pairs having the difference $[0, [0, Y, 0, 0], \beta, \gamma]$ where γ is non-zero and β represents every 255 difference value that can be obtained from the multiplication of M_1 with $[0, Y, 0, 0]^t$. Such a difference in the ciphertext pairs is observed with a probability of $1/2^{32} \cdot 255/2^{32} \cdot 255/2^{32} \cdot (2^{32} - 1)/2^{32} \approx 2^{-80}$. Therefore, $6013 \cdot 2^{K-8}$ pairs remain.

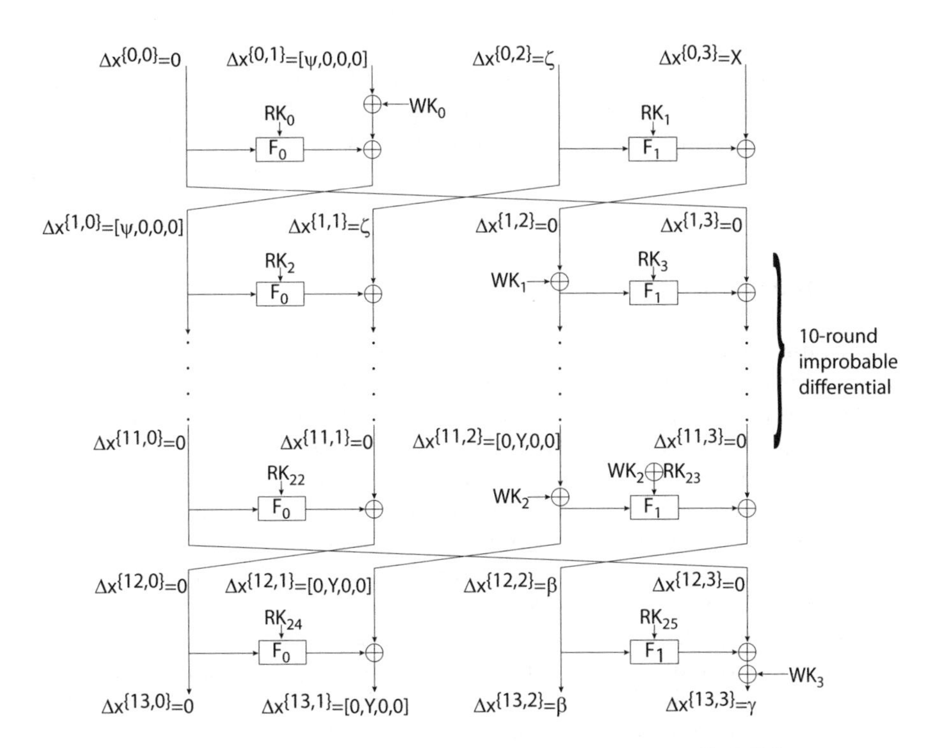

Fig. 5. Improbable differential attack on 13-round CLEFIA

Key Recovery. We keep counters for $RK_{23,1} \oplus WK_{2,1}|RK_{24}|RK_{25}$ for every guess of RK_1 and increase the corresponding counter when the improbable differential is obtained with a guessed key. Keys satisfying the improbable differential are obtained by differential table look-ups indexed on the input and the output differences of the 12th-round F_1 and 13th-round F_1. The probability of satisfying the improbable differential for a wrong key is $p = 2^{-40}$ from the average probabilities 2^{-8} and 2^{-32} for the 12th and 13th-round F_1 functions respectively. Therefore, the probability of obtaining the improbable differential for the correct key is $p_0 = p \cdot (1 - p') \approx 2^{-40.02}$.

During the attack we try to obtain the 104-bit round key, namely RK_1, $RK_{23,1} \oplus WK_{2,1}$, RK_{24}, RK_{25} and for the correct key to get the least number of hits, false alarm probability p_{fa} must be less than 2^{-104}. Feeding the Algorithm 1 with the inputs p, p_0, $p_{fa} = 2^{-105}$, and $p_{nd} = 1/100$ shows that when the threshold T is $673474 < 2^{20}$, $N_\infty \approx 2^{59.38}$ pairs are needed for the correct key to remain below the threshold and all of the wrong ones to remain above it with a success probability of 99%.

Attack Complexity. With the 2^{80} ciphertext filtering conditions, we need $2^{80} \cdot 2^{59.38} = 2^{139.38}$ pairs to perform the attack. Since we have 6013 choices for ψ, we need $2^K \approx 2^{54.83}$ structures so that $6013 \cdot 2^{72+K} = 2^{139.38}$. Hence, the data complexity of the attack is $2^{126.83}$ chosen plaintexts.

For every guess of RK_1 and RK_{24} and for every choice of ψ, we perform $2^{59.38}$ F-function computations which is $2^{64} \cdot 2^{59.38} \cdot 1/2 \cdot 1/13 \approx 2^{118.68}$ encryptions. However, the time complexity is $2^{126.83}$ encryptions for obtaining the ciphertexts.

The memory complexity of the attack comes from the 20-bit counters kept for the 104-bit round keys $RK_1|RK_{23,1} \oplus WK_{2,1}|RK_{24}|RK_{25}$, which require $20 \cdot 2^{104} \approx 2^{108.32}$ bits.

5 Conclusion

In previously defined statistical differential attacks on block ciphers, attacker's aim is to find an incident that is more probable for the correct key than a random key. Moreover, an impossible differential attack uses an impossible differential that is not possible when tried with the correct key. However, in this paper we introduced the improbable differential attack in which a given differential is less probable when tried with the correct key. Hence the impossible differential attack is just a special case of the improbable differential attack. We also modified the data complexity estimates given for statistical attacks by Blondeau et al. in order to use them in improbable differential attacks.

Moreover, we defined the almost miss in the middle technique for obtaining improbable differentials and we introduced a method for expanding impossible differentials to improbable differentials when suitable differentials that can be put on the top or the bottom of an impossible differential exist. Finally, we proposed improbable differential attacks on 13, 14, and 15-round CLEFIA by using this expansion method. To the best of our knowledge, these are the best cryptanalytic results on CLEFIA. Results of these improbable differential attacks and the impossible differential attacks of [10] on CLEFIA are summarized in Table 2.

In order to provide security against improbable attacks, block cipher designers should ensure that their designs contain no good improbable differentials. Since the almost miss in the middle technique uses two truncated differentials, providing upper bounds for truncated differentials may be used to provide security against improbable attacks.

Table 2. Results of the impossible differential attacks of [10] and improbable differential attacks on CLEFIA

#Rounds	Attack Type	Key Length	Data Complexity	Time Complexity	Memory (blocks)	Success Probability	Reference
12	Impossible	128, 192, 256	$2^{118.9}$	2^{119}	2^{73}	-	[10]
13	Improbable	128, 192, 256	$2^{126.83}$	$2^{126.83}$	$2^{101.32}$	%99	Sect. 4.2
13	Impossible	192, 256	$2^{119.8}$	2^{146}	2^{120}	-	[10]
14	Improbable	192, 256	$2^{126.98}$	$2^{183.17}$	$2^{126.98}$	%99	App. A
14	Impossible	256	$2^{120.3}$	2^{212}	2^{121}	-	[10]
15	Improbable	256	$2^{127.40}$	$2^{247.49}$	$2^{127.40}$	%99	App. B

Acknowledgments

I would like to express my deep gratitude to Ali Aydın Selçuk for his review and invaluable comments. I would like to thank Céline Blondeau for providing additional information about complexity estimates. I am grateful to Kerem Varıcı, Onur Özen and Meltem Sönmez Turan for their helpful comments. I also would like to thank anonymous referee for pointing out the earlier works [4] and [5].

References

1. Biham, E., Shamir, A.: Differential Cryptanalysis of DES-like Cryptosystems. J. Cryptology 4(1), 3–72 (1991)
2. Knudsen, L.R.: Truncated and higher order differentials. In: Preneel, B. (ed.) FSE 1994. LNCS, vol. 1008, pp. 196–211. Springer, Heidelberg (1995)
3. Biham, E., Biryukov, A., Shamir, A.: Cryptanalysis of Skipjack Reduced to 31 Rounds Using Impossible Differentials. J. Cryptology 18(4), 291–311 (2005)
4. Borst, J., Knudsen, L.R., Rijmen, V.: Two attacks on reduced IDEA. In: Fumy, W. (ed.) EUROCRYPT 1997. LNCS, vol. 1233, pp. 1–13. Springer, Heidelberg (1997)
5. Knudsen, L.R., Rijmen, V.: On the decorrelated fast cipher (DFC) and its theory. In: Knudsen, L.R. (ed.) FSE 1999. LNCS, vol. 1636, pp. 81–94. Springer, Heidelberg (1999)
6. Blondeau, C., Gérard, B.: On the data complexity of statistical attacks against block ciphers. In: Kholosha, A., Rosnes, E. (eds.) Workshop on Coding and Cryptography - WCC 2009, Ullensvang, Norway, pp. 469–488 (2009)
7. Blondeau, C., Gérard, B., Tillich, J.P.: Accurate Estimates of the Data Complexity and Success Probability for Various Cryptanalyses. To appear in Journal of Designs, Codes and Cryptography
8. Shirai, T., Shibutani, K., Akishita, T., Moriai, S., Iwata, T.: The 128-bit blockcipher CLEFIA (extended abstract). In: Biryukov, A. (ed.) FSE 2007. LNCS, vol. 4593, pp. 181–195. Springer, Heidelberg (2007)
9. Sony Corporation: The 128-bit Blockcipher CLEFIA, Security and Performance Evaluations, Revision 1.0, June 1 (2007), `http://www.sony.net/Products/cryptography/clefia/`
10. Tsunoo, Y., Tsujihara, E., Shigeri, M., Saito, T., Suzaki, T., Kubo, H.: Impossible differential cryptanalysis of CLEFIA. In: Nyberg, K. (ed.) FSE 2008. LNCS, vol. 5086, pp. 398–411. Springer, Heidelberg (2008)
11. Tsunoo, Y., Tsujihara, E., Shigeri, M., Suzaki, T., Kawabata, T.: Cryptanalysis of CLEFIA using multiple impossible differentials. In: International Symposium on Information Theory and Its Applications - ISITA 2008, December 7-10, pp. 1–6 (2008)
12. Zhang, W., Han, J.: Impossible differential analysis of reduced round CLEFIA. In: Yung, M., Liu, P., Lin, D. (eds.) Inscrypt 2008. LNCS, vol. 5487, pp. 181–191. Springer, Heidelberg (2009)
13. Cover, T.M., Thomas, J.A.: Elements of Information Theory. Wiley series in communications. Wiley, Chichester (1991)
14. Arratia, R., Gordon, L.: Tutorial on large deviations for the binomial distribution. Bulletin of Mathematical Biology 51, 125–131 (1989)
15. Blondeau, C.: Private communication (2009)

A Improbable Differential Attack on 14-Round CLEFIA

We expand our 13-round attack by one round on the ciphertext side to break 14-round CLEFIA for the key length of 192 or 256 bits. This attack captures 168 bits of the round keys, namely RK_1, $RK_{23,1}$, $RK_{24} \oplus WK_3$, $RK_{25} \oplus WK_2$, RK_{26}, and RK_{27}.

We move the whitening keys WK_1, WK_2, and WK_3 in the same way as in the 13-round attack.

Data Collection. We generate pairs in the same way as in the 13-round attack and we want 13th-round output difference to be $[[0, Y, 0, 0], \beta, \gamma, 0]$ to perform the attack. Consequently, we keep the ciphertext pairs satisfying the difference $[[0, Y, 0, 0], \beta', \gamma, \delta]$ where γ and δ are non-zero and β' is the XOR of β with the 255 possible values that can be obtained from the multiplication of M_0 with $[0, Y, 0, 0]^t$. Such a difference in ciphertext pairs is observed with a probability of $255/2^{32} \cdot ((255 \cdot 255)/2^{32} \cdot (2^{32} - 1)/2^{32} \cdot (2^{32} - 1)/2^{32} \approx 2^{-40}$. Therefore, $6013 \cdot 2^{K+32}$ pairs remain.

Key Recovery. We guess the second byte of RK_{24} and check if the second word of the output of 13th-round has difference β. The probability of this event is 2^{-8} and therefore, $6013 \cdot 2^{K+24}$ pairs remain. In order to check whether the 72-bit key $RK_{23,1}|RK_{25} \oplus WK_2|RK_{27}$ satisfies the improbable differential, we use differential tables indexed on the input and output differences of the 12th-round, 13th-round and 14th-round F_1 functions. The input values of these F_1 functions are obtained by the guesses of $RK_{24} \oplus WK_3$ and the first, third and fourth bytes of RK_{26}. The input of the 13th-round F_0 is obtained from RK_{27} candidates.

The probability of a candidate key to satisfy the improbable differential using three F_1 differential tables is $p = 2^{-72}$ from the average probabilities 2^{-8}, 2^{-32} and 2^{-32} for the 12th, 13th and 14th-round F_1 functions, respectively. Feeding the Algorithm 1 with the inputs p, p_0, $p_{fa} = 2^{-169}$, and $p_{nd} = 1/100$ shows that when the threshold T is $1022026 < 2^{20}$, $N_\infty \approx 2^{91.98}$ pairs are needed for the correct key to remain below the threshold and all of the wrong ones to remain above it with a success probability of 99%.

Keeping a key table for the attacked 168 key bits would require a memory that exceeds 2^{128} blocks where a block is 128 bits long. For this reason, we keep all of the $2^{126.98}$ plaintexts in a table, then guess RK_1 and choose the plaintext pairs for the attack.

Attack Complexity. We need $2^{91.98+40+8} = 2^{139.98}$ pairs in total to perform the attack. Since we have 6013 choices for ψ, we need $2^K \approx 2^{54.98}$ structures so that $6013 \cdot 2^{72+K} = 2^{139.98}$. Hence, the attack has data complexity of $2^{126.98}$ chosen plaintexts.

For every guess of RK_1, $RK_{24} \oplus WK_3$, and RK_{26}, we perform $2^{91.98}$ F-function computations which is $2^{96} \cdot 2^{91.98} \cdot 1/2 \cdot 1/14 \approx 2^{183.17}$ encryptions.

We keep 20-bit counters for the 72-bit keys $RK_{23,1}|RK_{25} \oplus WK_2|RK_{27}$ but the memory complexity is dominated by the ciphertext table of $2^{126.98}$ blocks.

B Improbable Differential Attack on 15-Round CLEFIA

We expand the 14-round improbable differential attack by one round on the ciphertext side to attack 15-round CLEFIA in which we exhaustively search for the 15th-round keys RK_{28} and RK_{29}. Our aim is to obtain the value of the 232-bit round key, namely RK_1, $RK_{23,1}$, RK_{24}, RK_{25}, $RK_{26} \oplus WK_3$, $RK_{27} \oplus WK_2$, RK_{28} and RK_{29}.

We move the whitening keys WK_1, WK_2, and WK_3 in the same way as in the 14-round attack.

For the inputs $p = 2^{-72}$, p_0, $p_{fa} = 2^{-233}$, and $p_{nd} = 1/100$, Algorithm 1 produces the outputs $N_\infty \approx 2^{92.40}$ and $T = 1361613 < 2^{21}$. Hence, the data complexity of the attack is $2^{127.40}$ chosen plaintexts and the memory complexity is $2^{127.40}$ blocks.

The time complexity of the attack comes from $2^{92.40}$ F-function computations for RK_1, RK_{24}, $RK_{26} \oplus WK_3$ guesses and the exhaustive search of RK_{28} and RK_{29}, which is $2^{92.40} \cdot 2^{96} \cdot 2 \cdot 2^{64} \cdot 1/2 \cdot 1/15 \approx 2^{247.49}$ encryptions.

C Practical Improbable Differential Attack on 6-Round CLEFIA

From the 9-round impossible differential used in Sect. 4, one can easily obtain the following 4-round impossible differential

$$[0_{(32)}, 0_{(32)}, 0_{(32)}, [\psi, 0, 0, 0]_{(32)}] \nrightarrow_{4r} [?_{(32)}, ?_{(32)}, ?_{(32)}, \psi'_{(32)}]$$

where ψ' is any difference other than ψ. We obtain a 5-round improbable differential by adding the following 1-round differential

$$[[\psi, 0, 0, 0]_{(32)}, \zeta_{(32)}, 0_{(32)}, 0_{(32)}] \rightarrow_{1r} [0_{(32)}, 0_{(32)}, 0_{(32)}, [\psi, 0, 0, 0]_{(32)}]$$

to the top of the 4-round impossible differential. The differential holds with probability $p' = 10/256$ if we choose $\psi = 08000000_x$ and $\zeta = 7EFCE519_x$.

In order to attack 6-round CLEFIA, we prepare plaintext pairs with the difference $[[\psi, 0, 0, 0]_{(32)}, \zeta_{(32)}, 0_{(32)}, 0_{(32)}]$. Then we guess RK_{11} and increase the counter for the guessed RK_{11} if $X_{5,3}$ has the difference $\psi'_{(32)}$. We expect the correct RK_{11} to have the smallest counter.

Feeding the Algorithm 1 with the inputs $p = 1 - 2^{-32}$, $p_0 = p \cdot (1 - 10/256)$, $p_{fa} = 2^{-33}$, and $p_{nd} = 1/100$ shows that when the threshold T is 184, $N_\infty = 185$ pairs are needed for the correct RK_{11} to remain below the threshold and all of the wrong ones to remain above it with a success probability of 99%.

Greedy Distinguishers and Nonrandomness Detectors

Paul Stankovski

Dept. of Electrical and Information Technology, Lund University,
P.O. Box 118, 221 00 Lund, Sweden

Abstract. We present the concept of greedy distinguishers and show how some simple observations and the well known greedy heuristic can be combined into a very powerful strategy (the Greedy Bit Set Algorithm) for efficient and systematic construction of distinguishers and nonrandomness detectors. We show how this strategy can be applied to a large array of stream and block ciphers, and we show that our method outperforms every other method we have seen so far by presenting new and record-breaking results for Trivium, Grain-128 and Grain v1.

We show that the greedy strategy reveals weaknesses in Trivium reduced to 1026 (out of 1152) initialization rounds using 2^{45} complexity – a result that significantly improves all previous efforts. This result was further improved using a cluster; 1078 rounds at 2^{54} complexity. We also present an 806-round distinguisher for Trivium with 2^{44} complexity.

Distinguisher and nonrandomness records are also set for Grain-128. We show nonrandomness for the full Grain-128 with its 256 (out of 256) initialization rounds, and present a 246-round distinguisher with complexity 2^{42}.

For Grain v1 we show nonrandomness for 96 (out of 256) initialization rounds at the very modest complexity of 2^{7}, and a 90-round distinguisher with complexity 2^{39}.

On the theoretical side we define the Nonrandomness Threshold, which explicitly expresses the nature of the randomness limit that is being explored.

Keywords: algebraic cryptanalysis, distinguisher, nonrandomness detector, maximum degree monomial, Trivium, Grain, Rabbit, Edon80, AES, DES, TEA, XTEA, SEED, PRESENT, SMS4, Camellia, RC5, RC6, HIGHT, CLEFIA, HC, MICKEY, Salsa, Sosemanuk.

1 Introduction

The output of a sensibly designed cipher should appear random to an external observer. Given a random-looking bit sequence, that observer should not be able to tell if the sequence is genuinely produced by the cipher in question or not. This simple idea is the core of cryptographic distinguishers and nonrandomness detectors.

G. Gong and K.C. Gupta (Eds.): INDOCRYPT 2010, LNCS 6498, pp. 210–226, 2010.

Recently we have seen several attempts at finding distinguishers and nonrandomness detectors and the best ones seem to be built using the maximum degree monomial test (see [27,11]) or some derivative of it. This test is superb for detecting nonrandomness, but it also provides a window into the internals of the cryptographic algorithm we are examining. The maximum degree monomial test can provide statements such as "The IV bits are not mixed properly", which can be invaluable to the algorithm designer.

The core of this test is a bit set, and the efficiency of the test is largely determined by how this bit set is selected. For this selection process, it seems that guesswork has been the most prominent ingredient. The reason for this may be that systematic methods have seemed too complicated to find or use, or simply that the importance of bit set selection has been underestimated. By far, the best systematic approach we have seen so far was due to Aumasson et al. [2]. They used a genetic algorithm to select a bit set, and this is a very reasonable approach for unknown and complex searchspaces. The complexity of the searchspace depends on the algorithm we are examining, but are they really so complex that we need to resort to such methods? In this paper we present a very simple deterministic and systematic approach that outperforms all other methods we have seen so far. We call it the Greedy Bit Set Algorithm.

Stream ciphers have an initialization phase, during which they "warm up" for a number of rounds before they are deemed operational. Block ciphers are not explicitly initialized in this way, but they do operate in rounds. For our purposes, this can be translated into an initialization phase.

How many rounds are needed to warm up properly? This is a question that every algorithm designer has been faced with, but we have not yet seen any satisfactory answer to this question. We make some observations that lead us to a definition of the Nonrandomness Threshold, which helps us to better understand the nature of the problem. The Greedy Bit Set Algorithm is a tool that can and should be used by designers to determine realistic lower bounds on the initialization period for their algorithm.

We go on to show how the Greedy Bit Set Algorithm performs against a wide variety of new and old stream and block ciphers, and we find new record-breaking results for Trivium, Grain-128 and Grain v1. We reveal weaknesses in Trivium reduced to 1026 out of 1152 initialization rounds in 2^{45} complexity, thereby significantly improving all previous efforts. By using a cluster we are able to improve this result even further to 1078 rounds at 2^{54} complexity. For Trivium we also present a new 806-round distinguisher of complexity 2^{44}. Both distinguishing and nonrandomness records are also set for Grain-128. We show nonrandomness in 256 (out of 256) initialization rounds, and present a 246-round distinguisher with complexity 2^{42}. For Grain v1 we show nonrandomness for 96 (out of 256) initialization rounds for a cost of only 2^7.

The paper is organized as follows. In Section 2 we give an overview of the black box model attack scenario and explain the maximum degree monomial test. We also briefly describe the software tools developed for this paper. In Section 3 we present our Greedy Bit Set Algorithm, comment on the importance of key weight and define the Nonrandomness Threshold. In Sections 4 and 5 we present and summarize our findings for the various algorithms. Finally, some concluding remarks are given in Section 6.

As a frame of reference, this article takes Filiol [12], Saarinen [27] and Englund et al. [11] as a starting point, and the most relevant previous work is due to Aumasson et al. [1,2] (see also Knudsen and Rijmen [19], Vielhaber [30], Dinur and Shamir [10] and Fischer, Khazaei and Meier [13]).

2 Background

2.1 The Black Box Model

Distinguishers may be built for block ciphers, stream ciphers, MACs, and so on, so adopting a black-box view of the cryptographic primitive is instructive. Consider the set-up in Fig. 1, dividing entities into potential input and output parameters to the left and right, respectively.

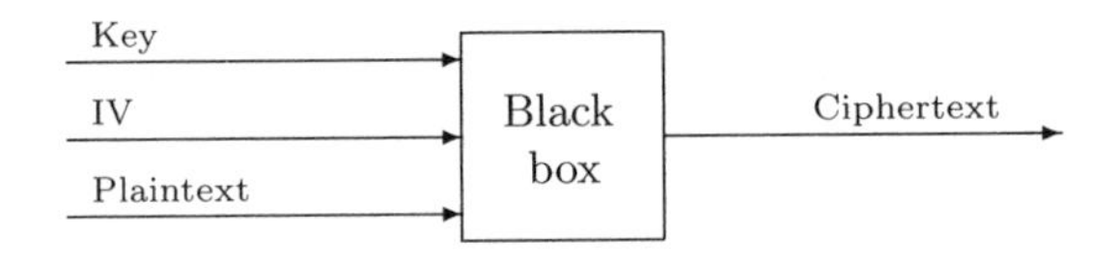

Fig. 1. Black box view of a cipher

A distinguisher attempts to determine if a given black box produces true random output or not. No cryptographic primitive produces truly random output, so the distinguisher can be thought of as a classifier. Given an output producing black box, the distinguisher answers "random" or "cipher", depending on its assertion. The distinguisher is said to be efficient if it significantly outperforms guessing, where the meaning of 'significantly' depends on the application.

For a distinguisher, the key is fixed and unknown. That is, the distinguisher may invoke the black box several times with different IVs, but the key is kept fixed. The IV bits constitute the input parameter bit space $B = \{0, 1\}^m$. The fixed key black box scenario is typical for real-world applications, and distinguishers are practical in the sense that they can be used in such a scenario.

Nonrandomness detectors are what we get when the input parameter bit space B includes key bits[1]. This renders them less useful in a real-world fixed key black box scenario, since they are related-key creatures by construction. Their merit, however, is that they can do a better job of detecting nonrandomness. This is invaluable for the cryptographic community, as we can get earlier indications on weaknesses in specific algorithms. Distinguishers show weaknesses in how IV bits

[1] We have not examined the effect of allowing plaintext bits in B, but this has the potential of working very well as these bits usually enter the state *after* both key and IV bits. This is true for block ciphers, but generally not for stream ciphers as encryption in that case usually works by simply XORing plaintext and keystream.

are handled, while nonrandomness detectors, in addition, can show weaknesses in how key bits are handled.

Explicitly summarizing the above, we have

Distinguisher: A {'random','cipher'}-classifier whose input parameter bit set B *does not* include key bits.

Nonrandomness detector: A {'random','cipher'}-classifier whose input parameter bit set B *does* include key bits.

Note, using a known or chosen key makes the {'random','cipher'}-classifier a nonrandomness detector, as we are then restricting the key space and effectively allowing key bits in B. A related discussion can be found in [19].

2.2 The Maximum Degree Monomial Signature

Algebraic techniques in general have recently been shown to be very powerful, and the maximum degree monomial (MDM) test stands out as a highly efficient randomness test. We have used this test in the following natural setting.

Consider a black box cipher that has been modified to produce output during its l initialization rounds. Choose a subset $S = \{0,1\}^n$ of the input variables B and regard the l-bit initialization round output of the black box as a Boolean vector function of the n variables $x_1, \cdots, x_n$ in S. Letting $f : \{0,1\}^n \rightarrow \{0,1\}^l$ denote the Boolean vector function, the sum

$$\sum_{x \in \{0,1\}^n} f(x)$$

produces a maximum degree monomial signature $\{0,1\}^l$ for the cipher. Note that this implicitly defines l (regular) Boolean functions $f_i, 1 \leq i \leq l$, one for each output bit. The ith signature bit is the coefficient of the maximum degree monomial $x_1 \cdots x_n$ in the algebraic normal form of f_i (see [27,2]).

An ideal cipher produces a random-looking MDM signature. That is, if a boolean function $g : \{0,1\}^n \rightarrow \{0,1\}$ is chosen uniformly at random from the universe of all such boolean functions, the maximum degree monomial exists in g with probability $\frac{1}{2}$.

The MDM signature for Trivium over the set consisting of every third IV bit, setting all other key and IV bits to zero, is

$$\underbrace{000\ldots000}_{930 \text{ zeros}}101\ldots.$$

The long sequence of leading zeros is very striking. We conclude that the sequence appears random-like close to where the first 1-bit appears, at round 931. We say that we have observed 930 zero rounds, and one interpretation of this is that 930 initialization rounds are not sufficient to properly mix the corresponding IV bits. Note that this is a nonrandomness result (chosen key).

Running the MDM test (producing the MDM signature and counting the number of initial zero rounds) over any given bit set S (permitting both key and IV bits) for an otherwise fixed key and fixed IV will produce a nonrandomness result. Fixed key nonrandomness detection over a bit set of size n has complexity 2^n and requires $O(l)$ space.

If the bit set S contains only IV bits, we have also implicitly produced a corresponding distinguisher. To assess the efficiency of this distinguisher, its performance needs to be sampled over *random* keys. Many different bias tests can be used here, but we have used MDM signature bit constantness (equal to zero) as measure, and two approaches stand out as simple, reasonable and typical.

Taking the *minimum* number of zero rounds over N randomly sampled keys assesses a distinguisher in $N \times 2^n$ time and $O(l)$ space. The time required for running this distinguisher is, however, only 2^n. Higher values for N increase the confidence level of the zero round number estimate.

Alternatively, taking the *maximum* number of zero rounds over N random keys assesses a distinguisher in $N \times 2^n$ time and $O(l)$ space. In the black box attack scenario, we need to examine N different black boxes before we find one that our distinguisher works for. The total running time for this distinguisher is therefore $N \times 2^n$. Taking the maximum costs us a factor N.

It is reasonable to take the maximum approach when the number of zero rounds varies heavily over the randomly selected keys. Without so much variation, it is more reasonable to take the minimum. This trades a few zero rounds for better time complexity. If complexity is less important, the highest zero round count is obtained by taking the maximum.

One key point is that the MDM test seems to be highly efficient and works very well in practice for some cryptographic algorithms. Another key point that makes the MDM tests attractive is that all output sequences can be successively XORed, so only a negligible amount of storage is required. Furthermore, one does not need to know anything at all about the internals of the algorithm that is being tested. The algorithm will quite politely but candidly reveal how susceptible any black box algorithm is to the MDM test.

2.3 Black Box Framework

A specialized cryptographic library that permits output of initialization data was put together for this paper. The library was written in C and supports bitsliced implementations and threading to make good use of multiple cores. This is something that the MDM test benefits from since it is spectacularly parallelizable. A unified interface makes it simple to author generic tests that can be used for all supported algorithms, and LaTeX-graphs of the results can be generated. This framework is an excellent tool for testing future generators. Interested researchers may find both ready-to-use executables and source code at [29]. The stream- and block ciphers used for this paper are listed in Table 1.

Table 1. Algorithms used to obtain the results in this paper

Stream ciphers		Block ciphers	
Trivium [8]	Rabbit [6]	AES-128 [24]	AES-256 [24]
Grain v1 [16]	Grain-128 [15]	DES [23]	PRESENT [7]
Edon80 [14]	MICKEY v2 [3]	RC5 [25]	RC6 [26]
HC-128 [34]	HC-256 [33]	TEA [31]	XTEA [32]
Salsa20/12 [5]	Sosemanuk [4]	SEED [20]	SMS4 [9]
		Camellia [22]	HIGHT [17]
		CLEFIA [28]	

3 The Algorithm and a Threshold

3.1 The Greedy Bit Set Algorithm

The trick to obtaining good results with the MDM test is to find an efficient bit set S for summation, a bit set that produces many zero rounds. The well known greedy heuristic provides a very simple but yet highly successful algorithm that outperforms all methods we have seen so far. The algorithm is made explicit in Fig. 2.

Algorithm GreedyBitSet
Input: key k, IV v, bit space B, desired bit set size n.
Output: Bit set S of size n.

$S \leftarrow \emptyset$;
repeat n times {
    $S \leftarrow$**GreedyAddOneBit**(k, v, B, S);
}
return S;

Fig. 2. The Greedy Bit Set Algorithm

Note that k and v are fixed, and that the bit space parameter B determines if key and/or IV bits may be used to build the resulting bit set. The subroutine GreedyAddOneBit is specified in Fig. 3.

Further note that the algorithm in Fig. 2 illustrates the straightforward greedy "add best bit"-strategy for building the resulting bit set S. GreedyBitSet can, by avoiding unnecessary recalculations, easily be implemented to sport a running time of precisely[2]

$$1 + \sum_{0 \le i < n} (m - i)2^i < m2^n$$

initializations for building a bit set of size n, where m is the size of the permissible bit space B.

[2] There are m choices for the first bit, $m - 1$ choices for the second bit, and so on.

```
Algorithm GreedyAddOneBit
Input: key k, IV v, bit space B, bit set S of size n.
Output: Bit set S' of size n + 1.

bestBit ← none;
max ← −1;
for all b ∈ B\S {
    zr ← numInitialZeroRounds(MDMsignature(k, v, S ∪ {b}));
    if (zr > max) {
        max ← zr;
        bestBit ← b;
    }
}
return S ∪ {bestBit};
```

Fig. 3. The **GreedyAddOneBit** subroutine

As a generalization one may allow other bit set building strategies, or a non-empty starting bit set. In this somewhat generalized form we denote an instance of the algorithm

$$\textbf{GreedyBitSet}(\text{strategy}, \text{starting bit set}, \text{primitive}, \text{bit space}, \text{key}, \text{IV}).$$

For example, running the Greedy Bit Set Algorithm with the "add best bit"-strategy on Trivium starting with an empty bit set, allowing only IV bits in the bit set, using the all-ones key and setting all remaining IV bits to zero may be denoted

$$\textbf{GreedyBitSet}(\text{Add1}, \emptyset, \text{Trivium}, \{\text{IV}\}, \mathbf{1}, \mathbf{0}).$$

Instead of starting with an empty bit set one may begin by computing a small optimal bit set and go from there. For most of our results below we have used optimal bit sets of sizes typically around five or six.

An alternative bit set building strategy is denoted "Add$N$". Add$N$ operates by adding the N bits that together produce the highest zero round count when added to the existing set. These bit sets should heuristically be better than the ones produced using the Add1 strategy as local optima are more likely to be avoided. The performances of the Add1 and Add2 strategies for Grain-128 are compared in Fig. 4, where the darker curve represents the Add2-strategy. GreedyBitSet with AddN strategy can be implemented with a running time of precisely[3]

$$1 + \sum_{0 \le i < k} \binom{m - iN}{N} 2^i < m^N 2^k$$

initializations for building a bit set of size kN.

We have standardized the graphs for uniform comparison between algorithms. Given a bit set, the portion of leading zero rounds in the initialization rounds is denoted 'bit set efficiency'.

[3] There are $\binom{m}{N}$ choices for the first bit, $\binom{m-N}{N}$ choices for the second bit, and so on.

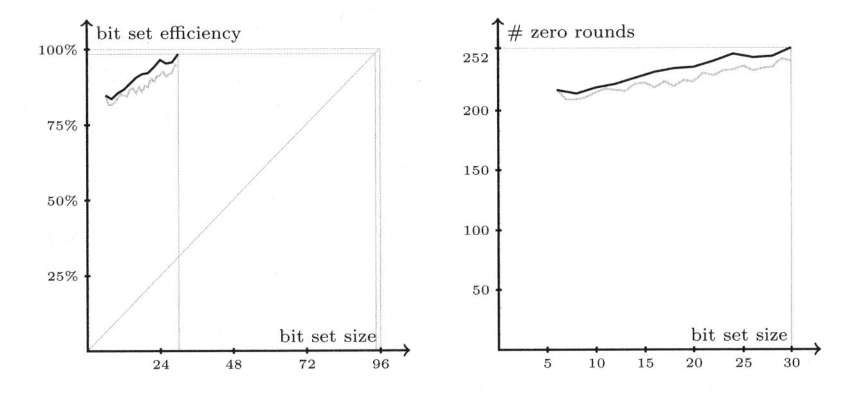

Fig. 4. Add1 (gray) vs. Add2 strategy (black) for Grain-128

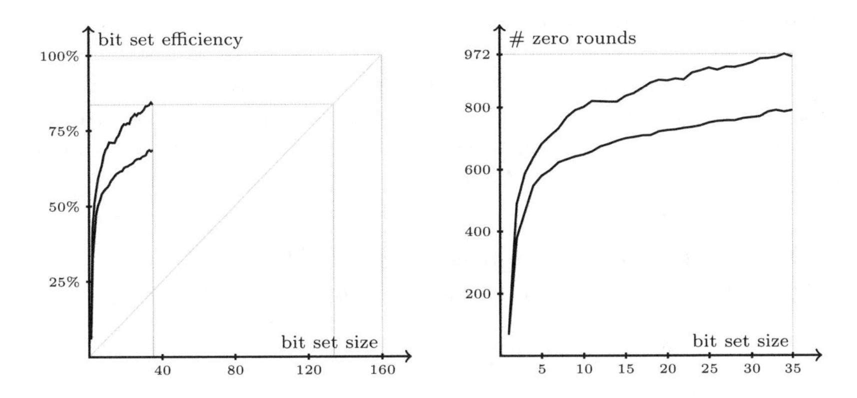

Fig. 5. The all-zeros key works better than the all-ones key for Trivium

For an ideal cipher, a bit set of size n produced by the Greedy Bit Set Algorithm will admit around $\lg(m - n)$ zero rounds.

3.2 Key Weight and the Nonrandomness Threshold

For some ciphers we have found that the result of the MDM test depends heavily on the weight of the key. A typical example of this is Trivium, for which the test seems to work best for the all-zeros key and worst for the all-ones key. Fig. 5 shows the efficiency of the bit sets produced by the Greedy Bit Set Algorithm for Trivium, starting with an empty set, using zero IV fill for these two keys.

For Trivium it seems that the all-zeros and all-ones keys are extreme cases. All other keys we have tried end up producing a curve that lies between these two, and a curve produced by averaging over several randomly chosen keys certainly falls between as well. So which value is most interesting: the maximum, minimum or the averaged one? Which zero round count should be reported? An attacker working on a deadline might be interested in the average performance over random keys, or possibly in the worst case performance if her deadline is really tight. But the algorithm designer may have quite other preferences.

Consider an algorithm analyst that needs to determine a reasonable number for how many initialization rounds that are needed for balancing initialization time and security in Trivium. Using the graphs in Fig. 5, the analyst can see that 1000 rounds will just barely withstand signs of improper mixing in this setting. At 972 rounds we start finding keys that allow us to prove that the bit mixing is inadequate. As we keep reducing the number of rounds, more and more keys show the same vulnerability. At 790 rounds, more or less all keys simultaneously chant "Inadequate mixing" in four-part harmony. The algorithm designer should, of course, in this case decide on an initialization round count well above 972. How much more is debatable.

Recall that we use a bit set $S = \{0, 1\}^n$ which is a subset of the entire bit space $B = \{0, 1\}^m$. The highest round count value 972 obtained above should really be viewed as a lower bound of a threshold - the Nonrandomness n-Threshold for bit sets of size n. That is the nature of the limit we are exploring here, a threshold for the existence of proof of inadequate bit mixing. Testing a specific bit set of size n over a single key and IV provides a lower bound for this threshold. The true threshold value is conceptually obtained by repeating the MDM Test several times taking the maximum over all possible keys, IVs and bit sets of size n for a total complexity of $\binom{m}{n}2^{m-n}$.

Definition 1. ***Nonrandomness n-Threshold***
The maximum number of zero rounds attainable according to

$$\max \ \text{numInitialZeroRounds}(\text{MDMsignature}(k, v, B, S)),$$

where the maximum is taken over $S \subseteq B$ with $|S| = n$, $k \in K$ and $v \in V$. B, K and V are the bit set-, key- and IV space, respectively.

4 Results

The algorithms are grouped according to susceptibility to the MDM test below, where particularly interesting algorithms are given room for elaboration. An algorithm is given a susceptibility rating high, significant, moderate or low according to its tendency to submit to the MDM test as the bit set size gets larger.

A direct comparison of our results to the previous best ones can be found in Table 2 in Section 5.

4.1 High Susceptibility

TEA and The Bit Flip Test. TEA is the top candidate. Starting with an empty bit set, we reach a full 100% zero round count after only two key bits have been added. It is the key bits that are the weak link, and this is a previously known deficiency in TEA (see [18]). The picture becomes quite different when one considers IV bit mixing. Allowing IV bits only results in a susceptibility that seems to be inherently low.

It seems that the shortcomings in key bit mixing have been properly dealt with for XTEA, as the Greedy Bit Set Algorithm cannot show anything beyond a low susceptibility level for any bit type. There is something we can learn from TEA. The TEA flaw is revealed by flipping two key bits, in which case the output does not change. We can devise an automated test for these simple symmetry faults. A Bit Flip Test can be defined by adding the two output sequences produced before and after flipping all bits in a given bit set. Trying all bit sets of small size will catch design flaws such as the one in TEA. The Bit Flip Test is, in fact, a MDM test for a bit set of size 1 with a prior change of basis. Instead of summing over a perfect cube, we sum over the "tilted" cube that is the result of a linear transformation of the basis.

Two such two-bit configurations are known for TEA, and we have verified that no other ones exist. We have also verified that none of the other algorithms we are considering here show any such bit flip weaknesses for small bit set sizes (five or so).

The Bit Flip Test should really be part of every algorithm designer's toolbox. This test, and many others, should be used routinely to check for errors or unexpected behavior.

Grain-128. For Grain-128, IV bits have a tendency to be more efficient than key bits and, as with Trivium, low weight keys work better than high weight keys. Running the Greedy Bit Set Algorithm on the all-zeros key with the Add2-strategy up to bit set size 40, IV bits only, we produced a nonrandomness detector for the full Grain-128 with its 256 initialization rounds. The successive development from the optimal 6-set to bit set size 40 is shown in Fig. 6.

We now turn our attention from nonrandomness detectors to distinguishers. The best previous distinguisher result on Grain-128 was due to Aumasson et al. [1]. Taking the maximum number of rounds over 64 random key trials, they found a 237-round distinguisher for a bit set of size 40.

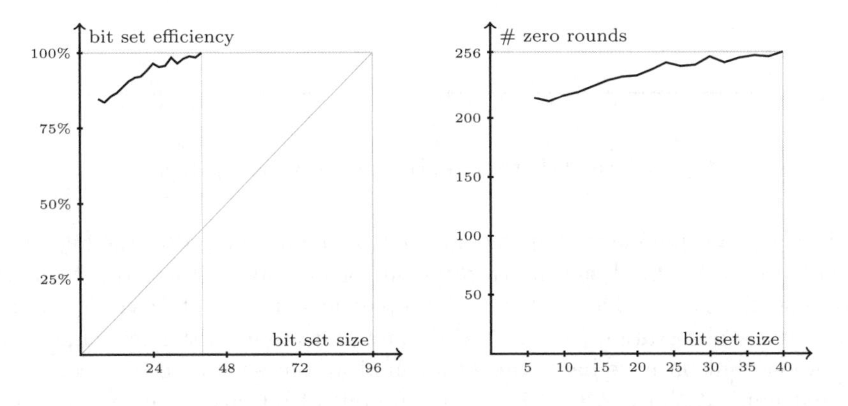

Fig. 6. Insufficient IV bit mixing in full Grain-128 (all 256 rounds)

Our greedy bit set of size 40 turns out to provide a 246-round distinguisher, measured by taking the maximum zero round count observed over 16 random key trials for a complexity of 2^{42}. The bit set is given below (zero indexed), and the order in which the bits have been added to the set has been preserved. The remaining IV bits were set to zero. The first six bit indices form the optimal 6-set.

34	59	63	64	67	69	55	61	25	85	35	58	2	73	30	38	5	6	10	44
24	50	3	77	91	95	12	13	41	72	19	29	15	79	7	37	21	45	8	71

To summarize the case for Grain-128, we have found one greedy nonrandomness detector showing that 256 (out of 256) rounds are insufficient for mixing the IV bits. This detector uses a bit set of size 40 and has complexity 2^{40}.

We also found a greedy 246-round distinguisher with complexity 2^{42}. This distinguisher uses the 38 first bits of the bit set above, taking the maximum zero round count over 16 random keys. The two last bits did not improve the distinguisher.

Trivium. There are several interesting observations for Trivium, apart from the importance of key weight that we have already established in Section 3.2. Key and IV bits are equally effective, but allowing both kinds in our bit set will take us much further. To see why this is not a contradiction, have a look at Fig. 7, which depicts the case where we allow only IV bits.

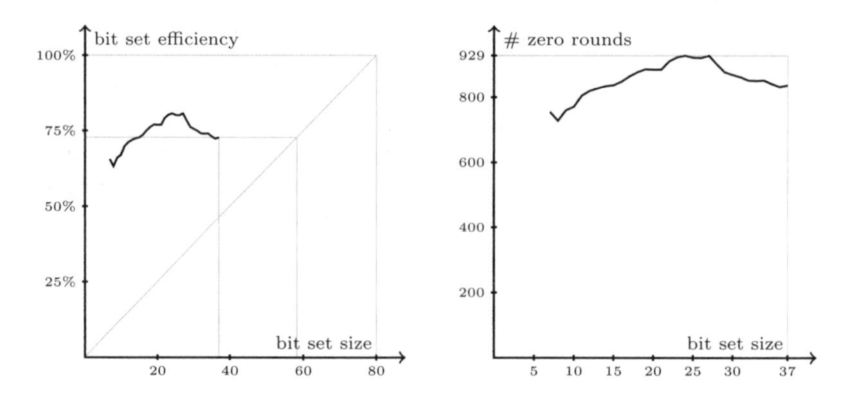

Fig. 7. GreedyBitSet(Add1, Opt7, Trivium, {IV}, **0**, **0**)

This graph is unique in that the curve drops significantly after the bit set has been built to size 27. Using *every third* bit for our bit set turns out to be the most effective choice. This is due to the threefold structure of Trivium, and this is not a new observation (see [21]). It doesn't seem to matter much which third we choose, but once we have started to build up our set we do best if we stick to that implicit third. After 27 bits we run out of bit space, but we can allow both key and IV bits.

Our best greedy nonrandomness detector using both key and IV bits takes us 1026 out of 1152 rounds. This is for the zero key, which we noted before was heavily biased. The greedy strategy was to start from the optimal 5-set and to use the Add2-strategy up to bit set size 29, via the Add1-strategy up to bit set size 37, to finally just guessing the last few bits for a total bit set size of 45. The resulting bit set is

Key bits	1 4 7 10 12 16 19 22 25 31 34 37
	40 43 46 49 52 55 58 61 64 70 73 76
IV bits	1 4 7 10 16 19 25 28 31 34 37 40
	43 46 49 52 55 58 64 67 70

The every-third-structure is evident in this bit set, so it would be interesting to measure the zero round performance of the corresponding 54-bit set with 27 key and 27 IV bits. Considering the bit set performance drop we saw in Fig. 7 above, it is reasonable to assume that we will see the same effect once we try to go beyond this supposedly near-optimal 54-bit set. More than one million core hours of computation on a cluster showed that we get 1078 zero rounds after 2^{54} encryptions.

We also present a distinguisher for 806-round Trivium. As noted before, one can use the internal structure of Trivium by using every third IV bit for the bit set. Unfortunately, we run out of bits after 27 of them have been added. We can, however, skip exploiting the threefold structure and, instead, just use the fact that multiplication is always performed between neighboring state variables. Using *every second* IV bit for the bit set will avoid fast initial term growth and take us 803 rounds over randomly selected keys. This was the *minimum* number of rounds obtained over 16 trials, so the resulting complexity is 2^{40}. Taking the maximum produces an 806-round distinguisher with complexity 2^{44}.

4.2 Significant Susceptibility

Grain v1 is definitely susceptible, as one can see from the direction of the curve in Fig. 8. The level of susceptibility seems limited, however, as the extrapolated greedy-curve will not hit the roof for any bit set of relevant size.

Key bits seem to work a little better than IV bits, in general, but our best nonrandomness result is 96 zero rounds for the all-zeros key with the optimal IV bit set of size 7 given by (zero-indexed)

1 22 26 37 45 47 55

A 90-round greedy distinguisher was derived from a bit set of size 35 by taking the maximum zero round count over 16 random keys for a complexity of 2^{39}. The zero-indexed IV bit set is

1 22 26 37 45 47 55 12 16 4 28 29 36 0 39 31 34 10
11 7 32 9 50 13 25 59 5 3 57 53 51 42 33 38 8

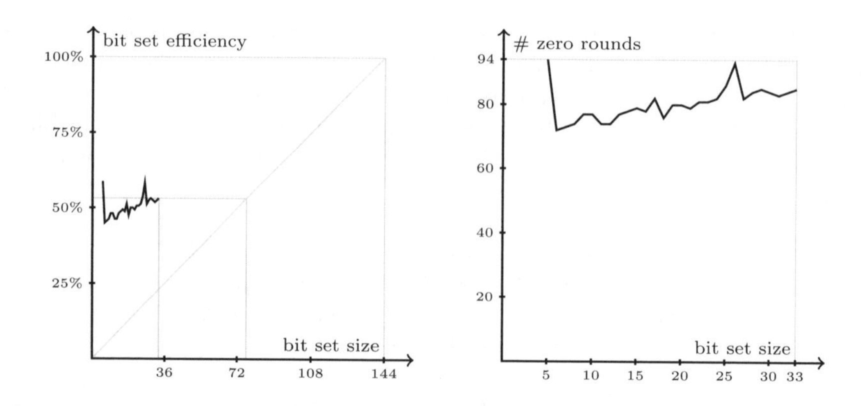

Fig. 8. **GreedyBitSet**(Add1, Opt5, Grain v1, {Key, IV}, **0**, **0**)

4.3 Moderate or Low Susceptibility

AES, DES, CLEFIA and HIGHT all start at and stay within a bit set efficiency in the range 25-50%. These algorithms show only very slight or no sign of budging as the bit set size increases.

The remaining ciphers have a bit set efficiency below 25%. Edon80 deviates from the norm by having a somewhat erratic curve, but it seems to stay within the 0-25% efficiency range. Sosemanuk does show a tendency to be affected by the MDM test, but all other algorithms seem to be more or less inherently non-susceptible.

It is interesting to see that the bit set efficiency for IV bits in RC5, and for IV bits in RC6 and key bits in XTEA to a lesser extent, show a *decreasing* tendency as the search progresses and bit set sizes increase. The curve for RC5 IV bits can be seen in Fig. 9.

HC-128 and HC-256 set a record of sorts at the low end by showing no significant susceptibility while producing an extremely large amount of initial data.

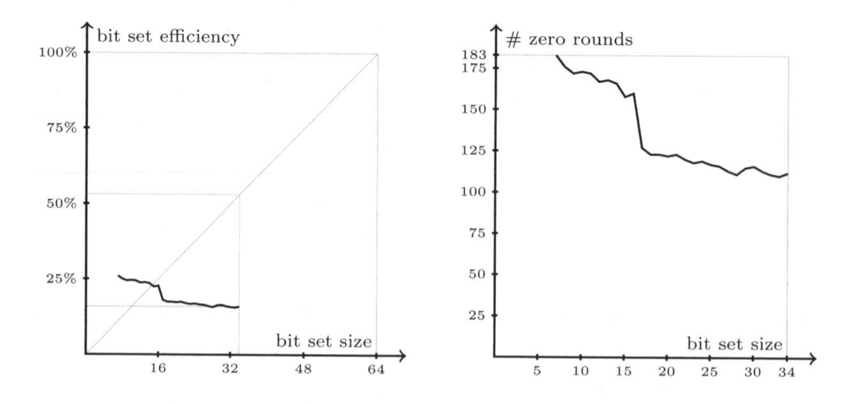

Fig. 9. IV bit sets for RC5 show decreasing efficiency

The yet unmentioned and remaining algorithms (Camellia, MICKEY v2, PRESENT, Rabbit, Salsa20/12, SEED and SMS4) all seem to be inherently non-susceptible.

5 Results Summary

We have shown how to find efficient bit sets in a systematic and deterministic way by using the Greedy Bit Set Algorithm. The record-breaking distinguishers and nonrandomness detectors derived from using the Greedy Bit Set Algorithm show that this algorithm outperforms all other bit set selection schemes we have seen so far. Table 2 compares the previous best results to ours for Trivium, Grain-128 and Grain v1.

We presented a nonrandomness detector showing that Grain-128 with full 256-round initialization does not behave sufficiently random. This detector uses an IV bit set of size 40 and has a complexity of 2^{40}. We also presented a 246-round distinguisher over random keys with complexity 2^{42}.

For Trivium we found a greedy 1026-round nonrandomness detector with complexity 2^{45}. Using a cluster, we went on to find a nonrandomness detector for 1078 out of 1152 rounds with 2^{54} comlexity. We also presented a 806-round distinguisher with 2^{44} complexity.

For Grain v1 we showed nonrandomness up to 96 rounds with complexity 2^7, and a 90-round distinguisher with complexity 2^{39}.

Table 2. Comparison to previous results

Algorithm	Attack type	Rounds	Time	Authors	Rounds	Time	Authors
Trivium	distinguisher	790	2^{30}	[2]	806	2^{44}	this paper
Trivium	nonrandomness	885	2^{27}	[2]	1078	2^{54}	this paper
Grain-128	distinguisher	237	2^{40}	[1]	246	2^{42}	this paper
Grain-128	nonrandomness	-	-	-	256	2^{40}	this paper
Grain v1	distinguisher	81	2^{24}	[1]	90	2^{39}	this paper
Grain v1	nonrandomness	-	-	-	96	2^7	this paper

6 Concluding Remarks

With the exception of TEA, all block ciphers we have tested seem reasonably resistant to the maximum degree monomial test. Due to differences in how zero rounds are measured in stream and block ciphers, one should, however, not immediately draw the conclusion that block ciphers are safer than stream ciphers.

The Greedy Bit Set Algorithm can be examined with more elaborate strategy variants, bit selection schemes, randomness tests, cryptographic algorithms, allowing plaintext bits in the bit set, and so on. The most urgent and constructive goal, however, would be to explain why the MDM test fails miserably for some algorithms. What minimal set of properties is guaranteed to render the MDM test useless?

Let us elaborate on the concept of "weak" bits, see [2,13]. Weak bits are such that they significantly increase the efficiency (the number of zero rounds) of a bit set if they are added to it. The first question one might ask is: Do weak bits exist at all? The Greedy Bit Set Algorithm answers this question and reveals some deeper insight into the concept of weakness. Our algorithm successively builds larger bit sets by repeatedly adding the weakest remaining single bit (Add1 strategy). For Trivium, bits at every third bit position eagerly reappear among the top ranked bits again and again as the bit set size steadily increases. The bits at other (off-third) positions do not show up as top ranked at all. This zero round distribution regularity is clear evidence that Trivium *has* weak bits. Other algorithms show no sign of weak bits. This does not prove their non-existence in any way, but we surmise that any bit selection strategy for a truly perfect algorithm should not perform much better than random choice. For Grain-128, there are signs of bit weakness, but they are much less conclusive than for Trivium.

The existence of weak bits is algorithm dependent. Also, when we use GreedyBitSet we successively expand a bit set with the *currently* weakest bit. This means that the existence of weak bits does not only depend on the choice of test, but also on the current state of the test. As for drawing conclusions on the existence of globally weak bits, defining how to measure bit weakness is only the first step into a rather non-trivial enterprise.

One consequence of this is that one cannot prove any general performance guarantees for GreedyBitSet stating that we will obtain a good bit with some supposedly high probability. As we have seen, for Trivium we do, for RC5 we don't.

Also, more intelligent analysis of the zero round distribution over the remaining bit space could lead to better practical assessment measures for bit weakness that could be used to improve The Greedy Bit Set Algorithm.

Automatic cryptanalysis can be performed on many cryptographic primitives. A toolbox of various tests, MDM-based and others, should be at the disposal of every algorithm designer. Such a toolbox can be used to reveal unexpected design weaknesses and to give better estimations on the required number of initialization rounds. The interested reader is referred to [29].

We wish to thank the anonymous reviewers for their insightful comments.

References

1. Aumasson, J.-P., Dinur, I., Henzen, L., Meier, W., Shamir, A.: Efficient FPGA Implementations of High-Dimensional Cube Testers on the Stream Cipher Grain-128 (2009), `http://eprint.iacr.org/2009/218/` (accessed June 17, 2009)
2. Aumasson, J.-P., Dinur, I., Meier, W., Shamir, A.: Cube Testers and Key Recovery Attacks on Reduced-Round MD6 and Trivium. In: Dunkelman, O. (ed.) Fast Software Encryption. LNCS, vol. 5665, pp. 1–22. Springer, Heidelberg (2009)
3. Babbage, S., Dodd, M.: The MICKEY Stream Ciphers. In: Robshaw, M.J.B., Billet, O. (eds.) New Stream Cipher Designs. LNCS, vol. 4986, pp. 191–209. Springer, Heidelberg (2008)

4. Berbain, C., Billet, O., Canteaut, A., Courtois, N., Gilbert, H., Goubin, L., Gouget, A., Granboulan, L., Lauradoux, C., Minier, M., Pornin, T., Sibert, H.: Sosemanuk, a Fast Software-Oriented Stream Cipher. In: Robshaw, M.J.B., Billet, O. (eds.) New Stream Cipher Designs. LNCS, vol. 4986, pp. 98–118. Springer, Heidelberg (2008)
5. Bernstein, D.J.: The Salsa20 Family of Stream Ciphers. In: Robshaw, M.J.B., Billet, O. (eds.) New Stream Cipher Designs. LNCS, vol. 4986, pp. 84–97. Springer, Heidelberg (2008)
6. Boesgaard, M., Vesterager, M., Pedersen, T., Christiansen, J., Scavenius, O.: Rabbit: A New High-Performance Stream Cipher. In: Johansson, T. (ed.) FSE 2003. LNCS, vol. 2887, pp. 307–329. Springer, Heidelberg (2003)
7. Bogdanov, A., Knudsen, L.R., Leander, G., Paar, C., Poschmann, A., Robshaw, M.J.B., Seurin, Y., Vikkelsoe, C.: PRESENT: An Ultra-Lightweight Block Cipher. In: Paillier, P., Verbauwhede, I. (eds.) CHES 2007. LNCS, vol. 4727, pp. 450–466. Springer, Heidelberg (2007)
8. De Cannière, C., Preneel, B.: Trivium. In: Robshaw, M.J.B., Billet, O. (eds.) New Stream Cipher Designs. LNCS, vol. 4986, pp. 244–266. Springer, Heidelberg (2008)
9. Diffie, W., Ledin, G.: SMS4 Encryption Algorithm for Wireless Networks. Version 1.03, (2008), `http://eprint.iacr.org/2008/329.pdf` (accessed February 18, 2010)
10. Dinur, I., Shamir, A.: Cube Attacks on Tweakable Black Box Polynomials. In: Joux, A. (ed.) EUROCRYPT 2009. LNCS, vol. 5479, pp. 278–299. Springer, Heidelberg (2010)
11. Englund, H., Johansson, T., Turan, M.S.: A framework for chosen IV statistical analysis of stream ciphers. In: Srinathan, K., Rangan, C.P., Yung, M. (eds.) INDOCRYPT 2007. LNCS, vol. 4859, pp. 268–281. Springer, Heidelberg (2007)
12. Filiol, E.: A new statistical testing for symmetric ciphers and hash functions. In: Deng, R.H., Qing, S., Bao, F., Zhou, J. (eds.) ICICS 2002. LNCS, vol. 2513, Springer, Heidelberg (2002)
13. Fischer, S., Khazaei, S., Meier, W.: Chosen IV Statistical Analysis for Key Recovery Attacks on Stream Ciphers. In: Vaudenay, S. (ed.) AFRICACRYPT 2008. LNCS, vol. 5023, pp. 236–245. Springer, Heidelberg (2008)
14. Gligoroski, D., Markovski, S., Knapskog, S.J.: The Stream Cipher Edon80. In: Robshaw, M.J.B., Billet, O. (eds.) New Stream Cipher Designs. LNCS, vol. 4986, pp. 152–169. Springer, Heidelberg (2008)
15. Hell, M., Johansson, T., Maximov, A., Meier, W.: A Stream Cipher Proposal: Grain-128. In: International Symposium on Information Theory—ISIT 2006, IEEE, Los Alamitos (2006)
16. Hell, M., Johansson, T., Meier, W.: Grain - a stream cipher for constrained environments. International Journal of Wireless and Mobile Computing, Special Issue on Security of Computer Network and Mobile Systems 2(1), 86–93 (2006)
17. Hong, D., Sung, J., Hong, S., Lim, J., Lee, S., Koo, B.-S., Lee, C., Chang, D., Lee, J., Jeong, K., Kim, H., Kim, J., Chee, S.: HIGHT: A New Block Cipher Suitable for Low-Resource Device. In: Goubin, L., Matsui, M. (eds.) CHES 2006. LNCS, vol. 4249, pp. 46–59. Springer, Heidelberg (2006)
18. Kelsey, J., Schneier, B., Wagner, D.: Related-key cryptanalysis of 3-WAY, Biham-DES, CAST, DES-X, NewDES, RC2, and TEA. In: Han, Y., Quing, S. (eds.) ICICS 1997. LNCS, vol. 1334, pp. 233–246. Springer, Heidelberg (1997)
19. Knudsen, L.R., Rijmen, V.: Known-Key Distinguishers for Some Block Ciphers. In: Kurosawa, K. (ed.) ASIACRYPT 2007. LNCS, vol. 4833, pp. 315–324. Springer, Heidelberg (2007)

20. Lee, H.J., Lee, S.J., Yoon, J.H., Cheon, D.H., Lee, J.I.: The SEED Encryption Algorithm. (2005), http://tools.ietf.org/html/rfc4269 (accessed February 18, 2010)
21. Maximov, A., Biryukov, A.: Two Trivial Attacks on Trivium. In: The State of the Art of Stream Ciphers, Workshop Record, SASC 2007, Bochum, Germany (January 2007)
22. NTT and Mitsubishi Electric Company. Specification of Camellia - A 128-bit Block Cipher. Version 2.0 (2000), http://info.isl.ntt.co.jp/crypt/eng/camellia/dl/01espec.pdf (accessed February 18, 2010)
23. U.S. Department of Commerce and NIST. Data Encryption Standard (DES). FIPS Publication 46-3 (1999)
24. U.S. Department of Commerce and NIST. Announcing the Advanced Encryption Standard (AES). FIPS Publication 197 (2001)
25. Rivest, R.L.: The RC5 encryption algorithm. In: Preneel, B. (ed.) FSE 1994. LNCS, vol. 1008, pp. 86–96. Springer, Heidelberg (1995)
26. Rivest, R.L., Robshaw, M.J.B., Sidney, R., Yin, Y.L.: The RC6 Block Cipher. Version 1.1 (1998), http://people.csail.mit.edu/rivest/Rc6.pdf (accessed February 18, 2010)
27. Saarinen, M.-J.O.: Chosen-IV statistical attacks on eSTREAM stream ciphers. eSTREAM, ECRYPT Stream Cipher Project, Report 2006/013 (2006), http://www.ecrypt.eu.org/stream
28. Shirai, T., Shibutani, K., Akishita, T., Moriai, S., Iwata, T.: The 128-Bit Blockcipher CLEFIA (Extended Abstract). In: Biryukov, A. (ed.) FSE 2007. LNCS, vol. 4593, pp. 181–195. Springer, Heidelberg (2007)
29. Stankovski, P.: Maximum Degree Monomial Toolkit with Source Code (2010), http://www.eit.lth.se/staff/paul.stankovski/phdprojects (accessed September 24, 2010)
30. Vielhaber, M.: Breaking ONE.FIVIUM by AIDA an Algebraic IV Differential attack (2007), http://eprint.iacr.org/2007/413/ (accessed June 17, 2009)
31. Wheeler, D.J., Needham, R.M.: TEA, a Tiny Encryption Algorithm, http://www.cix.co.uk/~klockstone/tea.pdf (accessed February18 , 2010)
32. Wheeler, D.J., Needham, R.M.: TEA extensions, http://www.cix.co.uk/~klockstone/xtea.pdf (accessed February18, 2010)
33. Wu, H.: A New Stream Cipher HC-256. In: Roy, B., Meier, W. (eds.) FSE 2004. LNCS, vol. 3017, pp. 226–244. Springer, Heidelberg (2004)
34. Wu, H.: The Stream Cipher HC-128. In: Robshaw, M.J.B., Billet, O. (eds.) New Stream Cipher Designs. LNCS, vol. 4986, pp. 39–47. Springer, Heidelberg (2008)

Polynomial Multiplication over Binary Fields Using Charlier Polynomial Representation with Low Space Complexity

Sedat Akleylek*, Murat Cenk, and Ferruh Özbudak**

Institute of Applied Mathematics, Middle East Technical University, Ankara, Turkey
{akleylek,mcenk,ozbudak}@metu.edu.tr

Abstract. In this paper, we give a new way to represent certain finite fields $GF(2^n)$. This representation is based on Charlier polynomials. We show that multiplication in Charlier polynomial representation can be performed with subquadratic space complexity. One can obtain binomial or trinomial irreducible polynomials in Charlier polynomial representation which allows us faster modular reduction over binary fields when there is no desirable such low weight irreducible polynomial in other representations. This representation is very interesting for NIST recommended binary field $GF(2^{283})$ since there is no ONB for the corresponding extension. We also note that recommended NIST and SEC binary fields can be constructed with low weight Charlier polynomials.

Keywords: Charlier polynomials, binary field representation, polynomial multiplication, subquadratic space complexity.

1 Introduction

Finite fields have many applications in coding theory, digital signal processing and cryptography [7], [8], [9]. Efficient arithmetic of finite field is an important factor for cryptographic applications. Before implementing cryptographic applications efficiently, several choices have to be made. The representation of finite field elements and choice of irreducible (reduction) polynomial have a crucial impact on the efficiency of field arithmetic [6]. These selections can be influenced by security considerations, application platform and constraints of the particular computing environment [1]. The measure of efficiency in hardware implementations is the number of $\#AND$ and $\#XOR$.

The binary extension field multiplication can be performed in two steps: polynomial multiplication over $GF(2)$ and modular reduction over $GF(2^n)$. As the complexity of finite field multiplication depends on the number of non-zero terms

* Sedat Akleylek is also with the Department of Computer Engineering, Ondokuz Mayıs University.

** Ferruh Özbudak is also with the Department of Mathematics, Middle East Technical University.

G. Gong and K.C. Gupta (Eds.): INDOCRYPT 2010, LNCS 6498, pp. 227–237, 2010.

in the reduction polynomials, it is desirable to use the reduction polynomials with as few non-zero terms as possible. Over binary fields, the use of trinomial or when trinomial does not exist for the corresponding extension, pentanomial is preferred since there is no irreducible binomial or quadranomial except for $x+1$ in $GF(2)[x]$.

There are mainly three types of representation of finite fields of characteristic 2, namely canonical (polynomial) basis, normal basis, redundant representation. Recently, Dickson polynomial representation has been proposed to obtain efficient binary field multiplication using low weight irreducible polynomial in [4] and [5]. Hasan and Negre formulate the multiplication of two elements in the field as a product of Toeplitz or Hankel matrix. Dickson polynomials seem interesting when no optimal normal basis (ONB) in any type exists for the field. This is the case for NIST recommended binary fields $GF(2^{163})$ and $GF(2^{283})$. By using Dickson polynomial representation, one can obtain irreducible Dickson binomials or trinomials. Depending on the choice of basis, binary field multiplication can be performed in different ways.

In this paper, we give a new way to represent certain finite fields $GF(2^n)$. This representation is based on Charlier polynomials. We show that multiplication in Charlier polynomial representation can be performed with subquadratic space complexity. One can obtain binomial or trinomial irreducible polynomials in Charlier polynomial representation which allows us faster modular reduction over binary fields when there is no desirable such low weight irreducible polynomial in other representations. This representation is very interesting for NIST recommended binary field $GF(2^{283})$ since there is no ONB for the corresponding extension. We also note that recommended NIST and SEC binary fields can be constructed with low weight Charlier polynomials such as $GF(2^{113})$, $GF(2^{131})$ and $GF(2^{233})$.

This paper is organized as follows: Section 2 describes Charlier polynomial and gives some general results on Charlier polynomials in $GF(2)[x]$. In section 3, we present the general method to multiply two polynomials in Charlier polynomial representation and give the total arithmetic complexity. We compare complexity of multipliers in view of $\#AND$ and $\#XOR$ gates in Section 4. We conclude the paper in Section 5.

2 Charlier Polynomials

In this section, we give preliminaries and describe a new representation of binary fields.

Charlier polynomials are the monic orthogonal polynomials associated with the inner product [2].

Definition 1. *The Charlier polynomials are $C_0(x) = 1$, $C_1(x) = x$ with the recursion*

$$C_n(x) = (x - n + 1)C_{n-1}(x)$$

for $n \geq 2$.

Table 1. Charlier Polynomials in $GF(2)[x]$

$C_0(x)$	1
$C_1(x)$	x
$C_2(x)$	x^2+x
$C_3(x)$	x^3+x^2
$C_4(x)$	x^4+x^2
$C_5(x)$	x^5+x^3
$C_6(x)$	$x^6+x^5+x^4+x^3$
$C_7(x)$	$x^7+x^6+x^5+x^4$
$C_8(x)$	x^8+x^4
$C_9(x)$	x^9+x^5
$C_{10}(x)$	$x^{10}+x^9+x^6+x^5$

Since we work in binary fields, we give the Charlier polynomials in $GF(2)[x]$ for $n \leq 10$ in Table 1. All values in Table 1 are computed by using Software for Algebra and Geometry Experimentation (Sage) [11].

2.1 Conversion of Coefficients from Polynomial Representation to Charlier Polynomial Representation

The polynomial basis $\{1, x, x^2, \cdots, x^{n-1}\}$, where x is a root of an irreducible polynomial of degree n over $GF(2)$ is usually preferred to represent the elements of $GF(2)[x]$. Let $a(x) = a'_{n-1}x^{n-1} + \cdots + a'_1 x + a'_0$, where $a'_i \in GF(2)$ be a polynomial with the standard (canonical) representation. Let $C_n(x) = \beta_n$ be the n-th Charlier polynomial in $GF(2)[x]$, where $n \geq 0$. $a(x)$ can be represented by using Charlier polynomials as $a = a_{n-1}\beta_{n-1} + \cdots + a_1\beta_1 + a_0\beta_0$, where $a_i \in GF(2)$ by using Algorithm 1.

Algorithm 1. Conversion of Coefficients From Polynomial Representation to Charlier Polynomial Representation

Input: $a(x) = \sum_{i=0}^{n-1} a'_i x^i$
Output: $(a_0, a_1, \cdots, a_{n-1})$, where $a = \sum_{i=0}^{n-1} a_i\beta_i$

```
1: T ← a
2: for i = n downto 1 do
3:   if deg(T) = i then
4:     a_i ← 1
5:     T ← T + β_i
6:   else
7:     a_i ← 0
8:   end if
9: end for
10: a_0 ← T
```

Note that since we are working in characteristic two, Algorithm 1 is self-inverse. One can obtain polynomial representation for Charlier representation by changing a_i with a'_i in Algorithm 1.

2.2 Charlier Basis

A basis for the finite field $GF(2^n)$ is a set of n elements $\{\beta_0, \beta_1, \cdots, \beta_{n-1}\} \in GF(2^n)$ such that every element of the binary field can be represented uniquely as a linear combination of basis elements. For a given $a \in GF(2^n)$, we can write

$$a = \sum_{i=0}^{n-1} a_i \cdot \beta_i$$

where $a_i \in GF(2)$ for $0 \leq i \leq n-1$.

Theorem 1. *Let f be an irreducible polynomial of degree n in $GF(2)[x]$. The set $\{\beta_0, \beta_1, \cdots, \beta_{n-1}\}$ forms a basis of $GF(2^n) \cong GF(2)[x]/(f)$.*

Proof. Consequences of Algorithm 1 show that the set $\{\beta_0, \beta_1, \cdots, \beta_{n-1}\}$ is linearly independent and each element in $GF(2^n)$ is uniquely expressed by using the set $\{\beta_0, \beta_1, \cdots, \beta_{n-1}\}$. Then, the set $\{\beta_0, \beta_1, \cdots, \beta_{n-1}\}$ forms a basis of $GF(2^n) \cong GF(2)[x]/(f)$. □

Theorem 2. *Let $C_n(x) = \beta_n$ be the n-th Charlier polynomial in $GF(2)[x]$, where $n \geq 0$. Then, for all $i, j \geq 0$ Charlier basis satisfies the following equation*

$$\beta_i \cdot \beta_j = \beta_{i+j} + \ell \cdot \beta_{i+j-1}$$

where $\ell \in GF(2)$. If i and j are both odd number, then $\ell = 1$. If i or j is an even number, then $\ell = 0$.

Proof. We will prove the theorem by induction on i and j. By using Table 1, the theorem is true for few terms. Assume that theorem is true for $i = n-1$. Then we need to show that it is true for $i = n$. We have four cases:

i. n is even and j is odd
ii. n is even and j is even
iii. n is odd and j is odd
iv. n is odd and j is even

Let $j < n-1$.

i. Let n be even and j be odd. Note that $\beta_1 = x$

$$\begin{aligned}\beta_n\beta_j &= (\beta_1\beta_{n-1} + \beta_{n-1})\beta_j \\ &= \beta_{n-1}(\beta_j(\beta_1 + 1)) \\ &= \beta_{n-1}(\beta_{j+1} + \beta_j + \beta_j) \\ &= \beta_{n+j}\end{aligned}$$

ii. Let n and j be even.

$$\begin{aligned}\beta_n\beta_j &= (\beta_1\beta_{n-1} + \beta_{n-1})\beta_j \\ &= \beta_{n-1}(\beta_j(\beta_1 + 1)) \\ &= \beta_{n-1}(\beta_{j+1} + \beta_j) \\ &= \beta_{n+j} + \beta_{n+j-1} + \beta_{n+j-1} \\ &= \beta_{n+j}\end{aligned}$$

iii. Let n and j be odd. Remember that addition of two odd integers is even.

$$\begin{aligned}\beta_n\beta_j &= (\beta_1\beta_{n-1})\beta_j \\ &= \beta_{n-1}(\beta_{j+1} + \beta_j) \\ &= \beta_{n+j} + \beta_{n+j-1}\end{aligned}$$

iv. Let n be odd and j be even.

$$\begin{aligned}\beta_n\beta_j &= (\beta_1\beta_{n-1})\beta_j \\ &= \beta_{n-1}\beta_{j+1} \\ &= \beta_{n+j}\end{aligned}$$

Note that if $j = n - 1$ or $j = n$, then this case can be proved by considering the factors of β_j or β_{j+1} as shown above. □

3 Polynomial Multiplication Using Charlier Polynomials over Binary Fields

In this section, we describe polynomial multiplication in Charlier polynomial representation for binary fields and give the total arithmetic complexity. Remember that multiplication in finite fields can be performed in two steps: multiplication over $GF(2)$ and modular reduction over $GF(2^n)$. Therefore, we divide this section into multiplication and reduction parts. Throughout this section, $M(n)$ and $A(n)$ denote the minimum number of multiplications and the minimum number of additions for corresponding algorithm for two n-term polynomials multiplication, respectively. The upper bounds of the required number of multiplications and additions to multiply polynomials in Charlier basis is given in the following theorem.

Theorem 3. *Let $a = a_{n-1}\beta_{n-1} + \cdots + a_0\beta_0$ and $b = b_{n-1}\beta_{n-1} + \cdots + b_0\beta_0$ be n-term polynomials over $GF(2)$ and $a \cdot b = c = c_{2n-2}\beta_{2n-2} + \cdots + c_0\beta_0$. Then, the coefficients of the polynomial, c are computed with*

$$M(n) + M(\left\lfloor \frac{n}{2} \right\rfloor) \text{ multiplications and}$$
$$A(n) + A(\left\lfloor \frac{n}{2} \right\rfloor) + 2\left\lfloor \frac{n}{2} \right\rfloor - 1 \text{additions}$$

by using any multiplication method.

Proof.

$$\begin{aligned}
c_0 &= a_0b_0 \\
c_1 &= a_0b_1 + a_1b_0 + a_1b_1 \\
c_2 &= a_0b_2 + a_2b_0 + a_1b_1 \\
c_3 &= a_0b_3 + a_3b_0 + a_1b_2 + a_2b_1 + a_1b_3 + a_3b_1 \\
&\vdots \\
c_{2n-3} &= a_{n-2}b_{n-1} + a_{n-1}b_{n-2} + \ell \cdot a_{n-1}b_{n-1} \\
c_{2n-2} &= a_{n-1}b_{n-1}
\end{aligned}$$

where $\ell = 1$ if $n-1$ is odd, otherwise, $\ell = 0$. There are extra terms when we compare this multiplication with ordinary multiplication. The extra terms can be expressed with $a_{2i+}b_{2j+1}$, where $0 \le i, j \le \lfloor \frac{n}{2} - 1 \rfloor - 1$. These elements correspond to multiplication of two $\lfloor \frac{n}{2} \rfloor$-term polynomials. Therefore, the total multiplication complexity is $M(n) + M(\lfloor \frac{n}{2} \rfloor)$. We need $2\lfloor \frac{n}{2} \rfloor - 1$ extra additions to combine these. Similarly, the total addition complexity is $A(n) + A(\lfloor \frac{n}{2} \rfloor) + 2\lfloor \frac{n}{2} \rfloor - 1$. □

Example 1. Let $n = p^j$, where p is a prime number and j is a positive integer. Let $a = a_{n-1}\beta_{n-1} + \cdots + a_0\beta_0$ and $b = b_{n-1}\beta_{n-1} + \cdots + b_0\beta_0$ be n-term polynomials over $GF(2)$ and $a \cdot b = c = c_{2n-2}\beta_{2n-2} + \cdots + c_0\beta_0$. Then, by using Karatsuba multiplication method [13],

1. If $p = 2$, the required number of multiplications is $n^{log_2 3} + \lfloor \frac{n}{2} \rfloor^{log_2 3}$ and the required number of additions is $8n^{log_2 3} - 11n + 3$.
2. If $p = 3$, the required number of multiplications is $n^{log_3 6} + \lfloor \frac{n}{2} \rfloor^{log_3 6}$ and the required number of additions is $\frac{116}{15}n^{log_3 6} - \frac{29}{5}n + \frac{7}{5}$.

We give an example to show that Theorem 3 is working.

Example 2. Let $a = a_3\beta_3 + a_2\beta_2 + a_1\beta_1 + a_0\beta_0$ and $b = b_3\beta_3 + b_2\beta_2 + b_1\beta_1 + b_0\beta_0$ be 4-term polynomials over $GF(2)$ in Charlier polynomial representation. Let $a \cdot b = c = c_6\beta_6 + \cdots + c_0\beta_0$. Then,

$$\begin{aligned}
c_0 &= a_0b_0 \\
c_1 &= a_0b_1 + a_1b_0 + \underline{a_1b_1} \\
c_2 &= a_0b_2 + a_2b_0 + a_1b_1 \\
c_3 &= a_0b_3 + a_3b_0 + a_1b_2 + a_2b_1 + \underline{a_1b_3 + a_3b_1} \\
c_4 &= a_1b_3 + a_3b_1 + a_2b_2 \\
c_5 &= a_2b_3 + a_3b_2 + \underline{a_3b_3} \\
c_6 &= a_3b_3
\end{aligned}$$

a_1b_1, $(a_1b_3 + a_3b_1)$ and a_3b_3 are the extra terms when we compare this multiplication with ordinary multiplication. The computation of these extra terms can be achieved by the following method:

Let $x_0 = a_1$, $y_0 = b_1$, $x_1 = a_3$ and $y_1 = b_3$. Then, the extra terms can be computed as follows:

$$m'_1 = x_0 y_0,$$
$$m'_2 = (x_0 + x_1)(y_0 + y_1) - m'_1 - m'_3,$$
$$m'_3 = x_1 y_1$$

The computation of extra terms requires 3 multiplications and 4 additions. One needs at most $9+3=12$ multiplications and $24+4+3=31$ additions by using Karatsuba method to compute $a \cdot b = c = c_6\beta_6 + \cdots + c_0\beta_0$.

Remark 1. Note that some or all elements of extra terms may be obtained without any cost, i.e. these are computed in n-term polynomials product. This, of course, depends your choice on multiplication method. Therefore, this reduces multiplication and addition complexity. However, in this paper, we give upper bounds.

Example 3. Let us consider Example 2. Consider two 4-term polynomials in standard representation $a(x) = \sum_{i=0}^{3} a_i x^i$ and $b(x) = \sum_{i=0}^{3} b_i x^i$. Karatsuba algorithm computes the product $c(x) = a(x)b(x) = \sum_{i=0}^{6} c_i x^i$ with the following 9 multiplications:

$$m_0 = a_0 b_0$$
$$m_1 = a_1 b_1$$
$$m_2 = a_2 b_2$$
$$m_3 = a_3 b_3$$
$$m_4 = (a_0 + a_1)(b_0 + b_1)$$
$$m_5 = (a_0 + a_2)(b_0 + b_2)$$
$$m_6 = (a_1 + a_3)(b_1 + b_3)$$
$$m_7 = (a_2 + a_3)(b_2 + b_3)$$
$$m_8 = (a_0 + a_1 + a_2 + a_3)(b_0 + b_1 + b_2 + b_3)$$

The extra terms in Charlier polynomial representation, i.e. $a_1 b_1$, $(a_1 b_3 + a_3 b_1)$ and $a_3 b_3$, are obtained without any cost:

$$m'_1 = m_0, m'_2 = m_6 + m_0 + m_3, m'_3 = m_3$$

Thus, one needs 9 multiplications and 24+3=27 additions by using Karatsuba method to compute $c = a \cdot b = c_6\beta_6 + \cdots + c_0\beta_0$

Remark 2. Let $a = a_{n-1}\beta_{n-1} + \cdots + a_0\beta_0$ be n-term polynomial over $GF(2)$ and $a^2 = c = c_{2n-2}\beta_{2n-2} + \cdots + c_0\beta_0$. Then,

$$c = a_{n-1}\beta_{2n-2} + a_{n-1} \cdot \ell\beta_{2n-3} + \cdots + 0\beta_3 + a_1\beta_2 + a_1\beta_1 + a_0\beta_0$$

Proof of this remark is very similar to Theorem 3. Note that the cost of squaring in Charlier polynomial representation is just reduction.

Now, we show how modular reduction process can be performed for irreducible Charlier polynomials.

3.1 Irreducible Charlier Binomials

Selected irreducible Charlier binomials are given in Table 2.

Table 2. Irreducible Charlier Binomials

$\beta_2 + \beta_0$	$\beta_3 + \beta_0$
$\beta_5 + \beta_0$	$\beta_7 + \beta_0$
$\beta_9 + \beta_0$	$\beta_{41} + \beta_0$
$\beta_{63} + \beta_0$	$\beta_{71} + \beta_0$
$\beta_{105} + \beta_0$	$\beta_{127} + \beta_0$
$\beta_{169} + \beta_0$	$\beta_{177} + \beta_0$

Reduction. By using irreducible Charlier binomial, one can perform reduction operation as follows:

Let $f = \beta_n + \beta_0$ be an irreducible polynomial of degree n over $GF(2)$. Let $n \leq i \leq 2n - 2$. Then,

$$\begin{aligned}
\beta_n\beta_{i-n} &= \beta_i + \beta_{i-1} \cdot \ell \\
\beta_0\beta_{i-n} &= \beta_i + \beta_{i-1} \cdot \ell \\
\beta_i &= \beta_{i-n} + \beta_{i-1} \cdot \ell \\
\beta_i &= \beta_{i-n} + (\beta_{i-n-1}\beta_n + \beta_{i-2} \cdot \ell_1) \cdot \ell \\
\beta_i &= \beta_{i-n} + \beta_{i-n-1} \cdot \ell
\end{aligned}$$

If $i - n$ and n or $i - n - 1$ and n are both odd, then $\ell = 1$ or $\ell_1 = 1$, respectively.. Otherwise, $\ell = 0$ or $\ell_1 = 0$. Note that $\beta_n = 1$. $\ell \cdot \ell_1 = 0$ since if $i - n$ is odd, then, $i - n - 1$ is even. Same trick is applicable for trinomial case.

3.2 Irreducible Charlier Trinomials

Table 3 tabulates selected irreducible Charlier trinomials. According to Table 3, it should be noted that recommended NIST or SEC binary fields such as $GF(2^{113})$, $GF(2^{131})$, $GF(2^{233})$ and $GF(2^{283})$ can be constructed with irreducible Charlier trinomials [10] and [12].

Reduction. By using irreducible Charlier trinomial, one can perform reduction operation as follows:

Let $f = \beta_n + \beta_k + \beta_0$ be an irreducible polynomial of degree n over $GF(2)$ and k be even. Let $n \leq i \leq 2n - 2$. Then,

$$\begin{aligned}
\beta_n\beta_{i-n} &= \beta_i + \beta_{i-1} \cdot \ell \\
(\beta_k + \beta_0)\beta_{i-n} &= \beta_i + \beta_{i-1} \cdot \ell \\
\beta_i &= \beta_{i-n+k} + \beta_{i-n} + \beta_{i-1} \cdot \ell \\
\beta_i &= \beta_{i-n+k} + \beta_{i-n} + (\beta_{i-n+k-1} + \beta_{i-n-1}) \cdot \ell
\end{aligned}$$

Table 3. Irreducible Charlier Trinomials

$\beta_{113}+\beta_{10}+\beta_0$	$\beta_{131}+\beta_{29}+\beta_0$
$\beta_{167}+\beta_2+\beta_0$	$\beta_{169}+\beta_3+\beta_0$
$\beta_{171}+\beta_4+\beta_0$	$\beta_{187}+\beta_3+\beta_0$
$\beta_{211}+\beta_{10}+\beta_0$	$\beta_{221}+\beta_2+\beta_0$
$\beta_{227}+\beta_3+\beta_0$	$\beta_{231}+\beta_{13}+\beta_0$
$\beta_{233}+\beta_{11}+\beta_0$	$\beta_{233}+\beta_{17}+\beta_0$
$\beta_{283}+\beta_3+\beta_0$	$\beta_{283}+\beta_{14}+\beta_0$
$\beta_{291}+\beta_{21}+\beta_0$	$\beta_{311}+\beta_3+\beta_0$
$\beta_{323}+\beta_{22}+\beta_0$	$\beta_{331}+\beta_{21}+\beta_0$
$\beta_{347}+\beta_{28}+\beta_0$	$\beta_{359}+\beta_{24}+\beta_0$
$\beta_{401}+\beta_{13}+\beta_0$	$\beta_{403}+\beta_{26}+\beta_0$
$\beta_{419}+\beta_{14}+\beta_0$	$\beta_{443}+\beta_9+\beta_0$
$\beta_{463}+\beta_{11}+\beta_0$	$\beta_{469}+\beta_{23}+\beta_0$
$\beta_{511}+\beta_{11}+\beta_0$	$\beta_{541}+\beta_3+\beta_0$
$\beta_{551}+\beta_{18}+\beta_0$	$\beta_{557}+\beta_{11}+\beta_0$

If $i-n$ and n are odd, then $\ell = 1$. Otherwise, $\ell = 0$.

Let $f = \beta_n + \beta_k + \beta_0$ be an irreducible polynomial of degree n over $GF(2)$ and k be odd. Let $n \leq i \leq 2n-2$. Then,

$$\begin{aligned}
\beta_n\beta_{i-n} &= \beta_i + \beta_{i-1}\cdot\ell \\
(\beta_k+\beta_0)\beta_{i-n} &= \beta_i + \beta_{i-1}\cdot\ell \\
\beta_i &= \beta_{i-n+k} + \beta_{i-n} + (\beta_{i+k-n-1} + \beta_{i-1})\cdot\ell \\
\beta_i &= \beta_{i-n+k} + \beta_{i-n} + \beta_{i-n-1}\cdot\ell
\end{aligned}$$

If $i-n$ and n are odd, then $\ell = 1$. Otherwise, $\ell = 0$.

3.3 Reduction Complexity

Table 4 shows reduction complexity for irreducible Charlier binomials and trinomials.

Table 4. Reduction Complexity

	Form	#XOR
Charlier Binomial	$\beta_n+\beta_0$	$\frac{3n}{2}$
Charlier Trinomial	$\beta_n+\beta_k+\beta_0$	$3n$

Remember that the cost of reduction process in polynomial basis representation strictly depends on the choice of reduction polynomial. Then, if one uses trinomial or pentanomial, reduction process requires $2n$ or $4n$ XOR gates, respectively.

4 Multiplication Complexity

In this section, we give modular multiplication complexity of multipliers in view of $\#AND$ and $\#XOR$ gates. Let $n = p^j$. Table 5 compares the space complexity and time complexity of selected multipliers. Note that this table is prepared by using Karatsuba multiplication method for Charlier basis [13]. According to Table 5, Charlier polynomial representation has better complexity than Dickson polynomial representation and ONB II. Therefore, binary fields can be constructed with low weight Charlier polynomials efficiently when there does not exist ONB for the corresponding extension. Remember that we give upper bounds for Charlier binomials and trinomials. The complexity of the field multiplication for Charlier polynomials can be further reduced by cleverly combining computed values (see Example 3). Therefore, for some cases, multiplication complexity for Charlier polynomial representation is also comparable with ONB I.

Table 5. Space Complexity Comparison of Selected Multipliers

	p	$\#AND$	$\#XOR$	Critical Delay
Charlier Binomial	2	$n^{log_2 3} + \lfloor \frac{n}{2} \rfloor^{log_2 3}$	$8n^{log_2 3} - 10n + 3$	$3log_2(n)T_X + T_A$
Charlier Binomial	3	$n^{log_3 6} + \lfloor \frac{n}{2} \rfloor^{log_3 6}$	$\frac{116}{15}n^{log_3 6} - \frac{29}{5}n + \frac{7}{5}$	$4log_3(n)T_X + T_A$
Charlier Trinomial	2	$n^{log_2 3} + \lfloor \frac{n}{2} \rfloor^{log_2 3}$	$8n^{log_2 3} - 8n + 3$	$3log_2(n)T_X + T_A$
Charlier Trinomial	3	$n^{log_3 6} + \lfloor \frac{n}{2} \rfloor^{log_3 6}$	$\frac{116}{15}n^{log_3 6} - \frac{14}{5}n + \frac{7}{5}$	$4log_3(n)T_X + T_A$
Polynomial Basis [13]	2	$n^{log_2 3}$	$6n^{log_2 3} - 8n + 2$	$(3log_2(n) - 1)T_X + T_A$
Polynomial Basis [13]	3	$n^{log_3 6}$	$\frac{29}{5}n^{log_3 6} - 8n + \frac{11}{5}$	$(4log_3(n) - 1)T_X + T_A$
Dickson Binomial [4]	2	$2n^{log_2 3}$	$11n^{log_2 3} - 11n$	$(2log_2(n) + 1)T_X + T_A$
Dickson Binomial [4]	3	$2n^{log_2 3}$	$\frac{48}{5}n^{log_3 6} - 11n + \frac{3}{5}$	$(3log_3(n) + 1)T_X + T_A$
Dickson Trinomial [4]	2	$2n^{log_3 6}$	$11n^{log_2 3} - 4n + 1$	$(2log_2(n) + 6)T_X + T_A$
Dickson Trinomial [4]	3	$2n^{log_3 6}$	$\frac{48}{5}n^{log_3 6} - 2n + \frac{1}{5}$	$(3log_3(n) + 6)T_X + T_A$
ONB I [3]	2	$n^{log_2 3} + n$	$\frac{11}{2}n^{log_2 3} - 4n - \frac{1}{2}$	$(2log_2(n) + 1)T_X + T_A$
ONB I [3]	3	$n^{log_3 6} + n$	$\frac{24}{5}n^{log_3 6} - 3n - \frac{4}{5}$	$(3log_3(n) + 1)T_X + T_A$
ONB II [3]	2	$2n^{log_2 3}$	$11n^{log_2 3} - 12n + 1$	$(2log_2(n) + 1)T_X + T_A$
ONB II [3]	3	$2n^{log_3 6}$	$\frac{48}{5}n^{log_3 6} - 10n - \frac{2}{5}$	$(3log_3(n) + 1)T_X + T_A$

Remark 3. NIST recommended binary field $GF(2^{283})$ can be constructed efficiently by using Charlier polynomials since there is no ONB for the corresponding extension.

5 Conclusion

In this paper, we give a new way to represent certain finite fields $GF(2^n)$. This representation is based on Charlier polynomials. We show that multiplication in

Charlier polynomial representation can be performed with subquadratic space complexity. One can obtain binomial or trinomial irreducible polynomials in Charlier polynomial representation which allows us faster modular reduction over binary fields when there is no desirable such low weight irreducible polynomial in other representations. This representation is very interesting for NIST recommended binary field $GF(2^{283})$ since there is no ONB for the corresponding extension. We also note that recommended NIST and SEC binary fields can be constructed with low weight Charlier polynomials.

Acknowledgment

The second and third authors are partially supported by TÜBİTAK under Grant No.TBAG-107T826 and and TBAG-109T672. The authors thank the anonymous referees for their detailed and very helpful comments.

References

1. Brown, M., Hankerson, D., Lopez, J., Menezes, A.: Software Implementation of the NIST Elliptic Curves over Prime Fields. In: Naccache, D. (ed.) CT-RSA 2001. LNCS, vol. 2020, pp. 250–265. Springer, Heidelberg (2001)
2. Godsil, C.D.: Algebraic Combinatorics. Chapman Hall/CRC Mathematics Series, Boca Raton (1993)
3. Fan, H., Hasan, M.A.: A New Approach to Subquadratic Space Complexity Parallel Multipliers for Extended Binary Fields. IEEE Trans. on Computers 56-2, 224–233 (2007)
4. Hasan, M.A., Negre, C.: Subquadratic Space Complexity Multiplication over Binary Fields with Dickson Polynomial Representation. In: von zur Gathen, J., Imaña, J.L., Koç, Ç.K. (eds.) WAIFI 2008. LNCS, vol. 5130, pp. 88–102. Springer, Heidelberg (2008)
5. Hasan, M.A., Negre, C.: Low Space Complexity Multiplication over Binary Fields with Dickson Polynomial Representation. IEEE Trans. on Computers (2010) (to appear) doi: 10.1109/TC.2010.132
6. Hankerson, D., Menezes, A., Vanstone, S.: Guide to Elliptic Curve Cryptography. Springer, Heidelberg (2004)
7. Lidl, R., Niederreiter, H.: Introduction to Finite Fields and Their Applications. Cambridge University, Cambridge (1997)
8. Menezes, A., Blake, I., Gao, X., Mullen, R., Vanstone, S., Yaghobian, T.: Applications of Finite Fields. Kluwer Academic, Boston (1993)
9. Mullen, G., Mummert, C.: Finite Fields and Applications. American Mathematical Society, Providence (2007)
10. National Institute of Standards and Technology, Recommended Elliptic curves for Federal Government Use (1999)
11. Sage: Open Source Mathematics Software, The Sage Group, `http://www.sagemath.org`
12. Standards for Efficient Cryptography Group (SECG), SEC 2: Recommended Elliptic Curve Domain Parameters (2010)
13. Weimerskirch, A., Paar, C.: Generalizations of the Karatsuba Algorithm for Efficient Implementations (2006), `http://eprint.iacr.org/2006/224`

Random Euclidean Addition Chain Generation and Its Application to Point Multiplication

Fabien Herbaut[1], Pierre-Yvan Liardet[2], Nicolas Méloni[3], Yannick Téglia[2], and Pascal Véron[3]

[1] Université du Sud Toulon-Var, IMATH, France
IUFM de Nice, Université de Nice
herbaut@unice.fr
[2] ST Microelectronics, Rousset, France
pierre-yvan.liardet@st.com, yannick.teglia@st.com
[3] Université du Sud Toulon-Var, IMATH, France
meloni@univ-tln.fr ,veron@univ-tln.fr

Abstract. Efficiency and security are the two main objectives of every elliptic curve scalar multiplication implementations. Many schemes have been proposed in order to speed up or secure its computation, usually thanks to efficient scalar representation [30,10,24], faster point operation formulae [8,25,13] or new curve shapes [2]. As an alternative to those general methods, authors have suggested to use scalar belonging to some subset with good computational properties [15,14,36,41,42], leading to faster but usually cryptographically weaker systems. In this paper, we use a similar approach. We propose to modify the key generation process using a small Euclidean addition chain c instead of a scalar k. This allows us to use a previous scheme, secure against side channel attacks, but whose efficiency relies on the computation of small chains computing the scalar. We propose two different ways to generate short Euclidean addition chains and give a first theoretical analysis of the size and distribution of the obtained keys. We also propose a new scheme in the context of fixed base point scalar multiplication.

Keywords: point multiplication, exponentiation, addition chain, SPA, elliptic curves.

1 Introduction

After twenty five years of existence, elliptic curve cryptography (ECC) is now one of the major public-key cryptographic primitives. Its main advantages, compared to it main competitor RSA, are its shorter keys and the lack of fast theoretical attacks. The recent factorization of an RSA modulus of 768 bits [17] is here to highlight the significant role that ECC will play during the next decade. In particular, it has been shown that ECC is suitable for cryptographic applications on devices with small resources. However, if 160-bit ECC is believed to remain secure, from a theoretical point of view, at least until 2020 [3], physical attacks on cryptographic devices have proved to be an immediate threat [22]. Thus,

G. Gong and K.C. Gupta (Eds.): INDOCRYPT 2010, LNCS 6498, pp. 238–261, 2010.

software or hardware ECC implementations have to deal with two apparently opposite requirements: efficiency and security. Indeed, protecting a device from physical attacks usually involves costly countermeasures.

In 2007, Méloni proposed a secure algorithm based on Euclidean addition chains [27]. As they only involve additions, they are naturally resistant to SCA. However, the efficiency of such a method relies on the existence of a small chain computing the scalar. It has been pointed out that finding such a chain becomes more and more difficult when the scalar grows in size. For cryptographic sizes, finding a good chain is costlier than the scalar multiplication itself. So, instead of proposing a new scalar multiplication scheme, **we propose to modify the key generation process**. More precisely, we show that it is possible to generate the key as a small Euclidean addition chain, allowing us to use Méloni's fast and secure scheme.

From a general point of view, generating keys in a specific shape is not a new idea. Various methods have been proposed through the years to generate scalars belonging to some subset of the set of all possible keys, with good computational properties [14,36,41,42]. However, this usually implies some serious security issues [33,34,9,38]. Some methods remain cryptographically secure in the context of a fixed base point but require large amount of stored data [4,15] (Coron et al. [15] suggest to store from 50 to 100 points for the same security level as that considered in the present work). Finally, some schemes use special endomorphisms, such as the Frobenius map, on Koblitz curves [20].

From that perspective, our approach is quite different. Méloni's scheme only efficiently applies on curve in Weierstrass form. Moreover, it is particularly suitable to small devices with low computational resources because of its low memory requirement (at most two stored points) and resistance to side channel attacks. Yet, generating random Euclidean addition chains leads to several problems. First, what is the size of the keys we can generate for a given chain length? In other words, is it possible to achieve a certain level of security with relatively small chains? The second problem is that of distribution. Indeed, many different chains of same length can compute the same integer. It is then important to ensure that generating keys this way does not weaken the discrete logarithm problem.

In this paper, we produce the first practical and theoretical results on random Euclidean addition chain generation. We also show that it can lead to efficient and SPA-secure scalar multiplication methods.

This work is organized as follows. In Section 2 we review Méloni's scheme. In Section 3 we recall some background about Euclidean addition chains and set notations. In Section 4 and 5, we describe two different families of Euclidean addition chains and give some results on their distribution (notice that in Section 4 two variants are described). Finally, in Section 6 we propose some comparisons with existing side-channel resistant scalar multiplication methods.

2 Scalar Multiplication Using Euclidean Addition Chains

For the sake of concision, we do not give details about scalar multiplication and side channel attacks. We invite the reader to refer to [7,12] for detailed overview of elliptic curve based cryptography.

This section is dedicated to a specific scalar multiplication algorithm based on Euclidean addition chains.

2.1 Euclidean Addition Chains

Definition 1. *An Euclidean addition chain (EAC) of length s is a sequence $(c_i)_{i=1...s}$ with $c_i \in \{0,1\}$. The integer k computed from this sequence is obtained from the sequence $(v_i, u_i)_{i=0..s}$ such that $v_0 = 1$, $u_0 = 2$ and $\forall i \geqslant 1$, $(v_i, u_i) = (v_{i-1}, v_{i-1}+u_{i-1})$ if $c_i = 1$ (small step), or $(v_i, u_i) = (u_{i-1}, v_{i-1}+u_{i-1})$ if $c_i = 0$ (big step). The integer k associated to the sequence $(c_i)_{i=1...s}$ is $v_s + u_s$, we will denote it by $\chi(c)$.*

EXAMPLE : From the EAC (10110) one can compute the integer 23 as follows : $(1,2) \rightarrow (1,3) \rightarrow (3,4) \rightarrow (3,7) \rightarrow (3,10) \rightarrow (10,13) \rightarrow 23 = \chi(10110)$.

2.2 Point Multiplication Using EAC

From any EAC c and any point P of an elliptic curve, it is shown in [27] that a new point Q can be computed using the following algorithm :

Algorithm 1. EAC_Point_Mul(c:EAC, P:point)

Require: P, $[2]P$
Ensure: $Q = \chi(c)P$
1: $(U_1, U_2) \leftarrow (P, [2]P)$
2: **for** $i = 1 \dots \text{length}(c)$ **do**
3: **if** $c_i = 0$ **then**
4: $(U_1, U_2) \leftarrow (U_2, U_1 + U_2)$
5: **else**
6: $(U_1, U_2) \leftarrow (U_1, U_1 + U_2)$
7: **end if**
8: **end for**
9: **return** $Q = U_1 + U_2$

Algorithm 2. Point multiplication

Require: P and an integer k
Ensure: $Q = kP$
1: $c \leftarrow$ Find_EAC(k)
2: **return** $Q =$ EAC_Point_Mul(c,P)

In [27], it is shown that any scalar multiplication can be performed using the preceding algorithm. It is achieved by finding an Euclidean addition chain computing the scalar. We will denote by `Find_EAC` , the algorithm which returns an addition chain for an integer k.

It has been shown in [27] that the `for loop` of algorithm 1 can be efficiently implemented on elliptic curves in Weierstrass form using Jacobian coordinates. This leads to a fast point multiplication method resistant to side channel attacks since at each step of the *for* loop, the same operation is used .

The efficiency of algorithm 2 directly depends on the length of the chains and the complexity of the algorithm `Find_EAC`. Although finding an Euclidean addition chain computing a given integer k is quite simple (it suffices to choose an integer g co-prime with k and apply the subtractive form of Euclid's algorithm) finding a short chain remains a hard problem. As an example, the average length of the computed chain for k with g uniformly distributed in the range $[1, k]$ is $O(\ln^2 k)$ [18]. For 160-bit scalars, experiments have shown that, on average, it is required to try more that 45,000 different g to find a relatively small chain using the Montgomery heuristic [32].

2.3 Our Approach

We propose in this paper to proceed differently. Instead of randomly choosing an integer k and then trying to find a suitable EAC c to finally compute the point kP, we propose to randomly generate a small length EAC c and then compute the associated point on the curve. More precisely, we will see in the next sections that the chains will be chosen in a subset of short chains for key distribution matters.

From definition 2.1, notice the random generation of a s-length EAC boils down to the random generation of a s-bit integer. As an example, algorithm 3 shows how to process Diffie-Helmann key exchange protocol with EAC.

Algorithm 3. Diffie-Helmann using EAC

Require: a point P
Ensure: Output a common key K
1: A randomly generates a short EAC c_1
2: B randomly generates a short EAC c_2
3: A computes Q_1 =EAC_Point_Mul(c_1, P) and send it to B
4: B computes Q_2 =EAC_Point_Mul(c_2, P) and send it to A
5: A computes K =EAC_Point_Mul(c_1, Q_2)
6: B computes K =EAC_Point_Mul(c_2, Q_1)

Using this approach we compute points kP for a subset S of all possible values for the integers k. Hence we do not deal any more with the classical Discrete Logarithm Problem but with the Constrained Discrete Logarithm Problem.

Name : CDLP
Input : p a prime number, g a generator of a group G, $S \subset \mathbb{Z}_p$, and g^x for $x \in S$.
Problem : Compute x.

The complexity of the discrete logarithm problem (DLP) over a generic group directly depends on the size of this group. It has been shown that the constrained version of this problem (CDLP), where only a subset S of the group is considered, has a similar complexity. Indeed, for a random set S, $\Omega(\sqrt{|S|})$ group operations [28] are required to solve the CDLP. As an example, if one naively chooses to generate random Euclidean chain of length 160 (which means there are 2^{160} of them), we will see, in section 3, that the biggest integer that can be generated is $F_{164} \simeq 2^{112.7}$. This means that the number of elements of the set S is at most 2^{113}, leading to a security of at most 66.5 bits.

In this paper, we propose to study two families of EAC providing good balance between security and efficiency. In practice, it implies that the chains are long enough so that the corresponding set S of generated keys is sufficiently large, but short enough so that the scalar multiplication remains fast.

3 Notations and Properties

We give in this section some notations and important results for the sequel of this paper.

Definition 2. *Let n and p be two integers, we define :*

- $\mathcal{M}$ *as the set of EAC and* $\mathcal{M}_n$ *as the set of EAC of length* $n > 0$,
- $\mathcal{M}_{n,p}$ *as the set of EAC of length* $n > 0$ *and Hamming weight* $p > 0$.
- χ *the map from* $\mathcal{M}$ *to* $\mathbb{N}$, *such that for* $m \in \mathcal{M}$, $\chi(m)$ *be the integer computed from the EAC* m,
- ψ *the map from* $\mathcal{M}$ *to* $\mathbb{N} \times \mathbb{N}$, *such that for* $m \in \mathcal{M}$, $\psi(m) = (v_s, u_s)$ *if* $m \in \mathcal{M}_s$,
- S_0 *the matrix* $\begin{pmatrix} 0 & 1 \\ 1 & 1 \end{pmatrix}$ *corresponding to a big step iteration,*
- S_1 *the matrix* $\begin{pmatrix} 1 & 1 \\ 0 & 1 \end{pmatrix}$ *corresponding to a small step iteration.*

With these notations, for $m = (m_1, \ldots, m_s) \in \mathcal{M}_s$, we have :

$$\psi(m) = (1,2) \prod_{i=1}^{s} S_{m_i} \text{ and } \chi(m) = \langle (1,2) \prod_{i=1}^{s} S_{m_i}, (1,1) \rangle.$$

Let r and s be two integers, we will denote by mm' the element of $\mathcal{M}_{r+s}$ obtained from the concatenation of $m \in \mathcal{M}_r$ and $m' \in \mathcal{M}_s$. This way, for $n > 0$, m^n is a word of $\mathcal{M}_{nr}$ if $m \in \mathcal{M}_r$.

For convenience, $\mathcal{M}_0$ will correspond to the set with one element e which is the identity element for the concatenation.

For m and m' two elements of $\mathcal{M}_r$ such that $\psi(m) = (v, u)$ and $\psi(m') = (v', u')$ we will say that $\psi(m) \leqslant \psi(m')$ if $v \leqslant v'$ and $u \leqslant u'$.

Proposition 1. *Let* $n > 0$, F_i *be the* i^{th} *Fibonacci number,* $\alpha_n = \frac{(1+\sqrt{2})^n + (1-\sqrt{2})^n}{2}$ *and* $\beta_n = \frac{(1+\sqrt{2})^n - (1-\sqrt{2})^n}{2\sqrt{2}}$ *:*

- $\psi(0^n) = (F_{n+2}, F_{n+3})$, $\psi(1^n) = (1, n+2)$, $\chi(0^n) = F_{n+4}$, $\chi(1^n) = n+3$,
- $\forall m \in \mathcal{M}_n$, $\chi(1^n) \leqslant \chi(m) \leqslant \chi(0^n)$, *and* $\psi(1^n) \leqslant \psi(m) \leqslant \psi(0^n)$,
- $S_0^n = \begin{pmatrix} F_{n-1} & F_n \\ F_n & F_{n+1} \end{pmatrix}$, $S_1^n = \begin{pmatrix} 1 & n \\ 0 & 1 \end{pmatrix}$,
- $(S_0 S_1)^n = \begin{pmatrix} \alpha_n - \beta_n & \beta_n \\ \beta_n & \alpha_n + \beta_n \end{pmatrix}$, $(S_1 S_0)^n = \begin{pmatrix} \alpha_n & 2\beta_n \\ \beta_n & \alpha_n \end{pmatrix}$.

Proof. The first property is straightforward. The other ones can be proved by induction.

Notice that from $S_0^{m+n} = S_0^m S_0^n$, we can recover a well known identity on Fibonacci sequence, namely :

$$F_{m+n} = F_{m-1}F_n + F_m F_{n+1}. \tag{1}$$

Proposition 2. *Let $n > 0$ and $m = (m_1, \ldots, m_n) \in \mathcal{M}_n$, then the map ψ is injective and $\chi(m_1, \ldots, m_n) = \chi(m_n, \ldots, m_1)$.*

Proof. We refer to [19] for standard link between EAC, Euclidean algorithm and continued fractions, which explains the second point. It is also explained that if $\psi(m) = (v, u)$ then $(u, v) = 1$ and the only chain which leads to (v, u) is obtained using the additive version of Euclidean algorithm.

4 A First Family of EAC

We will consider in this section $\mathcal{M}_n^0$ the subset of $\mathcal{M}_{2n}$ whose elements are EAC beginning with n zeros.

4.1 Some Properties of $\mathcal{M}_n^0$

From proposition 2 the restriction of χ to $\mathcal{M}_n$ is not injective because of the mirror symmetry property.

Proposition 3. *The restriction of χ to $\mathcal{M}_n^0$ is injective.*

Proof. Let x and y be two words of $\mathcal{M}_n^0$ such that $\chi(x) = \chi(y)$, and $m0^n$, $m'0^n$, be the words obtained when reading x and y from right to left. Using the symmetry property, we have $\chi(m0^n) = \chi(m'0^n)$. Let $(v, u) = \psi(m)$ and $(v', u') = \psi(m')$, then

$$\begin{aligned} & \chi(m0^n) = \chi(m'0^n) \\ \Leftrightarrow\ & F_n u + F_{n-1} v + F_{n+1} u + F_n v = F_n u' + F_{n-1} v' + F_{n+1} u' + F_n v' \\ \Leftrightarrow\ & F_{n+2}(u - u') = F_{n+1}(v' - v) \end{aligned}$$

Since $(F_{n+1}, F_{n+2}) = 1$, then F_{n+2} divides $v' - v$. Now from proposition 1, since v and v' are less or equal than F_{n+2} and nonzero, then $|v' - v| < F_{n+2}$ which implies that $v = v'$ and so $u = u'$. Hence $\psi(m) = \psi(m')$, so $m = m'$.

Proposition 4. $\chi(\mathcal{M}_n^0) \subset [(n+1)F_{n+2} + F_{n+3}, F_{2n+4}]$, *the lower (resp. the upper) bound being reached by $0^n 1^n$ (resp. 0^{2n}). The mean value is $\left(\frac{3}{2}\right)^n F_{n+4}$.*

Proof. Let $0^n x 1 y$ and $0^n x 0 y$ be two elements of $\mathcal{M}_n^0$ where x and y are chains of size a and $n - 1 - a$ with $a \in [0, n-1]$. From the definition of χ it follows that $\chi(0^n x 1 y) < \chi(0^n x 0 y)$. Hence the smallest integer is computed from the word $0^n 1^n$ and the greatest from $0^n 0^n$. Now $\chi(0^n 1^n) = \langle (1, 2) S_0^n S_1^n, (1, 1) \rangle$ and $\chi(0^{2n}) = \langle (1, 2) S_0^{2n}, (1, 1) \rangle$. From proposition 1, we deduce that $\chi(0^n 1^n) = (n + 1)F_{n+2} + F_{n+3}$ and $\chi(0^{2n}) = F_{2n+4}$.

To compute the mean value, let us consider n independent Bernoulli random variables $C_1, \ldots, C_n$ such that $\forall i \in [1, n]$, $\Pr(C_i = 0) = \Pr(C_1 = 1) = 1/2$. The mean value is $E(X)$ where $X = \chi(0^n C_1 \ldots C_n)$. Now

$$X = \langle (1,2) S_0^n \prod_{i=1}^{n} \left(\begin{smallmatrix} C_i & 1 \\ 1-C_i & 1 \end{smallmatrix}\right), (1,1) \rangle.$$

Notice that X is a polynomial of $\mathbb{Z}[C_1, \ldots, C_n]/(C_1^2 - C_1, \ldots, C_n^2 - C_n)$. As the C_i are independent then $\forall J \subset [1, n]$, $E(\prod_{j \in J} C_j) = \prod_{j \in J} E(C_j)$, hence

$$E(X) = \langle (1,2) S_0^n \prod_{i=1}^{n} \left(\begin{smallmatrix} E(C_i) & 1 \\ 1-E(C_i) & 1 \end{smallmatrix}\right), (1,1) \rangle = \langle (1,2) S_0^n \prod_{i=1}^{n} \left(\begin{smallmatrix} \frac{1}{2} & 1 \\ \frac{1}{2} & 1 \end{smallmatrix}\right), (1,1) \rangle.$$

The final result comes from proposition 1 and the equality :

$$\forall n \in \mathbb{N}^*, \begin{pmatrix} \frac{1}{2} & 1 \\ \frac{1}{2} & 1 \end{pmatrix}^n = (3/2)^{n-1} \begin{pmatrix} \frac{1}{2} & 1 \\ \frac{1}{2} & 1 \end{pmatrix}.$$

4.2 Application to Existing Standards

Using the set $\chi(\mathcal{M}_n^0)$, we can generate (with algorithm 1) 2^n distinct points $\chi(c)P$ for a point P whose order is greater than F_{2n+4}. Of course, when the order d of the point P is known, we have to choose the largest integer n such that $F_{2n+4} < d$.

Because of the results on the difficulty to solve the CDLP problem, we have to consider the set $\mathcal{M}_n^0$ only for $n \geqslant 160$. For $n = 160$, we have $\chi(\mathcal{M}_{160}^0) \subset [161 F_{162} + F_{163}, F_{324}]$, that is to say

$$\chi(\mathcal{M}_{160}^0) \subset [2^{118.6}, 2^{223.6}].$$

Such data fit well with the `secp224k1` and `secp224r1` parameters where the order of the point P is about $2^{223.99}$ [6]. Those parameters are consistent with ANSI X.962, IEEE P1363 and IPSec standards and are recommended for ANSI X9.63 and NIST standards.

4.3 A Variant for the `secp160k1` and `secp160r1` Recommended Parameters

In the `secp160k1` and `secp160r1` recommended parameters, the order d of the point P is around 2^{160}. Using the set $\mathcal{M}_n^0$, this leads us to choose $n = 114$ which only gives rise to 2^{114} distinct points. Hence the complexity of an attack is $\Omega(2^{57})$ which may expose this method to some attacks.

Notice that if we use an element c of $\mathcal{M}_{160}^0$ with the point P of order d then the algorithm 1 computes $(\chi(c) \bmod d)P$. If the values of $\{\chi(c) \bmod d, c \in \mathcal{M}_{160}^0\}$ are well distributed among $\mathbb{Z}/d\mathbb{Z}$, then we can use the above mentioned method, provided that computing $\chi(c)P$ with algorithm 1 be more efficient than computing kP with $k \in \mathbb{Z}/d\mathbb{Z}$, with a classical SPA-resistant method. This last point will be discussed in section 6, we will focus now on the problem of the distribution. To this end, we will adapt results on Stern sequences from [35].

Table 1. Distribution of $\chi(\mathcal{M}_n^0)$ modulo d

n \ t	0	1	2	3	4	5	6	7	8	9	10	11	12	13	$\frac{\#\chi(\mathcal{M}_0^n)}{d}$
10	294	312	214	70	12	4	1								0.66
15	10814	12044	6216	2026	436	76	15								0.66
20	266749	327372	194442	74219	20589	4344	789	100	18	1					0.7
25	6493638	8894037	5979946	2627531	850691	216285	44567	7832	1233	162	15	1	1		0.74
29	74024780	115132679	88980442	45585634	17436296	5315191	1347286	294399	56344	9674	1459	193	18	4	0.79

Theorem 1. *Let d be a prime number, $m \in \mathcal{M}_\ell$ such that $\psi(m) = (v, u)$. If $d \not| u$, and $d \not| v$ then there exist constants $c_d \in \mathbb{R}_+$ and $\tau_d \in [0, 1[$ so that for all $\alpha \in (\mathbb{Z}/d\mathbb{Z})^*$, $r \in \mathbb{N}$*

$$\left| \frac{\#\{x \in \mathcal{M}_r \mid \chi_d(mx) = \alpha\}}{2^r} - \frac{d}{d^2 - 1} \right| < c_d \tau_d^r, \text{ and}$$

$$\left| \frac{\#\{x \in \mathcal{M}_r \mid \chi_d(mx) = 0\}}{2^r} - \frac{1}{d + 1} \right| < c_d \tau_d^r.$$

Proof. See annex for the proof and the link with Stern sequences.

Taking m as the all zeros 160 bits vector, this asymptotic result let us think that the values $\chi(c)$ for $c \in \mathcal{M}_{160}^0$ are well distributed modulo d since $(F_{162}, d) = (F_{163}, d) = 1$. In order to illustrate this theoretical result, we made several numerical tests by generating all EAC of $\mathcal{M}_n^0$ (from $n = 10$ to $n = 29$) and reducing the corresponding integer modulo a n-bits prime number (recall that for a practical use, we consider elements of $\mathcal{M}_{160}^0$ and a point P of order d about 2^{160}). Table 1 seems to show that even for chains whose length is much smaller than d, the distribution is not so bad. For each value of n, a random n-bits prime d_n has been generated. We have then computed for each $\alpha \in \mathbb{Z}/d_n\mathbb{Z}$, the cardinality δ_α of $\chi^{-1}(\alpha)$ in $\mathcal{M}_n^0$. Let $\mathcal{T} = \{\delta_\alpha \mid \alpha \in \mathbb{Z}/d_n\mathbb{Z}\}$, for each t in $\mathcal{T}$ we have then computed how many integers in $[0, d-1]$ are exactly computed t times. As an example, for $t = 0$ we know how many integers are never reached when reducing modulo d the value $\chi(c)$ for $c \in \mathcal{M}_n^0$.

5 A Second Family of EAC and an Open Problem

In order to generate a 160 bits integer, the author of [26] gives numerical results which show that the search for a chain whose length be less than 260 needs about 2^{23} tests using a heuristic from Montgomery [32]. With such chains, Algorithm 1 needs less multiplications than the classical SPA-resistant algorithms. We investigated the problem of choosing shorter chains in order to speed up the performances of algorithm 1. Once the length ℓ is fixed we have to deal with two constraints :

- the number p of 1's in the chain must be chosen so that the greatest integer generated is as near as d ($\simeq 2^{160}$) as possible,

- because of the non-injectivity of χ , $\binom{\ell}{p}$ must be greater than 2^{160} in order to hope that the integers generated reach most of the elements of $[1, 2^{160}]$.

These two constraints lead us to study the sets $\mathcal{M}_{\ell,p}$ where $p < \ell/2$.

Theorem 2. *Let* $(p, \ell) \in \mathbb{N}^2$ *such that* $0 < 2p < \ell$. *Let* F_i *be the* i^{th} *Fibonacci,* $\alpha_p = \frac{(1+\sqrt{2})^p+(1-\sqrt{2})^p}{2}$, $\beta_p = \frac{(1+\sqrt{2})^p-(1-\sqrt{2})^p}{2\sqrt{2}}$, *then*

i) For all $m \in \mathcal{M}_{\ell,p}$ *we have,* $F_{\ell-p+4} + pF_{\ell-p+2} \leqslant \chi(m) \leqslant F_{\ell-2p+4}(\alpha_p + \beta_p) + \beta_p F_{\ell-2p+2}$.
ii) The lower bound is reached if and only if $m = 1^p 0^{\ell-p}$ *or* $m = 0^{\ell-p}1^p$.
iii) The upper bound is reached if and only if $m = (01)^p 0^{\ell-2p}$ *or* $m = 0^{\ell-2p}(10)^p$.

Proof. See annex.

To improve the performances of Algorithm 1, we choose to use chains whose length is 240. In this case $p = 80$ seems to be the best choice with respect to our two constraints. With such parameters, we can randomly generate about 2^{216} chains computing integers in the interval $[2^{117.7}, 2^{158.9}]$. Unfortunately, it seems to be a hard problem to compute the number of distinct integers generated in this way.

This naturally leads us to consider the set $\mathcal{M}_{3p,p}$. Numerical experiments for some values of p let us think that the cardinality of $\chi(\mathcal{M}_{3p,p})$ is near from 2^{2p}. Notice that from the preceding theorem, it can be proved that the upper bound for $\ell = 3p$ is equivalent to $\gamma\left(\frac{(1+\sqrt{2})(1+\sqrt{5})}{2}\right)^p$ where $\gamma = \frac{(1+\sqrt{5})^4}{32\sqrt{5}} + \frac{(1+\sqrt{5})^4}{32\sqrt{10}} + \frac{(1+\sqrt{5})^2}{8\sqrt{10}}$ which is close to $\gamma 2^{1.96p}$. We end this section with an open problem : What is the cardinality of $\chi(\mathcal{M}_{3p,p})$? The good performances of this method (see next section) make this problem of interest. Notice that a straightforward argument gives that the number of distinct integers generated (for the proposed parameters) is greater than 2^{106}. Indeed, it follows from proposition 3 that the chains $0^{120}c$, where $c \in \mathcal{M}_{120,80}$ give rise to distinct integers.

6 Comparisons

In this paper we have proposed three different ways to compute a point on the curve from an Euclidean addition chain :

Method 1: use a chain from $\mathcal{M}^0_{160}$ for curves of order about 2^{224}.
Method 2: use a chain from $\mathcal{M}^0_{160}$ for curves of order about 2^{160} even if $\chi(c)$ can be greater than the order, using the results on Stern sequences.
Method 3: use a chain from $\mathcal{M}_{240,80}$ for curves of order about 2^{160}.

The interest of Method 1 is that it is the only method for which we have a proved security. The main drawback is that it forces us to work on curves defined over larger fields ($\simeq 2^{224}$ elements instead of 2^{160}). However, we will see, in Section 6.2, that it might not be such a problem in practical implementations. Moreover, this method shows to be particularly relevant in the context of fixed base point scalar multiplication.

Method 2 allows us to reduce the size of the underlying fields by using Stern's results. Numerical samples tend to show that the generated keys are well distributed but we still lack a complete security proof.

Finally, in Method 3, we try to reduce both the size of the fields and the length of the chains. In that case, it becomes very complicated to analyze the distribution of the generated keys. In particular, the system becomes highly redundant, which might lead to a bias in the set of possible chains. Typically, this would make this method irrelevant for signature and key-exchange schemes.

We propose to compare our work to other side-channel resistant methods with a similar security level. We do not take into account special key generation methods as they usually provide lower security level [33,34]. One could try to increase the size of the underlying field (as we do with Method 1) to solve this issue, however this would make those approaches slower than general algorithms. In the fixed point scenario, special key generation methods usually provide enough security but require in the same time a large amount of stored data: from 50 to 1000 precomputed points [15,4] when we only require a table of two stored points.

In Table 2, we summarize the cost, in terms of field multiplications, of various SPA-resistant scalar multiplication schemes providing a security of 80 bits. In order to ease comparisons, we make the traditional assumption that the cost of a field squaring (S) is 80 percent of that of a field multiplication (M).

Random Base Point. In that scenario, a new base point P is computed for each new session. This implies that it is not possible to precompute offline multiples of P to speed up the process.

We consider the following SPA resistant methods:

- Dummy operations consist of adding a dummy point addition during the double-and-add algorithm, when the current bit is a 0.
- The Montgomery ladder is a SPA resistant algorithm from Peter Montgomery [31], performing one doubling and one addition for each bit of the scalar. It is only efficient on Montgomery curves.
- Unified formulae allow to perform doubling and addition with the same formulae on specific curve shapes. They can be then combined with the NAF representation for scalar multiplication. The cost of the unified operation is 11M, 12M and 14M on Edwards [2], Hessian [16] and Jacobi curves [23] respectively. We do not compare to Brier and Joye general unified formulae due to their quite high cost (18M) [5].
- Möller proposed a modified version of Brauer (2^w-ary) algorithm [29]. Using precomputations, its pattern is independent from the scalar itself. In that case, we used the latest and fastest point addition and point doubling formulae in Jacobian coordinates (see [1] for complete overview). One can also refer to the Left-to-Right recodings methods proposed in [39] whose performances are similar.

Fixed Base Point. In the context of a fixed base point we propose a new scalar multiplication scheme based on our chains generation method. Notice that in Methods 1 and 2, the 160 first steps of Algorithm 1 are fixed, independently of the scalar, and correspond to big steps. Hence we can precompute and store the points $F_{162}P$ and $F_{163}P$ and then generate random chains of length 160. We compared these methods to the classical Comb method.

Table 2. Cost of various SPA resistant scalar multiplication methods providing 80 bits of security

Point gen.	Method	curve	field size (bits)	# precomp. points	#Field Mult.
random	Dummy operations	general	160	1	3530
	Montgomery ladder	Montgomery	160	1	1463
	Unified formulae	Edwards	160	1	2335
	Unified formulae	Hessian	160	1	2548
	Unified formulae	Jacobi	160	1	2973
	Unified formulae	Edwards	160	7	2130
	Unified formulae	Hessian	160	7	2324
	Unified formulae	Jacobi	160	7	2711
	Möller's recoding	general	160	16	1843
fixed	Comb Method	general	160	2	1754
	Comb Method	general	160	4	1177
	Comb Method	general	160	8	866
	Comb Method	general	160	16	688
random	Method 1	general	224	1	**2104**
	Method 1 (x only)	general	224	1	**1790**
fixed	Method 1	general	224	2	**1048**
	Method 1 (x only)	general	224	2	**888**
random	Method 2	general	160	1	**2104**
	Method 2 (x only)	general	160	1	**1790**
	Method 3	general	160	1	**1576**
	Method 3 (x only)	general	160	1	**1336**
fixed	Method 2	general	160	2	**1048**
	Method 2 (x only)	general	160	2	**888**

To be completely fair, we evaluate in the next section the additional cost of working on larger fields with Method 1.

As for Method 2 and 3, Table 2 shows our different methods provide very good results. In the random point scenario, Methods 2 and 3 perform generally better that their counterparts. Only Montgomery's algorithm can be claim to be faster, but its use is restricted to Montgomery's curves. Besides, computing the x coordinate only with Method 3 leads to a faster scheme. In the fixed based point scenario, the Comb method requires at least 4 times more stored points to perform faster than our methods.

6.1 Comparison of Method 1 with Algorithms Working on 160-Bit Integers

Recall that Method 1 requires to work on fields of larger size (2^{224} elements). Hence to be fair in our comparisons, we need to evaluate the additional cost of performing multiplications in larger fields. To this end, we are going to consider three contexts for modular multiplication. For each of them, we will identify scenarios for which it is worth using our method.

a) The CIOS method [21]: the Coarsely Integrated Operand Scanning method is an efficient implementation of Montgomery's modular multiplication for a large class of processor. From [21] a modular multiplication between two integers stored as s words of w bits needs $2s^2+s$ w-bits multiplications, $4s^2+4s+2$ w-bits additions, $6s^2+7s+2$ w-bits read instructions and $2s^2+5s+1$ w-bits write instructions. Using these results, we can estimate the cost in terms of w-bits operations of the methods listed in Table 2.This leads us to the following remarks.

On a 32 bits processor, for a random point P, Method 1 with x-only is the most performant SPA resistant method which can be used on any curve and which only stores the point P. For a fixed point P, if only two points can be stored, this latter is better than Comb method. On a 64 bits processor, the preceding remarks remain true. Moreover, the fixed point method (without the x coordinate trick) is competitive with the Comb method when only two points are stored. On a 128 bits processor, since two words are needed to store 160 bits integer or 224 bits integer, we only have to compare the number of field multiplications in Table 2. This shows that our method in the fixed point context is better than the Comb method when storing 2 or 4 points. For random point context, if one needs an SPA resistant algorithm which works on any curve and stores no more than one point, then our method gives the best result.

b) The GNU multiprecision library: we provide benchmarks for fair comparisons between modular multiplications in the case of practical use with the library GMP. We compute several times 2^{28} modular multiplications (using

Table 3. Performances of Method 1 using CIOS method on 32 bits processor

	s	# Field mult.	32 bits ×	32 bits +	32 bits Read	32 bits Write
Method 1, x-only	7	1790	187950	404540	617550	239860
Dummy	5	3530	194150	430660	660110	268280
Method 1, fixed point, x-only	7	888	93240	200688	306360	118992
Comb method, 2 stored points	5	1754	96470	213988	327998	133304

Table 4. Performances of Method 1 using CIOS method on 64 bits processor

	s	# Field mult.	64 bits ×	64 bits +	64 bits Read	64 bits Write
Method 1, fixed point	4	1048	37728	85936	132048	55544
Comb method, 2 stored points	3	1754	36834	87700	135058	59636

Table 5. Time to compute 2^{28} modular multiplications with GnuMP

	p	32 bits Intel T2500 2.0Ghz	64 bits AMD Opteron 8382 2.6Ghz
160bits	$2^{160} - 2^{31} - 1$	$T_{\mathtt{min}}$ 135.96s $T_{\mathtt{max}}$ 140.49s $T_{\mathtt{average}}$ 137.88s	$T_{\mathtt{min}}$ 28.12s $T_{\mathtt{max}}$ 32.49s $T_{\mathtt{average}}$ 30.42s
224bits	$2^{224} - 2^{96} + 1$	$T_{\mathtt{min}}$ 198.22s $T_{\mathtt{max}}$ 201.86s $T_{\mathtt{average}}$ 200.51s	$T_{\mathtt{min}}$ 32.24s $T_{\mathtt{max}}$ 33.93s $T_{\mathtt{average}}$ 32.67s
$T_{\mathtt{average\ 224}}/T_{\mathtt{average\ 160}}$		1.45	1.07
$T_{\mathtt{max\ 224}}/T_{\mathtt{min\ 160}}$		1.48	1.17

Table 6. Performances of Method 1 with GnuMP

Point gen.	Method	curve	storage	Mult. (32 bits proc.)	Mult. (64 bits proc.)
random	Method 1	general	1	3114	2462
	Method 1 (x only)	general	1	2650	2095
fixed	Method 1	general	2	1552	1227
	Method 1 (x only)	general	2	1315	1039

`mpz_mul` and `mpz_mod`) on 32 and 64 bits processors. We consider reduction modulo a prime number p conformant to the ANSI X9.63 standard [6] (resp. FIPS186-3 standard [40]) for the 160 bits (resp. 224 bits) case. From these benchmarks, we deduce an average time to compute a modular multiplication as detailed in table 5. Let us now consider the ratio in the most pessimistic case : the cost of a 224 bits modular multiplication is 1.17 times (resp. 1.48 times) the cost of a 160 bits multiplication for 64 bits (resp. 32 bits) processor. Taking into account these results, we can give an estimate in terms of **160 bits multiplications** of Method 1. This shows the interest of Method 1, specially in the fixed point context (see table 2).

c) Hardware context: we considered in this section two kinds of components from the STMicroelectronics portfolio. The first embeds the Public Key 64 bits crypto processor from AST working at 200Mhz (CORE65LPHVT technology) and the second the 128 bits hardware smartcard cryptographic coprocessor Nescrypt working at 110Mhz. The AST crypto processor can compute about 2040816 160-bits modular multiplications and 1449275 224-bits modular multiplications per second. Taking into account these results, table 7 gives the time needed by the modular multiplications when computing a point multiplication. Once again, in the random point context, Method 1 obtains best performances if one needs an SPA resistant algorithm working on a general curve and storing only one point. In the fixed point context, Method 1 is faster than the Comb method with two points. Notice that we only consider the multiplications done by the cryptoprocessor for the times given in table 7. We do not take into account the overhead involved by the communications between the processor and the crypto processor.

Table 7. Time comparison for scalar multiplication methods in milliseconds

Point gen.	Method	curve	# precomp. points	msecs
random (160 bits)	Dummy operations	general	1	1.73
	Montgomery ladder	Montgomery	1	0.717
	Unified formulae	Edwards	1	1.144
	Unified formulae	Hessian	1	1.249
	Unified formulae	Jacobi	1	1.457
	Unified formulae	Edwards	7	1.044
	Unified formulae	Hessian	7	1.139
	Unified formulae	Jacobi	7	1.328
	Möller's recoding	general	16	0.903
fixed (160 bits)	Comb Method	general	2	0.859
	Comb Method	general	4	0.577
	Comb Method	general	8	0.424
	Comb Method	general	16	0.337
random (224 bits)	Method 1	general	1	**1.452**
	Method 1 (x only)	general	1	**1.235**
fixed (224 bits)	Method 1	general	2	**0.723**
	Method 1 (x only)	general	2	**0.613**

Nescrypt 128 bits crypto processor can compute about 339506 modular multiplications per second both for 160 bits and 224 bits integers. Indeed in both cases, only two 128 bits blocks are used to manipulate these integers. Hence we can do the same remarks as in the section about the CIOS method.

7 Conclusions

The goal of this paper was to describe subsets of integers k for which the computation of kP is faster, when dealing with the problem of SPA-secure exponentiation over an elliptic curve. We studied three such subsets and produced the first practical and theoretical results on random Euclidean addition chain generation. Table 2 shows that our methods provide good results in various situations when compared with the best SPA-secure methods.

We proved that the Method 1 we considered is secure and fast. In particular, in the context of a fixed base point, it is competitive with actual methods and faster when using similar amount of storage. We detailed several practical scenarios for which the method is relevant and improves efficiency : in CIOS context, in the context of GNU multiprecision library, and on some cryptoprocessors.

At last, we began the theoretical study of the other proposed methods. We made links between Method 2 and Stern sequences which enabled to obtain optimistic but asymptotic results. We managed to begin the study of the distribution of integers generated by Method 3. Both methods would improve the performances once again. For example Method 3 would be faster than Montgomery ladder, thus it would be worth studying it further. We have proved some results that may be useful for any further investigation.

References

1. Bernstein, D.J., Lange, T.: Explicit-formulas database, http://hyperelliptic.org/EFD
2. Bernstein, D.J., Lange, T.: Faster addition and doubling on elliptic curves. In: Kurosawa, K. (ed.) ASIACRYPT 2007. LNCS, vol. 4833, pp. 29–50. Springer, Heidelberg (2007)
3. Bos, J.W., Kaihara, M.E., Kleinjung, T., Lenstra, A.K., Montgomery, P.L.: On the security of 1024-bit rsa and 160-bit elliptic curve cryptography. Technical report, EPFL IC LACAL and Alcatel-Lucent Bell Laboratories and Microsoft Research (2009)
4. Boyko, V., Peinado, M., Venkatesan, R.: Speeding up discrete log and factoring based schemes via precomputations (1998)
5. Brier, E., Joye, M.: Weierstraß elliptic curves and side-channel attacks. In: Naccache, D., Paillier, P. (eds.) PKC 2002. LNCS, vol. 2274, pp. 335–345. Springer, Heidelberg (2002)
6. Certicom Research. Sec 2: Recommended elliptic curve domain parameters standards for efficient cryptography. Technical report, Certicom (2000)
7. Cohen, H., Frey, G., Avanzi, R., Doche, C., Lange, T., Nguyen, K., Vercauteren, F.: Handbook of Elliptic and Hyperelliptic Cryptography, Discrete Mathematics and its Applications, vol. 34. Chapman & Hall/CRC (2005)
8. Cohen, H., Miyaji, A., Ono, T.: Efficient elliptic curve exponentiation using mixed coordinates. In: Ohta, K., Pei, D. (eds.) ASIACRYPT 1998. LNCS, vol. 1514, pp. 51–65. Springer, Heidelberg (1998)
9. de Rooij, P.: On Schnorr's preprocessing for digital signature schemes. In: Helleseth, T. (ed.) EUROCRYPT 1993. LNCS, vol. 765, pp. 435–439. Springer, Heidelberg (1994)
10. Dimitrov, V., Imbert, L., Mishra, P.K.: Efficient and secure elliptic curve point multiplication using double-base chains. In: Roy, B. (ed.) ASIACRYPT 2005. LNCS, vol. 3788, pp. 59–78. Springer, Heidelberg (2005)
11. Graham, R.L., Knuth, D.E., Patashnik, O.: Concrete Mathematics: A Foundation for Computer Science. Addison-Wesley, Reading (1989)
12. Hankerson, D., Menezes, A., Vanstone, S.: Guide to Elliptic Curve Cryptography. Springer, Heidelberg (2004)
13. Hisil, H., Koon-Ho Wong, K., Carter, G., Dawson, E.: Twisted Edwards curves revisited. In: Pieprzyk, J. (ed.) ASIACRYPT 2008. LNCS, vol. 5350, pp. 326–343. Springer, Heidelberg (2008)
14. Hoffstein, J., Silverman, J.H.: Random small hamming weight products with applications to cryptography. Discrete Appl. Math. 130(1), 37–49 (2003)
15. M'Raïhi, D., Coron, J.-S., Tymen, C.: Fast generation of pairs (k, [k]p) for koblitz elliptic curves. In: Vaudenay, S., Youssef, A.M. (eds.) SAC 2001. LNCS, vol. 2259, pp. 151–174. Springer, Heidelberg (2001)
16. Joye, M., Quisquater, J.-J.: Hessian elliptic curves and side-channel attacks. In: Koç, Ç.K., Naccache, D., Paar, C. (eds.) CHES 2001. LNCS, vol. 2162, pp. 402–410. Springer, Heidelberg (2001)
17. Kleinjung, T., Aoki, K., Franke, J., Lenstra, A.K., Thomé, E., Bos, J.W., Gaudry, P., Kruppa, A., Montgomery, P.L., Osvik, D.A., te Riele, H., Timofeev, A., Zimmermann, P.: Factorization of a 768-bit rsa modulus. Technical report, EPFL IC LACAL and NTT and University of Bonn and INRIA CNRS LORIA and Microsoft Research and CWI (2010)

18. Knuth, D., Yao, A.: Analysis of the subtractive algorithm for greater common divisors. Proc. Nat. Acad. Sci. USA 72(12), 4720–4722 (1975)
19. Knuth, D.E.: The Art of Computer Programming: Fundamental Algorithms, 3rd edn, vol. 2. Addison Wesley, Reading (July 1997)
20. Koblitz, N.: CM-curves with good cryptographic properties. In: Feigenbaum, J. (ed.) CRYPTO 1991. LNCS, vol. 576, pp. 279–287. Springer, Heidelberg (1992)
21. Koc, C.K., Acar, T.: Analyzing and comparing montgomery multiplication algorithms. IEEE Micro 16, 26–33 (1996)
22. Kocher, P., Jaffe, J., Jun, B.: Differential power analysis. In: Wiener, M. (ed.) CRYPTO 1999. LNCS, vol. 1666, pp. 388–397. Springer, Heidelberg (1999)
23. Liardet, P.-Y., Smart, N.P.: Preventing SPA/DPA in ECC systems using the Jacobi form. In: Koç, Ç.K., Naccache, D., Paar, C. (eds.) CHES 2001. LNCS, vol. 2162, pp. 391–401. Springer, Heidelberg (2001)
24. Longa, P., Gebotys, C.: Setting speed records with the (fractional) multibase non-adjacent form method for efficient elliptic curve scalar multiplication. In: Jarecki, S., Tsudik, G. (eds.) Public Key Cryptography – PKC 2009. LNCS, vol. 5443, pp. 443–462. Springer, Heidelberg (2009)
25. Longa, P., Miri, A.: New composite operations and precomputation scheme for elliptic curve cryptosystems over prime fields. In: Cramer, R. (ed.) PKC 2008. LNCS, vol. 4939, pp. 229–247. Springer, Heidelberg (2008)
26. Meloni, N.: Arithmétique pour la Cryptographie basée sur les Courbes Elliptiques. PhD thesis, Université de Montpellier, France (2007)
27. Meloni, N.: New point addition formulae for ECC applications. In: Carlet, C., Sunar, B. (eds.) WAIFI 2007. LNCS, vol. 4547, pp. 189–201. Springer, Heidelberg (2007)
28. Mironov, I., Mityagin, A., Nissim, K.: Hard instances of the constrained discrete logarithm problem. In: Hess, F., Pauli, S., Pohst, M. (eds.) ANTS 2006. LNCS, vol. 4076, pp. 582–598. Springer, Heidelberg (2006)
29. Möller, B.: Securing elliptic curve point multiplication against side-channel attacks. In: Davida, G.I., Frankel, Y. (eds.) ISC 2001. LNCS, vol. 2200, pp. 324–334. Springer, Heidelberg (2001)
30. Möller, B.: Improved techniques for fast exponentiation. In: Lee, P.J., Lim, C.H. (eds.) ICISC 2002. LNCS, vol. 2587, pp. 298–312. Springer, Heidelberg (2003)
31. Montgomery, P.: Speeding the pollard and elliptic curve methods of factorization. Mathematics of Computation 48, 243–264 (1987)
32. Montgomery, P.L.: Evaluating recurrences of form $X_{m+n} = f(X_m, X_n, X_{m-n})$ via Lucas chains (1992), `http://ftp.cwi.nl/pub/pmontgom/Lucas.ps.gz`
33. Mui, J.A., Stinson, D.R.: On the low hamming weight discrete logarithm problem for nonadjacent representations. Applicable Algebra in Engineering, Communication and Computing 16, 461–472 (2006)
34. Nguyên, P.Q., Stern, J.: The hardness of the hidden subset sum problem and its cryptographic implications. In: Wiener, M. (ed.) CRYPTO 1999. LNCS, vol. 1666, pp. 31–46. Springer, Heidelberg (1999)
35. Reznick, B.: Regularity properties of the stern enumeration of the rationals. Journal of Integer Sequences 11 (2008)
36. Schnorr, C.P.: Efficient signature generation by smart cards. Journal of Cryptology 4, 161–174 (1991)
37. Stern, M.A.: über eine zahlentheoretische funktion. Journal für die reine und angewandte Mathematik 55, 193–220 (1858)
38. Stinson, D.R.: Some baby-step giant-step algorithms for the low hamming weight discrete logarithm problem. Mathematics of Computation 71, 379–391 (2000)

39. Theriault, N.: Spa resistant left-to-right integer recodings. In: Preneel, B., Tavares, S. (eds.) SAC 2005. LNCS, vol. 3897, pp. 345–358. Springer, Heidelberg (2006)
40. U.S. Department of Commerce and National Intitute of Standards and Technology. Digital signature standard (DSS). Technical report (2009)
41. Yacobi, Y.: Exponentiating faster with addition chains. In: Damgård, I.B. (ed.) EUROCRYPT 1990. LNCS, vol. 473, pp. 222–229. Springer, Heidelberg (1991)
42. Yacobi, Y.: Fast exponentiation using data compression. SIAM J. Comput. 28(2), 700–703 (1999)

Annex

Proof of Theorem 1

Definition 3. *Let* $(a,b) \in \mathbb{N}^2$*, the generalized Stern sequence* $(s_{a,b}(r,n))_{r\in\mathbb{N},n\in[0,2^r]}$ *is defined by* $s_{a,b}(0,0) = a$*,* $s_{a,b}(0,1) = b$*, and for* $r \geqslant 1$*,* $s_{a,b}(r,2n) = s_{a,b}(r-1,n)$*,* $s_{a,b}(r,2n+1) = s_{a,b}(r-1,n) + s_{a,b}(r-1,n+1)$*.*

In his original paper, Stern gave a practical description of his sequence using the following *diatomic array* [37] :

$$\begin{array}{ll}
(r=0) & a \quad b \\
(r=1) & a \;\; a+b \quad b \\
(r=2) & a \;\; 2a+b \;\; a+b \;\; a+2b \quad b \\
(r=3) & a \;\; 3a+b \;\; 2a+b \;\; 3a+2b \;\; a+b \;\; 2a+3b \;\; a+2b \;\; a+3b \;\; b \\
\vdots &
\end{array}$$

where each line r is exactly the sequence $s_{a,b}(r,n)$ for $n \in [0,2^r]$. Notice that to compute the row r, you just have to rewrite row $r-1$ and insert their sum between two elements. In the case $(a,b) = (1,1)$, the sequence is called the Stern sequence and has been well studied. For example, see the introduction of [35] or [11] for the link with the Stern Brocot array.

Now we will point deep connections between Stern sequences and the ψ and χ maps. These connections should not surprise us, because both are linked with continued fractions. As an example, if $(a,b) = (1,1)$, an easy induction enables us to prove that when $n \in [0,2^r[$, the sequence $(s_{1,1}(r+1,2n), s_{1,1}(r+1,2n+1))$ describes the set $\psi(\mathcal{M}_r)$.

$$\begin{array}{lll}
(r=0) & 1\,1 & \\
(r=1) & 1\,2\,1 & \\
(r=2) & 1\,3\,2\,3\,1 & (\mathcal{M}_2 = \{\ 00,\ \ 01,\ \ 10,\ \ 11\}) \\
(r=3) & 1\,4\,3\,5\,2\,5\,3\,4\,1 & (\psi(\mathcal{M}_2) = \{\ (3,5), (2,5), (3,4), (1,4)\}) \\
\vdots & &
\end{array}$$

Let us note $\Delta : \mathbb{N}^2 \to \mathbb{N}^2$ such that $\Delta(x,y) = (y,x)$. Another induction enables to prove that when $n \in [0,2^{r+1}-1]$, $(s_{1,1}(r+1,n), s_{1,1}(r+1,n+1))$ describes $\psi(\mathcal{M}_r) \cup \Delta(\psi(\mathcal{M}_r))$. For $\ell \in \mathbb{N}^*$ and $m \in \mathcal{M}_\ell$, let $A_r(m) = \{\psi(mx) \mid x \in \mathcal{M}_r\} \cup \{\Delta(\psi(mx)) \mid x \in \mathcal{M}_r\}$.

From now on, in order to simplify the notations, we will denote by $s(r,n)$ the value $s_{a,b}(r,n)$. Let us define the sequences $\left(S(r,n)\right)_{r\in\mathbb{N},n\in[0,2^r-1]}$ by $S(r,n) = (s(r,n), s(r,n+1))$ and $\left(S_d(r,n)\right)_{r\in\mathbb{N},n\in[0,2^r-1]}$ by $S_d(r,n) = (s(r,n) \mod d, s(r,n+1) \mod d)$. The following link between ψ and S can be proved by induction.

Lemma 1. *Let $r \geqslant 0$, $\ell > 0$ and let $m \in \mathcal{M}_\ell$ such that $\psi(m) = (a,b)$. Then $S(r+1,.)$ is a one to one map from $[0, 2^{r+1}-1]$ onto $A_r(m)$.*

It means that in our case, the values $\psi(c)$ and $\Delta(\psi(c))$ for $c \in \mathcal{M}_\ell^0$ correspond to the elements $S(\ell+1,n)$ for $n \in [0, 2^{\ell+1}-1]$ and $(a,b) = (F_{\ell+2}, F_{\ell+3})$. Recently, Reznick proved in [35] that, for $d \geqslant 2$, $(a,b) = (1,1)$, and r sufficiently large, the sequence $\{S_d(r,n)\}_{n\in\mathbb{N}}$ is well distributed among $\mathcal{S}_d := \{(i \mod d, j \mod d) \mid \gcd(i,j,d) = 1\}$. We need a similar result for any couple (a,b) in order to show that the values $\chi(c)$ are asymptotically well distributed modulo d. We will use similar notations and follow the arguments of [35] to prove the next theorem. We define :

- for $\gamma \in \mathcal{S}_d$, $B_d(r,\gamma) := \#\{n \in [0, 2^r-1] \,|S_d(r,n) = \gamma\}$,
- χ_d, the map such that $\chi_d(m) = \chi(m) \mod d$,
- ψ_d the map such that if $\psi(m) = (v,u)$ then $\psi_d(m) = (v \mod d, u \mod d)$,
- N_d the cardinality of $\mathcal{S}_d$.

Theorem 3. *Let $(a,b,d) \in \mathbb{N}^3$ such that d be prime and $(a,d) = (b,d) = 1$. There exist constants c_d and $\rho_d < 2$ so that if $m \in \mathbb{N}$ and $\alpha \in \mathcal{S}_d$, then for all $r \geqslant 0$,*

$$|B_d(r,\alpha) - \frac{2^r}{N_d}| < c_d \rho_d^r.$$

Proof. Due to the lack of space, we just give a short proof of this, following the arguments of section 4 in [35] and pointing out the differences. Since d is prime, we have $\mathcal{S}_d = \mathbb{Z}/d\mathbb{Z} \times \mathbb{Z}/d\mathbb{Z} \setminus \{(0,0)\}$, and $N_d = d^2-1$. We can define a graph $\mathcal{G}_d$ and the applications L and R in the same way, and also have $L^d = R^d = \text{id}$. In the proof of lemma 14, we have to give a slightly different proof that for each $\alpha = (x,y) \in \mathcal{S}_d$ there exists a way from (a,b) to α in the graph $\mathcal{G}_d$. Notice that since $(0,0) \notin \mathcal{S}_d$, then either $x \neq 0$ or $y \neq 0$. If $y \neq 0$, we notice that $R^{k'}(L^k(a,b)) = (a + k'(b+ka), b+ka)$. As $(a,d) = 1$, we can choose k such that $b + ka = y$. Thus, as $y \neq 0$, we can choose k' such that $a + k'(b+ka) = x$ and we are done. If $x \neq 0$, then we consider $L^{k'}(R^k(a,b)) = (a+kb, b+k'(a+kb))$: in the same way, we can choose (k,k') such that $a+kb = x$ and $b+k'(a+kb) = y$. In the first line of the proof of lemma 14, we also have to consider $(r_0,n_0) \in \mathbb{N}\times\mathbb{N}$ such that $S_d(r_0,n_0) = (0,1)$ rather than $S_d(0,0) = (0,1)$. Thus the adjacency matrix of the graph satisfies the same properties as in theorem 15 of [35]. So the conclusion remains true for B_d.

From Theorem 3 and Lemma 1 we can now give the proof of Theorem 1.

Proof. Let $\alpha \in (\mathbb{Z}/d\mathbb{Z})^*$, and $(\beta_i, \gamma_i)_{1 \leqslant i \leqslant d}$ the d elements of $\mathbb{Z}/d\mathbb{Z} \times \mathbb{Z}/d\mathbb{Z} \setminus \{(0,0)\}$ such that $\beta_i + \gamma_i = \alpha$, then

$$\#\{x \in \mathcal{M}_r \mid \chi_d(mx) = \alpha\} = \textstyle\sum_{i=1}^{d} \#\{x \in \mathcal{M}_r \mid \psi_d(mx) = (\beta_i, \gamma_i)\}.$$

If $n \in [0, 2^{r+1}]$ is such that $S_d(r+1, n) = (\beta_i, \gamma_i)$ then , thanks to Lemma 1, it corresponds to $x \in \mathcal{M}_r$ such that $\psi_d(mx) = (\beta_i, \gamma_i)$ or $\psi_d(mx) = (\gamma_i, \beta_i)$. In this last case, there exists an integer j, such that $(\gamma_i, \beta_i) = (\beta_j, \gamma_j)$. Hence,

$$\sum_{i=1}^{d} \#\{n \in [0, 2^{r+1}-1] \mid S_d(r+1, n) = (\beta_i, \gamma_i)\} = 2 \times \#\{x \in \mathcal{M}_r \mid \chi_d(mx) = \alpha\}.$$

Now by definition of B_d :

$$\sum_{i=1}^{d} \#\{n \in [0, 2^{r+1}-1] \mid S_d(r+1, n) = (\beta_i, \gamma_i)\} = \sum_{i=1}^{d} B_d(r+1, (\beta_i, \gamma_i)).$$

Thus,

$$\frac{\#\{x \in \mathcal{M}_r \mid \chi_d(mx) = \alpha\}}{2^r} - \frac{d}{d^2-1} = \sum_{i=1}^{d} \left(\frac{B_d(r+1, (\beta_i, \gamma_i))}{2^{r+1}} - \frac{1}{d^2-1} \right).$$

Then we can use Theorem 3 and the triangular inequality to prove the first inequality of the theorem. Using this inequality for $\alpha \neq 0$ we prove the second inequality.

Proof of Theorem 2

From now on, we will denote by $\overline{m}$, the value $\chi(m)$. Let us set $M = \sup\{\overline{m} \mid m \in \mathcal{M}_{\ell,p}\}$ and $I = \inf\{\overline{m} \mid m \in \mathcal{M}_{\ell,p}\}$. If $m \in \mathcal{M}_{\ell,p}$ is not one of the words of the points *ii)* (resp. *iii)*), we will propose $m' \in \mathcal{M}_{\ell,p}$ such that $\overline{m'} < \overline{m}$ (resp. $\overline{m'} > \overline{m}$). The lemmas in the two following subsections give the details about the words m' we use to compare.

We first look for $m \in \mathcal{M}_{\ell,p}$ such that $\overline{m} = I$. First suppose that two 1's in the word m are separated by one 0 or more. Then we can consider $(m, n, s) \in (\mathbb{N}^*)^3$ and $(a, b) \in \mathbb{N}^2$ and $(x, y) \in \mathcal{M}_a \times \mathcal{M}_b$ such that m is one of the words

$$1^m 0^n 1^s, \qquad x10^m 1^n 0y \qquad \text{or} \qquad y01^n 0^m 1x.$$

We won't consider the third case because it is the symmetric of the second one. The lemma 2 shows that $\overline{1^m 0^n 1^s} > \overline{1^{m+s} 0^n}$ and that $\overline{x10^m 1^n 0y} > \overline{x1^{n+1} 0^{m+1} y}$. So if $\overline{m} = I$, there are no 0 between two 1 of the word m, and so there are integers a and c such that $m = 0^a 1^p 0^c$. From lemma 3 we show that $a = 0$ (and so $c = \ell - p$) or $c = 0$ (and so $a = \ell - p$).

Now we look for $m \in \mathcal{M}_{\ell,p}$ such that $\overline{m} = M$. If there are two consecutive 1's in the word m, as $2p < \ell$ the word m will also have two consecutive 0's. We can consider the symmetry such that a subword 00 appears in m before a subword 11. In this case there exists $(a, b, n) \in \mathbb{N}^3$ and $(x, y) \in \mathcal{M}_a \times \mathcal{M}_b$ such that $m = x00(10)^n11y$. In this case, we have from lemma 4 that $\overline{m} < \overline{x(01)^{n+2}y}$. Now assume that there are no subword 11 in m. If two 1's in the word m are separated by two 0's or more, then there exists $(m, n, a, b) \in (\mathbb{N}^*)^2 \times \mathbb{N}^2$, $x \in \mathcal{M}_a$ and $y \in \mathcal{M}_b$ such that m is one of the words

$$x010^n(01)^m, \quad x010^n(01)^m0, \quad x010^n(01)^m0^2y,$$

$$10^n(01)^m, \quad 10^n(01)^m0, \quad \text{or} \quad 10^n(01)^m0^2y.$$

The Lemmas (5) and (6) give us in any case $m' \in \mathcal{M}_{\ell,p}$ such that $\overline{m} < \overline{m'}$. We have just proved that 11 is not a subword of m, and that two 1's of m are separated by exactly one 0. So the word m or its symmetric is $0^a(01)^p0^c$ where $(a, c) \in \mathbb{N}\times\mathbb{N}$. The lemma (7) shows that M is reached when $a = 0$ (so $c = \ell - 2p$) or when $c = 1$ (so $a = \ell - 2p - 1$). Then the point **iii)** is proved.

Now we just have to apply Proposition 1 to deduce *i)* from *ii)* and *iii)*.

Lemmas to Find the Lower Bound

Lemma 2. *Let* $(m, n, s) \in (\mathbb{N}^*)^3$ *and* $(a, b) \in \mathbb{N}^2$. *Let* $x \in \mathcal{M}_a$ *and* $y \in \mathcal{M}_b$. *We have*

i) $\overline{1^m0^n1^s} > \overline{1^{m+s}0^n}$ *and*
ii) $\overline{x10^m1^n0y} > \overline{x1^{n+1}0^{m+1}y}$.

Proof. For the point **i)** , we compute

$$\overline{1^m0^n1^s} = (1,2)\begin{pmatrix}1 & m\\ 0 & 1\end{pmatrix}\begin{pmatrix}F_{n-1} & F_n\\ F_n & F_{n+1}\end{pmatrix}\begin{pmatrix}1 & s\\ 0 & 1\end{pmatrix}\begin{pmatrix}1\\ 1\end{pmatrix}, \text{ so}$$

$$\overline{1^m0^n1^s} = (s+1)F_{n-1} + \big((s+1)(m+2)+1\big)F_n + (m+2)F_{n+1}. \tag{2}$$

$$\text{Also, } \overline{1^{m+s}0^n} = (1,2)\begin{pmatrix}1 & m+s\\ 0 & 1\end{pmatrix}\begin{pmatrix}F_{n-1} & F_n\\ F_n & F_{n+1}\end{pmatrix}\begin{pmatrix}1\\ 1\end{pmatrix}, \text{ so}$$

$$\overline{1^{m+s}0^n} = (2+m+s+1)F_{n+1} + (2+m+s)F_n. \tag{3}$$

The difference between (2) and (3) is msF_n, so it is positive in the conditions of the lemma. We can notice that there is equality when $s = 0$ or when $m = 0$, which can also be explained by the symmetry.

To prove the point *ii)*, let us set $(v, u) = \psi(x)$. We have

$$\psi(x10^m1^n0) = (v,u)\begin{pmatrix}1 & 1\\ 0 & 1\end{pmatrix}\begin{pmatrix}F_{m-1} & F_m\\ F_m & F_{m+1}\end{pmatrix}\begin{pmatrix}1 & n\\ 0 & 1\end{pmatrix}\begin{pmatrix}0 & 1\\ 1 & 1\end{pmatrix}$$

$$= \big((nF_m+F_{m+1})u+(nF_{m+1}+F_{m+2})v, F_{m+2}u+(F_{m+1}+(n+1)F_{m+2})v+(u-v)nF_m\big). \tag{4}$$

We also compute

$$\psi(x1^{n+1}0^{m+1}) = (v,u)\begin{pmatrix}1 & n+1\\ 0 & 1\end{pmatrix}\begin{pmatrix}F_m & F_{m+1}\\ F_{m+1} & F_{m+2}\end{pmatrix}$$

$$= (F_{m+1}u + (F_m + (n+1)F_{m+1})v, F_{m+2}u + (F_{m+1} + (n+1)F_{m+2})v). \quad (5)$$

As $u > v$, we can deduce the point *ii)* from the comparison components by components of the vectors (4) and (5).

Lemma 3. *Let* $(a, b, c) \in (\mathbb{N}^*)^3$*, we have* $\overline{0^a1^b0^c} > \overline{1^b0^{a+c}}$.

Proof. We first compute the left hand side

$$\overline{0^a1^b0^c} = (1,2)\begin{pmatrix} F_{a-1} & F_a \\ F_a & F_{a+1} \end{pmatrix}\begin{pmatrix} 1 & b \\ 0 & 1 \end{pmatrix}\begin{pmatrix} F_{c-1} & F_c \\ F_c & F_{c+1} \end{pmatrix}\begin{pmatrix} 1 \\ 1 \end{pmatrix}$$

$$= (F_{a+2}F_{c-1} + (bF_{a+2} + F_{a+3})F_c + F_{a+2}F_c + (bF_{a+2} + F_{a+3})F_{c+1} +)$$

$$= F_{a+2}F_{c+1} + F_{a+3}F_{c+2} + bF_{a+2}F_{c+2}. \quad (6)$$

We also compute the right hand side

$$\overline{1^b0^{a+c}} = (1,2)\begin{pmatrix} 1 & b \\ 0 & 1 \end{pmatrix}\begin{pmatrix} F_{a+c-1} & F_{a+c} \\ F_{a+c} & F_{a+c+1} \end{pmatrix}\begin{pmatrix} 1 \\ 1 \end{pmatrix}$$

$$= (F_{a+c-1} + (b+2)F_{a+c} + F_{a+c} + (b+2)F_{a+c+1})$$

$$= F_{a+c+4} + bF_{a+c+2}.$$

With eq. 1, page 242 we show that it is

$$F_{a+2}F_{c+1} + F_{a+3}F_{c+2} + bF_{a+c+2}. \quad (7)$$

The difference between (6) and (7) is $b(F_{a+c+2} - F_{a+2}F_{c+2}) = bF_cF_a$, which is positive in the conditions of the lemma. In the cases $c = 0$ or $a = 0$, there is equality which we already knew by the symmetry.

Lemmas to Compute the Upper Bound

Lemma 4. *Let* $(n, a, b) \in \mathbb{N}^3$*. For all* $x \in \mathcal{M}_a$ *and* $y \in \mathcal{M}_b$ *we have*

$$\overline{x00(10)^n11y} < \overline{x(01)^{n+2}y}.$$

Proof. We first compute

$$S_0^2\,(S_1S_0)^n\,S_1^2 = \begin{pmatrix} 1 & 1 \\ 1 & 2 \end{pmatrix}\begin{pmatrix} \alpha_n & 2\beta_n \\ \beta_n & \alpha_n \end{pmatrix}\begin{pmatrix} 1 & 2 \\ 0 & 1 \end{pmatrix} = \begin{pmatrix} \alpha_n + \beta_n & 3\alpha_n + 4\beta_n \\ \alpha_n + 2\beta_n & 4\alpha_n + 6\beta_n \end{pmatrix}.$$

We set $(v, u) = \psi(x)$, so we have

$$\begin{aligned} \psi(x00(10)^n11) &= (v,u)\begin{pmatrix} \alpha_n + \beta_n & 3\alpha_n + 4\beta_n \\ \alpha_n + 2\beta_n & 4\alpha_n + 6\beta_n \end{pmatrix} \\ &= (v(\alpha_n + \beta_n) + u(\alpha_n + 2\beta_n), v(3\alpha_n + 4\beta_n) + u(4\alpha_n + 6\beta_n)). \end{aligned} \quad (8)$$

From the other hand

$$S_0S_1\left(S_0S_1\right)^n S_0S_1 = \begin{pmatrix} 0 & 1 \\ 1 & 2 \end{pmatrix} \begin{pmatrix} \alpha_n - \beta_n & \beta_n \\ \beta_n & \alpha_n + \beta_n \end{pmatrix} \begin{pmatrix} 0 & 1 \\ 1 & 2 \end{pmatrix}$$

$$= \begin{pmatrix} \alpha_n + \beta_n & 2\alpha_n + 3\beta_n \\ 2\alpha_n + 3\beta_n & 5\alpha_n + 7\beta_n \end{pmatrix}, \text{ so}$$

$$\psi(x(01)^{n+2}) = (v(\alpha_n + \beta_n) + u(2\alpha_n + 3\beta_n), v(2\alpha_n + 3\beta_n) + u(5\alpha_n + 7\beta_n)). \quad (9)$$

As $v < u$ we can compare the vectors (8) and (9) components by components and then conclude.

Lemma 5. *Let* $(m, n, a, b) \in (\mathbb{N}^*)^2 \times \mathbb{N}^2$. *For all* $x \in \mathcal{M}_a$ *and* $y \in \mathcal{M}_b$, *we have*

i) $\overline{x010^n(01)^m} < \overline{x(01)^{m+1}0^n}$
ii) $\overline{x010^n(01)^m0} < \overline{x(01)^{m+1}0^{n+1}}$
iii) $\overline{x010^n(01)^m0^2y} < \overline{x(01)^{m+1}0^{n+2}y}$.

Proof. We compute

$$S_0S_1S_0^n = \begin{pmatrix} 0 & 1 \\ 1 & 2 \end{pmatrix} \begin{pmatrix} F_{n-1} & F_n \\ F_n & F_{n+1} \end{pmatrix} = \begin{pmatrix} F_n & F_{n+1} \\ F_{n+2} & F_{n+3} \end{pmatrix}.$$

We deduce

$$S_0S_1S_0^n\left(S_0S_1\right)^m = \begin{pmatrix} F_n & F_{n+1} \\ F_{n+2} & F_{n+3} \end{pmatrix} \begin{pmatrix} \alpha_m - \beta_m & \beta_m \\ \beta_m & \alpha_m + \beta_m \end{pmatrix} \quad \text{and so}$$

$$S_0S_1S_0^n\left(S_0S_1\right)^m = \begin{pmatrix} \alpha_m F_n + \beta_m F_{n-1} & \alpha_m F_{n+1} + \beta_m F_{n+2} \\ \alpha_m F_{n+2} + \beta_m F_{n+1} & \alpha_m F_{n+3} + \beta_m F_{n+4} \end{pmatrix}. \quad (10)$$

From the other hand, we can write $\left(S_0S_1\right)^{m+1} S_0^n = \left(S_0S_1\right)^m S_0S_1S_0^n$ so

$$\left(S_0S_1\right)^{m+1} S_0^n = \begin{pmatrix} \alpha_m - \beta_m & \beta_m \\ \beta_m & \alpha_m + \beta_m \end{pmatrix} \begin{pmatrix} F_n & F_{n+1} \\ F_{n+2} & F_{n+3} \end{pmatrix}, \text{ and then}$$

$$\left(S_0S_1\right)^{m+1} S_0^n = \begin{pmatrix} \alpha_m F_n + \beta_m F_{n+1} & \alpha_m F_{n+1} + \beta_m F_{n+2} \\ \alpha_m F_{n+2} + \beta_m (F_n + F_{n+2}) & \alpha_m F_{n+3} + \beta_m (F_{n+1} + F_{n+3}) \end{pmatrix} \quad (11)$$

Let us set $(v, u) = \psi(x)$.

$$\psi(x010^n(01)^m) = (v, u) \begin{pmatrix} \alpha_m F_n + \beta_m F_{n-1} & \alpha_m F_{n+1} + \beta_m F_{n+2} \\ \alpha_m F_{n+2} + \beta_m F_{n+1} & \alpha_m F_{n+3} + \beta_m F_{n+4} \end{pmatrix} \begin{pmatrix} 1 \\ 1 \end{pmatrix}$$

$$= v(\alpha_m F_{n+2} + \beta_m F_{n-1} + \beta_m F_{n+2}) + u(\alpha_m F_{n+4} + \beta_m F_{n+1} + \beta_m F_{n+4}). \quad (12)$$

We also have

$$\psi(x(01)^{m+1}0^n) = (v, u) \begin{pmatrix} \alpha_m F_n + \beta_m F_{n+1} & \alpha_m F_{n+1} + \beta_m F_{n+2} \\ \alpha_m F_{n+2} + \beta_m (F_n + F_{n+2}) & \alpha_m F_{n+3} + \beta_m (F_{n+1} + F_{n+3}) \end{pmatrix} \begin{pmatrix} 1 \\ 1 \end{pmatrix}$$

$$= v(\alpha_m F_{n+2} + \beta_m F_{n+1} + \beta_m F_{n+2}) + u(\alpha_m F_{n+4} + \beta_m F_{n+2} + \beta_m F_{n+4}). \quad (13)$$

The difference between (13) and (12) is $v\beta_m F_n + u\beta_m F_n$ so we deduce the first point.

To prove the point **ii)** we use (10) which gives

$$S_0 S_1 S_0^n (S_0 S_1)^m S_0 = \begin{pmatrix} \alpha_m F_{n+1} + \beta_m F_{n+2} & \alpha_m F_{n+2} + \beta_m F_{n-1} + \beta_m F_{n+2} \\ \alpha_m F_{n+3} + \beta_m F_{n+4} & \alpha_m F_{n+4} + \beta_m F_{n+1} + \beta_m F_{n+4} \end{pmatrix} \quad (14)$$

Let us set $(v, u) = \psi(x)$. We have

$$\begin{aligned} \overline{x010^n(01)^m0} = & \, v(\alpha_m F_{n+3} + \beta_m F_{n-1} + 2\beta_m F_{n+2}) \\ & + u(\alpha_m F_{n+5} + \beta_m F_{n+1} + 2\beta_m F_{n+4}). \end{aligned} \quad (15)$$

From (11) we deduce

$$(S_0 S_1)^{m+1} S_0^{n+1} = \begin{pmatrix} \alpha_m F_{n+1} + \beta_m F_{n+2} & \alpha_m F_{n+2} + \beta_m F_{n+3} \\ \alpha_m F_{n+3} + \beta_m F_{n+1} + \beta_m F_{n+3} & \alpha_m F_{n+4} + \beta_m F_{n+2} + \beta_m F_{n+4} \end{pmatrix} \quad (16)$$

So

$$\overline{x(01)^{m+1}0^{n+1}} = v(\alpha_m F_{n+3} + \beta_m F_{n+4}) + u(\alpha_m F_{n+5} + \beta_m F_{n+3} + \beta_m F_{n+5}). \quad (17)$$

The difference between (17) and (15) is $v\beta_m F_n$ so we have the positivity. To prove **iii)** we compute from (14) and (16)

$$S_0 S_1 S_0^n (S_0 S_1)^m S_0^2 = \begin{pmatrix} \alpha_m F_{n+2} + \beta_m F_{n-1} + \beta_m F_{n+2} & \alpha_m F_{n+3} + \beta_m F_{n-1} + 2\beta_m F_{n+2} \\ \alpha_m F_{n+4} + \beta_m F_{n+1} + \beta_m F_{n+4} & \alpha_m F_{n+5} + \beta_m F_{n+1} + 2\beta_m F_{n+4} \end{pmatrix}$$

and $(S_0 S_1)^{m+1} S_0^{n+2} = \begin{pmatrix} \alpha_m F_{n+2} + \beta_m F_{n+3} & \alpha_m F_{n+3} + \beta_m F_{n+4} \\ \alpha_m F_{n+4} + \beta_m F_{n+2} + \beta_m F_{n+4} & \alpha_m F_{n+5} + \beta_m F_{n+3} + \beta_m F_{n+5} \end{pmatrix}.$

We compare components by components and then deduce *iii)*

Lemma 6. *Let* $(m, n, b) \in (\mathbb{N}^*)^2 \times \mathbb{N}$ *and* $y \in \mathcal{M}_b$. *We have*

$$\begin{aligned} &\textbf{i)} \; \overline{10^n(01)^m} < \overline{(01)^{m+1}0^{n-1}}, \\ &\textbf{ii)} \; \overline{10^n(01)^m0} < \overline{(01)^{m+1}0^n}, \\ &\textbf{iii)} \; \overline{10^n(01)^m0^2y} < \overline{(01)^{m+1}0^{n+1}y}. \end{aligned}$$

Proof. We will use the computations of the proof of lemma 5. Let us consider that $\psi(x) = (1, 1)$. In this case $\psi(x0)$ would be $(1, 2)$, and we would have $\overline{x010^n(01)^m} = \overline{10^n(01)^m}$. So with (12) we have

$$\overline{10^n(01)^m} = \alpha_m F_{n+2} + \beta_m F_{n-1} + \beta_m F_{n+2} + \alpha_m F_{n+4} + \beta_m F_{n+1} + \beta_m F_{n+4}. \quad (18)$$

With (11) we have

$$\overline{(01)^{m+1}0^{n-1}} = \alpha_m F_{n+1} + \beta_m F_n + \beta_m F_{n+1} + 2\alpha_m F_{n+3} + 2\beta_m F_{n+1} + 2\beta_m F_{n+3}, \tag{19}$$

so $\overline{(01)^{m+1}0^{n-1}} - \overline{10^n(01)^m} = \alpha_m F_{n-1} + \beta_m F_{n+2}$, and then we have the point *i)* . In the same way, we also have

$\overline{10^n(01)^m0} = \alpha_m F_{n+3} + \beta_m F_{n-1} + 2\beta_m F_{n+2} + \alpha_m F_{n+5} + \beta_m F_{n+1} + 2\beta_m F_{n+4}$, and

$\overline{(01)^{m+1}0^n} = \alpha_m F_{n+2} + \beta_m F_{n+1} + \beta_m F_{n+2} + 2\alpha_m F_{n+4} + \beta_m F_{n+2} + \beta_m F_{n+4}$, so

$\overline{(01)^{m+1}0^n} - \overline{10^n(01)^m0} = F_n(\alpha_m + 2\beta_m)$ and then we deduce the point *ii)* .

Now, $\psi(10^n(01)^m0^2) = (1,1)S_0S_1S_0^n(S_0S_1)^mS_0^2$, and with (15) we find

$$\begin{aligned}\psi(10^n(01)^m0^2) = (\alpha_m(F_{n+2} + F_{n+4}) &+\beta_m(F_{n-1} + F_{n+5}),\\ &\alpha_m(F_{n+3} + F_{n+5}) + \beta_m(F_{n-1} + F_{n+5})).\end{aligned} \tag{20}$$

With (11) we compute

$$\begin{aligned}\psi((01)^{m+1}0^{n+1}) = (\alpha_m(F_{n+1}+ 2F_{n+3}) &+ \beta_m(F_{n+1} + 3F_{n+3}),\\ &\alpha_m(F_{n+2} + 2F_{n+4}) + \beta_m(F_{n+2} + 3F_{n+4}).\end{aligned} \tag{21}$$

The difference of the first component of (21) by the first component of (20) is $\alpha_m F_{n-2} + \beta_m F_{n+2}$, and the difference of the second components is $\alpha_m F_n + \beta_m(2F_{n+4} - F_{n-1} - F_{n+1})$ so the point *iii)* is proved.

Lemma 7. *Let* $(a,b,c) \in \mathbb{N} \times \mathbb{N}^* \times \mathbb{N}$. *If* $c \neq 1$ *and* $a \neq 0$ *then* $\overline{0^a(01)^b0^c} < \overline{(01)^b0^{a+c}}$.

Proof. We compute

$$\overline{0^a(01)^b0^c} = (1,2)\begin{pmatrix} F_{a-1} & F_a \\ F_a & F_{a+1}\end{pmatrix}\begin{pmatrix} \alpha_b - \beta_b & \beta_b \\ \beta_b & \alpha_b + \beta_b\end{pmatrix}\begin{pmatrix} F_{c-1} & F_c \\ F_c & F_{c+1}\end{pmatrix}\begin{pmatrix}1\\1\end{pmatrix}$$

$$= \alpha_b(F_{a+2}F_{c+1} + F_{a+3}F_{c+2}) + \beta_b(F_{a+1}F_{c+1} + F_{a+4}F_{c+2}). \tag{22}$$

On the other hand

$$\overline{(01)^b0^{a+c}} = (1,2)\begin{pmatrix} \alpha_b - \beta_b & \beta_b \\ \beta_b & \alpha_b + \beta_b\end{pmatrix}\begin{pmatrix} F_{a+c-1} & F_{a+c} \\ F_{a+c} & F_{a+c+1}\end{pmatrix}\begin{pmatrix}1\\1\end{pmatrix}$$

$$= \alpha_b F_{a+c+4} + \beta_b(F_{a+c+2} + F_{a+c+4}). \tag{23}$$

Using eq. 1, we prove that the difference between (23) by (22) is $\beta_b F_a(F_{c+1} - F_c)$. It is zero if and only if $a = 0$ or $c = 1$, and positive otherwise.

Attack on a Higher-Order Masking of the AES Based on Homographic Functions

Emmanuel Prouff[1] and Thomas Roche[1,2]

[1] Oberthur Technologies, 71-73, rue des Hautes Pâtures 92726 Nanterre, France
{e.prouff,t.roche}@oberthur.com
[2] University of Paris 8, Département de mathématiques, 2,
rue de la Liberté; 93526 Saint-Denis, France

Abstract. In the recent years, Higher-order Side Channel attacks have been widely investigated. In particular, 2nd-order DPA have been improved and successfully applied to break several masked implementations. In this context, the development of sound and practical countermeasures against attacks of arbitrary fixed order d is of crucial interest. Surprisingly, while many studies have been dedicated to the attacks, only a very few methods have been published that claim to provide security against dth-order side channel attacks whatever the order d. Among them, the one proposed by Courtois and Goubin at ICISC 2005 was especially interesting due to its great efficiency. In this paper we show that the method is however flawed and we exhibit several higher-order attacks that can defeat the countermeasure for any value of d.

1 Introduction

The observation of a device during its execution (*e.g.* through power consumption measurements) can give information on the internal values actually manipulated by the device. Based on this idea, a powerful attack targeting symmetric cipher implementations and called Differential Power Analysis (DPA for short) has been proposed by Kocher et al. in 1998 [1]. The main idea is to observe the device during the manipulation of key-dependent data (called *sensitive data* in the sequel), and to retrieve information about the key (and eventually the whole key) from this observation. Since the introduction of DPA, and more generally of *Side Channel Analysis* (SCA for short), many works have focused either on the enhancement of such attacks or on the search of sound countermeasures. In the latter area of research, masking techniques are currently the most promising type of countermeasure. The idea is to split any sensitive variable manipulated by the device into several shares such that the knowledge on a subpart of the shares does not give information on the sensitive value itself. When the number of shares is $d+1$, the countermeasure is usually called a *dth-order masking scheme*. When such a scheme is applied, the attacker has to retrieve information about the $d+1$ shares — *i.e.* to observe at least $d+1$ leakage points on the device — in order to gain knowledge about the targeted sensitive variable. Such an attack is called a $(d+1)$*th-order SCA attack* and it has been shown that

G. Gong and K.C. Gupta (Eds.): INDOCRYPT 2010, LNCS 6498, pp. 262–281, 2010.

its complexity increases exponentially with the order d [2]. While some 1st-order masking techniques have been successfully proved to be secure against 1st-order SCA attacks (see for instance [3, 4]), the practicality of 2nd-order attacks has been also demonstrated [5, 6, 7]. The construction of an efficient dth-order masking scheme thus became of great interest. The main difficulty resides in the handling of $d+1$ shares of a unique intermediate variable through a non-linear function (*i.e.* the cipher s-boxes). Several solutions to this so-called *higher-order masking problem* have been proposed in the literature: [8] and [9] only solve the problem for 2nd-order SCA attacks, [10] makes unrealistic assumptions on the adversary capabilities, assuming that some parts of the computation are unconditionally secured against SCA, and [11] has been broken for any order $d \geq 3$. Finally there are basically two types of constructions that claim (and are believed) to be secure against SCA for any arbitrary fixed order d: on one hand, the construction of Ishai *et al.* [12] and its recent adaptation by Rivain and Prouff [13] and, on the other hand, the construction of Courtois and Goubin [14] based on table re-computation techniques. In the present paper, it is argued that Courtois and Goubin scheme can actually always be broken by a 2nd-order SCA attack, making [12] and [13] the unique known solutions for the arbitrary higher-order masking problem.

The paper is organized as follows. In Sec. 2, the concept of Higher-order SCA (HO-SCA for short) and the Higher-order masking problem are formally defined. We also introduce some basics about homographic functions on which the construction proposed in [14] is based. In Sec. 3, Courtois and Goubin's scheme is presented. We then exhibit in Sec. 4 the construction weaknesses (or flaws) that make it vulnerable to 2nd-order SCA attacks (for any order of the masking scheme). We then evaluate the actual amount of information leakage resulting from each of the presented flaws and compare them with information leakage for the classical Boolean and affine masking schemes. These theoretical results are finally confirmed by simulations of SCA attacks in Sec. 5.

2 Preliminaries

2.1 Higher-Order Side-Channel Attacks and Higher-Order Security

SCA attacks exploit a dependency between a subpart of the secret key and the variations of a physical leakage as function of the plaintext. This dependency results from the manipulation of some sensitive variables by the implementation. We say that a variable is *sensitive* if it depends on both the plaintext and the secret key. For example, the x-or (denoted $\oplus$) between a key byte and a plaintext byte is a sensitive variable.

As recalled in introduction, the manipulation of sensitive data may be protected using a dth-order masking scheme. When such a scheme is applied, it is expected that no HO-SCA of order less than or equal to d can be successful. The next definition formalizes the notion of security with respect to dth-order SCA for a cryptographic algorithm. Note that this definition corresponds exactly to that involved in Courtois and Goubin's paper [14], along with numerous other papers (*e.g.* [15, 3, 16, 17, 18]).

Definition 1 (dth-Order Security). *A cryptographic algorithm $\mathcal{A}$ is secure against dth-order SCA if every family of at most d intermediate variables of $\mathcal{A}$ is independent of any sensitive variable.*

If a family of d intermediate variables depends on a sensitive variable then we say that the algorithm admits a *dth-order flaw*. A dth-order SCA aims at exploiting such a flaw.

In the rest of the paper, we will use a large letter (*e.g.* X) to denote a random variable, while a lowercase letter (*e.g.* x) will denote a particular value taken by a random variable.

2.2 Higher-Order Masking

Let M and K denote two random variables respectively associated with some plaintext subpart values m and a secret sub-key k manipulated by a cryptographic algorithm. Let us moreover denote by Z the sensitive variable $M \oplus K$. When dth-order aditive masking is involved to secure the manipulation of Z, the latter variable is randomly split into $d+1$ shares R_0, R_1, ..., R_d such that:

$$Z = R_0 \oplus R_1 \oplus \cdots \oplus R_d \ . \tag{1}$$

The R_i's, $i > 0$, are usually called *the masks* and are randomly generated. In this case, the share R_0, called the *masked variable*, plays a particular role and is defined such that $R_0 = Z \oplus R_1 \oplus \cdots \oplus R_d$.

To enable the application of a transformation S on a variable Z split in $d+1$ shares, as in (1), a so-called *dth-order masking scheme* (or simply a *masking scheme* if there is no ambiguity on d) must be designed. It leads to the processing of $d+1$ new shares $R'_0, R'_1, \cdots, R'_d$ such that:

$$S[Z] = R'_0 \oplus R'_1 \oplus \cdots \oplus R'_d \ . \tag{2}$$

Usually, the R'_i's, $i > 0$, are generated at random and the share R'_0 is defined such that $R'_0 = S[Z] \oplus R'_1 \oplus \cdots \oplus R'_d$. When S is non-linear, the critical point is to deduce R'_0 from $(R_i)_{i \geq 0}$ and $(R'_i)_{i>0}$ without compromising the security of the scheme against dth-order SCA. Several solutions have been proposed to deal with this issue (*e.g.* [12, 13, 14, 8, 11]). In this paper we pay particular attention to those based on the pre-processing of a *masked s-box* [14, 8, 11]. They involve so-called *(additive) masking functions* in their description. We give hereafter some basics about those functions:

- we denote by G_i the ith *input masking function* $X \mapsto X \oplus R_i$ and by G'_i the ith *output masking function* $X \mapsto X \oplus R'_i$,
- for every integer $j \geq 1$, we denote by $G_{j..1}$ and $G'_{j..1}$ the function $G_j \circ \cdots \circ G_1$ and $G'_j \circ \cdots \circ G'_1$ respectively,

– for every j, we have $G_{j..1}^{-1} = G_{j..1}$ and $G_{j..1}^{\prime -1} = G'_{j..1}$: masking functions are involutive over $\mathrm{GF}(2^n)$ and pairwisely commutative with respect to the composition law $\circ$.

When formulated in terms of masking functions, the masked variables R_0 and R'_0 respectively satisfy $R_0 = G_{d..1}(Z)$ and $R'_0 = G_{d..1}(S[Z])$, and the problem of the construction of a dth-order masking scheme can be stated as follows:

Problem 1 (Construction of a dth-order masking scheme). *Let S denote a non-linear function defined over the definition set of Z. Let $(G_i)_{i \leq d}$ (resp. $(G'_i)_{i \leq d}$) be respectively a family of d input (resp. output) additive masking functions and let $G_{d..1}(Z)$ be the dth-order masked representation of Z. Define a dth-order secure construction of the masked representation $(G'_{d..1} \circ S)(Z)$ of $S(Z)$ taking at inputs $G_{d..1}(Z)$ and the families $(G'_i)_{i \leq d}$ and $(G_i)_{i \leq d}$.*

Before starting the discussion about the masking scheme proposed in [14], we recall in the following section some basics about the so-called *homographic functions*, that are the core of the solution to Problem 1 proposed in [14].

2.3 Preliminaries on Homographic Functions

Let ∞ denote an element not included in $\mathrm{GF}(2^n)$ and let us denote by $\overline{\mathrm{GF}(2^n)}$ the set $\mathrm{GF}(2^n) \cup \{\infty\}$. Homographic functions over $\overline{\mathrm{GF}(2^n)}$ are functions which take the general form $X \mapsto \frac{aX+b}{cX+d}$ for $(a,b,c,d) \in \mathrm{GF}(2^n)^4$ and $X \in \mathrm{GF}(2^n) \setminus \{\infty, d/c\}$. They are defined in $X = \infty$ and $X = d/c$ by mapping those elements into a/c and ∞ respectively.

Homographic functions have a compact representation since every such function $A : X \mapsto \frac{aX+b}{cX+d}$ can be represented by a 2×2 matrix denoted by $\mathcal{M}_A$ and satisfying:

$$\mathcal{M}_A = \begin{pmatrix} a & b \\ c & d \end{pmatrix} . \tag{3}$$

The composition of two homographic functions is also homographic. Moreover, composing two homographic functions can be done efficiently by multiplying their respective matrix-representations: if $\mathcal{M}_\mathcal{A}$ and $\mathcal{M}_\mathcal{B}$ are the matrix representations of the homographic functions A and B, then the matrix representation $\mathcal{M}_{\mathcal{A} \circ \mathcal{B}}$ of $A \circ B$ satisfies:

$$\mathcal{M}_{\mathcal{A} \circ \mathcal{B}} = \mathcal{M}_\mathcal{A} \times \mathcal{M}_\mathcal{B} .$$

Eventually, the set of invertible homographic mappings (*i.e.* s.t. $ad \neq bc$) forms a group $\mathcal{H}$ under the composition law $\circ$. The neutral element is the identity function Id over $\overline{\mathrm{GF}(2^n)}$ and the matrix representation $\mathcal{M}_{A^{-1}}$ of the inverse A^{-1} — for any $A \in \mathcal{H}$ — can be easily deduced from $\mathcal{M}_A$:

$$\mathcal{M}_{A^{-1}} = \begin{pmatrix} d & b \\ c & a \end{pmatrix} , \tag{4}$$

where $\mathcal{M}_A$ satisfies (3).

3 Masking Scheme Based on the Pre-processing of a Homographic Masked s-Box

The method proposed by Courtois and Goubin in [14] to solve Problem 1 for any order d may be viewed as an adaptation of the generic dth-order table recomputation (recalled in Appendix A) to the particular case of the AES s-box. In the next section, we give the main outlines of Courtois and Goubin's masking scheme. The presentation is then completed in the following sections, where each step of the scheme is detailed.

3.1 Courtois and Goubin's Scheme: General Principle

To solve Problem 1, the table recomputation methods follow the hereafter approach:

1. From the look-up table representation of S, compute the look-up table representation of the following *masked s-box* $S^\star$:

$$S^\star = G'_d \circ \cdots \circ G'_2 \circ G'_1 \circ S \circ G_1 \circ G_2 \circ \cdots \circ G_d \ . \tag{5}$$

2. Evaluate $S^\star$ on $G_{d..1}(Z)$.

The processing of the first step must be done without ever manipulating a $(d-1)$-tuple $\mathbf{U}$ of intermediate variables such that $(\mathbf{U}, G_{d..1}(Z))$ is sensitive (*i.e.* depends on Z). This constraint implies several table recomputations during step 1 (see *e.g.* [11,8] and Appendix A). The core idea of Courtois and Goubin in [14] is to use homographic functions in such way that the costly table recomputations are replaced by 2-dimensional matrix products (as noticed in Sec. 2.3).

Homographic Extension. Courtois and Goubin approach requires to extend the functions S, $(G_i)_i$ and $(G'_i)_i$ defined over $\mathrm{GF}(2^n)$ to (homographic) functions $\overline{S}$, $(\overline{G}_i)_i$ and $(\overline{G}'_i)_i$ defined over $\overline{\mathrm{GF}(2^n)}$. As explained in Sec. 3.2, this is straightforward for the masking functions G_i's and G'_i's and possible for the AES s-box S due to its particular algebraic structure. Eventually, the method introduced in [14] enables the construction of the following function with only 2×2 matrix products:

$$\overline{S^\star} = \overline{G}'_d \circ \cdots \circ \overline{G}'_2 \circ \overline{G}'_1 \circ \overline{S} \circ \overline{G}_1 \circ \overline{G}_2 \circ \cdots \circ \overline{G}_d \ . \tag{6}$$

The core point is that the function $\overline{S^\star}$ exactly corresponds to the masked function $S^\star$ over the set $\overline{\mathrm{GF}(2^n)} \setminus \{\infty, G_{d..1}(0)\}$. In the rest of the paper, we will denote by $\overline{G}_{d..1}$ (resp. $\overline{G}'_{d..1}$) the function $\overline{G}_d \circ \cdots \circ \overline{G}_1$ (resp. $\overline{G}'_d \circ \cdots \circ \overline{G}'_1$).

The ∞-masking problem. Unfortunately, as noticed by the authors themselves, a direct application of $\overline{S^\star}$ to a masked variable $G_{d..1}(Z)$ would introduce a first-order flaw, called ∞-*masking problem*[1] in the following. Indeed, the masking

[1] Note that the ∞-masking problem is very close to the *zero-masking problem* (see for instance [19]).

functions $\overline{G}'_{d..1}$ and $\overline{G}_{d..1}$ do not mask the ∞ element ($\overline{G}_{d..1}(\infty) = \overline{G}'_{d..1}(\infty) = \infty$). To overcome this issue, Courtois and Goubin propose to reroute the ∞ element by involving two new families of d random functions $(\alpha_i)_i$ and $(\beta_i)_i$ in $\mathcal{H}$. This leads the authors to the construction of the following new masked s-box:

$$\overline{S^{\star\star}} = \alpha^{-1} \circ \overline{S^{\star}} \circ \beta^{-1} ,$$

where α and β respectively denote the functions $\alpha_1 \circ \cdots \circ \alpha_d$ and $\beta_1 \circ \cdots \circ \beta_d$.

We recall hereafter the different steps of Courtois and Goubin's masking scheme at order d:

1. Generate at random two families of d homographic permutations $(\alpha_i)_{1 \le i \le d}$ and $(\beta_i)_{1 \le i \le d}$.
2. Compute A and B such that:

$$A = \left(\alpha_d^{-1} \circ \cdots \circ \left(\alpha_2^{-1} \circ \left(\alpha_1^{-1} \circ \overline{G}'_1\right) \circ \overline{G}'_2\right) \circ \cdots \circ \overline{G}'_d\right)$$

and

$$B = \left(\overline{G}_1 \circ \cdots \circ \left(\overline{G}_{d-1} \circ \left(\overline{G}_d \circ \beta_d^{-1}\right) \circ \beta_{d-1}^{-1}\right) \circ \cdots \circ \beta_1^{-1}\right) ,$$

where the order of the computation (given by the brackets) is of importance to achieve dth-order security.
Note that we have $A = \alpha^{-1} \circ \overline{G}'_{d..1}$ and $B = \overline{G}_{d..1} \circ \beta^{-1}$.
3. Compute the representation of the homographic function

$$\overline{S^{\star\star}} = A \circ \overline{S} \circ B .$$

4. Compute step by step from right to left:

$$\alpha_1 \circ \cdots \circ \alpha_d \circ \overline{S^{\star\star}} \circ \beta_1 \circ \cdots \circ \beta_d(G_{d..1}(Z)) . \tag{7}$$

Note that this processing outputs $\overline{S^{\star}}(G_{d..1}(Z))$ that is, by definition of $\overline{S^{\star}}$, $G'_{d..1}(S(Z))$ for $Z \in \mathrm{GF}(2^n) \setminus \{0\}$. To enable computation on the 0 element, a pre-processing and a post-processing are applied on (7). Those steps, which involve *swapping functions*, are detailed in Sec.(s) 3.2 and 3.3.

3.2 Homographic Masked AES s-Box

We show here, how the masked s-box $S^{\star} = G'_{d..1} \circ S \circ G_{d..1}$ can be represented as a composition of Homographic functions. In the sequel, $\tau_{g,h}$ denotes the swapping function that exchanges the elements g and h and leaves all the other elements unchanged.

The AES s-box. The AES s-box is defined as the composition of an affine function with the inverse function over $\mathrm{GF}(2^n)^{\star}$ extended in 0 by setting $0^{-1} = 0$. To be compliant with the notations introduced in previous sections, we will denote this extended inverse function by S in the following. As noticed in [14], it can be extended over $\overline{\mathrm{GF}(2^n)}$ to a function $\tau_{0,\infty} \circ \overline{S}$ where $\overline{S}$ is a homographic function defined by:

$$\overline{S}(X) = \begin{cases} S(X) = X^{-1} \text{ if } X \notin \{0, \infty\} \\ 0 \mapsto \infty \\ \infty \mapsto 0 \end{cases} .$$

The masking functions. Every masking function G_i (resp. G'_i) can be straightforwardly extended to a homographic function $\overline{G_i}$ (resp. $\overline{G'_i}$) by setting $\overline{G_i}(\infty) = \infty$ (resp. $\overline{G'_i}(\infty) = \infty$). For every i, the matrix representation $\mathcal{M}_{\overline{G_i}}$ (resp. $\mathcal{M}_{\overline{G'_i}}$) is obtained by setting $a = d = 1, c = 0$ and $b = R_i$ (resp. $b = R'_i$) in (3).

As a consequence, the masked representation $S^\star$ of S defined in (5) can be considered as a restriction to $\mathrm{GF}(2^n)$ of the function

$$\tau_{\infty,\mathbf{R}'} \circ \overline{S^\star},$$

where $\mathbf{R}' = R'_1 \oplus \cdots \oplus R'_d$ and $\overline{S^\star}$ is a homographic function defined in (6). Let us remark that the swapped elements verify $\infty = \overline{G'_{d..1}}(\infty)$ and $\mathbf{R}' = \overline{G'_{d..1}}(0)$ respectively.

3.3 Courtois and Goubin's Scheme

We sum up in the following the masked s-box construction at order d proposed in [14]:

Algorithm 1. Masked s-box computation

INPUT: the random families $(\alpha_i^{-1})_{i\leq d}$ and $(\beta_i^{-1})_{i\leq d}$ and the masking functions $(\overline{G}_i)_{i\leq d}$ and $(\overline{G}'_i)_{i\leq d}$

OUTPUT: the masked s-box $\overline{S^{\star\star}}$ and the swapping elements $\{g, h\}$

1. Construct $A = \alpha^{-1} \circ \overline{G}'_{d..1}$:
   $\mathcal{M}_A \hookleftarrow \mathcal{M}_{Id}$
   for j **from** 1 **to** d **do**
   $\mathcal{M}_A \hookleftarrow \mathcal{M}_{\alpha_j^{-1}} \times \mathcal{M}_A \times \mathcal{M}_{\overline{G}'_j}$
2. Construct $B = \overline{G}_{d..1} \circ \beta^{-1}$:
   $\mathcal{M}_B \hookleftarrow \mathcal{M}_{Id}$
   for j **from** d **downto** 1 **do**
   $\mathcal{M}_B \hookleftarrow \mathcal{M}_{\overline{G}_j} \times \mathcal{M}_B \times \mathcal{M}_{\beta_j^{-1}}$
3. Construct $\overline{S^{\star\star}} \hookleftarrow A \circ \overline{S} \circ B$
   $\mathcal{M}_{\overline{S^{\star\star}}} \hookleftarrow \mathcal{M}_A \times \mathcal{M}_{\overline{S}} \times \mathcal{M}_B$
4. Evaluate $g = A(\infty)$
   $g \leftarrow \frac{A_1}{A_3}$ // where $\mathcal{M}_A = \begin{pmatrix} A_1 & A_2 \\ A_3 & A_4 \end{pmatrix}$
5. Evaluate $h = A(0)$
   $h \leftarrow \frac{A_2}{A_4}$ // where $\mathcal{M}_A = \begin{pmatrix} A_1 & A_2 \\ A_3 & A_4 \end{pmatrix}$

return $(\overline{S^{\star\star}}, g, h)$

In Algorithm 1, we used two different symbols $\hookleftarrow$ and $\leftarrow$ to highlight the difference between an instantiation of a matrix and an instantiation of a variable. It must however be noted that the operation $\hookleftarrow$ is itself composed of $\leftarrow$ operations which correspond to the manipulation of the matrix coordinates. Since the attacks - described in the next section - target the matrix instantiations, we give hereafter the details of the implementation of $\hookleftarrow$ in terms of $\leftarrow$.

Algorithm 2. Operation $\hookleftarrow$

INPUT: Two 2×2 matrices $\mathcal{M}_A = (a_{i,j})_{0 \le i,j \le 1}$ and $\mathcal{M}_B = (b_{i,j})_{0 \le i,j \le 1}$
OUTPUT: The matrix $\mathcal{M}_\mathcal{B}$ such that $\mathcal{M}_\mathcal{A} = \mathcal{M}_\mathcal{B}$

1. $b_{00} \leftarrow a_{00}$
2. $b_{01} \leftarrow a_{01}$
3. $b_{10} \leftarrow a_{10}$
4. $b_{11} \leftarrow a_{11}$

return $\mathcal{M}_\mathcal{B}$

It can be checked that for every $Z \in \mathrm{GF}(2^n)$ we have:

$$[G'_{d..1} \circ S](Z) = [\alpha \circ \tau_{g,h} \circ \overline{S^{\star\star}} \circ \beta] \left(G_{d..1}(Z)\right) \quad . \tag{8}$$

Namely, by applying the transformations β, $\overline{S^{\star\star}}$, $\tau_{g,h}$ and α to the masked representation $G_{d..1}(Z)$ of Z we get the masked representation $G'_{d..1}(S(Z))$ of $S(Z)$. Hence, once the homographic function $\overline{S^{\star\star}}$ and the elements g and h are generated by the means of Algorithm 1, the masked representation $G'_{d..1}(S(Z))$ of the AES s-box output is got by processing the following algorithm to the masked representation $G_{d..1}(Z)$ of Z.

Algorithm 3. Masked s-box evaluation

INPUT: the masked input $G_{d..1}(Z)$, the outputs of Alg. 1 ($\overline{S^{\star\star}}$ and $\{g, h\}$), and the random families $(\alpha_i)_i$ and $(\beta_i)_i$
OUTPUT: $G'_{d..1}(S(Z))$

1. $U \leftarrow G_{d..1}(Z)$
2. **for** j **from** d **downto** 1 **do**
3. $\quad U \leftarrow \beta_j(U)$
4. $V \leftarrow \overline{S^{\star\star}}(U)$
5. $W \leftarrow \tau_{g,h}(V)$
6. **for** j **from** d **downto** 1 **do**
7. $\quad W \leftarrow \alpha_j(W)$

return W

The completeness of Algorithm 3 is a straightforward consequence of (8) and the constructions performed in Algorithm 1.

4 The Flaws

The variable U manipulated at Step 4 of Algorithm 3 satisfies

$$U = \beta \circ \overline{G}_{d..1}(Z) \ ,$$

that is

$$U = B^{-1}(Z) \ , \tag{9}$$

where B is the homographic function designed at Step 2 of Algorithm 1. Hence, U can be viewed as the *homographically masked representation* of Z, where B^{-1} plays the role of an *homographic masking function*. Thus, as any kind of masking (*e.g.* Boolean [20], multiplicative [19] or affine [21]), the homographic masking function can be overcome by simultaneously targeting the manipulation of the masked variable and the manipulation of the masking material during the masked s-box construction (Algorithm 1). This remark leads to a straightforward flaw of order 5, that can be reduced to flaws of order 4, 3 and 2, and this, whatever the scheme order d. We exhibit them in Sec. 4.1. Additionally, we exhibit in Sec. 4.2 another kind of 2nd-order flaw related to the masking of the ∞ element. In Sec. 4.3, we eventually quantify the amount of information leakage resulting from those flaws when the scheme is implemented on a standard device.

4.1 First Category of Flaws

Let us denote by (B_1, B_2, B_3, B_4) the 4-tuple such that B satisfies:

$$\mathcal{M}_B = \begin{pmatrix} B_1 & B_2 \\ B_3 & B_4 \end{pmatrix} \ . \tag{10}$$

Equations (9) and (10) imply the following relation between Z, U and B:

$$Z = B(U) = \frac{B_1 U + B_2}{B_3 U + B_4} \ . \tag{11}$$

The above equation suggests us that information on U combined with information on the function B reveals information on the sensitive variable Z. Information on U can be retrieved during its manipulation at the end of the last iteration of the first loop (Step 2–Step 3) in Algorithm 3. Information on B can be obtained by observing the instantiation of the matrix-representation of B during the last iteration of the second loop in Algorithm 1. This processing is indeed done by calling Algorithm 2 and thus implies the manipulation of the 4 coordinates B_1, B_2, B_3 and B_4 of $\mathcal{M}_B$.

Due to (11), we have $\Pr[Z] \neq \Pr[Z \mid (U, B_1, B_2, B_3, B_4)]$ and we thus straightforwardly deduce the existence of the following 5th-order flaw on Z:

5th-order flaw: $$\{B_1, B_2, B_3, B_4, U\} \ . \tag{12}$$

At this point, it must be noted that the full description of B is not required to exhibit, together with U, a flaw on Z. Indeed, it can be easily checked (see Sec. 4.3) that subsets of $\{U, B_1, B_2, B_3, B_4\}$ are also dependent on Z. In other words, the scheme described in Sec. 3.3 admits flaws of order lower than 5. We list them hereafter:

4th-order flaws: $\{B_2, B_3, B_4, U\}, \{B_1, B_2, B_4, U\}, \{B_1, B_2, B_3, U\}$, (13)

3rd-order flaws: $\{B_1, B_2, U\}, \{B_2, B_3, U\}, \{B_2, B_4, U\}$, (14)

2nd-order flaw: $\{B_2, U\}$. (15)

A similar analysis could be done to exhibit a flaw on $\overline{S}(Z)$ by combining information on the intermediate variable $V = A \circ \overline{S}(Z)$ (Step 4 in Algorithm 3) with information on A (matrix instantiation at the end of the first loop in Algorithm 1). As a matter of fact, by construction of A, we have $\overline{S}(Z) = A^{-1}(V)$ when Z belongs to $\mathrm{GF}(2^n)^\star$. Again, the variable V can be viewed as the homographically masked representation of $\overline{S}(Z)$ by the homographic masking function A.

4.2 Second Category of Flaws

The masking scheme contains another 2nd-order flaw than the one exhibited in the previous section. This flaw is close to the *zero-masking problem* [19] or its projective version mentioned in [14]. To deal with the ∞-*masking problem*, the value $g = A(\infty)$ is computed in Algorithm 1. This variable g combined with the intermediate variable V occurring at Step 4 of Algorithm 3 reveals information on Z. Indeed, it can be checked that

$$(V = g) \text{ iff } \left(\overline{S}(Z) = \infty\right) \text{ iff } (Z = 0) .$$

We hence have $\Pr[Z] \neq \Pr[Z \mid (V, g)]$, that implies the following 2nd-order flaw:

2nd-order flaw: $\{V, g\}$. (16)

As we will see in Sec. 5, this flaw will lead to a powerful second-order SCA against Courtois and Goubin's Scheme. Actually, will show that this second-order flaw is very similar to the one related to affine masking [21] (see Remark 3, Sec. 4.3).

4.3 Information Theoretic Evaluation

In the previous section, we have shown that Courtois and Goubin's masking scheme has flaws of orders 2 to 5 whatever the order of the scheme. This is sufficient to claim that the countermeasure fails in achieving dth-order security for all d. However, the physical leakage of an implementation does not reveal the exact values of the variables manipulated but a noisy function of them. In this section, our purpose is to quantify the amount of information that each flaw reveals about the sensitive variable Z when the scheme is implemented on a classical embedded device. To achieve this goal, we first model the relationship between the physical leakage and the value of the variable manipulated at the time of the leakage. Secondly, we follow the approach introduced in [22] to evaluate theoretically the amount of information leaked with respect to each flaw.

Leakage Model. Every intermediate variable Y_i of an algorithm processing can be associated with a variable L_i representing the information leaking about Y_i through side channel. This leakage can be expressed as the sum of a deterministic *leakage function* φ of Y_i with an independent additive noise β_i:

$$L_i = \varphi(Y_i) + \beta_i \ . \tag{17}$$

In the following, we shall say that a d-tuple $\mathbf{L}$ of leakage variables L_i is a *dth-order leakage* if it corresponds to d different intermediate variables Y_i that jointly depend on some sensitive variable (*i.e.* is a dth-order flaw).

Thanks to (17), we can associate a dth-order leakage $\mathbf{L}$ to each d-tuple of intermediate variables listed in (12)–(16). We shall use the notation

$$\mathbf{L} \leftarrow (Y_1, \cdots, Y_d)$$

to specify the set of intermediate data manipulations that are related to $\mathbf{L}$.

Information Theoretic Evaluation. To evaluate the information revealed by each of the flaws exhibited in (12)–(16) with respect to HO-SCA, we follow the information theoretic approach introduced in [22]. Namely, we compute the mutual information between the sensitive variable Z and $\mathbf{L}$. For comparison purpose, we proceed similarly for Boolean and affine maskings (see for instance [20] and [21] for a detailed description of those maskings). We list hereafter the leakages we consider and the underlying leaking variables:

5th-order leakage of homographic masking: $\mathbf{L} \leftarrow (U, B_1, B_2, B_3, B_4)$. (18)

4th-order leakage of homographic masking: $\mathbf{L} \leftarrow (U, B_1, B_2, B_4)$. (19)

3rd-order leakage of homographic masking: $\mathbf{L} \leftarrow (U, B_1, B_2)$. (20)

2nd-order leakage of homographic masking: $\mathbf{L} \leftarrow (U, B_2)$. (21)

2nd-order leakage of homog. mask. (2nd cat.): $\mathbf{L} \leftarrow (V, g)$. (22)

2nd-order leakage of Boolean masking: $\mathbf{L} \leftarrow (Z \oplus B_2, B_2)$. (23)

2nd-order leakage of affine masking: $\mathbf{L} \leftarrow (B_1 \cdot Z \oplus B_2, B_2)$. (24)

3rd-order leakage of affine masking: $\mathbf{L} \leftarrow (B_1 \cdot Z \oplus B_2, B_2, B_1)$. (25)

Remark 1. The Boolean and affine maskings listed above can be viewed as particular cases of homographic masking by fixing respectively $(B_1, B_3, B_4) = (1, 0, 1)$ and $(B_3, B_4) = (0, 1)$ in (11).

Remark 2. For Flaws (13) and (14) only the best choice of leaking variables has been considered.

Remark 3. Flaws (22) and (24) are in fact very similar. Each of them can be interpreted as targeting a pair of intermediate variables (Y_1, Y_2) that depends on a sensitive variable Z in the following way:

$$(Y_1 = Y_2) \text{ iff } (Z = 0) \ .$$

In Flaw (22), we have $Y_1 = V$ and $Y_2 = g$ and we have $Y_1 = B_1 \cdot Z \oplus B_2$ and $Y_2 = B_2$ in Flaw (24).

For each kind of leakage, we computed the mutual information between Z and the tuple of leakages in the Hamming weight model with Gaussian noise: the leakage L_i related to a variable Y_i is distributed according to (17) with $\varphi = \mathrm{HW}$ and $\beta_i \sim \mathcal{N}(0, \sigma^2)$ (the different β_i's are also assumed independent). According to the assumptions made in [14] and recalled in Sec. 3, we assumed that the variable Z and the B_i's are uniformly distributed over their definition sets. In this context, the signal-to-noise ratio (SNR) of the leakage is defined as $\mathrm{Var}\left[\varphi(Y_i)\right] / \mathrm{Var}\left[\beta_i\right] = 2/\sigma^2$ (since the Y_is are 8-bits variables).

Figure 1 summarizes the information theoretic evaluation for each leakage[2]. As expected, when the noise is small, the amount of information given by the intermediate variables involved in Flaws (18) to (21) is an increasing function of the leakage order. When the noise increases, its impact on the information leakage is more important when the order of the flaw is high. Eventually, after some threshold, the ranking between the information leakages is reversed, as it can be observed between, *e.g.* , the 2O-Homographic (21), 3O-Homographic (20) and 4O-Homographic (19)[3].

5 Attacks Simulations and Comparisons

In Sec. 4.3, we have quantified the amount of sensitive information that each flaw exhibited in Sec.(s) 4.1 and 4.2 provides about Z. We will now see how those leakages can be exploited to perform HO-SCA that succeed in recovering the AES sub-keys.

To exploit the flaws exhibited in Sec.(s) 4.1 and 4.2, we have applied two types of HO-SCA: *higher-order CPA* (HO-CPA for short) such as introduced in [17] and improved in [23], and *higher-order template* attacks (HO-TA for short) [24,25]. Since each attack category involves adversaries with different capabilities (the one in HO-TA being much stronger than the one in HO-CPA), we think that

[2] Due to the cost of some mutual information computations, it was not possible to give results for all the flaws and all the noise standard deviations values. In particular, Flaw (18) could not be evaluated and evaluations of Flaws (19) and (20) are extrapolated on Figure 1 with dashed gray lines.

[3] The abrupt decrease of the mutual information for Flaws (19),(20) and (21) was not expected and does not correspond to the experimental results showed in Sec. 5. We believe these incoherences are due to a lack of precision of the estimation methods we involved.

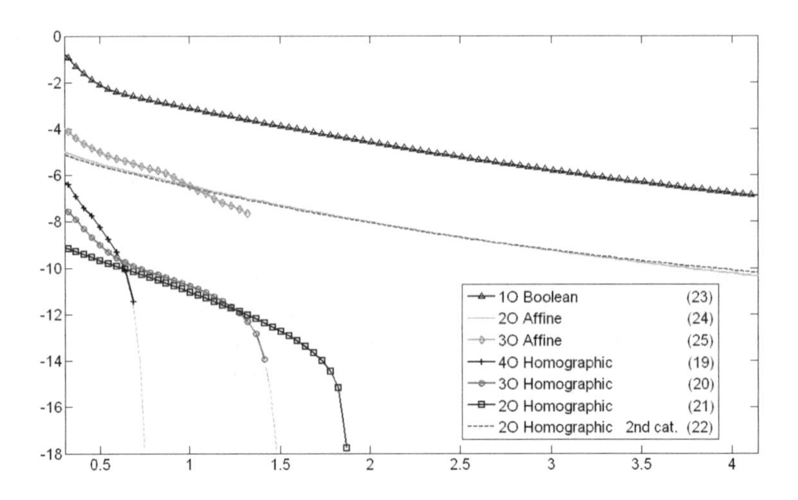

Fig. 1. Mutual information ($\log_{10}$) between the leakage and the sensitive variable over an increasing noise standard deviation (x-axis)

they together give a good overview of the scheme resistance against higher-order side channel attacks. We give in Appendix B the details of the attacks we have performed. Observations of the leakage vectors $\mathbf{L}$ in (18)–(25) have been simulated according to (17) by fixing φ equal to the Hamming weight function HW(). Due to the great computational complexity of template attacks for $n = 8$ and $d > 3$ (see Appendix B), we were not able to perform them for those parameters. To circumvent this issue and in order to be able to compare the efficiencies of all the (HO-CPA and HO-TA) attacks, we therefore choose to also perform our attacks simulations for $n = 4$. In this case, the flaws correspond to an s-box processing that does not exactly correspond to the non-linear part of the AES s-box (defined for $n = 8$), but it shares with it the same algebraic properties: it is the inverse function in $\mathrm{GF}(2^n)^\star$ extended in 0 by setting $0^{-1} = 0$. We first present the attack simulations over GF(16) and then, we present the simulations over GF(256). For comparison purposes, we also presented attack simulations against Boolean and affine maskings.

Each attack simulation has been performed 100 times for various SNR values ($+\infty$, 1, 1/2, 1/5 and 1/10), that is for several noise standard deviation values ($\sigma = 0, 1, \sqrt{2}, \sqrt{5}, \sqrt{10}$ for $n = 4$ and $\sigma = 0, \sqrt{2}, 2, \sqrt{10}, 2\sqrt{5}$ for $n = 8$). Tables 1 and 2 summarize the number of leakage measurements required to observe a success rate of 90% in retrieving k for the different attacks.

General Comments. As it can be observed in Tables 1 and 2, all the attacks recover the key with a success rate equal to 90% when the leakage is noise-free (SNR $= \infty$). The inefficiency of the 2O-CPA to exploit the flaw (21) for $n = 8$ was an expected result. Indeed, this flaw was quantified in Sec. 4.3 to be equal

Table 1. Attack Simulations over GF(16).

Masking	Attack	$\mathbf{L} \leftarrow ?$	SNR					
			∞	1	0.5	0.25	0.1	0.05
ICISC04 Flaw 1	2O-CPA	(21)	2600	9000	20 000	55 000	300 000	$> 10^6$
	2O-TA	(21)	360	4000	10 000	35 000	200 000	10^6
	3O-TA	(20)	130	2800	15 000	35 000	200 000	10^6
	4O-TA	(19)	120	2300	9000	30 000	250 000	$> 10^6$
	5O-TA	(18)	120	2600	10 000	40 000	$> 5.10^3$	$> 5.10^3$
ICISC04 Flaw 2	2O-CPA	(22)	210	800	1800	4000	25 000	80 000
	2O-TA	(22)	50	400	1100	4000	15 000	60 000
1O-Bool	2O-CPA	(23)	60	220	600	1200	8000	25 000
	2O-TA	(23)	20	180	330	1400	9000	30 000
Affine	2O-CPA	(24)	200	700	1700	5000	25 000	90 000
	2O-TA	(24)	50	500	1200	2800	15 000	60 000
	3O-TA	(25)	20	300	1000	2400	20 000	65 000

Table 2. Attack Simulations over GF(256)

Masking	Attack	$\mathbf{L} \leftarrow ?$	SNR				
			∞	1	0.5	0.25	0.1
ICISC04 Flaw 1	2O-CPA	(21)	10^7	$> 10^7$	$> 10^7$	$> 10^7$	$> 10^7$
	2O-TA	(21)	140 000	$> 10^6$	$> 10^6$	$> 10^6$	$> 10^6$
	3O-TA	(20)	70 000	$> 10^6$	$> 10^6$	$> 10^6$	$> 10^6$
	4O-TA	(19)	20 000	-	-	-	-
	5O-TA	(18)	-	-	-	-	-
ICISC04 Flaw 2	2O-CPA	(22)	7000	25 000	50 000	200 000	600 000
	2O-TA	(22)	1400	20 000	40 000	180 000	450 000
1O-Bool	2O-CPA	(23)	250	1100	2600	15 000	45 000
	2O-TA	(23)	20	500	1200	7000	20 000
Affine	2O-CPA	(24)	7000	25 000	50 000	200 000	550 000
	2O-TA	(24)	1200	20 000	40 000	200 000	550 000
	3O-TA	(25)	280	15 000	35 000	200 000	500 000

to at most $1.82 \cdot 10^{-5}$ and such a small mutual information cannot be exploited by a CPA in less than 10^6 measurements.

HO-TA *versus* HO-CPA. As expected, HO-TA are much more efficient than HO-CPA attacks. They succeed in exploiting all the information contained in the leakages related to the mask values B_1,..., B_4 (resp. g) to unmask the masked variable U (resp. V) and to finally recover the key. Moreover, the more information is exploited about the homographic masking function B (*i.e.* the higher the order of the template attack) the more efficient is the attack. We experimented that this was not the case with HO-CPA attacks: higher-order CPAs against homographic masking loss in efficiency when the order increases *i.e.* when information on B_2 is exploited simultaneously with that on other coordinates B_1, B_3 and B_4 of B. The good efficiency of template attacks in terms of number of leakage measurements must however be mitigated since their computation cost quickly increases with the order. This explains why we did not succeed in performing our attack simulations for SNRs greater than 1 when $n = 8$.

Comparison of The Different Flaws. As expected after our information theoretic evaluation campaign conducted in Sec. 4.3, attacks exploiting Flaw (16) are much more efficient than those exploiting Flaws (12)–(15). Furthermore, the similitude of efficiency between the HO-TAs against Flaws (16) and (24) (affine masking) confirms that they both provide the same security level (as stated in Remark 3, Sec. 4.3). If we do not take into account the second category of flaws, homographic masking seems to provide better resistance against HO-SCA than Boolean or affine masking. Indeed, even in the $n = 4$ scenario, HO-CPA and HO-TA against homographic masking are about 10 times less efficient than those against affine masking. Based on this observation, we think that this masking can be a good alternative to classical masking schemes if the second category of flaws can be patched.

6 Conclusion and Open Issues

In this paper, we have exhibited several weaknesses in the higher-order masking scheme proposed by Courtois and Goubin in [14]. Two kinds of flaws have been exhibited that render the construction vulnerable, in both cases, to 2nd-order SCA. It is a matter of fact that Courtois and Goubin's scheme was made particulary efficient by the use of homographic functions and then was attractive for high-order masked implementations of the AES. Our results demonstrate that this countermeasure must nevertheless be avoided if perfect resistance against dth-order SCA is expected. On the other hand, if the second category of flaws can be patched, then we think that the construction of Courtois and Goubin can be of interest to achieve satisfying resistance when the leakage signals contain enough noise. Indeed, we argued in this paper that the construction of Courtois and Goubin at any order d can be reduced to what we called a *homographic* masking scheme. Our theoretical and experimental evaluation of this masking shows that it offers better resistance to SCA than both Boolean and affine maskings. However, the study of the performances of a masking scheme based on homographic functions was out of the scope here and raises several non trivial implementation issues. We think that this study is an interesting avenue for further research on this subject.

References

1. Kocher, P., Jaffe, J., Jun, B.: Introduction to Differential Power Analysis and Related Attacks. Technical report, Cryptography Research Inc. (1998)

2. Chari, S., Jutla, C., Rao, J., Rohatgi, P.: A Cautionary Note Regarding Evaluation of AES Candidates on Smart-Cards. In: Second AES Candidate Conference – AES 2 (March 1999)
3. Blömer, J., Merchan, J.G., Krummel, V.: Provably Secure Masking of AES. In: Handschuh, H., Hasan, M.A. (eds.) SAC 2004. LNCS, vol. 3357, pp. 69–83. Springer, Heidelberg (2004)
4. Oswald, E., Mangard, S., Pramstaller, N., Rijmen, V.: A Side-Channel Analysis Resistant Description of the AES S-box. In: Gilbert, H., Handschuh, H. (eds.) FSE 2005. LNCS, vol. 3557, pp. 413–423. Springer, Heidelberg (2005)
5. Waddle, J., Wagner, D.: Toward Efficient Second-order Power Analysis. In: Joye, M., Quisquater, J.-J. (eds.) CHES 2004. LNCS, vol. 3156, pp. 1–15. Springer, Heidelberg (2004)
6. Peeters, E., Standaert, F.X., Donckers, N., Quisquater, J.J.: Improved Higher-order Side-Channel Attacks with FPGA Experiments. In: Rao, J.R., Sunar, B. (eds.) CHES 2005. LNCS, vol. 3659, pp. 309–323. Springer, Heidelberg (2005)
7. Oswald, E., Mangard, S., Herbst, C., Tillich, S.: Practical Second-order DPA Attacks for Masked Smart Card Implementations of Block Ciphers. In: Pointcheval, D. (ed.) CT-RSA 2006. LNCS, vol. 3860, pp. 192–207. Springer, Heidelberg (2006)
8. Rivain, M., Dottax, E., Prouff, E.: Block Ciphers Implementations Provably Secure Against Second Order Side Channel Analysis. Cryptology ePrint Archive, Report 2008/021 (2008), http://eprint.iacr.org/
9. Nikova, S., Rijmen, V., Schläffer, M.: Secure Hardware Implementation of Non-linear Functions in the Presence of Glitches. In: Lee, P.J., Cheon, J.H. (eds.) ICISC 2008. LNCS, vol. 5461, pp. 218–234. Springer, Heidelberg (2009)
10. Akkar, M.L., Bévan, R., Goubin, L.: Two Power Analysis Attacks against One-Mask Method. In: Roy, B., Meier, W. (eds.) FSE 2004. LNCS, vol. 3017, pp. 332–347. Springer, Heidelberg (2004)
11. Schramm, K., Paar, C.: Higher Order Masking of the AES. In: Pointcheval, D. (ed.) CT-RSA 2006. LNCS, vol. 3860, pp. 208–225. Springer, Heidelberg (2006)
12. Ishai, Y., Sahai, A., Wagner, D.: Private circuits: Securing hardware against probing attacks. In: Boneh, D. (ed.) CRYPTO 2003. LNCS, vol. 2729, pp. 463–481. Springer, Heidelberg (2003)
13. Prouff, E., Rivain, M.: Provable Secure Higher-Order Masking of AES. In: Mangard, S., Standaert, F.-X. (eds.) CHES 2010. LNCS, vol. 6225, Springer, Heidelberg (2010) (to appear)
14. Courtois, N., Goubin, L.: An Algebraic Masking Method to Protect AES against Power Attacks. In: Won, D.H., Kim, S. (eds.) ICISC 2005. LNCS, vol. 3935, pp. 199–209. Springer, Heidelberg (2006)
15. Coron, J.S.: A New DPA Countermeasure Based on Permutation Tables. In: Ostrovsky, R., De Prisco, R., Visconti, I. (eds.) SCN 2008. LNCS, vol. 5229, pp. 278–292. Springer, Heidelberg (2008)
16. Joye, M., Paillier, P., Schoenmakers, B.: On Second-order Differential Power Analysis. In: Rao, J.R., Sunar, B. (eds.) CHES 2005. LNCS, vol. 3659, pp. 293–308. Springer, Heidelberg (2005)
17. Messerges, T.: Using Second-order Power Analysis to Attack DPA Resistant Software. In: Paar, C., Koç, Ç.K. (eds.) CHES 2000. LNCS, vol. 1965, p. 238. Springer, Heidelberg (2000)
18. Piret, G., Standaert, F.X.: Security Analysis of Higher-Order Boolean Masking Schemes for Block Ciphers (with Conditions of Perfect Masking). IET Information Security (2008) (to Appear)

19. Golić, J., Tymen, C.: Multiplicative Masking and Power Analysis of AES. In: Kaliski Jr., B.S., Koç, Ç.K., Paar, C. (eds.) CHES 2002. LNCS, vol. 2523, pp. 198–212. Springer, Heidelberg (2003)
20. Goubin, L., Patarin, J.: DES and Differential Power Analysis – The Duplication Method. In: Koç, Ç.K., Paar, C. (eds.) CHES 1999. LNCS, vol. 1717, pp. 158–172. Springer, Heidelberg (1999)
21. Fumaroli, G., Martinelli, J., Prouff, E., Rivain, M.: Affine Masking against Higher-Order Side Channel Analysis. In: SAC 2010, Springer, Heidelberg (2010) (to appear)
22. Standaert, F.X., Malkin, T., Yung, M.: A unified framework for the analysis of side-channel key recovery attacks. In: Joux, A. (ed.) EUROCRYPT 2009. LNCS, vol. 5479, pp. 443–461. Springer, Heidelberg (2010)
23. Prouff, E., Rivain, M., Bévan, R.: Statistical Analysis of Second Order Differential Power Analysis. IEEE Trans. Comput. 58(6), 799–811 (2009)
24. Chari, S., Rao, J., Rohatgi, P.: Template Attacks. In: Kaliski Jr., B.S., Koç, Ç.K., Paar, C. (eds.) CHES 2002. LNCS, vol. 2523, pp. 13–29. Springer, Heidelberg (2003)
25. Oswald, E., Mangard, S.: Template Attacks on Masking Resistance is Futile. In: Abe, M. (ed.) CT-RSA 2007. LNCS, vol. 4377, pp. 243–256. Springer, Heidelberg (2006)
26. Rivain, M., Dottax, E., Prouff, E.: Block Ciphers Implementations Provably Secure Against Second Order Side Channel Analysis. In: Nyberg, K. (ed.) FSE 2008. LNCS, vol. 5086, pp. 127–143. Springer, Heidelberg (2008)
27. Messerges, T.: Securing the AES Finalists against Power Analysis Attacks. In: Schneier, B. (ed.) FSE 2000. LNCS, vol. 1978, pp. 150–164. Springer, Heidelberg (2001)
28. Akkar, M.L., Giraud, C.: An Implementation of DES and AES, Secure against Some Attacks. In: Koç, Ç.K., Naccache, D., Paar, C. (eds.) CHES 2001. LNCS, vol. 2162, pp. 309–318. Springer, Heidelberg (2001)
29. Coron, J.S., Prouff, E., Rivain, M.: Side Channel Cryptanalysis of a Higher Order Masking Scheme. In: Paillier, P., Verbauwhede, I. (eds.) CHES 2007. LNCS, vol. 4727, pp. 28–44. Springer, Heidelberg (2007)

A Generic dth-Order Table Recomputation Scheme

We recall here the outlines of a generic masking scheme which has been involved several times in the literature to solve Problem 1 (see *e.g.* [26, 11]). The core idea is to extend to higher orders a method, called *table re-computation*, which has been widely used to protect implementations against 1st-order SCA (see for instance [27, 28]). It essentially amounts to deduce from $G_{d..1}$ and $G'_{d..1}$ the look-up table representation of a new function $S^\star$ satisfying:

$$S^\star(G_{d..1}(Z)) = G'_{d..1} \circ S(Z) \ , \tag{26}$$

for every Z, or equivalently,

$$S^\star = G'_{d..1} \circ S \circ G_{d..1} \ , \tag{27}$$

since $G_{d..1}$ is involutive.

The tricky part in this approach is the concurrent construction of $S^\star$ which must not introduce any flaw with respect to dth-order SCA. An attempt to define a construction secure at any order d has been done in [11] where different table re-computation algorithms have been proposed. However, it has been shown in [29] that none of those methods were secure against dth-order attacks when d is greater than or equal to 3. We recall hereafter the most straightforward algorithm proposed in [11].

Algorithm 4. Generic dth-order Table recomputation Scheme

INPUT: the look-up table S and the families of masking functions $(G_i)_{1\leq i\leq d}$ and $(G'_i)_{1\leq i\leq d}$

OUTPUT: the look-up table $S^\star$

1. $S^\star \leftarrow S$
2. **for** j **from** 1 **to** d **do**
3. $\quad S_{tmp} \leftarrow S^\star$
4. $\quad$ **for** x **from** 0 **to** 255 **do**
   $\quad\quad$ //Construct $S^\star(x) = G'_j \circ S^\star \circ G_j(x)$
5. $\quad\quad tmp \leftarrow G_j(x)$
6. $\quad\quad tmp \leftarrow S_{tmp}[tmp]$
7. $\quad\quad S^\star[x] \leftarrow G'_j(tmp)$
8. $\quad$ **end**
9. **end**

B Attacks Description

Higher-Order Template Attacks. In HO-TA, the attacker owns some *templates* of the leakage that he previously acquired during a profiling phase (see for instance [24,25]). More precisely, for every possible value z he has some estimation of the probability density function (pdf for short) of the random variable $(\mathbf{L} \mid Z = z)$. We denote by $f_{\mathbf{L},z}$ those pdfs. Based on their estimations, a guess $\hat{k}$ on k is tested by estimating a *likelihood.*

Let ϕ_σ denotes the pdf of the Gaussian distribution $\mathcal{N}(0,\sigma)$ which satisfies $\phi_\sigma(x) = \frac{1}{\sqrt{2\pi}\sigma}\exp\left(-\frac{x^2}{2\sigma^2}\right)$ for every $x \in \mathbb{R}$. Since the coordinates of $\mathbf{L}$ are assumed to satisfy (17) in this paper, we can exhibit the exact expression of $f_{\mathbf{L},z}$:

– for the first category of flaws listed in (12)–(15), we have:

$$f_{\mathbf{L},z}(\ell) = \sum_{\substack{(b_1,\cdots,b_4)\\ \in \mathrm{GF}(2^n)^4}} \Pr\left[B_1{=}b_1,\cdots,B_4{=}b_4\right]\phi_\sigma\left(l_1\text{-}\varphi(u)\right)\prod_{i=2}^{d}\phi_\sigma\left(l_i\text{-}\varphi(b_{j_i})\right), \tag{28}$$

where l_i denotes the ith coordinate of the d-tuple ℓ and where $(u, b_{j_2}, \cdots, b_{j_d})$ is the value taken by the random variable $(U, B_{j_2}, \cdots, B_{j_d})$ defined such that $\mathbf{L} \leftarrow (U, B_{j_2}, \cdots, B_{j_d})$ (*i.e.* satisfies one of the Relations (18)–(21)).

- for the second category of flaws (16), we have:

$$f_{\mathbf{L},z}(\ell) = \sum_{\substack{(a_1,\cdots,a_4)\\ \in \mathrm{GF}(2^n)^4}} \Pr[A_1{=}a_1,\cdots,A_4{=}a_4]\,\phi_\sigma(l_1-\varphi(v))\,\phi_\sigma(l_2-\varphi(g)), \tag{29}$$

where we recall that v and g respectively denote the values taken by the random variables $V = A \circ \overline{S}(Z)$ and $A(\infty)$.

We give hereafter the main steps of the HO-TA we performed:

1. *[Measurements]* For N random plaintexts $(m_i)_i$, perform N leakage measurements $(\ell_i)_i = (l_1^i, .., l_d^i)_i$ at d different times.
2. *[Hypotheses Construction]* For every key hypothesis $\hat{k}$, and every plaintext m_i, compute the hypothesis $\hat{z}_i$ on the value taken by Z.
3. *[Discrimination]* Then, process the likelihood $\mathcal{L}(\hat{k}|(\ell_i,\hat{z}_i)_i)$ of the key guess $\hat{k}$:

$$\mathcal{L}(\hat{k}|(\ell_i,\hat{z}_i)_i) = \prod_{i=1}^{N} f_{\mathbf{L},\hat{z}_i}\ell_i\ . \tag{30}$$

4. *[Key-candidate Selection]* Select the key guess $\hat{k}$ that maximizes (30).

Higher-Order CPA attacks [17,23]. HO-CPA share the same two first stages with HO-TA but have a different discrimination process. The likelihood computation is replaced by a correlation estimation. Namely, the samples $(\ell_i)_i$ and $(\hat{z}_i)_i$ are used to estimate the correlation coefficient $\rho\left[\hat{\varphi}(\hat{Z}), \mathcal{C}(\mathbf{L})\right]$, where $\hat{Z}$ denotes the random variable associated to the sample $(\hat{z}_i)_i$, where $\mathcal{C}$ is a *combining function* that converts $\mathbf{L}$ into a 1-dimensional variable and where $\hat{\varphi}$ is a well-chosen *prediction function*. The guess $\hat{k}$ leading to the greatest correlation (in absolute value) is selected as key-candidate.

The analysis conducted in [23] states that a good choice for $\mathcal{C}$ is the *normalized product combining*:

$$\mathcal{C} : \mathbf{L} \mapsto \prod_i (L_i - \mathrm{E}[L_i]). \tag{31}$$

In [23], it is also shown that the best choice for $\hat{\varphi}$ given $\mathcal{C}$ is:

$$\hat{\varphi} : z \mapsto \mathrm{E}[\mathcal{C}(\mathbf{L})|Z = z]\ . \tag{32}$$

From now and until the end of the paper, we assume that $\mathcal{C}$ and $\hat{\varphi}$ respectively satisfy (31) and (32).[4]

If the leakage function φ in (17) is assumed to be known and if the noise is assumed to be independent with 0 mean, then (32) can be rewritten:

[4] As explained in [23], the attacker may not be able to evaluate $\hat{\varphi}$ without knowing the exact distribution of $\mathbf{L}$ given Z (as in a profiling attack scenario). In a security evaluation context, it however makes sense to assume that the attacker has this ability.

– for the first category of flaws:

$$\hat{\varphi} : z \mapsto \mathrm{E}\left[(U - \mathrm{E}[U]) \prod_{i=2}^{d} (B_{j_i} - \mathrm{E}[B_{j_i}]) \mid Z = z\right] , \tag{33}$$

where $(U, B_{j_2}, \cdots, B_{j_d})$ is defined s.t. $\mathbf{L} \leftarrow (U, B_{j_2}, \cdots, B_{j_d})$ (*i.e.* satisfies one of the Relations (18)–(21)).

– for the second category of flaws:

$$\hat{\varphi} : z \mapsto \mathrm{E}\left[(V - \mathrm{E}[V])(g - \mathrm{E}[g]) \mid Z = z\right] , \tag{34}$$

where V and g respectively denote the random variables $V = A \circ \overline{S}(Z)$ and $A(\infty)$.

Thanks to (33) and (34), the values $\hat{\varphi}(z)$ can be pre-computed and stored in a look-up table when φ is assumed to be known, which will be the case in our attack simulations reported in the next section.

About the Efficiency of HO-SCA. Since they are based on a maximum likelihood test, HO-TA need less observations N than HO-CPA do to recover k from a dth-order flaw. However, (28) shows that their processing involves $N \times d \times 2^{4n}$ evaluations of the function ϕ_σ to test each key hypothesis $\hat{k}$, whereas HO-CPA require less than $(10 + d) \times N$ elementary operations (products, additions and look-up table accesses) for this test. In the next section, we have followed the classical approach which consists in measuring the efficiency of an attack in terms of the number of leakage measurements N required to observe a success rate of 90% in retrieving the key. We did this choice to ease the comparison of our results with those published in the literature. However, we think that the computation cost should not be neglected when quantifying the efficiency of an attack. This factor explains in particular why we did not succeed in performing all our HO-TA simulations for $d \geq 3$.

Improved Impossible Differential Cryptanalysis of 7-Round AES-128

Hamid Mala[1], Mohammad Dakhilalian[1],
Vincent Rijmen[2,3,*], and Mahmoud Modarres-Hashemi[1]

[1] Cryptography & System Security Research Laboratory, Department of Electrical and Computer Engineering, Isfahan University of Technology, 8415683111 Isfahan, Iran
{hamid_mala@ec,mdalian@cc,modarres@cc}.iut.ac.ir
[2] COSIC, Dept. of EE, KULeuven and IBBT, Kasteelpark Arenberg 10, 3001 Heverlee, Belgium
[3] IAIK, Graz University of Technology, Inffeldgasse 16a, 8010 Graz, Austria
Vincent.Rijmen@iaik.tugraz.at

Abstract. Using a new 4-round impossible differential in AES that allows us to exploit the redundancy in the key schedule of AES-128 in a way more effective than previous work, we present a new impossible differential attack on 7 rounds of this block cipher. By this attack, 7-round AES-128 is breakable with a data complexity of about 2^{106} chosen plaintexts and a time complexity equivalent to about 2^{110} encryptions. This result is better than any previously known attack on AES-128 in the single-key scenario.

Keywords: AES, block cipher, cryptanalysis, impossible differential.

1 Introduction

The Advanced Encryption Standard (AES)[7] is a 128-bit block cipher with variable key lengths of 128, 192, and 256 bits, which are denoted as AES-128, AES-192 and AES-256, respectively. Since its selection as the standard by NIST in 2001, AES has drawn a great amount of attention from worldwide cryptology researchers. In this paper we reevaluate the security of AES-128 against impossible differential attacks.

Impossible differential cryptanalysis, an extension of the differential attack [4], was first introduced by Knudsen [11] and Biham [2] to analyze DEAL and Skipjack, respectively. Impossible differential attacks use differentials that hold

* This author's work was supported in part by the IAP Programme P6/26 BCRYPT of the Belgian State (Belgian Science Policy), and in part by the European Commission through the ICT programme under contract ICT-2007-216676 ECRYPT II. The information in this document reflects only the author's views, is provided as is and no guarantee or warranty is given that the information is fit for any particular purpose. The user thereof uses the information at its sole risk and liability.

G. Gong and K.C. Gupta (Eds.): INDOCRYPT 2010, LNCS 6498, pp. 282–291, 2010.

with probability zero to derive the right key by discarding the wrong keys which lead to the impossible differential.

Previous attacks, which are applicable up to seven rounds of AES-128 are as follows. The first attack is a Square attack which requires $2^{128} - 2^{119}$ chosen plaintexts and 2^{120} encryptions [9]. The second attack is a collision attack that requires 2^{32} chosen plaintexts and about 2^{128} encryptions [10]. The third attack is an impossible differential attack presented independently in [1] and [14] that requires $2^{115.5}$ chosen plaintexts and has a time complexity equivalent to 2^{119} 7-round encryptions. The most recently published impossible differential cryptanalysis on AES-128 was proposed in [12]. This attack is based on the attacks proposed in [1] and [14] and requires $2^{112.2}$ chosen plaintexts, and has a running time of $2^{117.2}$ memory accesses. Finally an improved meet in the middle attack on 7-rounds of AES-128 has been proposed in [8] which requires about 2^{80} chosen plaintexts and 2^{113} encryptions in the online stage, but at the expense of 2^{123} encryptions in the precomputation stage and 2^{122} blocks of memory. Since our proposed attack is in the single key scenario, here we do not elaborate on the related-key attacks on AES, but the reader is referred to [5] for the latest result which is a related-key boomerang attack on AES-128.

In this paper, we introduce a new 4-round impossible differential of AES, and exploit it to present a new attack on 7-round AES-128. In addition to the previous techniques including precomputation technique, early abort technique, and using structures of plaintexts and hash tables, our proposed attack uses some additional precomputation tables and key schedule considerations. The proposed 4-round impossible differential allows us to use the key schedule, in a way more lucrative than [12], to reduce the time complexity. This attack requires $2^{106.2}$ chosen plaintexts and has a time complexity equivalent to $2^{110.2}$ 7-round encryptions. We summarize our result along with previously known results on AES-128 in Table 1. In this table, data complexity is the number of the required chosen plaintexts (CP), and time complexity is measured in encryption (E) units or memory accesses (MA).

Table 1. Summary of previous single key attacks and our new attack on AES-128

Rounds	Data (CP)	Time	Memory (Byte)	Attack type	Source
5	$2^{29.5}$	2^{31} E	2^{42}	ID	[3]
6	$2^{91.5}$	2^{122} E	2^{93}	ID	[6]
7	$2^{128} - 2^{119}$	2^{120} E	2^{68}	S	[9]
7	2^{32}	2^{128} E	2^{100}	C	[10]
7	$2^{115.5}$	2^{119} E	2^{109}	ID	[1]
7	$2^{115.5}$	2^{119} E	2^{45}	ID	[14]
7	$2^{112.2}$	$2^{112.2}$E + $2^{117.2}$ MA $\approx 2^{112.3}$ E	$2^{93.2}$	ID	[12]
7	2^{80}	$2^{113} + 2^{123} \approx 2^{123}$ E	2^{126}	MitM	[8]
7	$2^{106.2}$	$2^{107.1}$E + $2^{117.2}$ MA $\approx 2^{110.2}$ E	$2^{94.2}$	ID	This work

ID: Impossible Differential, MitM: Meet in the Middle, S: Square, C: Collision.

The rest of this paper is organized as follows. Section 2 provides a brief description of AES. A 4-round impossible differential of AES is introduced in Section 3. We propose our improved impossible differential attack on 7-round AES-128 in Section 4. Finally, we conclude the paper in Section 5.

2 A Brief Description of AES

The AES [7] is an iterated block cipher with the SPN (substitution-permutation network) structure that supports key sizes of 128, 192, and 256 bits. The number of rounds for these three variants are 10, 12, and 14, respectively. To describe AES, a 128-bit plaintext is represented by a 4×4 matrix of bytes, where each byte represents a value in GF(2^8). An AES round applies the following four transformations to the state matrix:

- SubBytes (SB): a byte-wise nonlinear transformation that applies the same invertible S-box on each of the 16 bytes of the state.
- ShiftRows (SR): a linear transformation that rotates the i-th row of the state matrix by i bytes to the left (i=0, 1, 2, 3).
- MixColumns (MC): another linear transformation which is a multiplication of each column by a constant 4×4 matrix over the finite field GF(2^8).
- AddRoundKey (AK): a bit-wise XOR operation between the state matrix and the subkey of the current round.

In the first round, an additional AddRoundKey operation is applied (whitening) and in the last round, the MixColumns operation is omitted. Thus in our attack on 7 rounds of AES-128, we will take these two points into consideration.

The key schedule of AES-128 takes the 128-bit cipher key and transforms it into 11 subkeys of 128 bits each. The subkey words are represented by $W[0, ..., 43]$, where each 4-byte $W[i]$ forms a column of a round key. The first 4 words of $W[\cdot]$ are loaded with the cipher key, and the remaining key words are generated by the following algorithm:

- For $i = 4 : 43$, do
  - if $i = 0 \bmod 4$ then $W[i] = W[i-4] \oplus SB(W[i-1] \lll 8) \oplus RCON[i/4]$,
  - otherwise $W[i] = W[i-1] \oplus W[i-4]$,

where $RCON[\cdot]$ is an array of predetermined constants, and $\lll n$ is the rotation of a word by n bits to the left. In this paper, we will use the following notations: $x_i^I, x_i^{SB}, x_i^{SR}, x_i^{MC}$, and x_i^{AK} denote the input of round i and the intermediate values after the application of SB, SR, MC and AK operations of round i, respectively. The subkey of round i is denoted by k_i and the whitening key is denoted by k_0.

The byte located in the m-th row and n-th column of some intermediate state x_i (or a key k_i) is denoted by $x_{i,(4n+m)}$. For bytes located in positions $j_1, j_2, ...$ we use the notation $x_{i,(j_1,j_2,...)}$. We denote the z-th column of x_i by $x_{i,col(z)}$. We finally denote the bytes in x_i corresponding to the places after applying the ShiftRows operation on column z of x_i by $x_{i,SR(col(z))}$, e.g., $x_{i,SR(col(0))}$ is composed of bytes (0, 7, 10, 13). The concatenation of two bit strings a and b is denoted by $a|b$.

3 A 4-Round Impossible Differential of AES

In [3] a four-round impossible differential property of AES was presented which has been used in all the subsequent impossible differential attacks on AES. This property states that given a pair of $(x_i^I, x_i^{'I})$ with the difference $\Delta x_i^I = x_i^I \oplus x_i^{'I}$ which has only one non-zero byte, then the corresponding $(x_{i+3}^{SR}, x_{i+3}^{'SR})$ pair cannot have the difference $\Delta x_{i+3}^{SR} = x_{i+3}^{SR} \oplus x_{i+3}^{'SR}$ in which at least one of the four sets of bytes $SR(col(j)), j = 0, 1, 2, 3$ is equal to zero. It is known that this impossible differential is composed of two deterministic differentials which contradict in the middle.

Among the previous impossible differential attacks on 7-round AES-128 [1,12,14], only [12] exploits the redundancy in the key schedule of AES-128. In [12], first four attacks with four overlapping target subkey sets are applied separately, then the attacker checks the consistency of these four subkey sets to get a smaller number of suggestions for their union. We observed that by a modification in the impossible differential, we can use the key schedule with only one attack. The proposed impossible differential in one of its possible cases is illustrated in Fig. 1. One may consider this impossible differential as the previous work's impossible differential in which the two deterministic differentials have been swapped.

This impossible differential, in its general form, states that given a pair of $(x_i^{MC}, x_i^{'MC})$ which have zero differences in at least one of the four sets of bytes $SR^{-1}(col(j)), j = 0, 1, 2, 3$, then the corresponding $(x_{i+4}^{SR}, x_{i+4}^{'SR})$ pair cannot have zero differences in all bytes except one.

4 Impossible Differential Attack on 7 Rounds of AES-128

Based on the impossible differential property described in Section 3, we mount our new attack on 7-round AES-128. The attack is illustrated in Fig. 2. To reduce the size of the target key space, we change the order of the two linear transformations MixColumns and AddRoundKey in the 6th round. We use a collection of previously known techniques, including traditional precomputation tables [3], the early abort technique [13], as well as several different precomputation tables and key schedule considerations to reduce the complexity of the attack.

4.1 Precomputation Stage

In this section we prepare 4 (types of) precomputation tables $T_1, T_2, T_{3,i}, T_{4,j}$, for extracting the proper subkeys of $k_{7,SR(col(0))}$, $k_{0,SR^{-1}(col(2))}$, $k_{1,(10)}$, and $k_{1,(8)}$, respectively.

Table T_1: For all of the $2^{32} \times \binom{4}{3} \times (2^8 - 1) \approx 2^{42}$ possible pairs of $(x_{6,col(0)}^{AK}, x_{6,col(0)}^{'AK})$ which have the zero difference in exactly 3 out of the 4 bytes, compute the values of $(x_{7,(0,7,10,13)}^{SR}, x_{7,(0,7,10,13)}^{'SR})$. Store the obtained pairs in a hash table T_1 indexed by their difference. T_1 has 2^{32} rows and on average about $\frac{2^{42}}{2^{32}} = 2^{10}$ pairs lie in each of these rows.

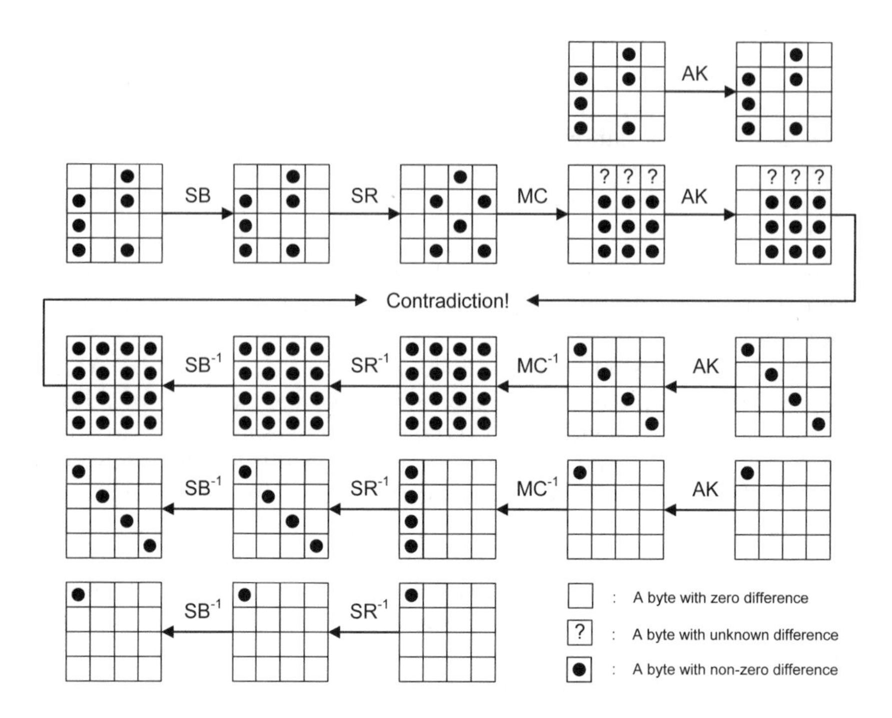

Fig. 1. One sample of the 4-round impossible differential of AES

Table T_2: For all of the about 2^{48} possible pairs of $(x^{MC}_{1,col(2)}, x'^{MC}_{1,col(2)})$ which have zero differences in bytes 9 and 11, and non-zero differences in bytes 8 and 10, compute the values of $(x^{I}_{1,(2,7,8,13)}, x'^{I}_{1,(2,7,8,13)})$. Store the obtained pairs in a hash table T_2 indexed by their difference. T_2 has 2^{32} rows and on average about $\frac{2^{48}}{2^{32}} = 2^{16}$ pairs lie in each row.

Tables $T_{3,i}, i = 0, 1, 2, 3$: For all of the about 2^{32} possible pairs of $(x^{AK}_{1,(0,10)}, x'^{AK}_{1,(0,10)})$ which have non-zero differences in these 2 bytes, compute the values of $\Delta x^{MC}_{2,col(0)}$. Then for $i = 0, 1, 2, 3$ choose the pairs $(x^{AK}_{1,(0,10)}, x'^{AK}_{1,(0,10)})$ whose corresponding difference $\Delta x^{MC}_{2,col(0)}$ is zero in byte i (we obtain about $2^{32} \times 2^{-8} = 2^{24}$ such pairs). Store the qualified pairs $(x^{AK}_{1,(10)}, x'^{AK}_{1,(10)})$ in a hash table $T_{3,i}$ indexed by the 24-bit parameter $x^{AK}_{1,(0)}|x'^{AK}_{1,(0)}|\Delta x^{AK}_{1,(10)}$. Each table $T_{3,i}$ has 2^{24} rows and on average about $\frac{2^{24}}{2^{24}} = 1$ pair lies in each row.

Tables $T_{4,j}, j = 8, 9, 10, 11$: For all of the about 2^{32} possible pairs of $(x^{AK}_{1,(2,8)}, x'^{AK}_{1,(2,8)})$ which have non-zero differences in these 2 bytes, compute the value of $\Delta x^{MC}_{2,col(2)}$. Then for $j = 8, 9, 10, 11$ choose the pairs $(x^{AK}_{1,(2,8)}, x'^{AK}_{1,(2,8)})$ whose corresponding difference $\Delta x^{MC}_{2,col(2)}$ is zero in byte j (we obtain about $2^{32} \times 2^{-8} = 2^{24}$ such pairs). Store the qualified pairs $(x^{AK}_{1,(8)}, x'^{AK}_{1,(8)})$ in a hash table $T_{4,j}$

indexed by the 24-bit parameter $x^{AK}_{1,(2)}|x'^{AK}_{1,(2)}|\Delta x^{AK}_{1,(8)}$. Each table $T_{4,j}$ has 2^{24} rows and on average about $\frac{2^{24}}{2^{24}} = 1$ pair lies in each row.

Since in the attack procedure each of the 4 tables $T_{4,j}, j = 8, 9, 10, 11$ can be used together with only one of the tables $T_{3,i}, i = 0, 1, 2, 3$, thus we say $T_{4,conj(i)}$ is the conjugate of $T_{3,i}$, where for $i = 0, 1, 2, 3$, the $conj(i)$ are 10, 11, 8, 9, respectively.

4.2 The Attack Procedure

The following attack algorithm gets 2^n structures of plaintexts and returns ϵ 14-byte joint subkeys $k_{0,(0,2,5,7,8,10,13,15)}|k_{1,(0,2,8,10)}|k_{7,(0,7,10,13)}$. Note that based on the key schedule of AES-128, the byte $k_{1,(0)}$ is determined by the two bytes $k_{0,(0)}$ and $k_{0,(13)}$, also $k_{1,(2)}$ is determined by the two bytes $k_{0,(2)}$ and $k_{0,(15)}$.

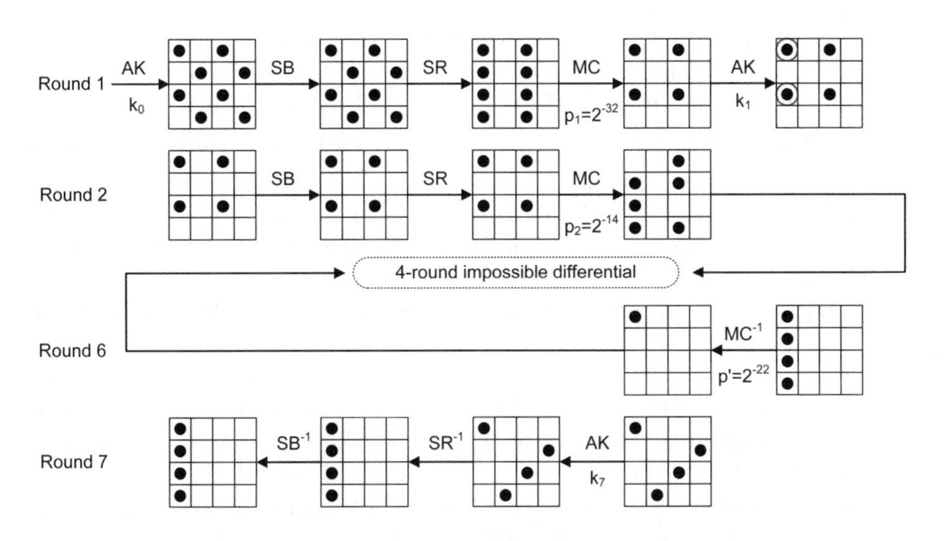

Fig. 2. New impossible differential attack on 7 rounds of AES-128

The attack algorithm is as follows (Algorithm 1):

1. Take 2^n structures of plaintexts such that each structure contains 2^{64} plaintexts that are fixed in the 8 bytes $(1, 3, 4, 6, 9, 11, 12, 14)$ and take all the possible values in other 8 bytes. It is obvious that $\binom{2^{64}}{2} \approx 2^{127}$ plaintext pairs can be obtained from each structure. Totally, we can collect about 2^{n+64} plaintexts and 2^{n+127} plaintext pairs (P, P') with the desired difference $\Delta P = P \oplus P'$ shown in Fig. 2.
2. To specify the ciphertext pairs that have non-zero differences only in bytes $SR(col(0)) = (0, 7, 10, 13)$, for each structure, first, obtain its 2^{64} ciphertexts and store them in a hash table H_1 indexed by their values in the 8 bytes $(1, 2, 3, 4, 5, 6, 8, 9)$. Then rearrange the ciphertexts of each row of H_1 having more than one text in another table H_2 indexed by their values in the 4

bytes $(11, 12, 14, 15)$. By following this 2-staged filtration of pairs, each two texts which lie in the same row of H_2 form a proper pair (i.e., a pair with the required plaintext and ciphertext differences). Store the proper pairs, then erase the hash table H_2 and repeat the process for the next row of H_1. After examining all rows of H_1, this table is refreshed for the next structure. As this procedure is considered a 96-bit filtration on all the 2^{n+127} plaintext pairs, at the end of this step the expected number of the remaining proper pairs from all the structures is $2^{n+127} \times 2^{-96} = 2^{n+31}$ pairs. Thus, to implement this step we need about 2^{64} blocks of memory for the hash table H_1, and $4 \times 2^{n+31}$ blocks of memory for storing the proper ciphertext pairs and their corresponding plaintext pairs. The size of H_2 is comparatively negligible.

3. In this step, we use a well-known property of invertible S-boxes: Given an input and an output difference of the SubBytes operation, there is on average one pair of actual values that satisfies these differences. Using this property, in the following way, will reduce the memory complexity of the attack. Since for any proper plaintext pair (P, P'), the difference $\Delta x_1^I = \Delta P$ is known, the knowledge of the difference $\Delta x_{1,(0,5,10,15)}^{SB}$ can be used to find the actual value of $(x_{1,(0,5,10,15)}^I, x_{1,(0,5,10,15)}^{'I})$, and consequently the value of $k_{0,(0,5,10,15)}$. There are only $(2^8 - 1)^2 \approx 2^{16}$ possible values of $\Delta x_{1,col(0)}^{MC}$ in which only the two bytes 0 and 2 are non-zero. Consequently, there are the same number of $\Delta x_{1,(0,5,10,15)}^{SB}$ for further analysis. Based on these considerations, perform the following substeps:
   (a) Initialize 2^{32} empty lists, each corresponding to a different value of $k_{0,(0,5,10,15)}$.
   (b) For each of the 2^{n+31} remaining proper pairs, and for each of the about 2^{16} possible differences in $\Delta x_{1,(0,5,10,15)}^{SB}$, compute the keys which lead this specific plaintext pair to this specific difference. Add this plaintext pair to the list corresponding to the obtained key value.

   For each of the 2^{n+31} proper pairs, about 2^{16} values of $\Delta x_{1,(0,5,10,15)}^{SB}$ are examined. Then, these $2^{n+31} \times 2^{16} = 2^{n+47}$ options are distributed in 2^{32} lists. Thus for a given subkey guess $k_{0,(0,5,10,15)}$, the expected number of stored pairs in the corresponding list is 2^{n+15} pairs.

   For each of the 2^{32} possible values of $k_{0,(0,5,10,15)}$, access the corresponding list and for each of the 2^{n+15} plaintext pairs (P, P') in that list, perform the following steps:
4. Access the row with index $\Delta P_{(2,7,8,13)}$ in table T_2. For each pair (y_1, z_1) in that row, select the value $P_{(2,7,8,13)} \oplus y_1$ as a candidate for $k_{0,(2,7,8,13)}$. Based on the structure of table T_2, the expected number of these candidates for $k_{0,(2,7,8,13)}$ will be about 2^{16}.
5. For each of the 2^{16} candidates for $k_{0,(2,7,8,13)}$ perform the following substeps.
   (a) Simply compute the two bytes $k_{1,(0,2)}$ as below:

$$k_{1,(0)} = k_{0,(0)} \oplus SB(k_{0,(13)}) \oplus c, \; k_{1,(2)} = k_{0,(2)} \oplus SB(k_{0,(15)}) \oplus c',$$

   where c and c' are constants known from the key schedule.

(b) Partially encrypt the plaintext pair to get $(x^{AK}_{1,(0,2)}, x'^{AK}_{1,(0,2)})$ and $(x^{MC}_{1,(8,10)}, x'^{MC}_{1,(8,10)})$.

(c) For $i = 0, 1, 2, 3$:

i. Access the row with index $x^{AK}_{1,(0)}|x'^{AK}_{1,(0)}|\Delta x^{AK}_{1,(10)}$ in table $T_{3,i}$. For each pair (y_2, z_2) in that row, select the value $x^{MC}_{1,(10)} \oplus y_2$ as a candidate for $k_{1,(10)}$. Based on the structure of table $T_{3,i}$, the expected number of these candidates will be about 1.

ii. Access the row with index $x^{AK}_{1,(2)}|x'^{AK}_{1,(2)}|\Delta x^{AK}_{1,(8)}$ in table $T_{4,conj(i)}$. For each pair (y_3, z_3) in that row, select the value $x^{MC}_{1,(8)} \oplus y_3$ as a candidate for $k_{1,(8)}$. Based on the structure of table $T_{4,conj(i)}$, the expected number of these candidates will be about 1.

6. Access the row with index $\Delta C_{(0,7,10,13)}$ in table T_1. For each pair (y_4, z_4) in that row, select the value $C_{(0,7,10,13)} \oplus y_4$. Based on the structure of table T_1, we expect to obtain about 2^{10} candidates $k_{7,(0,7,10,13)}$ from this table.
7. In this step, for each of the 2^{n+15} corresponding proper pairs, we know $2^{16} \times 4 \times 2^{10} = 2^{28}$ joint values of $k_{0,(2,7,8,13)}|k_{1,(0,2,8,10)}|k_{7,(0,7,10,13)}$ that result in the impossible differential. Remove these values from the list of all the 2^{80} possible values for these joint subkeys (Note that $k_{0,(0,5,10,15)}$ is previously guessed). After the trial of all of the pairs, if the list is not empty, announce the values in the list along with the guess of $k_{0,(0,5,10,15)}$ as the candidates for the correct 112-bit target subkey.

4.3 Complexity of the Attack

In step 7, for each of the 2^{32} values of $k_{0,(0,5,10,15)}$, and for each of the 2^{n+15} corresponding proper pairs, the attacker removes on average 2^{28} values out of the 2^{80} possible values from the key space. Thus the probability that a wrong subkey survives the elimination with one proper pair is $1 - \frac{2^{28}}{2^{80}} = 1 - 2^{-52}$. So, about $\epsilon = 2^{112}(1 - 2^{-52})^{2^{n+15}}$ values of the 112-bit target subkey remain as the output of the attack algorithm. If we take ϵ equal to 1, then n will be about 43.3. With this value for n, the attack requires $2^{n+64} = 2^{107.3}$ chosen plaintexts. The time complexities of different steps of the attack are calculated as functions of n in the second column of Table 2.

Based on the second column of Table 2, the time complexity of this attack is dominated by steps 2 and 7, which is equal to 2^{n+64} encryptions plus 2^{n+75} memory accesses.

Suppose Algorithm 1 outputs ϵ candidates for the 16-byte target subkeys $k_{0,(0,2,5,7,8,10,13,15)}|\ k_{1,(0,2,8,10)}|\ k_{7,(0,7,10,13)}$. For each of these values, two other bytes of k_0 are determined as below:

$$k_{0,(4)} = k_{1,(8)} \oplus k_{1,(0)} \oplus k_{0,(8)}$$

$$k_{0,(6)} = k_{1,(2)} \oplus k_{1,(10)} \oplus k_{0,(10)}$$

Thus only 6 bytes of k_0 are unknown. To find the whole key, the attacker must perform about 2^{48} trial encryptions for each of the ϵ output values of Algorithm

Table 2. Time complexity of the different steps of Algorithm 1

Step	Time Complexity	for $n = 42.2$
2	2^{n+64} E	$2^{106.2}$
3	$2^{n+31} \times 2^{16} = 2^{n+47}$ MA	$2^{89.2}$
4	$2^{32} \times 2^{n+15} \times 2^{16} = 2^{n+63}$ MA	$2^{89.2}$
5(a)	$2^{32} \times 2^{n+15} \times 2^{16} = 2^{n+63}$ MA	$2^{105.2}$
5(b)	$2^{32} \times 2^{n+15} \times 2^{16} \times \frac{4}{16} \times \frac{1}{7} = 2^{n+58.2}$ E	$2^{100.4}$
5(c)i	$2^{32} \times 2^{n+15} \times 2^{16} \times 4 = 2^{n+65}$ MA	$2^{107.2}$
5(c)ii	$2^{32} \times 2^{n+15} \times 2^{16} \times 4 = 2^{n+65}$ MA	$2^{107.2}$
6	$2^{32} \times 2^{n+15} \times 2^{10} = 2^{n+57}$ MA	$2^{99.2}$
7	$2^{32} \times 2^{n+15} \times 2^{16} \times 4 \times 2^{10} = 2^{n+75}$ MA	$2^{117.2}$

1. Thus the total time complexity to obtain the whole key is $2^{n+64} + \epsilon \times 2^{48}$ encryptions plus 2^{n+75} memory accesses. By choosing $\epsilon = 2^{58}$, equation $\epsilon = 2^{112}(1 - 2^{-52})^{2^{n+15}}$ yields $n = 42.2$. Thus the data complexity of the attack is $2^{n+64} = 2^{106.2}$, and for this value of n the time complexity of different steps of the attack is represented in the third column of Table 2. The dominant part of the time complexity is about $2^{n+64} + 2^{58} \times 2^{48} \approx 2^{107.1}$ encryptions plus $2^{n+75} = 2^{117.2}$ memory accesses. In a 32-bit implementation, each round of AES may be implemented by 20 memory accesses (16 table lookups for SB+MC+SR, and 4 table lookups for AK) [7], so one could declare the total complexity as $2^{107.1} + \frac{1}{7} \times \frac{1}{20} \times 2^{117.2} \approx 2^{110.2}$ encryptions.

The memory complexity of the attack is dominated by the memory required to produce 2^{32} lists of step 3, which is equal to $2^{n+31} \times 2^{16} \times 2 = 2^{90.2}$ blocks of memory, and the memory required as the list of removed candidates of $k_{0,(2,7,8,13)} \mid k_{1,(0,2,8,10)} \mid k_{7,(0,7,10,13)}$. For each guess of $k_{0,(0,5,10,15)}$, there exist only 2^{80} such candidates, and also for each guess of $k_{0,(0,5,10,15)}$ this list is refreshed, thus, we only need about 2^{80} 96-bit blocks of memory for step 7. The memory required for the pre-computation tables is negligible. Thus the memory complexity is about $2^{90.2}$ 128-bit blocks of memory.

5 Conclusion

In this paper, we proposed a new impossible differential attack on 7-round AES-128. Our proposed attack uses a different 4-round impossible differential that allows the attacker to use the redundancy in the key schedule of this cipher in a way more commodious than the previous work [12]. The attack requires about $2^{106.2}$ plaintexts, and has a time complexity equivalent to about $2^{110.2}$ encryptions which is better than previously published single key attacks on 7 rounds of AES-128.

Acknowledgment. The authors would like to thank Mohsen Shakiba and the anonymous reviewers of Indocrypt 2010 for their invaluable comments.

References

1. Bahrak, B., Aref, M.R.: Impossible differential attack on seven-round AES-128. IET Information Security 2, 28–32 (2008)
2. Biham, E., Biryukov, A., Shamir, A.: Cryptanalysis of Skipjack Reduced to 31 Rounds Using Impossible Differentials. In: Stern, J. (ed.) EUROCRYPT 1999. LNCS, vol. 1592, pp. 12–23. Springer, Heidelberg (1999)
3. Biham, E., Keller, N.: Cryptanalysis of reduced variants of Rijndael. In: The Third AES Candidate Conference (2000)
4. Biham, E., Shamir, A.: Differential Cryptanalysis of the Data Encryption Standard. Springer, Heidelberg (1993)
5. Biryukov, A., Nikolic, I.: Automatic Search for Related-Key Differential Characteristics in Byte-Oriented Block Ciphers: Application to AES, Camellia, Khazad and Others. In: Gilbert, H. (ed.) Advances in Cryptology – EUROCRYPT 2010. LNCS, vol. 6110, pp. 322–344. Springer, Heidelberg (2010)
6. Cheon, J.H., Kim, M., Kim, K., Lee, J., Kang, S.: Improved impossible differential cryptanalysis of Rijndael and Crypton. In: Kim, K. (ed.) ICISC 2001. LNCS, vol. 2288, pp. 39–49. Springer, Heidelberg (2002)
7. Daemen, J., Rijmen, V.: The design of Rijndael: AES– the Advanced Encryption Standard. Springer, Heidelberg (2002)
8. Demirci, H., Taşkin, İ., Çoban, M., Baysal, A.: Improved Meet-in-the-Middle Attacks on AES. In: Roy, B., Sendrier, N. (eds.) INDOCRYPT 2009. LNCS, vol. 5922, pp. 144–156. Springer, Heidelberg (2009)
9. Ferguson, N., Kelsey, J., Lucks, S., Schneier, B., Stay, M., Wagner, D., Whiting, D.: Improved Cryptanalysis of Rijndael. In: Schneier, B. (ed.) FSE 2000. LNCS, vol. 1978, pp. 213–230. Springer, Heidelberg (2001)
10. Gilbert, H., Minier, M.: A collision attack on 7 rounds of Rijndael. In: The Third AES Candidate Conference, pp. 230–241 (2000)
11. Knudsen, L.R.: DEAL – a 128-bit Block Cipher. Technical report, Department of Informatics, University of Bergen, Norway (1998)
12. Lu, J., Dunkelman, O., Keller, N., Kim, J.: New Impossible Differential Attacks on AES. In: Chowdhury, D.R., Rijmen, V., Das, A. (eds.) INDOCRYPT 2008. LNCS, vol. 5365, pp. 279–293. Springer, Heidelberg (2008)
13. Lu, J., Kim, J., Keller, N., Dunkelman, O.: Improving the Efficiency of Impossible Differential Cryptanalysis of Reduced Camellia and MISTY1. In: Malkin, T.G. (ed.) CT-RSA 2008. LNCS, vol. 4964, pp. 370–386. Springer, Heidelberg (2008)
14. Zhang, W., Wu, W., Feng, D.: New Results on Impossible Differential Cryptanalysis of Reduced AES. In: Nam, K.-H., Rhee, G. (eds.) ICISC 2007. LNCS, vol. 4817, pp. 239–250. Springer, Heidelberg (2007)

Cryptanalysis of a Perturbated White-Box AES Implementation

Yoni De Mulder[1], Brecht Wyseur[2], and Bart Preneel[1]

[1] Katholieke Universiteit Leuven
Dept. Elect. Eng.-ESAT/SCD-COSIC and IBBT,
Kasteelpark Arenberg 10, 3001 Heverlee, Belgium
{yoni.demulder,bart.preneel}@esat.kuleuven.be
[2] Nagravision S.A.
Route de Genève 22–24, 1033 Cheseaux-sur-Lausanne, Switzerland
brecht.wyseur@nagra.com

Abstract. In response to various cryptanalysis results on white-box cryptography, Bringer *et al.* presented a novel white-box strategy. They propose to extend the round computations of a block cipher with a set of random equations and perturbations, and complicate the analysis by implementing each such round as one system that is obfuscated with annihilating linear input and output encodings. The improved version presented by Bringer *et al.* implements the AEw/oS, which is an AES version with key-dependent S-boxes (the S-boxes are in fact the secret key). In this paper we present an algebraic analysis to recover equivalent keys from the implementation. We show how the perturbations and system of random equations can be distinguished from the implementation, and how the linear input and output encodings can be eliminated. The result is that we have decomposed the white-box implementation into a much more simple, functionally equivalent implementation and retrieved a set of keys that are equivalent to the original key. Our cryptanalysis has a worst time complexity of 2^{17} and a negligible space complexity.

Keywords: White-Box Cryptography, AES, Cryptanalysis, Structural Cryptanalysis.

1 Introduction

In the past decade, we have witnessed a trend towards the use of software applications with strong security requirements. Consider for example online banking and digital multimedia players. Building blocks to enable their security include cryptographic primitives such as the DES or the AES [13]. However, these building blocks are designed to be secure only when they are executed on a trustworthy system, which is typically no longer a valid assumption. White-box cryptography aims to address this issue – it aims to implement a given cryptographic cipher such that it remains 'secure' even when the adversary is assumed to have full access to the implementation and its execution environment (the white-box

G. Gong and K.C. Gupta (Eds.): INDOCRYPT 2010, LNCS 6498, pp. 292–310, 2010.

attack context). We refer to a white-box implementation as an implementation of a cipher to which these techniques are applied.

At SAC 2002, Chow *et al.* introduced the concept of white-box cryptography, applied to the AES [5], and to the DES in [6]. The main idea is to generate a network of re-randomized lookup tables that is functionally equivalent to a key-instantiated primitive. However, subsequent papers have shown that this strategy is prone to differential cryptanalysis [10,11,9,16] and algebraic cryptanalysis [2,12,15]. In [1], Billet and Gilbert proposed a traceable block cipher, by implementing the same instance of a cipher in many different ways. The security is based on the Isomorphisms of Polynomials (IP) problem [14]. Unfortunately, analysis of this IP problem [8] has defeated this approach. Based on the idea to introduce perturbations to reinforce the IP-based cryptosystems [7], Bringer *et al.* [3] reinforced the traceable block cipher, and presented a *perturbated* white-box AES implementation [4]. The main idea of the perturbated AES white-box implementations is to extend the AES rounds with a random system of equations and perturbation functions. The perturbations introduced at the first round are canceled out at the final round with a high probability; to guarantee correct execution, several such instances need to be implemented such that a majority vote can reveal the correct result. The random system of equations is discarded in the final round. As a challenging example, they apply their technique to the AEw/oS – a variant of the AES with non-standard, key-dependent S-boxes. These S-boxes are in fact the secret key. To the best of our knowledge, no attack against this white-box AEw/oS implementation has been proposed so far.

Our contribution. We developed a cryptanalysis of the perturbated white-box AEw/oS implementation; which extends naturally to perturbated white-box AES implementations. In a white-box attack context, the adversary will have access to each of the (obfuscated) rounds – these consist of the composition of random linear input and output encodings, the AES round operation with key-dependent S-boxes, and encompass the random system of equations and perturbated functions. The presence of the (unknown) linear encodings and the extra equations makes it hard to recover the secret information – the S-boxes – from the implementation.

In this paper, we describe the structural analysis of the white-box AEw/oS round operations. We show how to derive a set of equivalent S-boxes and linear encodings that describe a functionally equivalent implementation (due to the construction of the implementation, there are many candidate keys). As a result, we obtain a significantly simpler version of the white-box AEw/oS implementation, which is also invertible, thus defeating the security objective of the original implementation. Our cryptanalysis has a worst time complexity of 2^{17} and a negligible space complexity.

Organization of the paper. To support the cryptanalysis description, in Sect. 2, we present an overview of the perturbated white-box AEw/oS implementation as presented by Bringer *et al.* [4]. The cryptanalysis comprises three main steps, which are presented in Sect. 3.

2 The White-Box AEw/oS Implementation

In this section, we describe the perturbated white-box AEw/oS implementation, which is the Advanced Encryption Standard (AES) [13] with non-standard S-boxes – the choice of S-boxes is in fact the secret key, and there are 160 of them comprised in the entire implementation. Fig. 1 depicts the implementation, where X is the plaintext input, Z_r are the intermediate states (outputs of the perturbated round functions R'_r), where Z_{10} is the final output.

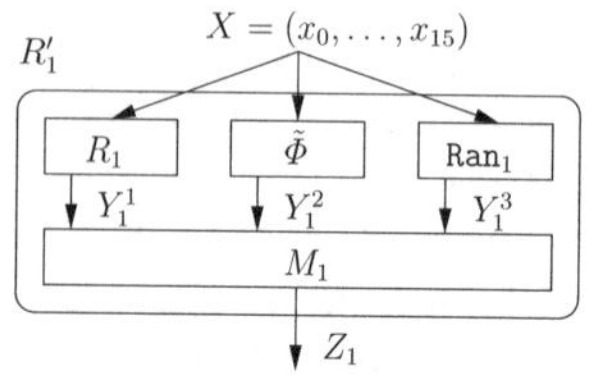

1.1: Perturbated first round R'_1.

(a) AddRoundKey K_0
(b) SubBytes $\{S_{1,0}, \ldots, S_{1,15}\}$
(c) ShiftRows
(d) MixColumns

1.2: Original first round R_1.

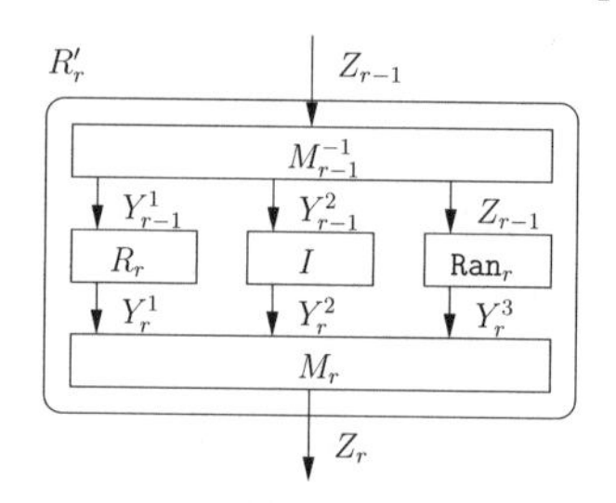

1.3: Perturbated intermediate rounds R'_r.

for r from 2 to 9:
(a) AddRoundKey K_{r-1}
(b) SubBytes $\{S_{r,0}, \ldots, S_{r,15}\}$
(c) ShiftRows
(d) MixColumns

1.4: Original intermediate rounds R_r.

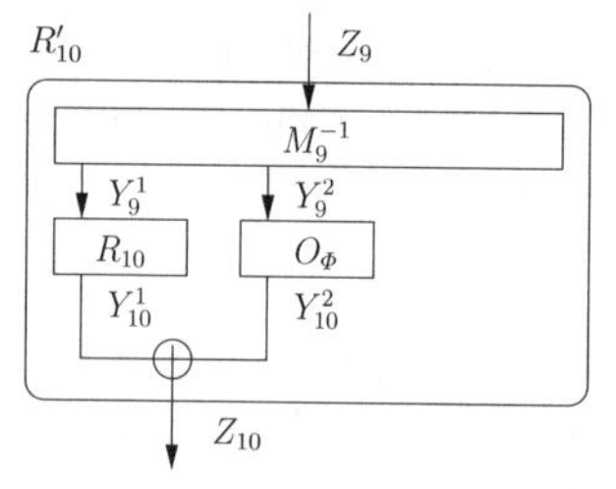

1.5: Perturbated final round R'_{10}.

(a) AddRoundKey K_9
(b) SubBytes $\{S_{10,0}, \ldots, S_{10,15}\}$
(c) ShiftRows
(d) AddRoundKey K_{10}

1.6: Original final round R_{10}.

Fig. 1. Description of the Perturbated White-box AEw/oS Implementation

The perturbated round functions. Each round R'_r of the perturbated AEw/oS is expressed as a system of 43 multivariate polynomials over GF (2^8); the final round as a system of 16 multivariate polynomials. Each system is defined over 43 variables (bytes), except for the initial round, which is defined over the 16 bytes of the plaintext. These extra variables and equations are due to the extension of the AES rounds with a perturbation system of 4 polynomials and a system of 23 random polynomials $\mathtt{Ran}_r$. The latter is introduced to dissimulate the perturbation and mask all internal operations.

The perturbation initialization system $\tilde{\Phi}$ is included in the first round R'_1 and comprises of 4 polynomials that "often" take the predefined value $(\varphi_1, \varphi_2, \varphi_3, \varphi_4)$ and is constructed as $\tilde{\Phi}(X) = (\tilde{0}(X) + \varphi_1, \tilde{0}(X) + \varphi_2, \tilde{0}(X) + \varphi_3, \tilde{0}(X) + \varphi_4)$, where the $\tilde{0}$-polynomial "often" vanishes.[1] The 4-byte output of $\tilde{\Phi}(X)$ is then carried through all intermediate rounds to ensure that all intermediate values Z_r are perturbated and all rounds are closely linked. These perturbations are canceled out at the final round R'_{10} by the perturbation cancelation system O_Φ – a function where $O_\Phi(\varphi_1, \varphi_2, \varphi_3, \varphi_4) = 0$. The result of this function is XOR-ed with the output of the original functionality, i.e. the ciphertext Y^1_{10}, to result into Z_{10}: $Z_{10} = Y^1_{10} \oplus Y^2_{10}$.

Linear Encodings $(M_r)_{1 \le r \le 9}$. Annihilating linear input and output encodings M_r over GF (2^8) between successive rounds ensure that all the variables are interleaved to make analysis hard – e.g. to prevent that an adversary is able to distinguish the system of random equations from the original functionality. These encodings M_r can be represented as a 43×43 diagonal block matrix constructed as follows:

$$M_r = \pi_r \circ \begin{pmatrix} A_r^{(1)} & & & \\ & \ddots & & \\ & & A_r^{(7)} & \\ & & & B_r \end{pmatrix} \circ \sigma_r \ ,$$

where (1) the $A_r^{(l)}|_{l=1,\ldots,7}$ are random invertible 5×5 matrices of which the inverse has exactly 2 non-zero coefficients in GF (2^8) on each row; (2) B_r is a random invertible 8×8 matrix of which the inverse has at least 7 non-zero coefficients in GF (2^8) on each row; and (3) π_r and σ_r are random permutations at byte level of $\{1, \ldots, 43\}$ defined such that the matrices $A_r^{(l)}|_{l=1,\ldots,7}$ mix the 16 original polynomials with 19 random polynomials, whereas B_r mixes the 4 perturbation polynomials with the remaining 4 random polynomials. We refer to [4] for determination of the constraints on the matrices. Our cryptanalysis exploits these characteristics of the linear encodings M_r.

Obtaining the Correct Result. Due to the introduction of the perturbation in the first round, there is a probability that the ciphertext is incorrect (when $\tilde{\Phi}(X) \neq (\varphi_1, \varphi_2, \varphi_3, \varphi_4)$ and thus $Y^2_{10} = O_\Phi(\tilde{\Phi}(X)) \neq 0$). Therefore, four correlated instances of the perturbated white-box AEw/oS implementation are generated. Each with a different perturbation function, constructed such that there are always two instances that give the correct result (ciphertext) while the other two result into different random values with an overwhelming probability. A majority vote can then be used to distinguish the correct result. We refer to [3,4] for a discussion on the correlation of the four $\tilde{0}$-polynomials. Our cryptanalysis requires only one instance of the perturbated white-box AEw/oS implementation.

[1] The construction of the $\tilde{0}$-polynomials is described in [3,4].

Summary. Putting everything together, the perturbated white-box AEw/oS implementation consists of four instances of implementations with different perturbations; each instance comprising of 10 rounds R'_r, defined as follows:

$$\begin{cases} M_1 \circ (R_1 \| \tilde{\Phi} \| \mathtt{Ran}_1) & \text{for } R'_1\,, \\ M_r \circ (R_r \| I \| \mathtt{Ran}_r) \circ M_{r-1}^{-1} & \text{for } R'_r|_{2 \le r \le 9}, \text{ where } I \text{ is the identity function}\,, \\ \bigoplus \circ (R_{10} \| O_{\Phi}) \circ M_9^{-1} & \text{for } R'_{10}\,. \end{cases}$$

Along the specifications of Bringer *et al.* [4], each instance accounts ≈142 MB, which brings the full size of the white-box implementation to ≈568 MB.

3 Cryptanalysis of the White-Box AEw/oS Implementation

In this section we describe our cryptanalysis, which comprises of the following three steps:

1. Analysis of the final round: distinguish the system of random equations and the perturbations from the AEw/oS round operations, and recover the input encoding M_9^{-1} up to an unknown constant factor s.t. the linear equivalent input of the original final round R_{10} can be observed.
2. Separate the output bytes of the S-boxes: eliminate the `MixColumns` operation from the penultimate round R_9 s.t. the unknown factors of the linear equivalent output of R_9 can be included into the secret S-boxes.
3. Full structural decomposition, i.e. recovering all linear input/output encodings up to an unknown constant factor and eliminating the `MixColumns` operation within all rounds. Recover linear equivalent key-dependent S-boxes.

Note that not all the information of the secret S-boxes and linear mappings can be extracted since there are many equivalent keys which yield the same white-box implementation. Indeed, we can multiply the input/output of an S-box with a fixed constant and compensate for it in the adjacent linear mapping. Our attack recovers an equivalent key, and hence the decomposed implementation can for example be used to decrypt arbitrary ciphertexts although the implementation was only intended to encrypt plaintexts.

Setup. Choose a random 16-byte plaintext X, encrypt it with the four correlated implementations, and select one of both instances that result into the correct ciphertext (using the majority vote). For that instance, store the intermediate and final states $Z_i|_{i=1,\ldots,10}$ (which are clearly readable in a white-box attack context), where the final state equals the ciphertext, i.e. $Z_{10} = Y_{10}^1$. Throughout our cryptanalysis, we will refer to these states as the initial unmodified states.

3.1 Analysis of the Final Round

The first phase of our cryptanalysis focuses on the perturbated final round R'_{10}, which is lossy since the system of random equations is discarded. We will recover a significant part of the linear input encoding, i.e., the first 16 rows of the linear input encoding M_9^{-1} up to an unknown 16×16 diagonal matrix Λ_9. This enables us to observe the linear equivalent input $\Lambda_9 Y_9^1$ of the original final round R_{10}. This phase consists of several consecutive steps.

Recover pairs of intermediate bytes in Z_9 generating each input byte of Y_9^1 of R_{10}. Due to the specific characteristics of M_9, i.e. the matrices $A_9^{(l)}|_{l=1,\ldots,7}$ mix the 16 original polynomials with 19 random polynomials, the concatenated 35-byte output of $A_9^{(l)-1}|_{l=1,\ldots,7}$ consists of the 16-byte input Y_9^1 of R_{10}, while the remaining 19 bytes are discarded in R'_{10}. Therefore, since these $A_9^{(l)-1}|_{l=1,\ldots,7}$ matrices in M_9^{-1} have exactly 2 non-zero coefficients on each row, each input byte $y_{9,i}^1$ of the original final round R_{10} is a linear combination in GF (2^8) of exactly two intermediate bytes of Z_9. The pair of intermediate bytes generating $y_{9,i}^1$ is denoted by (z_{9,i_1}, z_{9,i_2}).

In a white-box attack context, the adversary has access to the description of the (obfuscated) system of polynomials, and is able to manipulate the internal states. Hence, he can freely choose to modify bytes of Z_9 and observe the corresponding output Z_{10}. In this context, we present an algorithm to obtain the following sets:

$\mathscr{S}_{Z_9}(y_{10,i}^1)$: the set containing the pair of intermediate bytes (z_{9,i_1}, z_{9,i_2}) corresponding to each output byte $y_{10,i}^1$ of R_{10}. Due to the lack of the `MixColumns` step in the final round, R_{10} comprises of 16 one-to-one monovariate polynomials, and hence these sets can easily be assigned to the corresponding input byte $y_{9,i}^1$ by applying the inverse `ShiftRows` step;

$\mathscr{S}_{Z_9}(O_\Phi)$: the set containing those intermediate bytes of Z_9 which function as input bytes of the B_9^{-1} matrix in M_9^{-1} that only affects the 4-byte input Y_9^2 of O_Φ. This set contains at least 7 and at most 8 bytes.[2]

The setup of the algorithm is to generate for each output byte $z_{10,i}|_{i=0,\ldots,15}$ of the perturbated final round R'_{10} a set $\mathscr{S}_{Z_9}(z_{10,i})$ consisting of those intermediate bytes of Z_9 which influence $z_{10,i}$. Repeat the following steps for each intermediate byte $z_{9,i}$ of Z_9 one at a time:

Step 1: Make the intermediate byte $z_{9,i}$ active by introducing a non-zero difference $\Delta z_{9,i} \in \text{GF}(2^8) \setminus \{0\}$ while keeping all other bytes of Z_9 fixed to their initial value $(\forall l \neq i : \Delta z_{9,l} = 0)$;

Step 2: Compute R'_{10} and observe its output Z_{10} by comparing with the stored initial final state Y_{10}^1 (ciphertext): if the number of affected output bytes $z_{10,i}$ is larger than 5 bytes, then assign the current active intermediate byte $z_{9,i}$ directly to the set $\mathscr{S}_{Z_9}(O_\Phi)$. Else $z_{9,i}$ is assigned to each set

[2] Since B_9 mixes the 4 perturbation polynomials with 4 random polynomials, the 8-byte output of B_9^{-1} consists of the 4-byte input Y_9^2 of O_Φ and the other 4-byte output is discarded in the perturbated final round R'_{10}, hence we only focus on Y_9^2. If Y_9^2 only depends on 7 intermediate bytes of Z_9 instead of 8 (special case of 8×8 B_9^{-1} matrix), we are only able to identify 7 bytes.

$\mathscr{S}_{Z_9}(z_{10,i})$ of the output bytes $z_{10,i}$ it affects. In case that the number of affected output bytes is zero, $z_{9,i}$ is assigned to no set. Fig. 2 depicts the effect of one active intermediate byte $z_{9,i}$ on the output bytes of Z_{10} and explains the different cases.

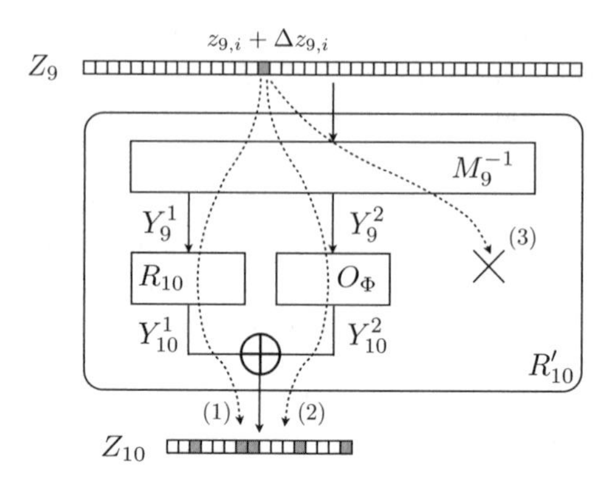

Fig. 2. In case that the active intermediate byte $z_{9,i}$ of Z_9 influences the input Y_9^1 of R_{10} through one of the $A_9^{(l)-1}$ matrices [case (1)], the maximum number of affected input bytes of R_{10} equals 5 since $A_9^{(l)-1}$ are 5×5 invertible matrices. This trivially translates to a maximum of 5 affected ciphertext bytes due to the lack of the `MixColumns` step in R_{10} and accordingly to a maximum of 5 affected output bytes $z_{10,i}$ in Z_{10}. So the case there are more than 5 affected output bytes $z_{10,i}$ only occurs when the active byte $z_{9,i}$ influences the input Y_9^2 of O_Φ through B_9^{-1} [case (2)], which causes the output Y_{10}^2 to change in more than 5 bytes. However, with a very low probability, only 5 or less bytes of Y_{10}^2 are affected which introduces false positives (see below). In case that the active intermediate byte $z_{9,i}$ only affects the input of the system of random polynomials [case (3)] - which has been discarded in R'_{10} - the number of affected output bytes is zero.

Concerning Step 2, false positives can occur, i.e. the incorrect assignment of $z_{9,i}$ to the sets $\mathscr{S}_{Z_9}(z_{10,i})$ instead of the set $\mathscr{S}_{Z_9}(O_\Phi)$. An active intermediate byte $z_{9,i}$ which influences the 4-byte input Y_9^2 of O_Φ through the B_9^{-1} matrix, modifies the initial value $(\varphi_1, \varphi_2, \varphi_3, \varphi_4)$. Since the 16-byte output Y_{10}^2 of O_Φ - which is zero for $Y_9^2 = (\varphi_1, \varphi_2, \varphi_3, \varphi_4)$ and random otherwise - is XOR-ed with the real ciphertext Y_{10}^1 to form the ouput Z_{10}, the probability that the number of affected output bytes $z_{10,i}$ is 5 or less is given by $\sum_{i=1}^{5} \binom{16}{i}(1/2^8)^{16-i}(1-1/2^8)^i \approx 1/2^{76}$. This corresponds also with the probability that $z_{9,i}$ is faulty assigned to each set $\mathscr{S}_{Z_9}(z_{10,i})$ of the affected output bytes $z_{10,i}$ (false positive).

Hence, at the end, the probability that each set $\mathscr{S}_{Z_9}(z_{10,i})|_{i=0,\ldots,15}$ contains exactly the pair of intermediate bytes (z_{9,i_1}, z_{9,i_2}) generating the corresponding input byte $y_{9,i}^1$ and that the set $\mathscr{S}_{Z_9}(O_\Phi)$ contains the 7 or 8 intermediate bytes functioning as input bytes of the B_9^{-1} matrix, equals $\approx (1 - 1/2^{76})^a$ with $a = 7$ or 8, which is ≈ 1. In that case $\mathscr{S}_{Z_9}(y_{10,i}^1) = \mathscr{S}_{Z_9}(z_{10,i})$.

The worst case scenario, i.e. the set $\mathscr{S}_{Z_9}(O_\Phi)$ contains less than 7 or 8 intermediate bytes and some or all sets $\mathscr{S}_{Z_9}(z_{10,i})|_{i=0,\ldots,15}$ contain next to the pair of intermediate bytes (z_{9,i_1}, z_{9,i_2}) also additional intermediate bytes which only influenced the input Y_9^2 of O_Φ (false positives), only occurs with a probability of $\approx 1-(1-1/2^{76})^a$ with $a=7$ or 8, which is ≈ 0. In that case, after the setup of the algorithm, the sets $\mathscr{S}_{Z_9}(z_{10,i})|_{i=0,\ldots,15}$ need to be reduced to the pair of intermediate bytes (z_{9,i_1}, z_{9,i_2}) while completing the set $\mathscr{S}_{Z_9}(O_\Phi)$. This case is fully handled in App. A.1, which is based on pairs of sets of plaintext-ciphertext together with all intermediate states for the same perturbated instance of the cipher we selected during the setup phase of our cryptanalysis.

Note: A side-effect of the above algorithm is the recovery of the set $\mathscr{S}_{Z_9}(O_\Phi)$. Keeping those 7 or 8 intermediate bytes fixed to their initial value ensures that the 4-byte input Y_9^2 of O_Φ remains unmodified, i.e. $(\varphi_1, \varphi_2, \varphi_3, \varphi_4)$, such that the output of O_Φ remains zero and hence we can always observe the real ciphertext: $Z_{10} = Y_{10}^1$. This allows us to circumvent the perturbations.

Decompose the linear input encoding M_9^{-1}. Each unknown input byte $y_{9,i}^1$ of the original final round R_{10} is a linear combination in $\mathrm{GF}\left(2^8\right)$ of a pair of intermediate bytes (z_{9,i_1}, z_{9,i_2}), which has been recovered in the previous step. Thus, there are two non-zero coefficients $c_{i,1}, c_{i,2} \in \mathrm{GF}\left(2^8\right)\setminus\{0\}$ on a row of one of the $A_9^{(l)-1}$ matrices in M_9^{-1} such that $z_{9,i_1} \bullet c_{i,1} + z_{9,i_2} \bullet c_{i,2} = y_{9,i}^1$, where $\bullet$ denotes multiplication in the Rijndael Galois Field [13]. In this step, we recover both coefficients up to an unknown factor $\alpha_{9,i}$, which enables us to observe the linear equivalent input byte $\alpha_{9,i} \bullet y_{9,i}^1$.

Both coefficients $c_{i,1}$ and $c_{i,2}$ can be expressed in terms of a single unknown parameter $\alpha_{9,i}$ as follows: (1) compute the output byte of Z_{10} corresponding to $y_{10,i}^1$ – knowing that $Z_{10} = Y_{10}^1$ (see note above) – where the relevant input bytes (z_{9,i_1}, z_{9,i_2}) are fixed to their initial value in Z_9, and (2) find another pair of values (z'_{9,i_1}, z'_{9,i_2}) by fixing $z'_{9,i_1} = z_{9,i_1} + \texttt{`01'}$ and searching for z'_{9,i_2} which yield the identical output byte in Z_{10}. Hence, since equal output bytes means equal input bytes for R_{10},

$$\begin{aligned} z_{9,i_1} \bullet c_{i,1} + z_{9,i_2} \bullet c_{i,2} &= y_{9,i}^1 \\ (z_{9,i_1} + \texttt{`01'}) \bullet c_{i,1} + z'_{9,i_2} \bullet c_{i,2} &= y_{9,i}^1 \,, \end{aligned}$$

from which we can derive that $c_{i,1} = \varepsilon_i \bullet c_{i,2}$, with $\varepsilon_i = z_{9,i_2} + z'_{9,i_2}$.

By assigning `01' to $c_{i,2}$, only the linear equivalent input byte $y_{9,i}^1$ can be recovered, i.e. $\alpha_{9,i} \bullet y_{9,i}^1 = \varepsilon_i \bullet z_{9,i_1} + z_{9,i_2}$ with $\alpha_{9,i}$ unknown. As a result, we retrieve an expression of the first 16 rows of the linear mapping M_9^{-1} up to (unknown) constants $\alpha_{9,i}$. That is, we obtain the following equation:

$$\begin{pmatrix} \alpha_{9,0} \bullet y_{9,0}^1 \\ \vdots \\ \alpha_{9,15} \bullet y_{9,15}^1 \end{pmatrix} = M_9^{-1}[0..15]' Z_9 = \begin{pmatrix} \alpha_{9,0} & & \\ & \ddots & \\ & & \alpha_{9,15} \end{pmatrix} \begin{pmatrix} L_{9,0} \\ \vdots \\ L_{9,15} \end{pmatrix} Z_9 \,, \qquad (1)$$

where $L_{9,i}$ denotes the i-th row of M_9^{-1} and contains the unknown coefficients $c_{i,1}, c_{i,2}$. The recovered submatrix is denoted by $M_9^{-1}[0..15]' = \Lambda_9 \circ M_9^{-1}[0..15]$

with $\Lambda_9 = diag(\alpha_{9,0}, \ldots, \alpha_{9,15})$, which transforms Z_9 into the linear equivalent input $\Lambda_9 Y_9^1$ of R_{10}. Each row of $M_9^{-1}[0..15]'$ is all '00' except an ε_i and '01' on the relevant columns, i.e. the columns corresponding to the pair of intermediate bytes (z_{9,i_1}, z_{9,i_2}).

3.2 Separate the S-Boxes

As a result of the first phase of the cryptanalysis, the adversary is able to derive the input bytes $y_{9,i}^1$ of the original final round R_{10} up to unknown coefficients $\alpha_{9,i}$, i.e. $\Lambda_9 Y_9^1$. Due to the annihilating nature of the linear encodings between successive perturbated rounds, this also corresponds to the linear equivalent output of the preceding, penultimate round R_9. Therefore, R_9 can be expressed as

$$\Lambda_9 \circ \texttt{MixColumns} \circ \texttt{ShiftRows} \circ \{S_{9,0}, \ldots, S_{9,15}\} \circ \bigoplus_{K_8} \circ M_8^{-1}[0..15]\,, \qquad (2)$$

where the set $\{S_{9,i}\}|_{i=0,\ldots,15}$ represents the 16 different invertible 8-to-8 bit original S-boxes of R_9, which together with the round key K_8 are part of the secret key.

The objective in this step of our cryptanalysis is to include the unknown factors of the linear equivalent output of R_9 into the secret S-boxes by separating the output bytes of the S-boxes, which can be achieved by eliminating the `MixColumns` operation from the round. However, due to the presence of the unknown values in Λ_9, this is not trivial since the `MixColumns` step is an invertible linear transformation which operates on four bytes. We address this problem in this section.

The main idea is to search for a transformation such that the matrix Λ_9 has the same factors α for each four bytes of a `MixColumns` operation. Even though this factor remains unknown, such a diagonal matrix can be swapped with the `MixColumns` operation (multiplication with a diagonal matrix with all the same elements is a commutative operation in the group of square matrices). As a result, the `MixColumns` operation is the final operation and can be eliminated by multiplying the result with the inverse `MixColumns` operation.

In total there are four parallel `MixColumns` steps $\texttt{MC}_i|_{i=0,\ldots,3}$ since each step $\texttt{MC}_i$ operates on the output bytes of four different S-boxes. Accordingly, Λ_9 can be divided into four 4×4 diagonal submatrices $\Lambda_{9,i}$, each containing those unknown factors α corresponding to the four output bytes of each $\texttt{MC}_i$: $\Lambda_{9,i} = diag(\alpha_{9,i}, \alpha_{9,i+4}, \alpha_{9,i+8}, \alpha_{9,i+12})$ for $i = 0, \ldots, 3$. Hence out of the requirement $\Lambda_{9,i} \circ \texttt{MC}_i = \texttt{MC}_i \circ \Lambda'_{9,i}$, we seek a 4×4 diagonal matrix $\Lambda'_{9,i} = \texttt{MC}_i^{-1} \circ \Lambda_{9,i} \circ \texttt{MC}_i$. This is the case when all diagonal entries of $\Lambda_{9,i}$ are identical, e.g. $\Lambda_{9,i} = diag(\alpha_{9,i}, \alpha_{9,i}, \alpha_{9,i}, \alpha_{9,i})$, and moreover $\Lambda'_{9,i} = \Lambda_{9,i}$. Here we present an algorithm - which has been successfully implemented in `C++` and confirmed by computer experiments - such that:

Given black-box access to the structure as shown in (2)*, a white-box adversary is able to ensure that all diagonal entries of* $\Lambda_{9,i}$ *become identical and hence* *construct* $\Lambda'_{9,i} = diag(\alpha_{9,i}, \alpha_{9,i}, \alpha_{9,i}, \alpha_{9,i})$ *for each of the four parallel* MixColumns *steps* $MC_i|_{i=0,\dots,3}$ *such that* (2) *becomes:*

$$\texttt{MixColumns} \circ \Lambda'_9 \circ \texttt{ShiftRows} \circ \{S_{9,0}, \dots, S_{9,15}\} \circ \bigoplus_{K_8} \circ M_8^{-1}[0..15] \ . \qquad (3)$$

Since in (3) the ShiftRows step is just a permutation on byte level, the unknown diagonal entries in Λ'_9 can be included into the secret S-boxes by applying the inverse ShiftRows step.

The setup of the algorithm is to generate for each $\mathtt{MC}_i|_{i=0,\dots,3}$ a set $\mathscr{S}_{Z_8}(\mathtt{MC}_i)$ consisting of those intermediate bytes of Z_8 which influence the input of $\mathtt{MC}_i$. This is done by making each intermediate byte of Z_8 (the input of (2)) one at a time active and observing the corresponding active output bytes in $\Lambda_9 Y_9^1$ (the output of (2)). Since each input byte of $\mathtt{MC}_i$ depends on a pair of intermediate bytes of Z_8 and due to the special structure of the $A_8^{(l)-1}|_{l=1,\dots,7}$ matrices in M_8^{-1}, modifying one of the bytes in $\mathscr{S}_{Z_8}(\mathtt{MC}_i)$ results in making one or two of the four input bytes of $\mathtt{MC}_i$ active in most cases. However in the special case when the four input bytes of $\mathtt{MC}_i$ are controlled by two distinct pairs of intermediate bytes of Z_8, only exactly two input bytes of $\mathtt{MC}_i$ can be made active.

Repeat the following steps for each MixColumns step $\mathtt{MC}_i|_{i=0,\dots,3}$:

Step 1: Given the initial unmodified value of the intermediate state Z_8, store the corresponding 4-byte output of $\mathtt{MC}_i$ in $\Lambda_9 Y_9^1$, denoted by $Y_{\mathtt{MC}_i}$;[3]

Step 2: Modify one byte in $\mathscr{S}_{Z_8}(\mathtt{MC}_i)$ and store the corresponding 4-byte output of $\mathtt{MC}_i$ in $\Lambda_9 Y_9^1$, denoted by $Y'_{\mathtt{MC}_i}$. In the case when less than three bytes between $Y_{\mathtt{MC}_i}$ and $Y'_{\mathtt{MC}_i}$ have become active and hence at least three of the four input bytes of $\mathtt{MC}_i$ have become active,[4] we discard this case and continue with Step 4;

Step 3: Given the pair $(Y_{\mathtt{MC}_i}, Y'_{\mathtt{MC}_i})$, keep the first factor $\alpha_{9,i}$ fixed while varying the other three factors $(\alpha_{9,i+4}, \alpha_{9,i+8}, \alpha_{9,i+12})$ over GF $(2^8)\backslash\{0\}$ by multiplying the second, third and fourth byte within both values $(Y_{\mathtt{MC}_i}, Y'_{\mathtt{MC}_i})$ with respectively $\beta, \gamma, \delta \in \mathrm{GF}\ (2^8) \setminus \{0\}$. For each combination $(\beta,\gamma,\delta)_j$ with the corresponding pair $(Y^{(j)}_{\mathtt{MC}_i}, Y'^{(j)}_{\mathtt{MC}_i})$ with $j = 1, \dots, (2^8-1)^3$, invert the MixColumns step, i.e. $(Y^{(j)}_{\mathtt{MC}_i^{-1}}, Y'^{(j)}_{\mathtt{MC}_i^{-1}}) = (\mathtt{MC}^{-1}(Y^{(j)}_{\mathtt{MC}_i}), \mathtt{MC}^{-1}(Y'^{(j)}_{\mathtt{MC}_i}))$, and construct the following solution set by comparing both values:

$$\mathscr{S} = \{(\beta,\gamma,\delta)_j \mid \text{one or two active bytes between } (Y^{(j)}_{\mathtt{MC}_i^{-1}}, Y'^{(j)}_{\mathtt{MC}_i^{-1}})\} \ ;$$

Step 4: Repeat Step 2 and Step 3 for each byte in $\mathscr{S}_{Z_8}(\mathtt{MC}_i)$ one at a time;

Step 5: At the end, a solution set $\mathscr{S}$ has been constructed for each modified byte in $\mathscr{S}_{Z_8}(\mathtt{MC}_i)$. The triplet $(\beta,\gamma,\delta)_i$ for $\mathtt{MC}_i$ is derived as the intersection between all solution sets.

[3] $Y_{\mathtt{MC}_i} = (\alpha_{9,i} \bullet y^1_{9,i}, \alpha_{9,i+4} \bullet y^1_{9,i+4}, \alpha_{9,i+8} \bullet y^1_{9,i+8}, \alpha_{9,i+12} \bullet y^1_{9,i+12})$.

[4] The branch number of the MixColumns step equals 5.

As can be observed in Step 3, the algorithm only keeps track of single and double active input bytes[5] to each MixColumns step for each modified byte in $\mathscr{S}_{Z_8}(\texttt{MC}_i)$ and each combination $(\beta,\gamma,\delta)_j$. When modifying one byte in $\mathscr{S}_{Z_8}(\texttt{MC}_i)$, we distinguish the following two cases:

1. one or two of the four input bytes of $\texttt{MC}_i$ have become active. The resulting solution set $\mathscr{S}$ obtained in Step 3 is considered *valid* and contains the triplet $(\beta,\gamma,\delta)_j$ for which the same one or two bytes have become active between $(Y^{(j)}_{\texttt{MC}_i^{-1}}, Y'^{(j)}_{\texttt{MC}_i^{-1}})$, which only occurs when the triplet made all diagonal entries of $\Lambda_{9,i}$ identical (i.e. all equal to $\alpha_{9,i}$) such that $\Lambda_{9,i}$ could be swapped with the MixColumns step $\texttt{MC}_i$. This triplet is contained within all *valid* solution sets;
2. at least three of the four input bytes of $\texttt{MC}_i$ have become active and the case has not been discarded in Step 2. Hence the resulting solution set $\mathscr{S}$ obtained in Step 3 is considered *invalid*.

Hence, in Step 5 only one intersection occurs between all solution sets, i.e. between *valid* sets since there is no intersection with *invalid* sets.

The triplet $(\beta,\gamma,\delta)_i$ as outcome of the above algorithm applied to each $\texttt{MC}_i$ are the factors needed to ensure that all diagonal entries of $\Lambda_{9,i}$ become identical to the first factor $\alpha_{9,i}$, i.e. $diag(\texttt{'01'},\beta_i,\gamma_i,\delta_i)\circ\Lambda_{9,i} = diag(\alpha_{9,i},\alpha_{9,i},\alpha_{9,i},\alpha_{9,i}) = \Lambda'_{9,i}$. So by multiplying each set of four rows corresponding to each $\texttt{MC}_i$ of the recovered submatrix $M_9^{-1}[0..15]'$ (see (1)) with the derived quartet $(\texttt{'01'},\beta_i,\gamma_i,\delta_i)$, the adversary is able to construct Λ'_9 in which the diagonal entries corresponding to each $\texttt{MC}_i$ are identical:

$$\begin{pmatrix} I_{4\times4} & 0 & 0 & 0 \\ 0 & B & 0 & 0 \\ 0 & 0 & \Gamma & 0 \\ 0 & 0 & 0 & \Delta \end{pmatrix} M_9^{-1}[0..15]' = \begin{pmatrix} D_9 & 0 & 0 & 0 \\ 0 & D_9 & 0 & 0 \\ 0 & 0 & D_9 & 0 \\ 0 & 0 & 0 & D_9 \end{pmatrix} \begin{pmatrix} L_{9,0} \\ \vdots \\ L_{9,15} \end{pmatrix}, \tag{4}$$

where $B = diag(\beta_0,\beta_1,\beta_2,\beta_3)$, $\Gamma = diag(\gamma_0,\gamma_1,\gamma_2,\gamma_3)$, $\Delta = diag(\delta_0,\delta_1,\delta_2,\delta_3)$, $D_9 = diag(\alpha_{9,0},\alpha_{9,1},\alpha_{9,2},\alpha_{9,3})$. The obtained submatrix of (4) is denoted by $M_9^{-1}[0..15]'' = \Lambda'_9 \circ M_9^{-1}[0..15]$ with $\Lambda'_9 = diag(D_9,D_9,D_9,D_9)$.

3.3 Decomposing the Rounds

The final phase of our cryptanalysis presents the full decomposition of the perturbated white-box AEw/oS implementation and shows how to obtain a set of candidate S-boxes (the secret key). We present an algorithm to recover all remaining linear encodings up to a constant factor and to eliminate the MixColumns operation, when applied to all perturbated intermediate rounds $R'_r|_{2\le r\le 9}$ from the bottom up.

[5] Keeping track of double active input bytes is necessary due to the special case mentioned in the setup of the algorithm.

Given black-box access to a perturbated intermediate round R'_r which linear output encoding $M_r^{-1}[0..15]'' = \Lambda'_r \circ M_r^{-1}[0..15]$ has already been recovered s.t. the linear equivalent output $\Lambda'_r Y_r^1$ of the original intermediate round R_r can be observed and the MixColumns *step can be separated from the S-boxes, i.e.:*

$$\texttt{MixColumns} \circ \Lambda'_r \circ \texttt{ShiftRows} \circ \{S_{r,0}, \ldots, S_{r,15}\} \circ \bigoplus_{K_{r-1}} \circ M_{r-1}^{-1}[0..15]\,,$$

a white-box adversary is able to derive the linear input encoding up to an unknown factor, i.e. $M_{r-1}^{-1}[0..15]'' = \Lambda'_{r-1} \circ M_{r-1}^{-1}[0..15]$, s.t. the linear equivalent input $\Lambda'_{r-1} Y_{r-1}^1$ of R_r can be observed and the MixColumns *step of the preceding round R_{r-1} can be separated from the S-boxes, and hence the structure of R_r becomes:*

$$\texttt{MixColumns} \circ \Lambda'_r \circ \texttt{ShiftRows} \circ \{S_{r,0}, \ldots, S_{r,15}\} \circ \bigoplus_{K_{r-1}} \circ \Lambda'^{-1}_{r-1}\,.$$

As a result of the previous two phases of the cryptanalysis – i.e. the recovery of the linear output encoding $M_9^{-1}[0..15]''$ up to a constant factor (see (4)) and the elimination of the MixColumns step (see (3)) of the penultimate round R_9 – the above algorithm first applies to R'_9 and then to the remaining perturbated intermediate rounds $R'_r|_{2 \le r \le 8}$ from the bottom up since each linear input encoding matches the linear output encoding of the preceding round. The main steps of the algorithm are:

1. Assign to each input byte $y_{r-1,i}^1$ of the original round R_r a pair of intermediate bytes $(z_{r-1,i_1}, z_{r-1,i_2})$ of Z_{r-1} contained within the set $\mathscr{S}_{Z_{r-1}}(y_{r-1,i}^1)$;
2. Decompose the linear input encoding M_{r-1}^{-1}: recover the first 16 rows up to a 16×16 diagonal linear bijection $\Lambda_{r-1} = diag(\alpha_{r-1,0}, \ldots, \alpha_{r-1,15})$, i.e. $M_{r-1}^{-1}[0..15]' = \Lambda_{r-1} \circ M_{r-1}^{-1}[0..15]$;
3. Eliminate the MixColumns step in the preceding round R_{r-1} by converting Λ_{r-1} into Λ'_{r-1} where the diagonal entries corresponding to each $\texttt{MC}_i|_{i=0,\ldots,3}$ are identical: $M_{r-1}^{-1}[0..15]'' = \Lambda'_{r-1} \circ M_{r-1}^{-1}[0..15]$ with $\Lambda'_{r-1} = diag(D_{r-1}, D_{r-1}, D_{r-1}, D_{r-1})$, where $D_{r-1} = diag(\alpha_{r-1,0}, \alpha_{r-1,1}, \alpha_{r-1,2}, \alpha_{r-1,3})$.

We refer to App. A.2 for a detailed description of each step, which are very similar to the ones stated in Sect. 3.1 and 3.2. However, the algorithm for separating the MixColumns step from the S-boxes applied to the first round R_1, i.e. the case when $r = 2$, is simplified since the perturbated round R'_1 lacks an input encoding.

As a result of the algorithm mentioned above, the white-box adversary has black-box access to the following structures of each round $R_r|_{r=1,\ldots,10}$:

$$\begin{cases} \texttt{SR} \circ \bigoplus_{K'_{10}} \circ \{S_{10,i}\}|_{i=0,\ldots,15} \circ \bigoplus_{K_9} \circ \Lambda'^{-1}_9 & \text{for } R_{10}\,, \\ \texttt{MC} \circ \texttt{SR} \circ \Lambda''_r \circ \{S_{r,i}\}|_{i=0,\ldots,15} \circ \bigoplus_{K_{r-1}} \circ \Lambda'^{-1}_{r-1} & \text{for } R_r|_{2 \le r \le 9}\,, \\ \texttt{MC} \circ \texttt{SR} \circ \Lambda''_1 \circ \{S_{1,i}\}|_{i=0,\ldots,15} \circ \bigoplus_{K_0} & \text{for } R_1\,, \end{cases} \quad (5)$$

where each set $\{S_{r,i}\}|_{i=0,\ldots,15;r=1,\ldots,10}$ represents the 16 different invertible 8-to-8 bit original S-boxes of R_r, which together with round keys $K_r|_{r=0,\ldots,10}$ are part of the secret key. By altering $\Lambda'_r|_{r=1,\ldots,9}$, its order with the ShiftRows step can be reversed, i.e. $\Lambda''_r = \texttt{SR}^{-1} \circ \Lambda'_r \circ \texttt{SR} = diag(D_r, D_r^{\gg 1}, D_r^{\gg 2}, D_r^{\gg 3})$ where $D_r^{\gg i}$ denotes the matrix $D_r = diag(\alpha_{r,0}, \alpha_{r,1}, \alpha_{r,2}, \alpha_{r,3})$ whose diagonal entries are cyclically shifted i-times to the right. Note that the first and final rounds respectively lack a linear input and output encoding.

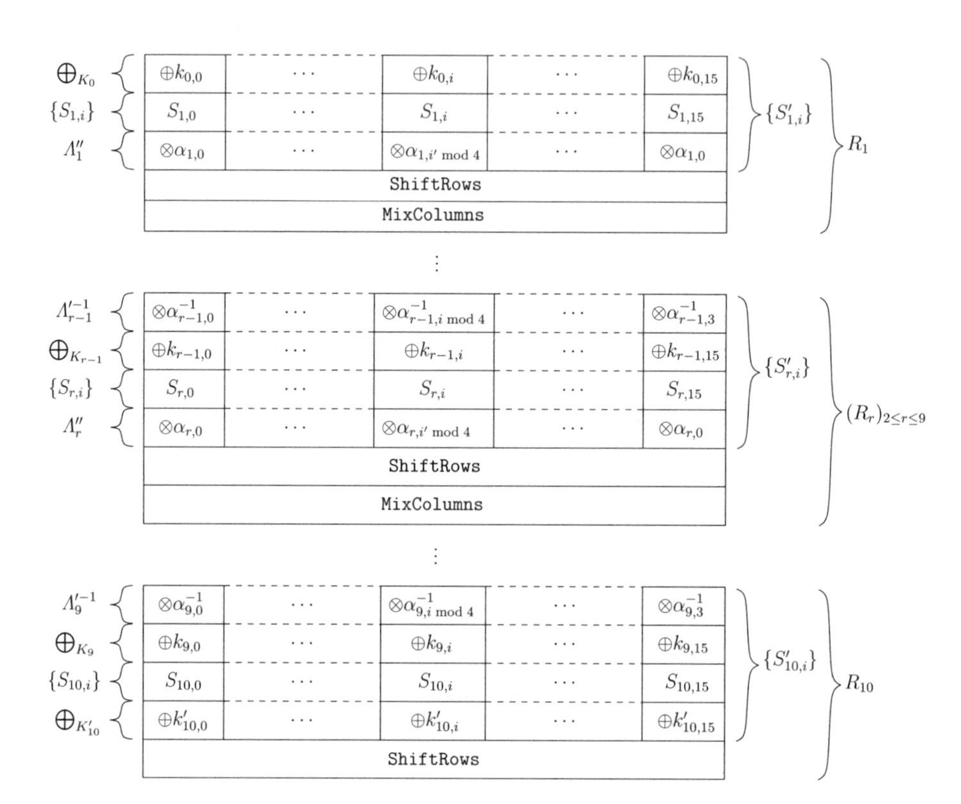

Fig. 3. Our Invertible Functionally Equivalent AEw/oS Implementation

In the structures of (5), only the ShiftRows and MixColumns steps are known to the adversary. Since all unknown linear bijections $\Lambda'^{-1}_r|_{r=1,\ldots,9}$ and $\Lambda''_r|_{r=1,\ldots,9}$ are 16×16 diagonal matrices, the unknown factors α as diagonal entries can easily be included into respectively the input and output of the secret S-boxes. Hence the linear equivalent key-dependent S-boxes, denoted by $S'_{r,i}$, have the following form for each round $R_r|_{r=1,\ldots,10}$:

$$\begin{aligned} S'_{10,i}|_{i=0,\ldots,15} &= \bigoplus_{k'_{10,i}} \circ S_{10,i} \circ \bigoplus_{k_{9,i}} \circ \bigotimes_{\alpha^{-1}_{9,i \bmod 4}} , \\ S'_{r,i}|_{i=0,\ldots,15;r=2,\ldots,9} &= \bigotimes_{\alpha_{r,i' \bmod 4}} \circ S_{r,i} \circ \bigoplus_{k_{r-1,i}} \circ \bigotimes_{\alpha^{-1}_{r-1,i \bmod 4}} , \\ S'_{1,i}|_{i=0,\ldots,15} &= \bigotimes_{\alpha_{1,i' \bmod 4}} \circ S_{1,i} \circ \bigoplus_{k_{0,i}} . \end{aligned} \tag{6}$$

Although all components in the S-boxes of (6) are unknown to the adversary, each S-box $S'_{r,i}|_{i=0,\ldots,15;r=1,\ldots,10}$ can be defined by varying its input byte $\alpha_{r-1,i \bmod 4} \bullet y^1_{r-1,i}$ (in the case of R_1: the i-th plaintext byte x_i) over $GF(2^8)$ and record the corresponding output byte in $\mathtt{SR}^{-1}(\mathtt{MC}^{-1}(\Lambda'_r Y^1_r))$ (in the case of R_{10}: in $\mathtt{SR}^{-1}(Z_{10})$). In order to vary the input, we keep one of the pair bytes z_{r-1,i_1} fixed and vary the other byte z_{r-1,i_2}.

Fig. 3 depicts an overview of the full decomposition of the perturbated white-box AEw/oS implementation in order to obtain an invertible, functionally equivalent version.

4 Conclusion

In this paper we presented a structural cryptanalysis of the perturbated white-box AEw/oS implementation, presented by Bringer *et al.* [4]. Our attack has a worst time complexity of 2^{17} and negligible space complexity (see App. B). Our cryptanalysis trivially extends to perturbated white-box AES implementations as well.

The technique decomposes the obfuscated round structure of the white-box implementation. After eliminating the system of random equations and perturbations, we show how to distinguish the output bytes of individual S-boxes – by eliminating the `MixColumns` from the round functions. This elimination step is crucial in our cryptanalysis, and a proof of concept has been implemented in `C++`. From the obtained structure, a definition for each S-box can be derived. These S-boxes are linear equivalent to the original (secret) key that was chosen to construct the implementation. Indeed, there are several candidate keys possible that yield the same white-box implementation. This is embodied by the factors α that we meet in our cryptanalysis – these can take any value in $\mathrm{GF}\left(2^8\right) \setminus \{0\}$.

Each equivalent key consists of 160 bijective key-dependent 8-bit S-boxes, which can be used to construct a simpler, functionally equivalent version of the white-box AEw/oS implementation (as depicted in Fig. 3 in Sect. 3.3). The S-boxes occupy a total storage space of $\approx$ 41 kB; hence the total size of the implementation is significantly reduced from several hundred MB to just a few tens of kB. On top of this, the implementation becomes invertible, which renders it useless for many practical implementation where white-box cryptography would be of value.

The cryptanalysis is independent of the definition of the perturbation functions that are introduced in the first round; we exploit the characteristics of the input/output linear encodings and some properties of the AEw/oS block ciphers (such as the `MixColumns` operation). Modifying some specifications in an attempt to mitigate our cryptanalysis, such as the number of non-zero elements on the rows of the $A_r^{(l)-1}$ may turn the white-box implementation useless (its size will increase exponentially).

Although our cryptanalysis is specific to the particular structure of the implementation, some algorithms are of independent interest. In particular for research in structural cryptanalysis. Future research may include research to extend these algorithms to more generic constructions, e.g., where the `MixColumns` operations are also key dependent.

Acknowledgments. This work was supported in part by the IWT-SBO project on Security of Software for Distributed Applications (SEC SODA) and by the IAP Programme P6/26 BCRYPT of the Belgian State (Belgian Science Policy). Yoni De Mulder was supported in part by IBBT (Interdisciplinary institute for BroadBand Technology) of the Flemish Government and by a research grant of the Katholieke Universiteit Leuven.

References

1. Billet, O., Gilbert, H.: A Traceable Block Cipher. In: Laih, C.-S. (ed.) ASIACRYPT 2003. LNCS, vol. 2894, pp. 331–346. Springer, Heidelberg (2003)
2. Billet, O., Gilbert, H., Ech-Chatbi, C.: Cryptanalysis of a White Box AES Implementation. In: Handschuh, H., Hasan, M.A. (eds.) SAC 2004. LNCS, vol. 3357, pp. 227–240. Springer, Heidelberg (2004)
3. Bringer, J., Chabanne, H., Dottax, E.: Perturbing and Protecting a Traceable Block Cipher. In: Leitold, H., Markatos, E.P. (eds.) CMS 2006. LNCS, vol. 4237, pp. 109–119. Springer, Heidelberg (2006)
4. Bringer, J., Chabanne, H., Dottax, E.: White box cryptography: Another attempt. Cryptology ePrint Archive, Report 2006/468 (2006), `http://eprint.iacr.org/`
5. Chow, S., Eisen, P.A., Johnson, H., van Oorschot, P.C.: White-Box Cryptography and an AES Implementation. In: Nyberg, K., Heys, H.M. (eds.) SAC 2002. LNCS, vol. 2595, pp. 250–270. Springer, Heidelberg (2003)
6. Chow, S., Eisen, P.A., Johnson, H., van Oorschot, P.C.: A white-box DES implementation for DRM applications. In: Feigenbaum, J. (ed.) DRM 2002. LNCS, vol. 2696, pp. 1–15. Springer, Heidelberg (2003)
7. Ding, J.: A New Variant of the Matsumoto-Imai Cryptosystem through Perturbation. In: Bao, F., Deng, R., Zhou, J. (eds.) PKC 2004. LNCS, vol. 2947, pp. 305–318. Springer, Heidelberg (2004)
8. Faugère, J.-C., Perret, L.: Polynomial Equivalence Problems: Algorithmic and Theoretical Aspects. In: Vaudenay, S. (ed.) EUROCRYPT 2006. LNCS, vol. 4004, pp. 30–47. Springer, Heidelberg (2006)
9. Goubin, L., Masereel, J.-M., Quisquater, M.: Cryptanalysis of White Box DES Implementations. In: Adams, C., Miri, A., Wiener, M. (eds.) SAC 2007. LNCS, vol. 4876, pp. 278–295. Springer, Heidelberg (2007)
10. Jacob, M., Boneh, D., Felten, E.W.: Attacking an Obfuscated Cipher by Injecting Faults. In: Feigenbaum, J. (ed.) DRM 2002. LNCS, vol. 2696, pp. 16–31. Springer, Heidelberg (2003)
11. Link, H.E., Neumann, W.D.: Clarifying Obfuscation: Improving the Security of White-Box DES. In: Proceedings of the International Conference on Information Technology: Coding and Computing (ITCC 2005), Washington, DC, USA, vol. 1, pp. 679–684. IEEE Computer Society, Los Alamitos (2005)
12. Michiels, W., Gorissen, P., Hollmann, H.D.L.: Cryptanalysis of a Generic Class of White-Box Implementations. In: Avanzi, R.M., Keliher, L., Sica, F. (eds.) SAC 2008. LNCS, vol. 5381, pp. 414–428. Springer, Heidelberg (2009)
13. National Institute of Standards and Technology. Advanced encryption standard. FIPS publication 197 (2001), `http://csrc.nist.gov/publications/fips/fips197/fips-197.pdf`
14. Patarin, J.: Hidden Fields Equations (HFE) and Isomorphisms of Polynomials (IP): Two New Families of Asymmetric Algorithms. In: Maurer, U.M. (ed.) EUROCRYPT 1996. LNCS, vol. 1070, pp. 33–48. Springer, Heidelberg (1996)

15. Wyseur, B.: White-Box Cryptography. PhD thesis, Katholieke Universiteit Leuven (2009)
16. Wyseur, B., Michiels, W., Gorissen, P., Preneel, B.: Cryptanalysis of White-Box DES Implementations with Arbitrary External Encodings. In: Adams, C., Miri, A., Wiener, M. (eds.) SAC 2007. LNCS, vol. 4876, pp. 264–277. Springer, Heidelberg (2007)

A Algorithms

A.1 Algorithm: Recover Pairs of Intermediate Bytes in Z_9 Generating Each Input Byte of Y_9^1 of R_{10} - Worst Case Scenario

The worst case scenario, i.e. the set $\mathcal{S}_{Z_9}(O_\Phi)$ contains less than 7 or 8 intermediate bytes and some or all sets $\mathcal{S}_{Z_9}(z_{10,i})|_{i=0,\ldots,15}$ contain next to the pair of intermediate bytes (z_{9,i_1}, z_{9,i_2}) also additional intermediate bytes which only influenced the input Y_9^2 of O_Φ (false positives), occurs with a probability of $\approx 1-(1-1/2^{76})^a$ with $a = 7$ or 8, which is ≈ 0. In that case, after the setup of the algorithm, the sets $\mathcal{S}_{Z_9}(z_{10,i})|_{i=0,\ldots,15}$ need to be reduced to the pair of intermediate bytes (z_{9,i_1}, z_{9,i_2}) while completing the set $\mathcal{S}_{Z_9}(O_\Phi)$.

Being able to do so, another triplet consisting of plaintext-ciphertext $\{X, Z_{10} = Y_{10}^1\}$ together with the intermediate state Z_9 is required for the same perturbated instance of the cipher we selected in the setup phase of the cryptanalysis (see Sect. 3). We introduce the index m to refer to each of both triplets, i.e. $\{X_m, (Z_9)_m, (Z_{10})_m\}$ with $m = 1, 2$, where $m = 1$ concerns the triplet stored during the setup phase of the cryptanalysis. By applying the setup of the algorithm mentioned in Sect. 3.1 as well to the newly computed triplet – and in case the worst case scenario occurs again – we obtain the sets $\mathcal{S}_{Z_9}^m(z_{10,i})$ and $\mathcal{S}_{Z_9}^m(O_\Phi)$ for both triplets $m = 1, 2$, where the set $\mathcal{S}_{Z_9}^m(O_\Phi)$ is incomplete and some sets $\mathcal{S}_{Z_9}^m(z_{10,i})$ contain more than 2 bytes for both $m = 1, 2$.

First combine the retrieved information by taking the union $\bigcup_{m=1,2} \mathcal{S}_{Z_9}^m(O_\Phi)$, denoted by $\mathcal{S}_{Z_9}^{\cup}(O_\Phi)$, and removing in all sets $\mathcal{S}_{Z_9}^m(z_{10,i})|_{m=1,2;i=0,\ldots,15}$ those intermediate bytes $z_{9,i} \in \mathcal{S}_{Z_9}^{\cup}(O_\Phi)$. Then repeat the following steps for each output byte $z_{10,i}$ one at a time:

Step 1: Take the intersection $\bigcap_{m=1,2} \mathcal{S}_{Z_9}^m(z_{10,i})$, denoted by $\mathcal{S}_{Z_9}^{\cap}(z_{10,i})$:
If $|\mathcal{S}_{Z_9}^{\cap}(z_{10,i})| = 2$, then this set identifies the pair of intermediate bytes (z_{9,i_1}, z_{9,i_2}) corresponding to the output byte $y_{10,i}^1$ and is assigned to the set $\mathcal{S}_{Z_9}(y_{10,i}^1)$. Go to Step 4.
Else take all possible combinations when choosing two bytes out of $\mathcal{S}_{Z_9}^{\cap}(z_{10,i})$, with a total number of $\binom{|\mathcal{S}_{Z_9}^{\cap}(z_{10,i})|}{2}$.

Step 2: Repeat the following for each combination and for both triplets $\{X_m, (Z_9)_m, (Z_{10})_m\}$ with $m = 1, 2$: construct $(Z_9)_m|_{m=1,2}$ in which the chosen two bytes are replaced by '00' while the others remain fixed to their original value. Compute the perturbated final round R_{10}' and compare both output bytes $(z_{10,i})_m|_{m=1,2}$;

Step 3: **If** only one collision occurs in Step 2 for all possible combinations, then the two bytes chosen out of $\mathscr{S}^{\cap}_{Z_9}(z_{10,i})$ in the combination for which the collision occurred, identify the pair of intermediate bytes (z_{9,i_1}, z_{9,i_2}) corresponding to the output byte $y^1_{10,i}$ and are assigned to the set $\mathscr{S}_{Z_9}(y^1_{10,i})$. Go to Step 4.

Else in the case more than one collision occurs (hence for more than one combination), go back to Step 1 and continue with the next output byte. At the end, repeat all these steps again for the output bytes for which more than one collision occurred. In case multiple collisions continue to occur, start over again with a new and different triplet $m = 2$.

Step 4: Assign all intermediate bytes of both sets $\mathscr{S}^{m}_{Z_9}(z_{10,i})|_{m=1,2}$ which are not an element of $\mathscr{S}_{Z_9}(y^1_{10,i})$ to the set $\mathscr{S}^{\cup}_{Z_9}(O_\Phi)$ and remove in all remaining sets $\mathscr{S}^{m}_{Z_9}(z_{10,j})|_{m=1,2;i<j}$ those intermediate bytes $z_{9,i} \in \mathscr{S}^{\cup}_{Z_9}(O_\Phi)$.

At the end, we obtained the sets $\mathscr{S}_{Z_9}(y^1_{10,i})|_{i=0,\ldots,15}$ and trivially $\mathscr{S}_{Z_9}(O_\Phi) = \mathscr{S}^{\cup}_{Z_9}(O_\Phi)$.

Concerning Step 3, always at least one collision occurs, i.e. for the combination in which the two bytes chosen out of $\mathscr{S}^{\cap}_{Z_9}(z_{10,i})$ are in fact the pair of intermediate bytes (z_{9,i_1}, z_{9,i_2}) producing the input byte $y^1_{9,i}$ of the original final round R_{10}. Hence when both bytes are replaced by '`00`' in $(Z_9)_m|_{m=1,2}$ while the other bytes remain fixed to their initial value, the 8-byte input of B_9^{-1} remains unmodified. The latter ensures that the 4-byte input Y_9^2 of O_Φ is unchanged, i.e. $(\varphi_1, \varphi_2, \varphi_3, \varphi_4)$, such that the output of O_Φ remains zero and hence we can observe the real ciphertext for both $m = 1, 2$: $(Z_{10})_m = Y^1_{10} \oplus O_\Phi(\varphi_1, \ldots, \varphi_4) = Y^1_{10}$. The former ensures that the input byte $y^1_{9,i}$ becomes '`00`' for both $m = 1, 2$ which causes that the corresponding output bytes $y^1_{10,i}$ in the ciphertext collide.

The probability that only one collision occurs, equals $(\frac{2^8-1}{2^8})^{[\binom{|\mathscr{S}^{\cap}_{Z_9}(z_{10,i})|}{2}-1]}$. In the worst case, i.e. when $|\mathscr{S}^{\cap}_{Z_9}(z_{10,i})| = 10$, the probability becomes ≈ 0.84 . Hence it is assumed that the algorithm succeeds in the worst case with only 1 or 2 additional triplets.

A.2 Algorithm: Recover the Linear Input Encoding Up to an Unknown Constant Factor of Each Perturbated Intermediate Round $R'_r|_{r=2\leq r\leq 9}$

The adversary is able to observe the linear equivalent output $\Lambda'_r Y^1_r$ of the original intermediate round R_r due to the knowledge of the linear output encoding up to an unknown constant factor, i.e. $M_r^{-1}[0..15]'' = \Lambda'_r \circ M_r^{-1}[0..15]$. Moreover, due to the fact that the `MixColumns` step is separated from the S-boxes, he can apply its inverse and observe $Y_r^{1\prime} = \mathtt{MC}^{-1}(\Lambda'_r Y^1_r) = \Lambda'_r \circ \mathtt{MC}^{-1}(Y^1_r)$. Hereby, the diffusion property of the `MixColumns` step is lost and hence there is again a one-to-one correspondence between single active input and output bytes, which is the `ShiftRows` step. Now we describe each step of the algorithm:

Assign a pair of intermediate bytes $(z_{r-1,i_1}, z_{r-1,i_2})$ *of* Z_{r-1} *to each input byte* $y^1_{r-1,i}$ *of* R_r. By making each intermediate byte $z_{r-1,i}$ one at a time active and keeping track of the corresponding active bytes in $Y_r^{1'}$, a pair $(z_{r-1,i_1}, z_{r-1,i_2})$ is assigned to each output byte $y_{r,i}^{1'}$ due to the specific predefined structure of the $A_{r-1}^{(l)-1}|_{l=1,\dots,7}$ matrices in M_{r-1}^{-1}. By inverting the ShiftRows step, these pairs are reassigned to the corresponding input bytes $y_{r-1,i}^1$ of R_r.

Decompose the linear input encoding M_{r-1}^{-1}. This step is completely similar as the one described in Sect. 3.1, but then applied to R_r. Hence search for two different values of the intermediate bytes, i.e. $\{(z_{r-1,i_1}, z_{r-1,i_2}), (z_{r-1,i_1} \oplus \text{'01'}, z'_{r-1,i_2})\}$, which both map onto the same value of the input byte $y_{r-1,i}^1$ by indirectly observing the corresponding output byte $y_{r,i}^{1'}$ in $Y_r^{1'}$. As a result, a relation between both coefficients is derived, i.e. $c_{i,1} = \varepsilon_i \bullet c_{i,2}$, with $\varepsilon_i = z_{r-1,i_2} + z'_{r-1,i_2}$. By assigning '01' to $c_{i,2}$, only the linear equivalent input byte $y_{r-1,i}^1$ can be recovered, i.e. $\alpha_{r-1,i} \bullet y_{r-1,i}^1 = \varepsilon_i \bullet z_{r-1,i_1} + z_{r-1,i_2}$ where $\alpha_{r-1,i} \in \text{GF}(2^8) \setminus \{0\}$ is unknown.

As a result, we obtain the first 16 rows of the linear input encoding M_{r-1}^{-1} up to an unknown 16×16 diagonal linear bijection Λ_{r-1}, i.e. $M_{r-1}^{-1}[0..15]' = \Lambda_{r-1} \circ M_{r-1}^{-1}[0..15]$ with $\Lambda_{r-1} = diag(\alpha_{r-1,0}, \dots, \alpha_{r-1,15})$. Each row of $M_{r-1}^{-1}[0..15]'$ is all '00' except a '01' and ε_i on the relevant columns.

Eliminate the MixColumns step in the preceding round R_{r-1}. The algorithm to convert Λ_{r-1} into Λ'_{r-1} in which the unknown diagonal entries corresponding to each MixColumns step $\text{MC}_i|_{i=0,\dots,3}$ are identical s.t. the order between Λ'_{r-1} and the MixColumns step can be reversed, is identical to the one applied to the penultimate round R_9 in Sect. 3.2. However there is one special case, i.e. to generate Λ'_1 when applied to the first round R_1. In contrast to all intermediate rounds, the first round lacks a linear input encoding, and thus it is possible to make only one of the four input bytes to the i-th MixColumns step MC_i active by modifying the corresponding byte in the plaintext X. So for $r = 2$, each set $\mathscr{S}_X(\text{MC}_i)$ contains exactly four plaintext bytes. Moreover, when constructing the solution sets $\mathscr{S}$ in Step 3 for each modified plaintext byte in $\mathscr{S}_X(\text{MC}_i)$, the algorithm only needs to keep track of single active input bytes. This simplifies the algorithm and increases the performance in this special case.

As a result, we recover the new submatrix $M_{r-1}^{-1}[0..15]'' = \Lambda'_{r-1} \circ M_{r-1}^{-1}[0..15]$, where $\Lambda'_{r-1} = diag(D_{r-1}, D_{r-1}, D_{r-1}, D_{r-1})$ with $D_{r-1} = diag(\alpha_{r-1,0}, \alpha_{r-1,1}, \alpha_{r-1,2}, \alpha_{r-1,3})$, which transforms the 43-byte intermediate value Z_{r-1} into the linear equivalent input $\Lambda'_{r-1} Y_{r-1}^1$ of R_r.

B Complexity

The time complexity of our cryptanalysis is expressed in the number of perturbated round evaluations (parsing system of equations over $\text{GF}(2^8)$). With respect to the round operations, other computations in the attack are negligible and have been ommitted for simplicity.

- *Setup phase of the cryptanalysis:* encryption of one randomly chosen plaintext by all four correlated instances of the perturbated white-box AEw/oS implementation requires $4 \cdot 10$ round executions;
- *Analysis of the final round:* construction of both sets $\mathscr{S}_{Z_9}(y^1_{9,i})$ and $\mathscr{S}_{Z_9}(O_\Phi)$ for the final round needs (1) without the worst case scenario 43 round evaluations or (2) with the worst case scenario – which occurs with a negligible probability – an additional $2 \cdot (8 \cdot 4 \cdot 10 + 43 + 2 \cdot 5 \cdot \binom{10}{2})$ (under the assumption that only eight plaintexts need to be encrypted in order to find an additional triplet with probability $1 - (1/2)^8 \approx 0.996$); the recovery of the coefficients of the linear combination up to an unknown factor α for each pair intermediate bytes demands $16 \cdot 2^8$ round operations;
- *Separate the S-boxes:* elimination of the `MixColumns` step in the penultimate round R_9 requires $43 + 4 \cdot 8$ round evaluations;
- *Decomposing the rounds:* construction of the set $\mathscr{S}_{Z_{r-1}}(y^1_{r-1,i})$ for all intermediate rounds $R_r|_{2 \leq r \leq 9}$ needs $8 \cdot 43$ round evaluations; the recovery of the coefficients of the linear combination up to an unknown factor α for each pair of intermediate bytes for the intermediate rounds $R_r|_{2 \leq r \leq 9}$ demands $8 \cdot 16 \cdot 2^8$ round operations; the elimination of the `MixColumns` step in the rounds $R_r|_{1 \leq r \leq 8}$ requires $7 \cdot (43 + 4 \cdot 8) + (4 \cdot 4)$ round evaluations; finally, in order to define the linear equivalent key, $10 \cdot 16 \cdot 2^8$ perturbated rounds need to be executed.

This brings the total worst time complexity down to $80493 = 2^{16.2966}$ perturbated round evaluations.

The space complexity of our cryptanalysis is negligible.

A Program Generator for Intel AES-NI Instructions

Raymond Manley* and David Gregg**

Lero@TCD
School of Computer Science and Statistics
Trinity College Dublin, Ireland
manleyr@tcd.ie,David.Gregg@cs.tcd.ie

Abstract. Recent Intel processors provide hardware instructions that implement a full AES round in a single instruction. Existing libraries use hand-tuned assembly language to overlap the execution of multiple AES instructions and extract maximum performance. We present a program generator that creates optimized AES code automatically from a simple, annotated C version of the code. We show how this generator can be used to rapidly create highly optimized versions of several AES modes. The resulting code generated has performance that is equal to, or up to 7% faster than the hand-tuned assembly libraries from Intel.

Keywords: AES, AES-NI, program generator, code generation.

1 Motivation

Intel has recently extended its popular x86 architecture to support the Advanced Encryption Standard (AES) [1,2]. The new instructions perform a full AES round in a single instruction. Making good use of these instructions is not simple because they have a latency of several cycles. Good performance depends on the execution of multiple AES instructions being overlapped in the processor's pipeline, while also carefully managing other resources such as registers. Existing compilers do not do this well, so Intel provides a library of highly-optimized assembly routines implementing various AES modes [3,4].

Assembly code is both difficult and expensive to understand, maintain or modify. The need for assembly language programming is a major barrier to experimenting with new variations of the code. New versions of architectures often require changes in the assembly. Maintaining multiple versions of the same basic piece of assembly code is a costly software engineering problem. As newer versions of architectures appear, sub-optimal code may be used, simply to avoid creating yet another version.

* Supported by the Irish Research Council for Science Engineering and Technology (IRCSET) and Intel Ireland.

** This work was additionally supported, in part, by Science Foundation Ireland grant 03/CE2/I303_1 to Lero – The Irish Software Engineering Research Centre (www.lero.ie).

G. Gong and K.C. Gupta (Eds.): INDOCRYPT 2010, LNCS 6498, pp. 311–327, 2010.

Assembly programming also makes it difficult to combine multiple algorithms into a single piece of code. For example, it is often faster to do encryption and authentication together rather than separately, by interleaving the code from each algorithm [5]. But if the code for each algorithm consists of hand-tuned assembly library routines, manually creating a combined version is a cumbersome engineering task. In contrast, our generator can automatically intermingle two pieces of code with little programmer effort.

This paper describes a program generator that takes an annotated C version of AES code, generates many different variants and automatically finds a variant of the code that runs well on the target processor. The system is flexible enough that the cryptographer can specify multiple different ways of varying the code using several strategies. These strategies are also described. The generator uses an iterative, feedback-directed approach to finding efficient code for the particular architecture. The main contributions of this paper are:

- We automate the application of low-level optimizations used in the Intel library.
- We automate the search for the best variation of optimizations, and show that the resulting code can be equal to or faster than hand-tuned assembly.
- We show that CTR mode caching can give good speedups on AES-NI.
- We show that exploiting the properties of `xor` can speed up CBC mode.
- We show the ease of optimizing when using function-stitching with GCM.

The remainder of this paper is organized as follows: Section 2 provides background on Intel's AES instruction set extension. Our implementation of the program generator is described in Section 3. A breakdown of optimizations and features are in Sections 4—6, with the flexibility of the generator with AES counter (CTR) mode discussed in Section 4, making algorithmic variations with AES cipher-block-chaining (CBC) mode in Section 5, and exploiting instruction level parallelism with AES Galois/Counter Mode (GCM) in Section 6. An analysis of our experimental results appears in Section 7 with related and future work mentioned in Sections 8 and 9, respectively. Our conclusions are offered in Section 10.

2 Intel AES-NI Instructions

The Intel Advanced Encryption Standard New Instructions (AES-NI) are six new x86 instructions to support AES encryption, decryption and key expansion which debuted on the Intel Westmere architecture [4]. The `aesenc` instruction performs a full "round" of encryption in a single instruction. The AES algorithm consists of 10, 12 or 14 "rounds of encryption", which corresponds to using key sizes of 128, 192, or 256 bits, respectively. The code for encrypting a single block consists of a chain of 10, 12, or 14 `aesenc` instructions. The output of each round becomes the input to the next round, so each round must complete before the subsequent one can start. This can be seen in the loop of Figure 1, which shows the C code implementation of our round 1 AES counter mode (CTR).

```
#include <wmmintrin.h>
#include "aes-table.h"
void AES_CTR_Encrypt(__m128i* plaintext, __m128i* ciphertext,
                     __m128i* key, long long ivec, long nonce,
                                                int blocks){
  __m128i result, result0, result1;
  __m128i fake_key, saved_r1, table_mask;
  __m128i counter_block = _mm_setzero_si128();
  unsigned scalar_key0, scalar_result0, scalar_result1;
  unsigned my_counter, scalar_counter = 0;
  int i = 0;

  counter_block = _mm_insert_epi64(counter_block, ivec, 1);
  counter_block = _mm_insert_epi32(counter_block, nonce, 1);
  counter_block = _mm_srli_si128(counter_block, 4);

  result0 = _mm_xor_si128(counter_block, key[0]);
  scalar_key0 = _mm_extract_epi32(key[0], 3);
  result1 = _mm_insert_epi32(result0, scalar_key0 & 0xFFFFFF, 3);
  saved_r1 = _mm_aesenc_si128(result1, key[1]);
  table_mask = _mm_cvtsi32_si128(table3[0]);
  saved_r1 = _mm_xor_si128(saved_r1, table_mask);

  for(i = 0; i < blocks; i++){
    my_counter = scalar_counter++;
    scalar_result0 = ((_bswap(scalar_key0)) & 0xFF) ^ my_counter;
    table_mask =  _mm_cvtsi32_si128(table3[scalar_result0]);
    result1 = _mm_xor_si128(saved_r1, table_mask);

    result = _mm_aesenc_si128(result1, key[2]);
    result = _mm_aesenc_si128(result, key[3]);
    result = _mm_aesenc_si128(result, key[4]);
    result = _mm_aesenc_si128(result, key[5]);
    result = _mm_aesenc_si128(result, key[6]);
    result = _mm_aesenc_si128(result, key[7]);
    result = _mm_aesenc_si128(result, key[8]);
    result = _mm_aesenc_si128(result, key[9]);

    fake_key = _mm_xor_si128(key[10], plaintext[i]);
    result = _mm_aesenclast_si128(result, fake_key);
    ciphertext[i] = result;
  }
}
```

Fig. 1. AES CTR (round 1) encryption in C using AES-NI instructions

In the current hardware implementation of AES-NI, the `aesenc` instruction has a latency of six cycles [6]. Thus, encrypting a 16-byte block using a 128 bit key requires ten rounds, or at least 60 cycles. However, a new `aesenc` instruction can be started every two cycles. Therefore, with parallel modes of operation, such as AES CTR, the encryption of multiple blocks can be overlapped.

Intel provides a library of hand-tuned assembly routines which overlap the execution of AES instructions [4]. The library uses standard techniques for exploiting instruction-level parallelism (ILP), such as manually unrolling loops and manually interleaving the instructions from the unrolled iterations. For example, the AES CTR version unrolls the loop four times and interleaves the resulting instructions. The assembly code is carefully written to make good use of processor resources and achieves excellent performance.

3 The AES-GEN Generator

Our generator (AES-GEN) automates many of the optimization techniques used in the Intel library and adds some additional ones. The main difference is that the system of creating variants of the AES code is general. So it can be applied to any of the AES modes or similar code, and can find efficient code for any processor with the AES-NI instruction set.

Our generator builds code using two types of variants. The first type relates to choices in the basic source code that is generated. For example, you can encourage key values to be stored in registers by assigning them to variables. We call these *algorithmic variants.* The second type of variant is in the ordering of the code that is generated. For example, in CTR, the main loop may be unrolled several times and the instructions from the different iterations can be interleaved. These types of variants are aimed at better exploiting instruction-level parallelism, so we call them *ILP variants.*

Figure 2 shows the structure of the AES-GEN system. In the first phase it considers algorithmic variants, such as assigning key values to registers, using streaming stores, and `xor` options. This is done with a pass through a Python-powered template engine called Cheetah [7]. The generator then applies optimizations to the post-templated code to exploit ILP, such as loop unrolling and instruction scheduling to build the ILP variants. The resulting optimized C code is compiled and, based on performance, the tuning script modifies the parameters to the generator and a new version of the code is created. This iterative process is repeated thousands of times, until an efficient variant of the code is found.

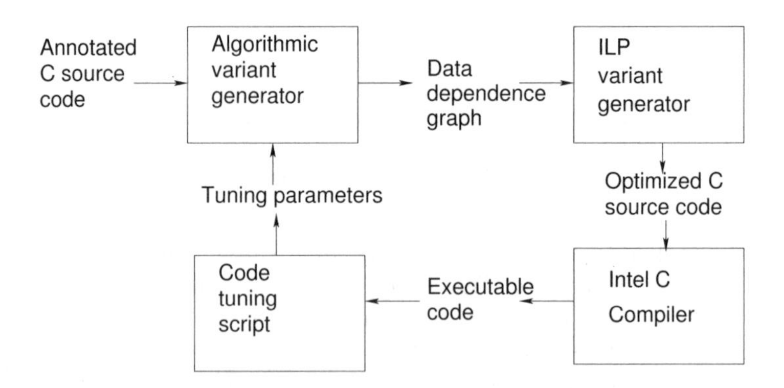

Fig. 2. Structure of the AES-GEN generator

The input to the generator is annotated C code which implements the encryption loop. The annotations can be used to express algorithmic variants and other information. After converting the code to a simple version of static single assignment (SSA) form [8], a data dependency graph is built to feed into the scheduler.

AES-GEN uses two scheduling strategies. The basic block scheduler applies a standard list scheduling algorithm [9] to the data dependence graph to generate code that can be executed in parallel. For example, if the loop is unrolled with an unrolling factor of two, and there are no data dependencies between iterations, the basic block scheduler will build a data dependence graph that includes statements from both iterations. The instructions from both iterations will be interleaved in the generated code, which will usually allow them to execute in parallel. The generator uses variable renaming to eliminate false data dependencies.

The ILP variants that AES-GEN creates are based on standard compiler optimizations for VLIW architectures [10]. Like a standard VLIW compiler, we model data dependencies and build schedules of statements based on these data dependencies. However, unlike a standard VLIW compiler, we do not attempt to accurately model processor resources. We assume that we do not have a clear idea of the available resources and generate a large number of different variants of the code instead. We use iterative feedback and machine learning to find a variant of the code that fits the actual machine resources. The result is that our generator should does a reasonable job of finding good code for any current or future processors supporting the Intel AES instructions.

Nonetheless, we need some way to affect the amount of ILP that is exposed by the generated code. Exposing more ILP usually results in more code growth, more variable renaming and greater register pressure. We can adjust the ILP in several ways, while limiting code growth. We often use modulo scheduling [11] instead of loop unrolling. Modulo scheduling is a form of software pipelining where the length of the loop kernel is fixed in advance and instructions are scheduled based on a measure of their dependency modulo value.

Each statement in the source code has a latency value, which is a notional number of cycles that it will take to theoretically execute. AES-GEN uses these notional latencies when building the schedule. By increasing or decreasing the latencies of statements, the ordering of the statements in the schedule can be changed, which in turn affects the amount of ILP in the schedule. For basic block scheduling, the ordering of statements is determined entirely by the latencies.

Our modulo scheduler can also control the amount of exposed ILP by varying the *initiation interval* (II). The II is the number of notional cycles that elapse between starting successive iterations of the loop. Lower II values result in higher amounts of instruction-level parallelism that exist in the generated code. By varying both the II and the latency of the instructions in the loop, a very large space of possible modulo schedules for the loop can be explored. Figure 3 shows a possible transformation of the CTR loop (from Figure 1) when latencies are set to 2 cycles per instruction and an II of 6.

Part of the motivation for building AES-GEN came from results obtained from an earlier and less robust generator we created that built only counter and cipher-block-chaining code [12]. These algorithms were hard-coded into a generator and left us with little flexibility to try slightly different algorithmic variants to achieve faster AES implementations.

```
/* pre modulo-scheduled code */
  for(i = 0; i < blocks - 4; i += 1 ){
    sp0_result = _mm_aesenc_si128(sp0_result, key[9]);
    sp1_result = _mm_aesenc_si128(sp1_result, key[6]);
    sp2_result = _mm_aesenc_si128(sp2_result, key[3]);
    sp3_table_mask = _mm_cvtsi32_si128(table3[sp3_scalar_result0]);
    sp4_scalar_counter =  scalar_counter++;
    sp4_fake_key = _mm_xor_si128(key[10], plaintext[i + 4]);
    sp0_result = _mm_aesenclast_si128(sp0_result, sp0_fake_key);
    sp1_result = _mm_aesenc_si128(sp1_result, key[7]);
    sp2_result = _mm_aesenc_si128(sp2_result, key[4]);
    sp3_result1 = _mm_xor_si128(saved_r1 , sp3_table_mask);
    sp4_my_counter = sp4_scalar_counter;
    ciphertext[i] = sp0_result8;
    sp1_result = _mm_aesenc_si128(sp1_result, key[8]);
    sp2_result = _mm_aesenc_si128(sp2_result, key[5]);
    sp3_result = _mm_aesenc_si128(sp3_result1, key[2]);
    sp4_scalar_result0 = ((_bswap(scalar_key0)) & 0xFF) ^ sp4_my_counter;
    // variable copy cleanup
  }
/* post modulo-scheduled code */
```

Fig. 3. An example modulo-scheduled body of the CTR (round 1) encryption loop

AES-GEN uses annotated C code as input. Therefore, prototype implementations of AES code can be fed into the generator immediately to give us quick feedback on its performance on the target architecture. This allows an exploratory approach to thinking about new ways to write the AES code. With an assembly language implementation, the cost of experimenting with new ways of writing the code is very high. Every change made may require a completely different schedule and register allocation. AES-GEN allows us to write a straightforward C implementation of the code and then it handles all low-level details associated with finding a good schedule.

4 Generator Flexibility with CTR

The first algorithm we ported to AES-GEN was AES CTR. Apart from the counter increment, this algorithm has no cyclic data dependences, so a lot of ILP is available.

The CTR code AES-GEN builds shows that the same modulo scheduled code works well for various input sizes of both 1K and 32K. When loops are unrolled, a second loop is necessary to encrypt any remaining blocks. This second loop is slow, which is a problem for very small inputs which spend a large proportion of time in the second loop. With modulo scheduling there is a single loop, which flows smoothly from one iteration to another.

With the ease of generating code variants from different source files, we are able to explore optimizations very specific to each AES mode without significant rewrites as would be necessary with hand-tuned assembly. Specifically to CTR, the 16-byte counter value must be incremented for each block. If the loop is unrolled, counter increments are scheduled together as they are the first statements of each iteration. However, this creates a data dependency from one increment to the next, which slightly limits ILP. Splitting the counter into multiple variables is a possible solution [4], but this increases register pressure. With modulo scheduling, the statements from multiple iterations are overlapped seamlessly and there is only a single copy of each statement within the generated loop body. For a loop such as the one for CTR-128, where an iteration of the loop takes around 20 cycles, the cyclic data dependence from the counter increment is too small to affect the schedule.

Faster Than "Optimal" CTR

We used this flexibility to create implementations of AES CTR mode that are faster than the "optimal" code. The Intel AES instructions implement a full round of AES in a single instruction. On the Intel Westmere microarchitecture, the AES instructions have a throughput of one instruction every two cycles. Therefore, if an implementation of AES CTR completes each round in an average of a little over 2 cycles per round, we consider it close to "optimal". Table 1 shows the cycles/round for our generated versions of AES-CTR. These implementations are pretty close to "optimal", especially for large inputs.

Table 1. Performance of AES-GEN CTR code in cycles per AES round

Encryption Mode	**AES-128**		**AES-256**	
	1K buffer	4080B buffer	1K buffer	4080B buffer
CTR (plain)	2.194	2.007	2.142	2.002
CTR (round 1)	2.106	1.956	2.017	1.905
CTR (round 2)	2.230	2.077	1.982	1.854

Our implementations of AES achieve good performance by keeping the hardware AES unit running at maximum capacity. However, while the AES unit is saturated, the rest of the processor core is largely idle. If we could use these other idle resources to implement some of the AES CTR rounds, we might be able to exceed the speed of the "optimal" code.

We explored a new version of the AES CTR code that mixes Intel AES instructions and a traditional AES implementation based on table lookups. We used counter-mode caching [13] to reduce the number of table lookups for the first two rounds. We created two versions of the code: one where round 1 of AES is implemented using table lookups rather than the Intel AES instruction (shown in Figure 1), and another where both rounds 1 and 2 use table lookups.

These versions of AES CTR work only for inputs of up 4080 bytes (255 blocks). Despite the size limitation, the AES CTR code using table lookups is valuable. AES is commonly used to encrypt data packets in wireless networks, which are typically less than 2K in length. Table 2 show results of these implementations with input sizes of 1024 bytes and 4080 bytes.

The results in Table 2 show that replacing round 1 with scalar instructions and table lookups is faster than using the AES-NI equivalent. With a 1K buffer,

Table 2. Performance of AES-GEN CTR code measured in cycles per byte

Encryption Mode	AES-128		AES-256	
	1K buffer	4080B buffer	1K buffer	4080B buffer
CTR (plain)	1.371	1.255	2.142	1.752
CTR (round 1)	1.316	1.222	1.765	1.667
CTR (round 2)	1.398	1.298	1.734	1.622
CTR (Intel assembly)	1.38		1.88	

the 128 and 256 bit implementations of AES using this method achieves a 5% and 7% speedup, respectively (see Table 5). The maximum possible speedup from eliminating the AES instruction for one of fourteen rounds is about 7.1%, so a 7% speedup is an excellent result. Replacing 2 rounds with table lookups performs slightly worse than Intel's implementation with AES-128, but achieves over an 8% speedup when using a 256 bit key. We believe that the results in Tables 1 and 2 demonstrate that it is possible for AES CTR to run faster that the "optimal" possible when using just Intel AES instructions. These are also the first implementations—to our knowledge—that complete an AES round in under two cycles.

5 Algorithmic Variants with CBC

Unlike counter mode, cipher-block-chaining (CBC) mode does not pipeline well. AES CBC encrypts the input plaintext directly and in order to eliminate patterns in the ciphertext, CBC combines the output of encrypting one 16-byte block with the plaintext for the next block, using an `xor` operation. This results in a large cyclic data dependency in CBC code, which prevents pipelining of the encryption. This dependence makes CBC slow and makes optimization difficult. On the other hand, the difficulty of optimizing CBC makes any speedup very welcome.

Figure 4(a) shows a C translation of the assembly code in the Intel hand-optimized high-performance AES encryption library [4]. The code is a direct implementation of standard descriptions of CBC mode [14]. It combines the result (variable `r` in the code) from the previous block of encryption with plaintext of the current block, using an `xor` operation. It then applies the standard AES encryption algorithm to the result. The cyclic dependency in the loop includes all statements except two: the first statement in the loop, which loads the plaintext and the last statement, which stores the encrypted block to memory. This long cyclic dependency prevents any significant exploitation of instruction-level parallelism in the CBC loop.

It is possible to slightly shorten the length of the cyclic dependence in the CBC loop. The first two statements in the cyclic data dependency are `xor` operations. As the `xor` operator is both commutative and associative, these expressions can be rewritten to increase the amount of exploitable ILP.

Neither the value in `plain` nor the value in `key0` depends on any value computed within the cyclic data dependence. Figure 4(b) shows a version of the code

```
while(i < size) {
 p = plain[i];
 r = xor(r, p);
 r = xor(r, key0);
 r = aesenc(r, key1);
 r = aesenc(r, key2);
 /* more rounds */
 r = aesenc(r, key8);
 r = aesenc(r, key9);
 r = enclast(r, key10);
 cipher[i] = r;
 i++;
}
```

(a)

```
while(i < size-1) {
 r = xor(r, tmp);
 r = aesenc(r, key1);
 r = aesenc(r, key2);
 /* more rounds */
 r = aesenc(r, key8);
 p = plain[i+1];
 tmp = xor(p, key0);
 r = aesenc(r, key9);
 r = enclast(r, key10);
 cipher[i] = r;
 i++;
}
```

(b)

```
while(i < size-1) {
  r = aesenc(r, key1);
  r = aesenc(r, key2);
  /* more rounds */
  r = aesenc(r, key8);
  p = plain[i+1];
  tmp = xor(p, key0);
  fake = xor(tmp, key10);
  r = aesenc(r, key9);
  r = enclast(r, fake);
  correct = xor(r, tmp);
  cipher[i] = correct;
  i++;
}
```

(c)

Fig. 4. AES CBC algorithm implementations

with the load of the plaintext and the first `xor` operation scheduled to execute in parallel with the encryption code for the previous block.

With very careful programming, it is possible to further reduce the length of the cyclic dependency. Recall that the final round of AES encryption involves three steps: (1) substitute bytes, (2) shift rows, and (3) add round key. The add round key step is an `xor` operation, where the result of the previous steps are combined with the key using `xor`. All three steps are implemented with the single x86 `aesenclast` instruction. The result of the `aesenclast` is used to store the ciphertext to memory and is also fed into the next round using an `xor` operation. Given that the last stage of `aesenclast` is an `xor` computation, we can combine the `aesenclast` and `xor` statements.

Performing this transformation removes the second `xor` operation from the cyclic data dependence. However, a problem occurs because the result of the `aesenclast` operation is not just used in the encryption of the next block. The result is also stored to memory as the ciphertext. It is possible to repair the result coming from the `aesenclast` operation, at the cost of adding another `xor` operation to the code.

Figure 4(c) shows code implementing this strategy. This version of the code contains an additional `xor` operation in comparison with all the other versions, so more work must be done on each iteration of the loop. However, both original `xor` operations have been removed from the cyclic data dependence. To our knowledge, we are the first to propose this way of writing AES CBC code.

CBC XOR Performance

We compared the four different versions of the AES CBC code, each with their own strategy for dealing with the `xor` operations. Table 3 shows the performance of each of the four versions, alongside performance numbers on from Intel's high performance, hand-tuned assembly version [4].

Table 3. Performance of AES CBC versions found in Figure 4 in cycles per byte

Encryption Mode	AES-128		AES-256	
	1K buffer	32K buffer	1K buffer	32K buffer
CBC version (a)	4.539	5.315	6.042	6.815
CBC version (b)	4.156	4.067	5.656	5.567
CBC version (c)	3.851	3.759	5.351	5.259
Intel assembly	4.15		5.65	

As these results show, removing the `xor` operations from the cyclic dependence chain makes a significant difference to performance. The version that removes both instructions from the cyclic dependence gives the best performance, even though it increases code size. Moving the load of the plaintext and the first `xor` operation forward affects the performance of our CBC implementations.

The influence of moving the load operation forward is particularly visible for the 32K inputs as version (a) of Figure 4 does not move the load and cache misses cause poor performance. Although Intel's hand-tuned assembly version of the code uses the same approach as in Figure 4(a), it performs faster. Our version in Figure 4(c) is nonetheless faster than the assembly, because it uses a fundamentally more efficient approach.

A final question is whether we could further improve the performance of our code in Figure 4(c) by more careful programming or writing in assembly. To address this question we compute the number of cycles per round of AES encryption. Given that the code is dominated by a large cyclic dependency of AES instructions, the best we can hope to do is perform the encryption in 6 cycles per round. Additional CBC results in Table 5 show the performance of each version of our CBC code in cycles per round completed and suggest that our CBC version (c) is already quite close to optimal and that any possible remaining speedups from careful coding are likely to be small.

Applicability of XOR Optimizations to AES CTR

Note that in AES CTR mode the result of the last AES round is combined with the plaintext using an xor operation. The technique we use in Figure 4(c) can also be used to reverse the order of these two operations in AES CTR. This slightly reduces the length of the long chain of data dependences in AES CTR. Given that AES CTR is fully parallelizable, there is no large-scale execution speed benefit. However, shortening the long dependence chain in CTR mode has three benefits.

First, it reduces the execution time of very short inputs. With short inputs the execution time is dominated by latency of one execution rather than throughput of many blocks. Secondly, where loop unrolling is used to implement AES-CTR, there needs to be a final sequential loop which deals with any iterations that are not an even multiple of the unrolling degree (this is one of the major reasons we do not advocate loop unrolling). Shortening the long dependence chain speeds up this loop. Finally, when creating modulo schedules, the number of iterations overlapped in the schedule depends on the length of the dependence chain that is being pipelined. A slightly shorter dependence chain means slightly less overlapping of iterations, which means slightly less code growth and register pressure.

6 Exploiting ILP by Function-Stitching with GCM

Many applications require both encryption and authentication of the same data. In AES, Galois/Counter Mode (GCM) [15] combines both the already-mentioned counter mode (CTR) of encryption with Galois authentication. The GCM algorithm operates by applying a GHASH function after a block of data has been encrypted in CTR mode. Due to this behaviour, there are opportunities to get speedups by overlapping the execution of both using a process called *function stitching* [5].

Writing code that schedules statements from two algorithms together in the same loop is tedious and time-consuming. Often the two algorithms contain different mixes of instructions, which create good opportunities for exploiting ILP between the algorithms. However, manual scheduling is always difficult. AES-GEN simplifies this problem greatly. We simply provide sequential code for the two algorithms in the same loop. The generator builds the data dependency graph and attempts to find a good software pipeline that overlaps the execution of the two algorithms.

Intel provides C code listings of GCM code [16] and we have adapted both the one- (1x) and four-at-a-time (4x) implementations for use in our generator. GCM encrypts data using AES CTR and authenticates it by applying the GHASH to the encrypted blocks. The main loop body of our GCM 1x and 4x implementation contains both the encryption and fully inlines the GHASH function and our modulo scheduler mixes statements from both algorithms in the schedule. With larger input sizes, we are able to achieve speedups by applying AES-GEN to function-stitched code. This can be seen in Table 4.

The results in Table 4 show that our generator is able to overlap the execution of the AES encryption code and the Galois field multiplication using modulo scheduling. As GCM is essentially CTR encryption and authentication, we can safely say that the GHASH function takes the majority of run time. The GHASH

Table 4. Performance of AES GCM 128 in various modes in cycles/byte

Encryption Mode	**AES-128**		
	1K buffer	4K buffer	16K buffer
GCM 1x (AES-GEN)	5.175	4.892	4.883
GCM 1x (Intel C)	5.49	5.36	5.33
GCM 4x (AES-GEN)	3.964	3.597	3.505
GCM 4x (Intel C)	4.16	3.88	3.70
GCM 4x (Intel assembly)	3.85	3.60	3.54

function has a large dependency chain and despite this, speedups are achieved in both 1x and 4x versions by pulling out several instructions from GHASH and replacing them with constants.

We also used associating rules to replace sequential chains of `xor` operations with balanced trees of `xor` operations. Even these improvements, it is difficult to exploit ILP with GCM 4x. Each loop iteration contains 40 to 56 long-latency AES-NI instructions. This is why we see speedups of less than 1% in 4K and 16K and perform slower than the assembly version when using 1K of input data. We suspect that the biggest reason for not being able to match the hand-tuned assembly version is due to the compiler. We suspect that low-level optimizations are not being exploited in full. This is a basis for one of several ideas to improve our GCM results as we mention in Section 9.

7 Experimental Results

We tested our generated code on an Intel 3.2GHz Core i5 650 machine (cpu family 6, model 37), with all power management disabled, lightly loaded, and running 64-bit Linux. All code is compiled with the Intel C Compiler (icc) with -O3. We call our generated code from a simple timing harness that provides a random input of plaintext, a random key schedule, and a random initialization vector. We use the processor's time stamp counter to measure timings report the median time of 10 million runs. Our timings include only the encryption itself, not the key expansion.

Results for CTR, CBC (1 stream), and GCM (1x) in previous sections show our generator produces variants that are as fast or faster than their Intel hand-tuned counterparts. To significantly improve upon the assembly numbers in both CTR and CBC, we make algorithmic variations to the annotated C code before we pass it to AES-GEN. This shows the usefulness and versatility of AES-GEN in general. With CTR, we re-write the counter from a vector to a scalar value; with CBC, additional encryption streams requires minimal code additions; with GCM, we collapse the `xor` chains into trees. These high-level changes, written in C, are easy to make and AES-GEN creates further algorithmic variants—as described in Section 3—automatically.

More specifically detailing AES-GEN's performance requires a look at Table 5. This shows the execution time of our generated code in cycles per byte and cycles per completed AES round. For comparison and when available, we also include results from Intel [4,16] for inputs of 1K. Discussion on these results have been mentioned in greater detail in previous sections.

The main aim of AES-GEN is to automatically generate code that makes good use of the AES-NI instructions, rather than having to write the assembly code by hand. If the generated code is as efficient as the assembly language, then we get the benefits of maintainability, retargetability and flexibility without sacrificing performance. The results in Table 5 show that we have largely achieved that goal. In all cases but one, our generated code achieves performance that is at least as good as the assembly implementations. The exception is GCM, where

Table 5. Performance of generated code in cycles/byte and cycles/round. (*) Results for CTR (round 1) and CTR (round 2) reference 4080B, all others reference 32K.

	AES-128					AES-256				
	Intel	cycles/byte		cycles/round		Intel	cycles/byte		cycles/round	
Encryption Mode	1K	1K	32K*	1K	32K*	1K	1K	32K*	1K	32K*
CTR	1.38	1.371	1.258	2.194	2.013	1.88	1.875	1.756	2.142	2.007
CTR (round 1)*		1.316	1.222	2.106	1.956		1.765	1.667	2.017	1.905
CTR (round 2)*		1.398	1.298	2.230	2.077		1.734	1.622	1.982	1.854
CBC – 1 stream	4.15	3.852	3.753	6.162	6.005	5.65	5.352	5.253	6.117	6.004
CBC – 2 streams		1.930	1.878	3.087	3.004		2.682	2.628	3.065	3.003
CBC – 3 streams		1.293	1.256	2.069	2.009		1.792	1.762	2.048	2.013
CBC – 4 streams	1.33	1.298	1.270	2.077	2.033		1.819	1.771	2.079	2.024
CBC – 5 streams		1.290	1.257	2.064	2.011		1.798	1.756	2.055	2.007
GCM 1x	5.49	5.175	4.825	8.280	7.720		5.574	5.262	8.918	6.014
GCM 4x	3.85	3.964	3.527	6.342	5.643		4.543	4.037	5.192	4.614

our code is slightly slower for 1K inputs, but faster for larger input sizes (see Table 4).

Why Not Optimize with a Standard Compiler?

The techniques we use to optimize the AES implementations are mostly standard compiler optimizations to exploit instruction level parallelism. Most of these techniques have been implemented in compilers, particularly compilers for VLIW architectures [10]. In principle, a standard compiler could do most of these optimizations. However, exploiting instruction level parallelism is not a priority for compilers that target out-of-order architectures.

Even a quick glance at Table 6 shows the large performance gap between even aggressive optimization with a standard compiler compared to AES-GEN. If we decided to only pursue avenues of optimization based on the results with a standard compiler (icc in this case), we would never consider using CTR (round 2). It runs slower than both the standard and round 1 versions of CTR in both 128- and 256-bit modes. However, from Table 5, we see that CTR (round 2) with AES-GEN gives us our best counter results with a 256 bit keysize. We see similar behaviour with CBC results in Table 6—it's not clear that using more streams is better. This makes AES-GEN an important tool to fully experiment with different AES implementations.

Implementing efficient AES code would be much more straightforward if a standard compilers had good modulo schedulers. However, adding such a scheduler and other sophisticated optimizations needed for a compiler to reorder memory operations would be a significant engineering task. Out-of-order pipelines are sufficiently complex that a compiler can rarely predict, in advance, exactly which code ordering is likely to be best, so an approach that uses experimentation is valuable. AES-GEN was built to be flexible enough to explore specific optimizations for general problems.

Table 6. Standard compiler vs. AES-GEN in cycles/byte. (*) Results for CTR (round 1) and CTR (round 2) reference 4080B, all others reference 32K.

	AES-128				AES-256			
	1K		32K*		1K		32K*	
Encryption Mode	icc	gen	icc	gen	icc	gen	icc	gen
CTR	2.000	1.371	1.880	1.258	3.273	1.875	3.132	1.756
CTR (round 1)*	1.828	1.316	1.722	1.222	2.902	1.765	2.788	1.667
CTR (round 2)*	2.148	1.398	2.043	1.298	3.285	1.734	3.173	1.622
CBC – 1 stream	4.280	3.852	5.065	3.753	5.781	5.352	6.565	5.253
CBC – 2 streams	2.257	1.930	2.193	1.878	3.029	2.682	2.953	2.628
CBC – 3 streams	1.897	1.293	1.852	1.256	2.935	1.792	2.896	1.762
CBC – 4 streams	1.893	1.298	1.990	1.270	2.916	1.819	2.922	1.771
CBC – 5 streams	1.926	1.290	2.081	1.257	2.926	1.798	3.140	1.756
GCM 1x	7.820	5.175	7.571	4.825	9.340	5.574	9.260	5.262
GCM 4x	4.300	3.964	3.844	3.527	4.777	4.543	4.283	4.037

8 Related Work

The most closely related to work to ours has been from various outlets at Intel, including its hand-tuned assembly language library [4,16] for the AES-NI instructions. Both this library and our generator use standard techniques for optimizing assembly code in the presence of long-latency instructions. The main difference is that our generator automates the process of applying these and other optimizations. Although our performance is similar, the generated code is actually very different. We use modulo scheduling to execute multiple iterations together, whereas the Intel code uses loop unrolling to achieve the same goal. The Intel library code also achieves significant code size savings by combining the AES code for different key sizes.

Akdemir *et al.* present performance results for CBC using AES-NI on multi-core processors [6]. To achieve speedups with CBC, they utilize multiple cores and multiple threads. They also present a method for overlapping the execution of the key expansion with the encryption. Our implementations do not consider this technique, but would be worthwhile looking into as our attempts with function stitching with GCM provided excellent results.

Gopal *et al.* [5] proposed using function stitching as cryptographic applications often process pairs of independent algorithms. In AES GCM, for example, both encryption and authentication algorithms are serially executed. In their work, they present several different stitching techniques, among them stitching CBC with SHA-1 [17] which are executed in parallel within a single composite function to achieve better speeds—often at the speed of only the longer executing portion. Gopal *et al.* [18] optimized GCM further in a white paper released while our paper was under review by improving the GHASH function is by treating constants differently while encrypting four blocks at a time. We've implemented a similar technique as an algorithmic variant to assign any permutation of constants used in the GHASH function to registers.

Like AES-GEN, there are code generators that focus on a particular subset of problems as they are ideal for optimizing a small piece of code that use large amounts of processing time and which the best optimizations are not obvious. Code generators have been very successfully applied to several domains such as Spiral, which generates code for linear transformations [19] and FFTW, which generates code for computing the discrete Fourier transform [20].

9 Future Work

Our GCM 1K results are slightly slower than those published by Gueron [16]. Results published by Gopal *et al.* in [18] are also significantly better than our GCM results with higher input sizes. However, this comparison isn't exactly like for like. Gopal's measurements include the use of hyper-threading with two threads of GCM running simultaneously—our numbers do not. We believe that their GCM results are nonetheless significantly faster than ours and will investigate ways to improve performance. We want to augment our current testing framework to accommodate hyper-threading so we can better compare results.

Our program generation process takes input source code, makes algorithmic changes, is then sent to the variant generator, then compiled to exploit possible low-level optimizations. We'd like to include an additional step to further improve the compiled code. We have observed the post-compiler assembly code and believe that there are additional optimization opportunities that are not taken by the compiler.

We are currently working on expanding our benchmark suite with several other modes of encryption and authentication. While CTR, CBC, and GCM modes all have unique algorithm properties that test AES-GEN's flexibility, more AES benchmarks would further show its usefulness in a more general case. It is our belief that other loop driven algorithms that have many long-latency instructions would benefit from our tool. Additionally, with other researchers expressing interest in using AES-GEN, we will also work towards to a stable version that could be released.

10 Conclusion

AES-GEN is used to generate code that uses the Intel AES instructions. The generator reads in annotated source code and uses iterative methods to try to find a variant of the source code which executes fast on the target hardware. The generator is also a useful experimental tool for programmers. The ability to make small, exploratory changes to an algorithm that can be easily scheduled differently and quickly evaluated is extremely valuable.

Our results show that AES-GEN creates faster code without a massive increase in code size, due to its good modulo scheduler. We have implemented several algorithmic variants of AES CTR, CBC and GCM modes. Our standard implementations of AES counter perform almost exactly the same as the hand-tuned assembly code, which is a good result. However, we also discuss additional implementations that are faster than anything published by Intel; to our knowledge, we have the fastest cycle per round implementations of AES CTR and AES CBC on the Westmere architecture.

With AES-GEN, it is easy for us to generate CBC code that operates on any number of input streams, as we simply make those changes in high-level C code and re-run the generator. We also presented dependency reduction strategies for the cyclic-dependant algorithm by exploiting the properties of the `xor` operation. These strategies yielded good results and testing these changes would have been a time consuming task if one were to make the changes in assembly.

Similarly with GCM, we were able to try a number of techniques in an attempt to improve GCM code generation with minimal effort. Trying to optimize multiple algorithms running in the same loop body as GCM does would be tedious in assembly. AES-GEN's modulo scheduler is very effective at building good "function-stitched" code.

Our goal was to maximize the performance of the Intel AES-NI instructions. We believe that we have done that. AES-GEN generates many variations of the AES algorithm and a good solution is found quickly that is tuned for any particular architecture that supports AES-NI.

Acknowledgments. Thanks to Mike O'Hanlon and Vinodh Gopal of Intel for suggestions, feedback and comments on our generator research.

References

1. Specification for the Advanced Encryption Standard (AES) (2001)
2. Daemen, J., Rijmen, V.: The design of Rijndael: AES — the Advanced Encryption Standard. Springer, Heidelberg (2002)
3. Gueron, S.: Intel's New AES Instructions for Enhanced Performance and Security. In: Dunkelman, O. (ed.) Fast Software Encryption. LNCS, vol. 5665, pp. 51–66. Springer, Heidelberg (2009)
4. Gueron, S.: Intel Advanced Encryption Standard (AES) Instructions Set (White Paper). Intel Corp. (2010), `http://software.intel.com/file/24917`
5. Gopal, V., Feghali, W., Guilford, J., Ozturk, E., Wolrich, G., Dixon, M., Locktyukhin, M., Perminov, M.: Fast Cryptographic Computation on Intel Architecture Via Function Stitching (White Paper). Intel Corp. (2010), `http://download.intel.com/design/intarch/PAPERS/323686.pdf`
6. Akdemir, K., Dixon, M., Feghali, W., Fay, P., Gopal, V., Guilford, J., Ozturk, E., Wolrich, G., Zohar, R.: Breakthrough AES Performance with Intel AES New Instructions (White Paper). Intel Corp. (2010), `http://software.intel.com/file/27067`
7. Rudd, T.: Cheetah - The Python-Powered Template Engine (2007), `http://www.cheetahtemplate.org/`
8. Cytron, R., Ferrante, J., Rosen, B.K., Wegman, M.N., Zadeck, F.K.: Efficiently computing static single assignment form and the control dependence graph. ACM Trans. Program. Lang. Syst. 13(4), 451–490 (1991)
9. Skiena, S.S.: The Algorithm Design Manual. Springer, New York (1998)

10. Fisher, J.A.: Very Long Instruction Word architectures and the ELI-512. In: ISCA 1983: Proceedings of the 10th Annual International Symposium on Computer Architecture, pp. 140–150. ACM, New York (1983)
11. Rau, B.R.: Iterative modulo scheduling: an algorithm for software pipelining loops. In: MICRO 27: Proceedings of the 27th Annual International Symposium on Microarchitecture, pp. 63–74. ACM, New York (1994)
12. Manley, R., Gregg, D.: Code Generation for Hardware Accelerated AES. In: 21st IEEE International Conference on Application-specific Systems, Architectures and Processors (Poster Session), ASAP 2010 (2010)
13. Bernstein, D.J., Schwabe, P.: New AES Software Speed Records. In: Chowdhury, D.R., Rijmen, V., Das, A. (eds.) INDOCRYPT 2008. LNCS, vol. 5365, pp. 322–336. Springer, Heidelberg (2008)
14. Ehrsam, W.F., Meyer, C.H.W., Powers, R.L., Smith, J.L., Tuchman, W.L.: Product block cipher system for data security. Patent, US 3962539 (June 1976)
15. McGrew, D.A., Viega, J.: The Galois/Counter Mode of Operation, GCM (2004), `http://csrc.nist.gov/CryptoToolkit/modes/proposedmodes/gcm/gcm-spec.pdf` **GCM**
16. Gueron, S., Kounavis, M.E.: Intel Carry-Less Multiplication Instruction and its Usage for Computing the GCM Mode (White Paper). Intel Corp. (2010), `http://software.intel.com/file/24918`
17. Eastlake, D.E., Jones, P.E.: US Secure Hash Algorithm 1, SHA1 (2001), `http://www.ietf.org/rfc/rfc3174.txt?number=3174`
18. Gopal, V., Ozturk, E., Feghali, W., Guilford, J., Wolrich, G., Dixon, M.: Optimized Galois-Counter-Mode Implementation on Intel Architecture Processors. Intel Corp. (2010), `http://download.intel.com/design/intarch/PAPERS/324194.pdf`
19. Püschel, M., Moura, J.M.F., Johnson, J., Padua, D., Veloso, M., Singer, B., Xiong, J., Franchetti, F., Gacic, A., Voronenko, Y., Chen, K., Johnson, R.W., Rizzolo, N.: SPIRAL: Code Generation for DSP Transforms. Proceedings of the IEEE, special issue on Program Generation, Optimization, and Adaptation 93(2), 232–275 (2005)
20. Frigo, M., Steven, Johnson, G.: The Design and Implementation of FFTW3. Proceedings of the IEEE, 216–231 (2005)

ECC2K-130 on NVIDIA GPUs*

Daniel J. Bernstein[1], Hsieh-Chung Chen[2], Chen-Mou Cheng[3], Tanja Lange[4], Ruben Niederhagen[3,4], Peter Schwabe[4], and Bo-Yin Yang[2]

[1] Department of Computer Science
University of Illinois at Chicago
851 S. Morgan Street, Chicago, IL 60607-7053, USA
djb@cr.yp.to

[2] Institute of Information Science
Academia Sinica
128, Section 2, Academia Road, Taipei 11529, Taiwan
{kc,by}@crypto.tw

[3] Department of Electrical Engineering
National Taiwan University
1, Section 4, Roosevelt Road, Taipei 10617, Taiwan
doug@crypto.tw, ruben@polycephaly.org

[4] Department of Mathematics and Computer Science
Eindhoven University of Technology
Den Dolech 2, 5600 MB Eindhoven, Netherlands
tanja@hyperelliptic.org, peter@cryptojedi.org

Abstract. A major cryptanalytic computation is currently underway on multiple platforms, including standard CPUs, FPGAs, PlayStations and Graphics Processing Units (GPUs), to break the Certicom ECC2K-130 challenge. This challenge is to compute an elliptic-curve discrete logarithm on a Koblitz curve over $\mathbb{F}_{2^{131}}$. Optimizations have reduced the cost of the computation to approximately 2^{77} bit operations in 2^{61} iterations.

GPUs are not designed for fast binary-field arithmetic; they are designed for highly vectorizable floating-point computations that fit into very small amounts of static RAM. This paper explains how to optimize the ECC2K-130 computation for this unusual platform. The resulting GPU software performs more than 63 million iterations per second, including 320 million $\mathbb{F}_{2^{131}}$ multiplications per second, on a $500 NVIDIA GTX 295 graphics card. The same techniques for finite-field arithmetic and elliptic-curve arithmetic can be reused in implementations of larger systems that are secure against similar attacks, making GPUs an interesting option as coprocessors when a busy Internet server has many elliptic-curve operations to perform in parallel.

---

* Permanent ID of this document: 1957e89d79c5a898b6ef308dc10b0446. Date of this document: 2010.09.25. This work was sponsored in part by the National Science Foundation under grant ITR–0716498, in part by Taiwan's National Science Council under grant NSC-96-2221-E-001-031-MY3, and under grant NSC-96-2218-E-001-001, also through the Taiwan Information Security Center under grant NSC-97-2219-E-001-001, and under grant NSC-96-2219-E-011-008, and in part by the European Commission through the ICT Programme under Contract ICT–2007–216676 ECRYPT II and the ICT Programme under Contract ICT–2007–216499 CACE.

G. Gong and K.C. Gupta (Eds.): INDOCRYPT 2010, LNCS 6498, pp. 328–346, 2010.

Keywords: Graphics Processing Unit (GPU), Elliptic Curve Cryptography, Pollard rho, qhasm.

1 Introduction

The elliptic-curve discrete-logarithm problem (ECDLP) is the number-theoretic problem behind elliptic-curve cryptography (ECC): the problem of computing a user's ECC secret key from his public key. Pollard's rho method solves this problem in $O(\sqrt{\ell})$ iterations, where ℓ is the largest prime divisor of the order of the base point. A parallel version of the algorithm by van Oorschot and Wiener [18] provides a speedup by a factor of $\Theta(N)$ when running on N computers, if ℓ is larger than a suitable power of N. In several situations a group automorphism of small order m provides a further speedup by a factor of $\Theta(\sqrt{m})$. No further speedups are known for any elliptic curve chosen according to standard security criteria; this is the end of the story.

However, these asymptotic iteration counts ignore many factors that have an important influence on the cost of an attack. Understanding the hardness of a specific ECDLP requires a more thorough investigation. The publications summarized on `www.keylength.com`, giving recommendations for concrete cryptographic key sizes, all extrapolate from such investigations. To reduce extrapolation errors it is important to use as many data points as possible, and to push these investigations beyond the ECDLPs that have been carried out before.

Certicom published a list of ECDLP prizes in 1997 [11] in order to "increase the cryptographic community's understanding and appreciation of the difficulty of the ECDLP". These challenges range from very easy exercises, solved in 1997 and 1998, to serious cryptanalytic challenges. The last Certicom challenge that was publicly broken was a 109-bit ECDLP in 2004. Certicom had already predicted the lack of further progress: it had stated in [11, page 20] that the subsequent challenges were "expected to be infeasible against realistic software and hardware attacks, unless of course, a new algorithm for the ECDLP is discovered."

Since then new hardware platforms have become available to the attacker. Processor design has moved away from increasing the clock speed and towards increasing the number of cores. This means that implementations need to be parallelized in order to make full use of the processor. Running a serial implementation on a recent processor might take longer than 5 years ago, because the average clock speed has decreased, but if this implementation can be parallelized and occupy the entire processor then the implementation will run much faster. An extreme example of this high parallelism at reduced clock speed is the NVIDIA GTX 295 graphics card. This card contains two G200b Graphics Processing Units (GPUs); each GPU contains 30 cores; each core contains 8 ALUs; each ALU is capable of performing a 32-bit operation every cycle. Each ALU operates at just 1.242 GHz, but 480 ALUs together have a tremendous amount of computational power.

Certicom's estimate a decade ago was that ECC2K-130, the first "infeasible" challenge, would require (on average) 2700000000 "machine days" of computation.

Our main result is that a cluster of just 534 graphics cards running our software would solve ECC2K-130 in just 24 months.

In this paper we explain how we use Pollard's rho algorithm to compute ECDLP on GPUs. In particular, we give details on how we implement binary-field multiplication, a problem that at first seems quite poorly suited to GPUs, and on how we implement a complete ECDLP iteration function. Some of our optimizations are specific to ECC2K-130, but most of our implementation techniques can be reused in larger ECDLP computations. Furthermore, the binary-field arithmetic operations that we optimize are also the primary bottlenecks in various cryptographic computations; the same implementation techniques open up the interesting possibility of using GPUs as high-performance coprocessors to offload binary-field ECC computations from busy Internet servers.

This paper is part of a large collaborative project that has optimized ECDLP computations for several different platforms and that aims to break ECC2K-130. See [1] and `http://www.ecc-challenge.info` for more information about the project. All of the platforms use the same ECC2K-130 iteration function, allowing an objective comparison of the power of different platforms and putting our GPU speeds in perspective: finishing the computation in two years would require

- 1595 standard PCs (3.2 GHz AMD Phenom II X4 955 CPU) [1]; or
- 1231 PlayStation 3 computers (Cell CPU with 6 usable SPEs and 1 PPE) [10]; or
- 534 GTX 295 graphics cards, as shown in this paper; or
- 308 XC3S5000-4FG676 FPGAs [12];

or any combination of the above.

2 The GTX 295 Graphics Card

The most impressive feature of GPUs is their theoretical floating-point performance. Each of the 480 ALUs on a GTX 295 can dispatch a single-precision floating-point multiplication (with a free addition) every cycle. There are also 120 "special-function units" that can each dispatch another single-precision floating-point multiplication every cycle, for a total of 745 billion floating-point multiplications per second.

The most useful GPU arithmetic instructions for the ECC2K-130 computation are 32-bit logical instructions (ANDs and XORs) rather than floating-point multiplications, but 596 billion 32-bit logical instructions per second are still much more impressive than (e.g.) the 28.8 billion 128-bit logical instructions per second performed by a typical 2.4 GHz Intel Core 2 CPU with 4 cores and 3 128-bit ALUs per core.

However, the GPUs also have many bottlenecks that make most applications run slower, often one or two orders of magnitude slower, than the theoretical throughput figures would suggest. The most troublesome bottlenecks are discussed in the remainder of this section and include a heavy divergence penalty, high instruction latency, low SRAM capacity, high DRAM latency, and relatively low DRAM throughput per ALU.

2.1 The Dispatcher

The 8 ALUs in a GPU core are fed by a single *dispatcher*. The dispatcher cannot issue more than one new instruction to the ALUs every 4 cycles. The dispatcher can send this one instruction to a *warp* containing 32 separate *threads* of computation, applying the instruction to 32 pieces of data in parallel and keeping all 8 ALUs busy for all 4 cycles; but the dispatcher cannot direct some of the 32 threads to follow one instruction while the remaining threads follow another.

Branching is allowed, but if threads in a warp take different branches ("diverge") then the threads taking one branch will no longer operate in parallel with the threads in the other branch. For example, if 32 threads are split among all three branches in the code `if(x) if(y) aaa else bbb else ccc`, then the first threads will execute `aaa` while the other threads remain idle; next the second threads will execute `bbb` while the other threads remain idle; and finally the third threads will execute `ccc` while the other threads remain idle. The total time is the sum of the times taken by `aaa`, `bbb`, `ccc`, rather than the maximum of those times.

2.2 Instruction Latency

Each thread follows its instructions strictly in order. NVIDIA does not document the exact pipeline structure but recommends running at least 192 threads (6 warps) on each core to hide arithmetic latency. If 8 ALUs are fully occupied with 192 threads then each thread runs every 24 cycles; evidently the latency of an arithmetic instruction is below 24 cycles.

One might think that a single warp of 32 threads can keep the 8 ALUs fully occupied, if the instructions in each thread are scheduled for 24-cycle arithmetic latency (i.e., if an arithmetic result is not used until 6 instructions later). However, our experiments show that the minimum number of threads that can keep the 8 ALUs fully occupied is 128. Our experiments also show a drop in the ALU utilization for 128 threads when more than 25% of the instructions are complex instructions that include memory access, or when complex instructions are adjacent.

NVIDIA also recommends running many more than 192 threads to hide memory latency. This does not mean that one can achieve the best performance by simply running the maximum number of threads that fit into the core. Threads share several critical resources, as discussed below, so increasing the number of threads means reducing the resources available to each thread. The ECC2K-130 computation puts extreme pressure on shared memory, as discussed later in this paper; to minimize this pressure we ended up using just 128 threads, skirting the edge of severe latency problems.

2.3 SRAM: Registers and Shared Memory

Each core has 16384 32-bit registers divided among the threads running on the core. For example, if the core is running 256 threads, then each thread is assigned

64 registers. If the core is running 128 threads, then each thread is assigned 128 registers, although the "high" 64 registers are somewhat limited: the architecture does not allow a high register as the second operand of an instruction, and does not allow access to the last high register.

The core also has 16384 bytes of *shared memory* that provide variable array indexing and communication between threads. This memory is split into 16 banks, each of which can dispatch one 32-bit read or write every two cycles. If the 16 threads in a half-warp read from the same location or 16 different banks then there are no bank conflicts, but if the threads all read from different locations of the same bank then they take 16 times as long.

Threads also have fast access to an 8192-byte *constant cache*. This cache can broadcast a 32-bit value from one location to every thread simultaneously, but it cannot read more than one location per cycle.

2.4 DRAM: Global Memory and Local Memory

The CPU makes data available to the GPU by copying it into the DRAM on the graphics card outside the GPU. The cores on the GPU can then load data from this *global memory* and store results in global memory to be retrieved by the CPU. Global memory is also a convenient temporary storage area for data that does not fit into shared memory. However, global memory is limited to a throughput of just one 32-bit load from each GPU core per cycle, with a latency of 400–600 cycles.

Each thread also has access to *local memory*. The name "local memory" might suggest that this storage is fast, but in fact it is another area of DRAM, as slow as global memory. Instructions accessing local memory automatically incorporate the thread ID into the address being accessed, effectively partitioning the local memory among threads without any extra address-calculation instructions.

There are no hardware caches for global memory and local memory. Programmers can, and must, set up their own schedules for copying data to and from global memory.

2.5 Choice of GPU

We decided to focus on the GTX 295 because it provides an excellent price-performance ratio. The price of a GTX 295 was only about $500 when we began this project. We built several $2000 PCs, each containing two GTX 295s and a standard CPU, following the advice of Bernstein, Chen, Chen, Cheng, Hsiao, Lange, Lin, and Yang in [6] and [5]; each PC is (considering the extra cost of power and cooling) more than twice as expensive as a PC containing a CPU alone, but it gives us several times better performance.

Similar GPUs are also widely available in computing clusters, which typically use self-contained 1U rackmount Tesla S1060 units. Each S1060 contains 4 G200 (or G200b) GPUs and is microarchitecturally identical to a pair of slightly over-clocked GTX 295s. The United States TeraGrid network includes two such GPU clusters, namely Lincoln (384 GPUs) and Longhorn (512 GPUs).

There are two other GPU architectures of note: AMD's Evergreen (R8xx) GPUs, as in the Radeon HD 5970, and NVIDIA's very new Fermi (GF1xx) line of GPUs, as in the GTX 480. All of these GPUs pose similar parallelization challenges, but there are many differences in the details. The current dominance of G200-based Tesla GPUs in computer clusters makes these GPUs the most attractive target for now. We focus on the GTX 295 throughout this paper.

3 The ECDLP and Parallel Pollard Rho

The standard method for solving the ECDLP in prime-order subgroups is Pollard's rho method [15]. For large instances of the ECDLP, one usually uses a parallelized rho method due to van Oorschot and Wiener [18]. This section briefly reviews the ECDLP and the parallel rho method.

The ECDLP is the following problem: Given an elliptic curve E over a finite field $\mathbb{F}_q$ and two points $P \in E(\mathbb{F}_q)$ and $Q \in \langle P \rangle$, find an integer k such that $Q = [k]P$.

Let ℓ be the order of P, and assume in the following that ℓ is prime. The parallel rho method is a client-server approach in which each client does the following:

1. Generate a pseudo-random starting point R_0 as a known linear combination of P and Q: $R_0 = a_0P + b_0Q$;
2. apply a pseudo-random iteration function f to obtain a sequence $R_{i+1} = f(R_i)$, where f is constructed to preserve knowledge about the linear combination of P and Q;
3. for each i, after computing R_i, check whether R_i belongs to an easy-to-recognize set D, the set of distinguished points, a subset of $\langle P \rangle$;
4. if at some moment a distinguished point R_i is reached, send (R_i, a_i, b_i) to the server and go to step 1.

The server receives all the incoming triples (R, a, b) and does the following:

1. Search the entries for a *collision*, i.e., two triples $(R, a, b), (R', a', b')$ with $R = R'$ and $b \not\equiv b' \pmod{\ell}$;
2. obtain the discrete logarithm of Q to the base P as $k = \frac{a'-a}{b-b'}$ modulo ℓ.

The expected running time of this parallel version of Pollard's rho algorithm is approximately $\sqrt{\pi\ell/2}$ calls to the iteration function f, assuming perfectly random behavior of f.

4 ECC2K-130 and the Iteration Function

The specific ECDLP addressed in this paper is given in the Certicom challenge list [11] as Challenge ECC2K-130. The given elliptic curve is the Koblitz curve $E : y^2 + xy = x^3 + 1$ over the finite field $\mathbb{F}_{2^{131}}$; the two given points P and Q have order ℓ, where ℓ is a 129-bit prime.

This section reviews the definition of distinguished points and the iteration function used in [1]. For a more detailed discussion, an analysis of communication costs, and a comparison to other possible implementation choices, the interested reader is referred to [1].

4.1 Definition of the Iteration Function

A point $R \in \langle P \rangle$ is *distinguished* if $\mathrm{HW}(x_R)$, the Hamming weight of the x-coordinate of R in normal-basis representation, is smaller than or equal to 34. The iteration function is defined as

$$R_{i+1} = f(R_i) = \sigma^j(R_i) + R_i,$$

where σ is the Frobenius endomorphism and

$$j = ((\mathrm{HW}(x_{R_i})/2) \bmod 8) + 3.$$

The restriction of σ to $\langle P \rangle$ corresponds to scalar multiplication with a particular easily computed scalar r. For an input $R_i = a_i P + b_i Q$, the output of f will be $R_{i+1} = (r^j a_i + a_i)P + (r^j b_i + b_i)Q$.

Each walk starts by picking a random 64-bit seed s which is then expanded deterministically into a linear combination $R_0 = a_0 P + b_0 Q$. To reduce bandwidth and storage requirements, the client does not report a distinguished triple (R, a, b) to the server but instead transmits only s and a 64-bit hash of R. On the occasions that a hash collision is found, the server recomputes the linear combinations in P and Q for $R = aP + bQ$ and $R' = a'P + b'Q$ from the corresponding seeds s and s'. This has the additional benefit that the client does not need to keep track of the coefficients a and b or counters for how often each Frobenius power is used. This speedup is particularly beneficial for highly parallel architectures such as GPUs, which otherwise would need a conditional addition to each counter in each step.

4.2 Computing the Iteration Function

The main task for each client is to repeatedly compute the function f defined above. Each iteration starts with a point $R = (x, y)$, computes a normal-basis Hamming weight $\mathrm{HW}(x)$ and thus $j = ((\mathrm{HW}(x)/2) \bmod 8) + 3$, computes $\sigma^j(R) = (x^{2^j}, y^{2^j})$, and adds (x, y) to (x^{2^j}, y^{2^j}) on E.

The addition on E requires 2 multiplications, one squaring, 6 additions, and 1 inversion in $\mathbb{F}_{2^{131}}$ in affine Weierstrass coordinates; see, e.g., [7]. Inversions are significantly more expensive than multiplications but Montgomery's trick [14] reduces these expenses in a batch of inversions: it replaces m separate inversions with $3(m-1)$ multiplications and 1 inversion. For example, a 10-way batched inversion takes $3 \cdot (10 - 1) = 27$ multiplications and 1 inversion; each of the original inversions is thus replaced by 0.1 inversions and 2.7 multiplications.

The obvious way to compute x^{2^j} for $3 \le j \le 10$ is to first square 3 times, obtaining x^{2^3}, and then square repeatedly, at most 7 more times, until reaching

x^{2^j}. However, this involves several expensive branches depending on the value of j. A branch-free strategy stated in [8] is to compute $r = x^{2^3}, s = r+b_0(r^2+r), t = s + b_1(s^4 + s), u = t + b_2(t^{16} + t)$ where $j = 3 + b_0 + 2b_1 + 4b_2$, i.e., where $\mathrm{HW}(x) = 2b_0 + 4b_1 + 8b_2 + \cdots$. The computation $s = r + b_0(r^2 + r)$ is a *conditional squaring*, computing $s = r = x^{2^3}$ if $b_0 = 0$ or $s = r^2 = x^{2^4}$ if $b_0 = 1$, i.e., $s = x^{2^{3+b_0}}$; similarly $t = x^{2^{3+b_0+2b_1}}$ and $u = x^{2^{3+b_0+2b_1+4b_2}} = x^{2^j}$.

In total, each iteration in a batch of 10 iterations takes 4.7 multiplications, 0.1 inversions, 1 squaring, 2 3-squarings (where a 3-squaring means a sequence of 3 squarings), 2 conditional squarings, 2 conditional 2-squarings, 2 conditional 4-squarings, and 1 normal-basis Hamming-weight computation. As the batch size increases, the number of multiplications per iteration converges to 5 while the number of inversions per iteration converges to 0.

If each inversion is replaced by computing the $(2^{131} - 2)$nd power using the sequence of multiplications and squarings shown in [8, Figure 5.3] then each iteration in a batch of 10 iterations takes 5.5 multiplications, 1.3 squarings, 2.6 m-squarings for various m, 2 conditional squarings, 2 conditional 2-squarings, 2 conditional 4-squarings, and 1 normal-basis Hamming-weight computation.

4.3 Bitslicing

The next step in optimizing multiplications in $\mathbb{F}_{2^{131}}$, squarings in $\mathbb{F}_{2^{131}}$, etc. is to decompose these arithmetic operations into the operations available on the target platform.

Modern microprocessors operate on words of many bits, say n bits. The fast XOR logical instruction reads two n-bit words, computes n bit XORs (additions in $\mathbb{F}_2$) in parallel, and produces an n-bit output word. Similar comments apply to other logical instructions: AND (multiplication in $\mathbb{F}_2$), OR, etc.

If one n-bit word can be viewed as n independent elements of $\mathbb{F}_2$, then k n-bit words can be viewed as n independent elements of the vector space $\mathbb{F}_2^k$, for example representing n independent elements of the field $\mathbb{F}_{2^k}$ on a suitable basis. If an operation in $\mathbb{F}_{2^k}$ can be carried out with $T(k)$ additions and multiplications in $\mathbb{F}_2$ then the same operation can be carried out in parallel on n elements in $\mathbb{F}_{2^k}$ with $T(k)$ XOR instructions and AND instructions.

This technique is called *bitslicing*. It was introduced by Biham [9] in an implementation of the Data Encryption Standard. The speedups that can be achieved from bitslicing binary-field arithmetic on a modern CPU were demonstrated last year by Bernstein in [3]. We decided to use bitslicing on a GPU as it allows very efficient use of logical operations in implementing binary-field operations. The final cost of our optimized bitsliced multiplier turned out to be smaller than a lower bound for the cost of a traditional non-bitsliced table-based multiplier.

The most obvious challenge in efficient bitsliced arithmetic is to optimize $T(k)$, the number of bit operations required for an operation in $\mathbb{F}_{2^k}$. However, the requirement to work on nk bits in parallel creates additional challenges that are particularly troublesome for GPUs: an n-bit GPU instruction is efficient only if n is very large (32 bits times $\geq$128 threads), but if n is very large then the cost of accessing nk bits becomes a bottleneck.

The rest of this section reviews the techniques that produce the smallest number of bit operations known for the ECC2K-130 iteration function. The rest of this paper explains how we made the iteration function run quickly on the GPU: Section 5 analyzes 131-bit polynomial multiplication, and Section 6 analyzes the complete iteration function.

4.4 Choice of Basis

There is no irreducible trinomial of degree 131 in the polynomial ring $\mathbb{F}_2[z]$. The standard representation of $\mathbb{F}_{2^{131}}$ is as $\mathbb{F}_2[z]$ modulo an irreducible pentanomial, such as $z^{131} + z^{13} + z^2 + z + 1$. Multiplication on the basis $1, z, z^2, \ldots, z^{130}$ of $\mathbb{F}_2[z]/(z^{131} + z^{13} + z^2 + z + 1)$ involves one 131-bit polynomial multiplication and just 455 extra bit operations.

However, the ECC2K-130 iteration involves not only several multiplications but also also many squarings, including m-squarings, making normal bases particularly attractive. A squaring in normal basis corresponds to a simple cyclic shift of the coefficients. An m-squaring corresponds to a cyclic shift by m positions. Working with x in normal basis also removes the need to convert x from pentanomial basis to normal basis.

The main problem with normal bases is the cost of multiplication. However, $\mathbb{F}_{2^{131}}$ has a type-II optimal normal basis, and Shokrollahi's type-II multiplier [16] (see also [13]) has a surprisingly small overhead above the cost of polynomial multiplication. Bernstein and Lange [8] recently reduced this overhead further and showed that combining the optimal normal basis with an *optimal polynomial basis* achieves a significantly lower cost than a pentanomial basis for the ECC2K-130 iteration.

Our final GPU implementation uses this multiplier, as described in Section 6. Here we give a short summary of the mathematics necessary for understanding the implementation. For full details see [8].

Field elements such as x and y are represented on the permuted optimal normal basis $(\zeta+\zeta^{-1}, \zeta^2+\zeta^{-2}, \zeta^3+\zeta^{-3}, \ldots, \zeta^{131}+\zeta^{-131})$ of $\mathbb{F}_{2^{131}}$, where $\zeta \in \mathbb{F}_{2^{262}}$ is a primitive 263rd root of 1. Squaring is a permutation of coefficients in this basis, although no longer a simple cyclic shift. Multiplication has four steps:

- Convert each input to the *optimal polynomial basis* $(\zeta+\zeta^{-1}, (\zeta+\zeta^{-1})^2, (\zeta+\zeta^{-1})^3, \ldots, (\zeta+\zeta^{-1})^{131})$. A streamlined recursive conversion takes just 325 bit operations.
- Apply 131-bit polynomial multiplication, obtaining a product of the form $a_2(\zeta+\zeta^{-1})^2 + a_3(\zeta+\zeta^{-1})^3 + \cdots + a_{131}(\zeta+\zeta^{-1})^{131} + \cdots + a_{262}(\zeta+\zeta^{-1})^{262}$.
- Apply a double-size inverse conversion, obtaining the same product as a linear combination of $\zeta+\zeta^{-1}, \zeta^2+\zeta^{-2}, \zeta^3+\zeta^{-3}, \ldots, \zeta^{262}+\zeta^{-262}$. This transformation takes just 779 bit operations.
- Reduce the product to the original basis $(\zeta+\zeta^{-1}, \zeta^2+\zeta^{-2}, \zeta^3+\zeta^{-3}, \ldots, \zeta^{131}+\zeta^{-131})$ using the identities $\zeta^{262}+\zeta^{-262} = \zeta+\zeta^{-1}$, $\zeta^{261}+\zeta^{-261} = \zeta^2+\zeta^{-2}$, etc. This transformation takes just 130 bit operations.

We use the name `multprep` for the initial conversion of an input from permuted optimal normal basis to optimal polynomial basis. We use the name `ppn` for the remaining steps, taking two inputs in optimal polynomial basis and delivering an output in normal basis.

Sometimes the output of a multiplication is used again as input to another multiplication, and is not used for any squarings. In such cases Bernstein and Lange keep the lower half of the polynomial product in polynomial-basis representation and use the conversion routine only to compute the polynomial reduction, ending up in polynomial-basis representation and skipping a subsequent `multprep`. We use the name `ppp` for this operation, taking two inputs in optimal polynomial basis and delivering an output in optimal polynomial basis.

The costs of either type of field multiplication, `ppn` or `ppp`, are dominated by the costs of the 131-bit polynomial multiplication. The next section optimizes polynomial multiplication for the GPU, and Section 6 optimizes the entire iteration function for the GPU.

5 Polynomial Multiplication on the GPU

With optimal polynomial bases (see Section 4.4), each iteration involves slightly more than five 131-bit polynomial multiplications and only about 10000 extra bit operations. We are not aware of any 131-bit polynomial multiplier using fewer than 11000 bit operations; in particular, Bernstein's multiplier [3, Section 2] uses 11961 bit operations.

These figures show that polynomial multiplication consumes more than 80% of the bit operations in each iteration. We therefore placed a high priority on making multiplication run quickly on the GPU, preferably below 200 cycles in a single core. This section explains how we achieved this goal.

5.1 The Importance of Avoiding DRAM

We began by exploring an embarrassingly vectorized approach: T threads in a core work on $32T$ independent multiplication problems in bitsliced form. The $32T \times 2$ inputs are stored as 262 vectors of $32T$ bits, and the $32T$ outputs are stored as 261 vectors of $32T$ bits.

The main difficulty with this approach is that, even if the outputs are perfectly overlapped with the inputs, even if no additional storage is required, the inputs cannot fit into SRAM. For $T = 128$ the inputs consume 134144 bytes, while shared memory and registers together have only 81920 bytes. Reducing T to 64 (and tolerating 50% GPU utilization) would fit the inputs into 67072 bytes, but would also make half of the registers inaccessible (since each thread can access at most 128 registers), reducing the total capacity of shared memory and registers to 49152 bytes.

There is more than enough space in DRAM, even with very large T, but DRAM throughput then becomes a serious bottleneck. A single pass through the input vectors, followed by a single pass through the output vectors, keeps the

DRAM occupied for $523T$ cycles (i.e., more than 16 cycles per multiplication), and any low-memory multiplication algorithm requires many such passes.

We implemented and optimized several multiplication algorithms and complete iteration functions using this approach, but our best result using this approach was only 26 million iterations per second on a GTX 295. The remainder of this section describes a faster approach.

5.2 How to Fit into Shared Memory

The SIMD programming model of GPUs highly relies on the exploitation of data-level parallelism. However, data-level parallelism does not require having each thread work on a completely independent computation: parallelism is also available within computations. For example, the addition of two 32-way-bitsliced field elements is nothing but a sequence of 131 32-bit XOR operations; it naturally contains 131-way data-level parallelism. Similarly, there are many ways to break 131-bit binary-polynomial multiplication into several smaller-degree polynomial multiplications that can be carried out in parallel.

Registers do not communicate between threads, so having several threads cooperate on a single computation requires the active data for the computation to fit into shared memory. On the other hand, registers have more space than shared memory; during multiplication we use some registers as spill locations for data not involved in the multiplication, reversing the traditional direction of data spilling from registers to memory.

Our final software carries out 128 independent 131-bit multiplications (i.e., four 32-way bitsliced 131-bit multiplications) inside shared memory and registers, with no DRAM access. This means that each multiplication has to fit within 1024 bits of shared memory. This would not have been a problem for schoolbook multiplication, but it was a rather tight fit for the fast Karatsuba-type multiplication algorithm that we use (see below); more simultaneous multiplications would have meant compromises in the multiplication algorithm.

We decided to use 128 threads. This means that 32 threads are cooperating on each of the four 32-way bitsliced 131-bit multiplications. We expected, and ran experiments to confirm, that this would be enough threads to hide almost all latencies in the most time-consuming parts of the iteration function, particularly multiplication. Our 131-bit multiplication algorithm allows close to 32-way parallelization, as discussed below, although the parallelization is not perfect.

We would have had fewer latency problems from 192 or 256 threads, but the overall benefit is small and overwhelmed by increased parallelization requirements within each multiplication. In the opposite direction, we could have reduced the parallelization requirements by running 96 or 64 threads, but below 128 threads the GPU performance drops almost linearly.

5.3 Vectorized 128-Bit Multiplication

Our main task is now to multiply 131-bit polynomials, at each step using using 32 parallel bit operations to the maximum extent possible. We repeat the resulting algorithm on 128 independent inputs to obtain what the code actually does with 128 threads: namely, 128 separate multiplications of 131-bit polynomials, stored in bitsliced form as $4 \cdot 131$ 32-bit words, using 128 parallel 32-bit operations to the maximum extent possible.

We begin with the simpler task of multiplying 128-bit polynomials. Here we apply three levels of Karatsuba expansion. Each level uses $2n$ XORs to expand a $2n$-bit multiplication into three n-bit multiplications, and then $5n - 3$ XORs to collect the results (with Bernstein's "refined Karatsuba" from [4]).

Three levels of Karatsuba result in 27 16-bit polynomial multiplications. The inputs to these multiplications occupy a total of 864 bits, consuming most but not all of the 1024 bits of shared memory available to each 131-bit multiplication. We scheduled the code from [4] for a 16-bit polynomial multiplication, fitting it into 67 registers, and applied it to these 27 multiplications in parallel, leaving 5 threads idle out of 32. In total 108 16-bit polynomial multiplications coming from the four independent 131-bit polynomial multiplications are carried out by 108 threads in the `27xmult16` subroutine. In this subroutine each thread executes 413 instructions (350 bit operations and 63 load/store instructions) resulting in a total of 44604 instructions.

For comparison, applying two levels of Karatsuba would produce 9 32-bit multiplications. Our best register allocation for the 32-bit multiplier from [4] would require 161 registers, requiring additional spills to shared memory.

The initial expansion is trivially parallelizable. We merged it across all three levels, operating on blocks of 16 bits per operand and using 8 loads, 19 XORs, and 27 stores per thread.

Collection is more work. On the highest level (level 3), each block of 3 16-bit results is combined into a 32-bit intermediate result for level 2. This takes 5 loads (2 of these conditional), 3 XORs and 3 stores per thread on each of the 9 blocks. On level 2 we operate on blocks of 3 32-bit intermediate results leading to 3 64-bit blocks of intermediate results for level 1. This needs 6 loads and 5 XORs for each of the 3 blocks. The 3 blocks of intermediate results of this step do not need to be written to shared memory and remain in registers for the following final step on level 1. On level 1 the remaining three blocks of 64 bits are combined to the final 128-bit result by 12 XORs per thread.

5.4 Vectorized 131-Bit Multiplication

To multiply 131-bit polynomials we split each input into a 128-bit low part and a 3-bit high part, handling the 3×3-bit and 3×128-bit products separately.

The 3×3-bit multiplication is carried out almost for free by an otherwise idle 16×16 multiplication thread. The 3×128-bit multiplication is implemented straightforwardly by 6 loads for the 3 highest bits of each input, $3 \cdot 4$ combined loads and ANDs per input, and 24 XORs.

Overall our multiplier uses 13087 bit operations, and about 40% of our ALU cycles are spent on these bit operations rather than on loads, stores, address calculations, and other overhead. We lose an extra factor of about 1.1 from 32-way parallelization, since the 32 threads are not always all active. For comparison, the Toom-type techniques from [4] use only 11961 bit operations, saving about 10%, but appear to be more difficult to parallelize.

6 ECC2K-130 Iterations on the GPU

Recall that polynomial multiplication, the topic of the previous section, consumes more than 80% of the bit operations in the ECC2K-130 computation. This does not mean that the 20% overhead can be ignored! Imagine, for example, that the polynomial-multiplication code is carrying out useful bit operations in 40% of its cycles, while the remaining code fits much less smoothly into the GPU and is carrying out useful bit operations in only 5% of its cycles. The total time would then be *triple* the polynomial-multiplication time.

This section discusses several aspects of the overhead in the ECC2K-130 computation. Our main goal, as in the previous section, is to identify 32-way parallelism in the bit operations inside each 131-bit operation. This is often more challenging for the "overhead" operations than it is for multiplication, and in some cases we change algorithms to improve parallelism.

6.1 Basis Conversion (`multprep`)

As explained in Section 4.4 we keep most elements of $\mathbb{F}_{2^{131}}$ in (permuted) normal basis. Before those elements are multiplied we convert them from normal basis to polynomial basis.

Consider an element a of $\mathbb{F}_{2^{131}}$ in (permuted) normal basis:

$$a = a_0(\zeta + \zeta^{-1}) + a_1(\zeta^2 + \zeta^{-2}) + \cdots + a_{130}(\zeta^{131} + \zeta^{-131}).$$

On the first two levels of the basis conversion algorithm the following sequence of operations is executed on bits $a_0, a_{62}, a_{64}, a_{126}$:

$$\begin{aligned} a_{62} &\leftarrow a_{62} + a_{64}, && \text{then} \\ a_0 &\leftarrow a_0 + a_{126}, && \text{then} \\ a_{64} &\leftarrow a_{64} + a_{126}, && \text{then} \\ a_0 &\leftarrow a_0 + a_{62}. \end{aligned}$$

Meanwhile the same operations are performed on bits $a_1, a_{61}, a_{65}, a_{125}$; on bits $a_2, a_{60}, a_{66}, a_{124}$; and so on through $a_{30}, a_{32}, a_{94}, a_{96}$. We assign these 31 groups of bits to 32 threads, keeping almost all of the threads busy.

Merging levels 2 and 3 and levels 4 and 5 works in the same way. This assignment keeps 24 out of 32 threads busy on levels 2 and 3, and 16 out of 32 threads busy on levels 4 and 5. This assignment of operations to threads also avoids almost all memory-bank conflicts (see Section 2).

6.2 Multiplication with Reduction (ppp and ppn)

Recall that a ppp operation produces a product in polynomial basis, suitable for input to a subsequent multiplication. A ppn operation produces a product in normal basis, suitable for input to a squaring.

The main work in ppn, beyond polynomial multiplication, is a conversion of the product from polynomial basis to normal basis. This conversion is almost identical to multprep above, except that it is double-size and in reverse order. The main work in ppp is a more complicated double-size conversion, with similar parallelization.

6.3 Squaring and m-Squaring (sq, msq and sqseq)

Squaring (subroutine sq) and m-squaring (subroutine msq) are simply permutations in normal basis, costing 0 bit operations, but this does not mean that they cost 0 cycles.

The obvious method for 32 threads to permute 131 bits is for them to pick up the first 32 bits, store them in the correct locations, pick up the next 32 bits, store them in the correct locations, etc.; each thread performs 5 loads and 5 stores, with most of the threads idle for the final load and store. The addresses determined by the permutation for different m-squarings can be kept in constant memory. However, this approach triggers two GPU bottlenecks.

The first bottleneck is shared-memory bank throughput. Recall from Section 2 that threads in the same half-warp cannot simultaneously store values to the same memory bank. To almost completely eliminate this bottleneck we wrote a greedy search tool that decides on a good order to pick up 131 bits, trying to avoid all memory bank conflicts for both the loads and the stores. For almost all values of m, including the most frequently used ones, this tool found a conflict-free assignment. For two values of m the assignment involves a few bank conflicts, but these values are used only in inversion, not in the main loop.

The second bottleneck is constant-cache throughput. If thread i loads from a constant array at position i then the constant cache serves only one thread per cycle. To eliminate this bottleneck we move these loads out of the main loop and dedicate 10 registers per thread to hold 20 load and 20 store positions for the 4 most-often used m-squarings, packing 4 1-byte positions in one 32-bit register. Unpacking the positions costs just one shift and one mask instruction for the two middle bytes, a mask instruction for the low byte, and a shift instruction for the high byte.

6.4 Hamming-weight Computation (hamming and below)

The hamming subroutine computes the Hamming weight of x in a parallel manner: multiple bits are added per clock cycle. The basic operation in this subroutine is a *full adder*, picking up 3 bits b_1, b_2, b_3 and storing 2 bits c_0, c_1 such that $b_1 + b_2 + b_3 = 2c_1 + c_0$.

In the first addition round 96 bits are added into 64 result bits, of which 32 are on level 0 (corresponding to 2^0) and the other 32 belong to level 1 (corresponding to 2^1). In the next round, the 32 bits on level 0 are added to 34 of the remaining $131 - 96 = 35$ bits in groups of 3 giving 22 bits on level 0 and 22 bits on level 1; simultaneously 30 of the 32 bits on level 1 are added producing 10 bits on level 1 and 10 on level 2. Note that $22 + 10 = 32$ and thus these operations can be handled simultaneously. The following steps are analogous: in every round the maximal multiple of 3 bits per level is handled and produces bits on that level and the one above. Each output is carefully positioned to simplify address calculations in subsequent steps. Eventually there is only one bit per level and the Hamming weight can be read off. Note that after the second round fewer than 32 bits are used at once, but we have achieved the shortest length possible.

Once the Hamming weight is computed the subroutine `below` tests whether the weight is below 34, i.e., whether the point is distinguished.

6.5 Kernel-Launch Overhead, Register Spills, etc

GPU code is organized into *kernels* called from the CPU. Calling (*launching*) a kernel takes several microseconds on top of any time needed to copy data between global memory and the CPU. To eliminate these costs we run a single kernel for several seconds. The kernel consists of a loop around a complete iteration; it performs the iteration repeatedly without contacting the CPU. Any distinguished points are masked out of subsequent updates; distinguished points are rare, so negligible time is lost computing unused updates.

We stream a batch of iterations in a simple way between global memory and shared memory; this involves a small number of global-memory copies in each iteration. We avoid spilling any additional data to DRAM; in particular, we avoid all use of local memory.

All of this sounds straightforward but in fact required completely redesigning our programming environment. To explain this we briefly review NVIDIA's standard programming environment and the problems that we encountered with it.

Non-graphical applications for NVIDIA GPUs are usually programmed in CUDA, a C-like language designed by NVIDIA. NVIDIA's `nvcc` compiler translates a `*.cu` CUDA file into a `*.ptx` file in a somewhat machine-independent language called PTX. NVIDIA's `ptxas` compiler translates this `*.ptx` file into a machine-specific binary `*.cubin` file. The `*.cubin` file is loaded onto the GPU and run.

NVIDIA's register allocators were designed to handle small kernels consisting of hundreds of instructions; their memory-use scaling appears to be quadratic with the kernel size, and their time scaling appears to be even worse. For medium-size kernels we found NVIDIA's compilers intolerably slow; even worse, the resulting code involved frequent spills to local memory, dropping performance by an order of magnitude. For larger kernels the compilers ran out of memory and crashed.

We experimented with writing code in the PTX language, but this language still requires compilation by `ptxas`; even though `ptxas` is labelled as an "assembler" it turns out to be the culprit in NVIDIA's register-allocation problems. To control register allocation we eventually resorted to the reverse-engineered assembler `cudasm` by van der Laan [17]. We fixed various bugs in `cudasm` and, to improve usability, extended Bernstein's `qhasm` programming language [2] to support `cudasm`. We used `qhasm` on top of `cudasm` to implement the whole kernel: more than 90000 instructions after macro processing for batch size 32. We then designed our own assembly-level function-call convention and merged large stretches of instructions into functions to reduce instruction-cache misses.

6.6 Overall Results

Table 1 reports timings of all major building blocks in our software. For example, in the `multprep` row of the table, the cycles/iteration column is 52, indicating that `multprep` was responsible for 52 cycles in each iteration. The calls/iteration column is 4.12, indicating that an average iteration involved 4.12 calls to `multprep`; in fact, a batch of 128 iterations involved 527 calls to `multprep`. The cycles/call column is 12.

We collected these numbers by performing the following experiment. On a typical pass through the main loop (specifically, the 10000th pass), inside each thread, we checked the GTX 295's hardware cycle half-counter before and after each use of `multprep`, and tallied the cycles spent inside `multprep`. We repeated this experiment 20 times, with 128 threads in each experiment, yielding a total of $20 \cdot 128 = 2560$ cycle counts. We divided the cycle counts by 16384 because each pass through the main loop performs $16384 = 128 \cdot 128$ iterations: our implementation always handles 128 iterations in parallel, and on top of this we chose a batch size of 128. The average of these cycle counts was 52. Table 1 also reports standard deviations: e.g., 52 ± 1.

After measuring one operation we removed the cycle counters and placed them around each occurrence of another operation. Only one operation is measured at a time. Cycle counting is not free, so the sum of times measured for the individual operations is slightly more than the time measured for the entire main loop. The operations shown in Table 1 are as follows:

- `ppp` multiplies two field elements in polynomial basis, producing output in polynomial basis.
- `ppn` multiplies two field elements in polynomial basis, producing output in normal basis.
- `multprep` converts from normal basis to polynomial basis.
- `sqseq` is the sequence of squarings used to compute x^{2^j}: a 3-squaring, a conditional squaring, a conditional 2-squaring, and a conditional 4-squaring.

Table 1. Timings of major building blocks within an iteration

operation	calls/iteration	cycles/call	cycles/iteration	
complete iteration			1164.43 ± 12.05	100.00%
ppp	2.9766	158.08 ± 0.23	470.55 ± 0.68	40.41%
ppn	2.0625	159.54 ± 1.64	329.05 ± 3.38	28.26%
sqseq	2.0000	44.99 ± 0.83	89.97 ± 1.67	7.73%
readfen	7.9844	9.03 ± 0.10	72.12 ± 0.82	6.19%
multprep	4.1172	12.63 ± 0.05	52.00 ± 0.21	4.47%
writefen	5.9844	7.91 ± 0.27	47.35 ± 1.63	4.07%
hamming	1.0000	41.60 ± 0.11	41.60 ± 0.11	3.57%
add	7.0000	4.01 ± 0.06	28.04 ± 0.40	2.41%
readbit	5.0000	2.90 ± 0.84	14.52 ± 4.22	1.25%
cadd	2.0000	6.81 ± 0.77	13.62 ± 1.53	1.17%
below	1.0000	11.26 ± 0.16	11.26 ± 0.16	0.97%
msq	1.0703	9.60 ± 0.12	10.27 ± 0.13	0.88%
copy	2.0000	3.73 ± 0.87	7.45 ± 1.74	0.64%
writebit	4.0000	1.25 ± 0.02	5.00 ± 0.09	0.43%
131-bit poly mult	5.0391	119.11 ± 0.13	600.21 ± 0.65	51.55%
27xmult16	5.0391	65.02 ± 0.19	327.66 ± 0.98	28.14%
DRAM access			139.01 ± 5.73	11.94%
inversion			13.74 ± 0.16	1.18%

- `msq` is an m-squaring for any $m \in \{1, 2, 3, 4, 8, 16, 32, 65\}$.
- `add` adds two field elements.
- `cadd` is conditional field addition, i.e., addition masked by an extra bit.
- `hamming` computes Hamming weight.
- `below` checks for a distinguished point.
- `readfen` copies a field element from global memory to shared memory.
- `writefen` copies a field element the other way.
- `readbit` copies a bit from global memory to shared memory.
- `writebit` copies a bit the other way.
- `copy` copies a field element within shared memory.

There are some additional rows in the table showing total time spent in 131-bit polynomial multiplication; total time spent in the `27xmult16` subroutine; total time spent on global-memory access; and total time spent in inversion.

Each of the instructions in our software handles 128 iterations, but is also followed by 128 threads, keeping the GPU core busy for at least 16 cycles. The number of cycles spent per iteration is therefore at least 0.125 times the number of instructions. For example, `27xmult16` occurs 5.04 times per iteration and involves 413 instructions, accounting for $0.125 \cdot 5.04 \cdot 413 \approx 260$ cycles in each iteration. The actual number of cycles spent on `27xmult16` is about 25% higher than this; the gap is explained in part by instruction-cache misses and in part by the delays for complex instructions discussed in Section 2.

The complete kernel uses 1164 cycles per iteration on average on a single core on a GTX 295 graphics card. Therefore we achieve 63 million iterations per second on a single card (60 cores, 1.242 GHz). The whole ECC2K-130 computation would be finished in two years (on average; the rho method is probabilistic) using 534 GTX 295 graphics cards.

References

1. Bailey, D.V., Batina, L., Bernstein, D.J., Birkner, P., Bos, J.W., Chen, H.-C., Cheng, C.-M., van Damme, G., de Meulenaer, G., Dominguez Perez, L.J., Fan, J., Güneysu, T., Gurkaynak, F., Kleinjung, T., Lange, T., Mentens, N., Niederhagen, R., Paar, C., Regazzoni, F., Schwabe, P., Uhsadel, L., Van Herrewege, A., Yang, B.-Y.: Breaking ECC2K-130. Cryptology ePrint Archive, Report 2009/541 (2009), `http://eprint.iacr.org/2009/541`
2. Bernstein, D.J.: qhasm: tools to help write high-speed software, `http://cr.yp.to/qhasm.html`
3. Bernstein, D.J.: Batch binary Edwards. In: Halevi, S. (ed.) Advances in Cryptology - CRYPTO 2009. LNCS, vol. 5677, pp. 317–336. Springer, Heidelberg (2009) Document ID: 4d7766189e82c1381774dc840d05267b, `http://cr.yp.to/papers.html#bbe`
4. Bernstein, D.J.: Minimum number of bit operations for multiplication (2009), `http://binary.cr.yp.to/m.html` (accessed 2009-12-07)
5. Bernstein, D.J., Chen, H.-C., Chen, M.-S., Cheng, C.-M., Hsiao, C.-H., Lange, T., Lin, Z.-C., Yang, B.-Y.: The billion-mulmod-per-second PC. In: Workshop Record of SHARCS 2009: Special-purpose Hardware for Attacking Cryptographic Systems, pp. 131–144 (2009), `http://www.hyperelliptic.org/tanja/SHARCS/record2.pdf`
6. Bernstein, D.J., Chen, T.-R., Cheng, C.-M., Lange, T., Yang, B.-Y.: ECM on graphics cards. In: Joux, A. (ed.) EUROCRYPT 2009. LNCS, vol. 5479, pp. 483–501. Springer, Heidelberg (2009) Document ID: 6904068c52463d70486c9c68ba045839, `http://eprint.iacr.org/2008/480/`
7. Bernstein, D.J., Lange, T.: Explicit-formulas database, `http://www.hyperelliptic.org/EFD/` (accessed 2010-09-25)
8. Bernstein, D.J., Lange, T.: Type-II optimal polynomial bases. In: Anwar Hasan, M., Helleseth, T. (eds.) WAIFI 2010. LNCS, vol. 6087, pp. 41–61. Springer, Heidelberg (2010) Document ID: 90995f3542ee40458366015df5f2b9de, `http://binary.cr.yp.to/opb-20100209.pdf`
9. Biham, E.: A fast new DES implementation in software. In: Biham, E. (ed.) FSE 1997. LNCS, vol. 1267, pp. 260–272. Springer, Heidelberg (1997)
10. Bos, J.W., Kleinjung, T., Niederhagen, R., Schwabe, P.: ECC2K-130 on Cell CPUs. In: Bernstein, D.J., Lange, T. (eds.) Progress in Cryptology – AFRICACRYPT 2010. LNCS, vol. 6055, pp. 225–242. Springer, Heidelberg (2010) Document ID: bad46a78a56fdc3a44fcf725175fd253, `http://eprint.iacr.org/2010/077`
11. Certicom. Certicom ECC challenge (1997), `http://www.certicom.com/images/pdfs/cert_ecc_challenge.pdf`
12. Fan, J., Bailey, D.V., Batina, L., Güneysu, T., Paar, C., Verbauwhede, I.: Breaking elliptic curves cryptosystems using reconfigurable hardware. In: 20th International Conference on Field Programmable Logic and Applications (FPL 2010), Milano, Italy, August 31–September 2 (2010)

13. von zur Gathen, J., Shokrollahi, A., Shokrollahi, J.: Efficient multiplication using type 2 optimal normal bases. In: Carlet, C., Sunar, B. (eds.) WAIFI 2007. LNCS, vol. 4547, pp. 55–68. Springer, Heidelberg (2007)
14. Montgomery, P.L.: Speeding the Pollard and elliptic curve methods of factorization. Mathematics of Computation 48, 243–264 (1987)
15. Pollard, J.M.: Monte Carlo methods for index computation (mod p). Mathematics of Computation 32, 918–924 (1978)
16. Shokrollahi, J.: Efficient implementation of elliptic curve cryptography on FPGAs. PhD thesis, Rheinische Friedrich-Wilhelms Universität (2007), Dissertation, http://nbn-resolving.de/urn:nbn:de:hbz:5N-09601
17. van der Laan, W.J.: Cubin utilities (2007), http://wiki.github.com/laanwj/decuda
18. van Oorschot, P.C., Wiener, M.J.: Parallel collision search with cryptanalytic applications. Journal of Cryptology 12(1), 1–28 (1999)

One Byte per Clock: A Novel RC4 Hardware

Sourav Sen Gupta[1], Koushik Sinha[2],
Subhamoy Maitra[1], and Bhabani P. Sinha[1]

[1] Indian Statistical Institute, 203 B T Road, Kolkata 700 108, India
[2] Honeywell Technology Solutions Lab, Bangalore 560 076, India
sg.sourav@gmail.com, sinha_kou@yahoo.com, {subho,bhabani}@isical.ac.in

Abstract. RC4, the widely used stream cipher, is well known for its simplicity and ease of implementation in software. In case of a special purpose hardware designed for RC4, the best known implementation till date is 1 byte per 3 clock cycles. In this paper, we take a fresh look at the hardware implementation of RC4 and propose a novel architecture which generates 1 keystream byte per clock cycle. Our strategy considers generation of two consecutive keystream bytes by unwrapping the RC4 cycles. The same architecture is customized to perform the key scheduling algorithm at a rate of 1 round per clock.

Keywords: Fast Implementation, Hardware, RC4, Stream Cipher.

1 Introduction

RC4 is one of the widely used software stream ciphers that is mostly implemented in software. This cipher is used in network protocols such as SSL, TLS, WEP and WPA. As well the cipher finds applications in Microsoft Windows, Lotus Notes, Apple AOCE, Oracle Secure SQL etc. Though several other efficient and secure stream ciphers have been discovered after RC4, it is still the most popular stream cipher algorithm due to its simplicity, ease of implementation, speed and efficiency. The algorithm can be stated in a few lines, yet after two decades of analysis, its strengths and weaknesses are of great interest to the community. In spite of several cryptanalysis attempts on RC4 (see [1,2,4,9,10,11,13,14,15,16,18] and references therein), the cipher stands secure if used properly.

In this paper we present a novel hardware design of RC4 for fast generation of keystream. To motivate our contribution, we need to discuss the basic framework of RC4 first. A short note on RC4 follows.

1.1 RC4 Algorithm

The RC4 stream cipher has been designed by Ron Rivest for RSA Data Security in 1987. It uses an S-Box $S = (S[0], \ldots, S[N-1])$ of length N, each location storing one byte. Typically, $N = 256$, and S is initialized as the identity permutation, i.e., $S[y] = y$ for $0 \leq y \leq N-1$. A secret key k of size l bytes (typically, $5 \leq l \leq 16$) is used to scramble this permutation. An array

G. Gong and K.C. Gupta (Eds.): INDOCRYPT 2010, LNCS 6498, pp. 347–363, 2010.

$K = (K[0], \ldots, K[N-1])$ is used to hold the secret key, where the key is repeated in K at key length boundaries. i.e., $K[y] = k[y \bmod l]$, for $0 \leq y \leq N-1$.

RC4 has two components, namely, the Key Scheduling Algorithm (KSA) and the Pseudo-Random Generation Algorithm (PRGA). The KSA uses the secret key K to generate a pseudo-random permutation S of $0, 1, \ldots, N-1$ and the PRGA uses this pseudo-random permutation to generate pseudo-random keystream bytes. The two pieces of the RC4 algorithm are as shown in Algorithm 1 and Algorithm 2. Any addition used related to the RC4 is in general addition modulo N unless specified otherwise. The keystream output byte Z is XOR-ed with the message byte to generate the ciphertext byte at the sender end, and is XOR-ed with the ciphertext byte to get back the message byte at the receiver end. The software implementation of RC4 is simple. Detailed comparison of the software performance of eStream portfolio and RC4 is given in [19].

Input: Secret Key K.
Output: S-Box S generated by K.

for $i = 0, \ldots, N-1$ **do**
  $S[i] = i$;
end

Initialize counter: $j = 0$;

for $i = 0, \ldots, N-1$ **do**
  $j = j + S[i] + K[i]$;
  Swap $S[i] \leftrightarrow S[j]$;
end

Algorithm 1. KSA

Input: S-Box S, output of KSA.
Output: Random stream Z generated from S.

Initialize the counters: $i = j = 0$;

while *TRUE* **do**
  $i = i + 1$;
  $j = j + S[i]$;
  Swap $S[i] \leftrightarrow S[j]$;
  Output $Z = S[S[i] + S[j]]$;
end

Algorithm 2. PRGA

1.2 Our Contribution

A 3-clock efficient implementation of RC4 on a custom pipelined hardware was proposed in Kitsos et al. [6] in 2003. Though there are already a few attempts to propose efficient hardware implementation [3,5,7,12] of RC4, the basic issue remained ignored that the design motivation should be initiated from the question that "In how many clocks a byte can be generated in an RC4 hardware?" To the best of our knowledge, this line of thought has never been studied and exercised in a disciplined manner in the existing literature, which in fact, is quite surprising.

In this paper we present a novel hardware for RC4 from this specific design motivation. Our model is a generic circuit based on simple ideas of combinational and sequential logic design. The main contribution of our work is to take a new look at RC4 by combining consecutive pairs of cycles in a pipelined fashion, and to read off the values of one state of the S-box from previous or later rounds of the cipher. To the best of our knowledge, the unwrapping of RC4 cycles to extract S-box information from previous or later stages is an idea which has never been exploited in designing an efficient hardware for RC4. The comprehensive design strategy and analysis of the circuit is presented in the following sections.

1.3 Organization of the Paper

Section 2: In this section, we propose the basic idea for our hardware implementation. This includes the modifications in the RC4 algorithm required for our hardware, and the circuits to perform each step of the modified algorithm.

Section 3: This section deals with a comprehensive timing analysis of the proposed architecture and presents a timing diagram for a few illustrative clock cycles in details.

Section 4: Here we present the complete circuit for our hardware, and prove our claim ('one byte per clock') as a formal theorem. In this section, we also discuss the minor modifications required to perform the KSA round using the same circuit (see Section 4.1), and compare our proposed architecture with the existing models for RC4 hardware, as found in the literature (see Section 4.2).

Section 5: This section concludes the paper by suggesting platforms for the practical implementation of the proposed hardware. We also present some ideas regarding parallelization of the circuit.

2 Hardware Implementation

We consider the generation of two consecutive values of Z together, for the two consecutive message bytes to be encrypted. Assume that the initial values of the variables i, j and S are i_0, j_0 and S_0, respectively. After the first execution of the PRGA loop, these values will be i_1, j_1 and S_1, respectively and the output byte is Z_1, say. Similarly, after the second execution of the PRGA loop, these will be i_2, j_2, S_2 and Z_2, respectively. Thus, for the first two loops of execution to complete, we have to perform the operations shown in Table 1.

Table 1. Two consecutive loops of RC4 Stream Generation

Steps	First Loop	Second Loop
1	$i_1 = i_0 + 1$	$i_2 = i_1 + 1 = i_0 + 2$
2	$j_1 = j_0 + S_0[i_1]$	$j_2 = j_1 + S_1[i_2] = j_0 + S_0[i_1] + S_1[i_2]$
3	Swap $S_0[i_1] \leftrightarrow S_0[j_1]$	Swap $S_1[i_2] \leftrightarrow S_1[j_2]$
4	$Z_1 = S_1[S_0[i_1] + S_0[j_1]]$	$Z_2 = S_2[S_1[i_2] + S_1[j_2]]$

For hardware realization, to store the S value, we use a bank of 8-bit registers, 256 in total. The output lines of any one of these 256 registers can be accessed through a 256 to 1 Multiplexer (MUX), with its control lines set to the required address i_1, j_1, i_2 or j_2. Thus, we need 4 such 256 to 1 MUX units to simultaneously read $S[i_1]$, $S[i_2]$, $S[j_1]$ and $S[j_2]$. Before that, let us study how to compute the increments of i and j at each level.

2.1 Step 1: Calculation of i_1 and i_2

We first note that incrementing i_0 by 1 and 2 can be done by the same clock pulse applied to two synchronous 8-bit counters both initialized with the value i_0; in one counter the clock pulse is applied to the input of all the flip-flops, and in the other, it is applied to all the flip-flops except the one at the LSB position, as shown in Fig. 1. Also, we need to load these counters with new values every other clock cycle (alternate with incrementing), and hence these must be counters with parallel load mechanism, so that the loading takes a single clock pulse.

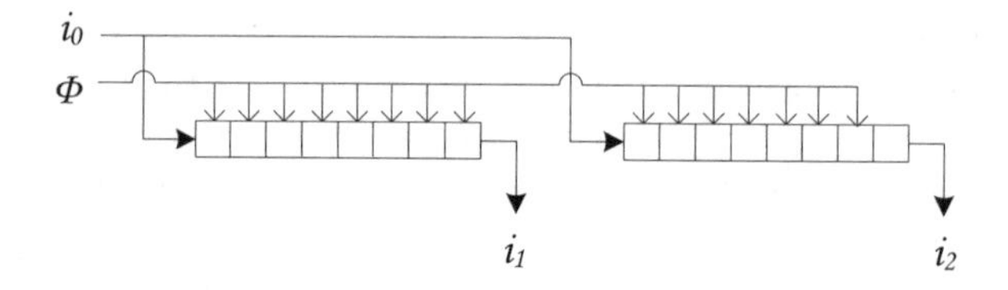

Fig. 1. [Circuit 1] Circuit to compute i_1 and i_2

Note that it is possible to simplify the implementation even further. We know that i_1 and i_2 will always be equal but for the LSB, which is always 0 for i_1 and 1 for i_2. Hence one needs only to implement a 7-bit counter for the 7 common MSBs of i_1, i_2, and append appropriate LSB.

2.2 Step 2: Calculation of j_1 and j_2

The values of j_1 and j_2 will be computed and stored in two 8-bit registers. To compute j_1, we need a 2-input parallel adder unit. It may be one using a carry lookahead adder, or one using *scan* operation as proposed by Sinha and Srimani [17], or one using *carry-lookahead-tree* as proposed by Lynch and Swarzlander, Jr. [8]. For computing j_2, there are two special cases:

$$j_2 = j_0 + S_0[i_1] + S_1[i_2] = \begin{cases} j_0 + S_0[i_1] + S_0[i_2] & \text{if } i_2 \neq j_1, \\ j_0 + S_0[i_1] + S_0[i_1] & \text{if } i_2 = j_1. \end{cases}$$

Note that the only change from S_0 to S_1 is the swap $S_0[i_1] \leftrightarrow S_0[j_1]$, and hence we need to check if i_2 is equal to either of i_1 or j_1. Now, i_2 can not be equal to i_1 as they differ only by 1 modulo 256. Therefore, $S_1[i_2] = S_1[j_1] = S_0[i_1]$ if $i_2 = j_1$, and $S_1[i_2] = S_0[i_2]$ otherwise. In both the cases, three binary numbers are to be added. Let us denote the kth bit of $j_0, S_0[i_1]$ and $S_1[i_2]$ (either $S_0[i_2]$ or $S_0[i_1]$) by a_k, b_k and c_k, respectively, where $0 \leq k \leq 7$. We first construct two 9-bit vectors R and C, where the kth bits ($0 \leq k \leq 8$) of R and C are given by

$$R_k = \mathsf{XOR}(a_k, b_k, c_k) \quad \text{for } 0 \leq k \leq 7, \quad R_8 = 0, \quad \text{and}$$
$$C_0 = 0, \quad C_k = a_{k-1}b_{k-1} + b_{k-1}c_{k-1} + c_{k-1}a_{k-1} \quad \text{for } 1 \leq k \leq 8.$$

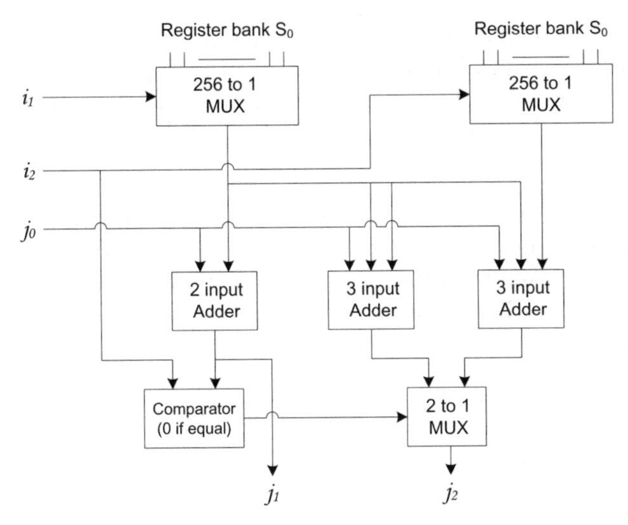

Fig. 2. [Circuit 2] Circuit to compute j_1 and j_2. (All connectors are 8-line bus.)

Note that in case of RC4, all additions are done modulo 256. Hence, we can discard the 9th bit ($k = 8$) of the vectors R, C while adding them together, and carry out normal 8-bit parallel addition considering $0 \leq k \leq 7$. Therefore, one may add R and C by a parallel full adder as used for j_1. The circuit to compute j_1 and j_2 is as shown in Fig. 2.

2.3 Step 3: Swapping the S Values

Step 3 in Algorithm 2 consists of one of the following 8 possible data transfer requirements among the registers of the S-register bank, depending on the different possible values of i_1, j_1, i_2 and j_2. We have to check if i_2 and j_2 can be equal to i_1 or j_1 (we only know that $i_2 \neq i_1$). All the cases in this direction can be listed as in Table 2. A more detailed explanation for each case is presented in Appendix A.

After the swap operation is completed successfully, one obtains S_2 from S_0. From the point of view of the receiving registers (in the S-register bank) in case of the above mentioned register-to-register transfers, we can summarize the cases as follows.

Table 2. Different cases for the Register-to-Register transfers in the swap operation

#	Condition	Register-to-Register Transfers
1	$i_2 \neq j_1$ & $j_2 \neq i_1$ & $j_2 \neq j_1$	$S_0[i_1] \rightarrow S_0[j_1]$, $S_0[j_1] \rightarrow S_0[i_1]$, $S_0[i_2] \rightarrow S_0[j_2]$, $S_0[j_2] \rightarrow S_0[i_2]$
2	$i_2 \neq j_1$ & $j_2 \neq i_1$ & $j_2 = j_1$	$S_0[i_1] \rightarrow S_0[i_2]$, $S_0[i_2] \rightarrow S_0[j_1] = S_0[j_2]$, $S_0[j_1] \rightarrow S_0[i_1]$
3	$i_2 \neq j_1$ & $j_2 = i_1$ & $j_2 \neq j_1$	$S_0[i_1] \rightarrow S_0[j_1]$, $S_0[i_2] \rightarrow S_0[i_1] = S_0[j_2]$, $S_0[j_1] \rightarrow S_0[i_2]$
4	$i_2 \neq j_1$ & $j_2 = i_1$ & $j_2 = j_1$	$S_0[i_1] \rightarrow S_0[i_2]$, $S_0[i_2] \rightarrow S_0[i_1] = S_0[j_1] = S_0[j_2]$
5	$i_2 = j_1$ & $j_2 \neq i_1$ & $j_2 \neq j_1$	$S_0[i_1] \rightarrow S_0[j_2]$, $S_0[j_2] \rightarrow S_0[j_1] = S_0[i_2]$, $S_0[j_1] \rightarrow S_0[i_1]$
6	$i_2 = j_1$ & $j_2 \neq i_1$ & $j_2 = j_1$	$S_0[i_1] \rightarrow S_0[j_1] = S_0[i_2] = S_0[j_2]$, $S_0[j_1] \rightarrow S_0[i_1]$
7	$i_2 = j_1$ & $j_2 = i_1$ & $j_2 \neq j_1$	Identity permutation, no data transfer.
8	$i_2 = j_1$ & $j_2 = i_1$ & $j_2 = j_1$	Impossible, as it implies $i_1 = i_2 = i_1 + 1$.

- $S_2[i_1]$ can receive data from any one of $S_0[i_1], S_0[j_1]$ or $S_0[i_2]$,
- $S_2[j_1]$ can receive data from any one of $S_0[i_1], S_0[j_1], S_0[i_2]$ or $S_0[j_2]$,
- $S_2[i_2]$ can receive data from any one of $S_0[i_1], S_0[j_1], S_0[i_2]$ or $S_0[j_2]$,
- $S_2[j_2]$ can receive data from any one of $S_0[i_1], S_0[i_2]$ or $S_0[j_2]$.

In view of the above discussions, the input data (1 byte) for each of the 256 registers in the S-register bank will be taken from the output of an 8 to 1 MUX unit, whose data inputs are taken from $S_0[i_1], S_0[j_1], S_0[i_2], S_0[j_2]$, and the control inputs are taken from the outputs of three comparators comparing (i) i_2 and j_1, (ii) j_2 and i_1, (iii) j_2 and j_1. The circuit to realize the swap operation is shown in Fig. 3.

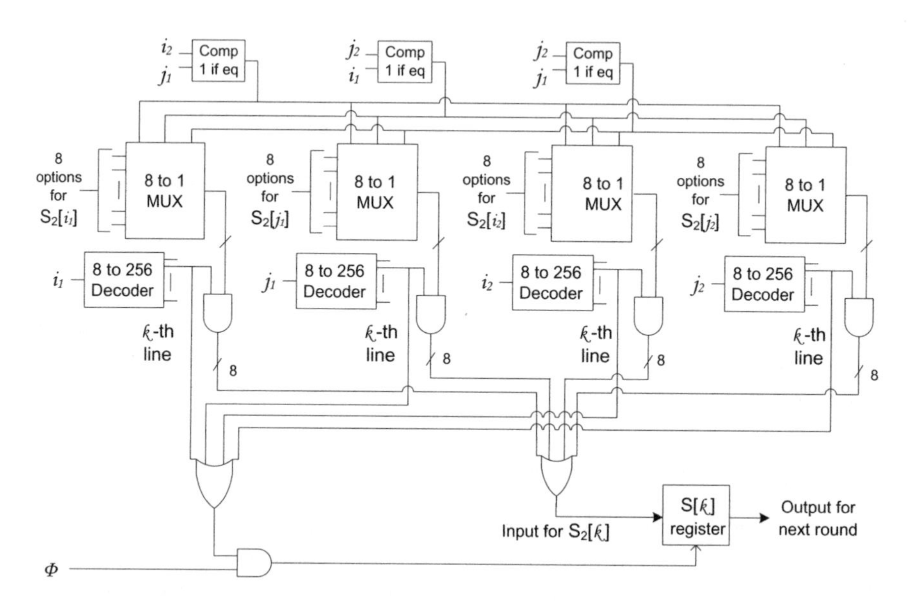

Fig. 3. [Circuit 3] Circuit to swap S values. (Data lines shown only for a fixed k.)

For simultaneous data transfers during the swap operation, we propose that the S-registers in the register bank be constituted by Master-Slave J-K flip-flops. The details of such flip-flops are presented in Appendix B. By using this type of registers, one can perform all the required register-to-register transfer operations in a single clock cycle.

2.4 Step 4: Calculation of Z_1 and Z_2

In step 4 of Algorithm 2, we have $S_1[i_1] + S_1[j_1] = S_0[j_1] + S_0[i_1]$, and

$$Z_1 = S_1\left[S_0[j_1] + S_0[i_1]\right] = \begin{cases} S_2[i_2] & \text{if } S_0[j_1] + S_0[i_1] = j_2, \\ S_2[j_2] & \text{if } S_0[j_1] + S_0[i_1] = i_2, \\ S_2[S_0[j_1] + S_0[i_1]] & \text{otherwise.} \end{cases}$$

Thus, the computation of Z_1 involves adding $S_0[i_1]$ and $S_0[j_1]$ first, which can be done using a 2-input parallel adder. The 256 to 1 MUX, which is used to extract appropriate data from S_2, will be controlled by another 4 to 1 MUX. This 4 to 1 MUX is in turn controlled by the outputs of two comparators comparing (i) $S_0[j_1] + S_0[i_1]$ and i_2, and (ii) $S_0[j_1] + S_0[i_1]$ and j_2. The circuit to compute Z_1 is as illustrated in Fig. 4.

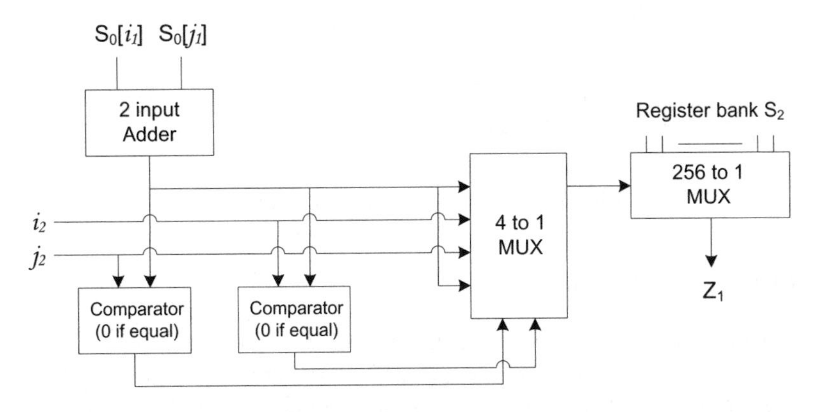

Fig. 4. [Circuit 4] Circuit to compute Z_1

Computation of Z_2, however, involves adding $S_1[i_2], S_1[j_2]$, as follows:

$$Z_2 = S_2\left[S_2[i_2] + S_2[j_2]\right] = S_2\left[S_1[j_2] + S_1[i_2]\right].$$

Note that we never actually store configuration S_1 of the register bank. This is because we move from state S_0 to S_2 directly to make the algorithm more efficient. Hence, we have to unwrap one cycle of RC4 and gather the values of $S_1[i_2]$ and $S_1[j_2]$ from the S_0 state. $S_1[i_2]$ and $S_1[j_2]$ receive the values from the appropriate registers of S_0 as given below, depending on the following conditions:

- $i_2 \neq j_1$ and $j_2 \neq i_1$ and $j_2 \neq j_1$: $S_1[i_2] = S_0[i_2]$ and $S_1[j_2] = S_0[j_2]$
- $i_2 \neq j_1$ and $j_2 \neq i_1$ and $j_2 = j_1$: $S_1[i_2] = S_0[i_2]$ and $S_1[j_2] = S_0[i_1]$
- $i_2 \neq j_1$ and $j_2 = i_1$ and $j_2 \neq j_1$: $S_1[i_2] = S_0[i_2]$ and $S_1[j_2] = S_0[j_1]$
- $i_2 \neq j_1$ and $j_2 = i_1$ and $j_2 = j_1$: $S_1[i_2] = S_0[i_2]$ and $S_1[j_2] = S_0[j_1]$
- $i_2 = j_1$ and $j_2 \neq i_1$ and $j_2 \neq j_1$: $S_1[i_2] = S_0[i_1]$ and $S_1[j_2] = S_0[j_2]$
- $i_2 = j_1$ and $j_2 \neq i_1$ and $j_2 = j_1$: $S_1[i_2] = S_0[i_1]$ and $S_1[j_2] = S_0[i_1]$
- $i_2 = j_1$ and $j_2 = i_1$ and $j_2 \neq j_1$: $S_1[i_2] = S_0[i_1]$ and $S_1[j_2] = S_0[j_1]$

These conditions can be realized using an 8 to 1 MUX unit controlled by the outputs of three comparators comparing (i) i_2 and j_1, (ii) j_2 and i_1, (iii) j_2 and j_1. Note that we can use the same control lines as in case of the swapping operation. The computation can be performed using the circuit as shown in Fig. 5.

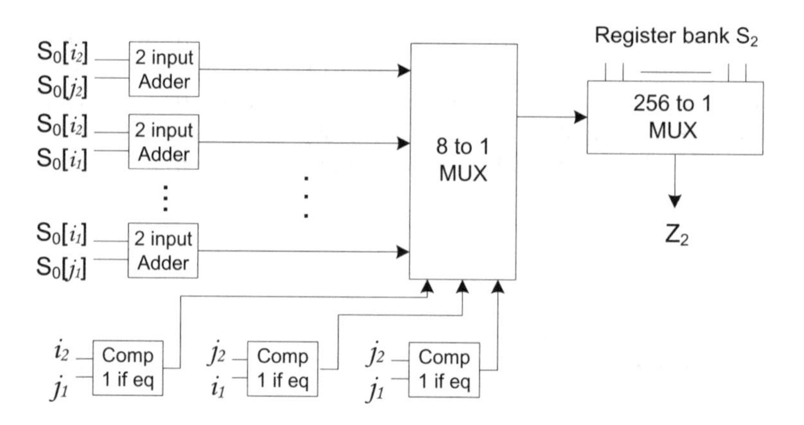

Fig. 5. [Circuit 5] Circuit to compute Z_2

3 Timing Analysis

Let us denote the operational clock by ϕ, and its ith cycle by ϕ_i for all $i \geq 0$. We denote the initial clock cycle by ϕ_0, which triggers the PRGA circuit. Based on this notation, the timing analysis for the complete PRGA circuit (shown in Fig. 7) is as follows. We analyze the first two iterations of our model, that is, the generation of Z_1, Z_2 and Z_3, Z_4, and the analysis for the subsequent iterations fall along similar lines.

Initialization: The initial inputs to the circuit are $i_0 = 0$, $j_0 = 0$ and S_0 from the output of KSA. The two registers in Fig. 1 are preloaded with i_0, i.e., the values are set to 0. The S-registers in the register bank are all set to store the corresponding 8-bit values of S generated by the KSA.

Clock cycle ϕ_0: At the trailing edge of ϕ_0, the registers containing i_0 increment to produce i_1 and i_2. Also, the inputs to the four 256 to 1 MUX units are read from S_0.

Clock cycle ϕ_1: At the start of this cycle, we already have i_1, i_2 and S_0. Hence, one can obtain: (i) $S_0[i_1]$ and $S_0[i_2]$ from the first two 256 to 1 MUX, controlled by i_1, i_2, and (ii) j_1 and j_2 by combining j_0, $S_0[i_1]$ and $S_0[i_2]$, using the circuit in Fig. 2.

The values $i_1, i_2, j_1, j_2, S_0[i_1]$ and $S_0[i_2]$ are latched at the leading edge of ϕ_1. At the trailing edge of ϕ_1, the latched values of j_1, j_2 are accessed to control the last two 256 to 1 MUX.

Simultaneously, the first register of Fig. 1 is loaded (in parallel) with the value of i_2 at the trailing edge of ϕ_1, replacing the previous entry i_1. The second one already contains i_2.

Clock cycle ϕ_2: As we have latched j_1, j_2 released, we obtain $S_0[j_1]$ and $S_0[j_2]$ from the last two 256 to 1 MUX controlled by j_1, j_2. These two values from S_0 are latched at the leading edge of ϕ_2.

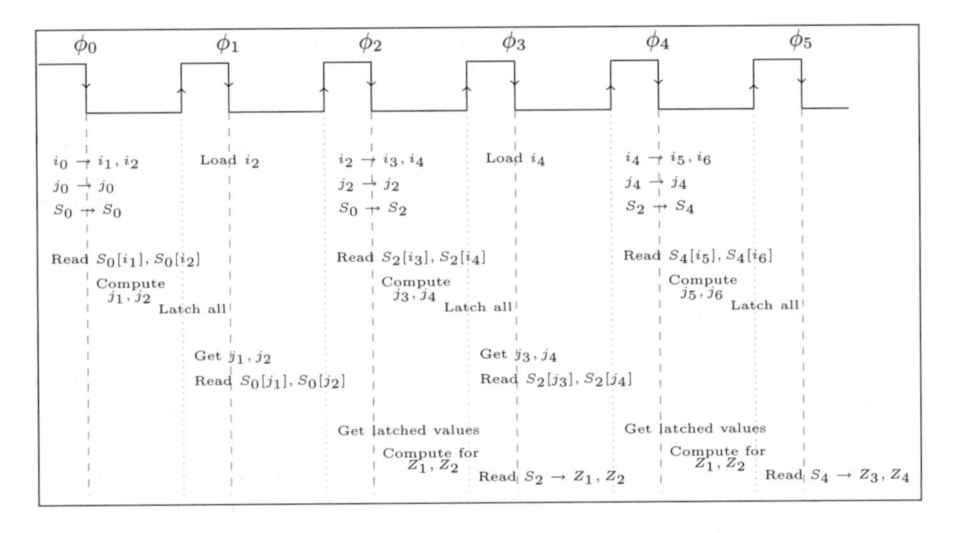

Fig. 6. Timing diagram for the complete PRGA circuit

At the trailing edge of ϕ_2, the *swap* operation takes place among the appropriate registers of S_0. One has all the values required for the swap module shown in Fig. 3, and hence can obtain the new configuration S_2 from S_0.

Simultaneously, the latched values for i_1, i_2, j_1, j_2, $S_0[i_1]$ and $S_0[i_2]$ are accessed, and the registers in Fig. 1 are incremented to i_3 and i_4 at the trailing edge of ϕ_2.

Clock cycle ϕ_3: At the start of this cycle, one has i_1, i_2, j_1, j_2, $S_0[i_1]$, $S_0[i_2]$, $S_0[j_1]$ and $S_0[j_2]$, released from the latches. Moreover, the register-bank configuration is S_2 at this stage. Hence, one can compute Z_1 using circuit in Fig. 4, and Z_2 using circuit in Fig. 5. The combinational logic of these circuits operate during the cycle and the bytes from S_2 are read at the trailing edge of ϕ_3.

Simultaneously, during cycle ϕ_3, the indices i_3, i_4 control the first two 256 to 1 MUX units to produce $S_0[i_3]$ and $S_0[i_4]$, similar to cycle ϕ_1. These values are used to produce j_3, j_4 and they are latched at the leading edge of ϕ_3. At the trailing edge of ϕ_3, the latches for j_3, j_4 are accessed.

Clock cycle ϕ_4: Similar to cycle ϕ_2, here we obtain $S_2[j_3], S_2[j_4]$, and latch these values at the leading edge of ϕ_4. Another swap operation is performed at the trailing edge of ϕ_4 to produce S_4 from S_2. Simultaneously, the first register in Fig. 1 is loaded with i_4 (replacing i_3), and incremented to i_5, i_6 at the trailing edge of ϕ_4.

Clock cycle ϕ_5: Similar to cycle ϕ_3, we compute the combinational logic for Z_3, Z_4 in this cycle (using circuits in Fig. 4 and Fig. 5), and the final values of Z_3 and Z_4 are read from S_4 at the trailing edge of ϕ_5.

The process continues similarly over the clock cycles and we obtain a timing diagram as shown in Fig. 6. The combinational logics operate between the clock pulses and the latches operate with the edges of the pulses (loaded at the leading edge and released at the trailing edge). All read, swap and increment operations are done at the trailing edges of the clock pulses.

One may observe that the first two bytes Z_1, Z_2 of the output are obtained at the end of the third clock cycle ϕ_3 and the next two bytes Z_3, Z_4 are obtained at the fifth clock cycle ϕ_5. A formal statement to summarize the observation is given in the next section.

4 The Complete Circuit

The complete circuit diagram for the PRGA stage of our RC4 hardware is shown in Fig. 7. Here, L_i denotes the latches operated by the trailing edge of ϕ_{2n+i}, i.e., the $(2n+i)^{th}$ cycle of the master clock ϕ where $n \geq 0$. For example, the latches labeled L_1 (four of them) are released at the trailing edge of $\phi_1, \phi_3, \phi_5, \ldots$ and the latches labeled L_2 (eight of them) are released at the trailing edge of $\phi_2, \phi_4, \phi_6, \ldots$ etc. We can now generalize our previous observation to state the following.

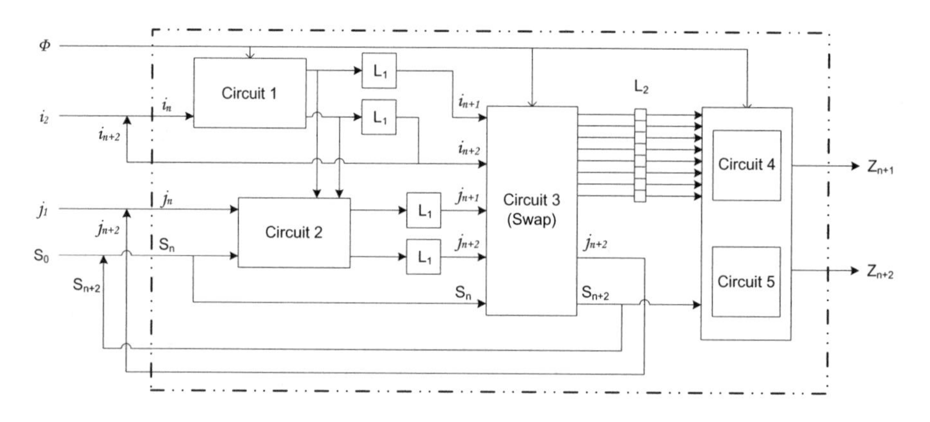

Fig. 7. Circuit for PRGA stage of RC4

Theorem 1. *The hardware proposed for the PRGA stage of RC4, as shown in Fig. 7, produces "one byte per clock" after an initial delay of two clock cycles.*

Proof. Let us call the stage of the PRGA circuit shown in Fig. 7 the nth stage. This actually denotes the nth iteration of our model, which produces the output bytes Z_{n+1} and Z_{n+2}.

The first block (Circuit 1) operates at the trailing edge of ϕ_n, and increments i_n to i_{n+1}, i_{n+2}. During cycle ϕ_{n+1}, the combinational part of Circuit 2 operates to produce j_{n+1}, j_{n+2}. The trailing edge of ϕ_{n+1} releases the latches of type L_1, and activates the swap circuit (Circuit 3). The combinational logic of the swap circuit functions during cycle ϕ_{n+2} and the actual swap operation takes place at the trailing edge of ϕ_{n+2} to produce S_{n+2} from S_n. Simultaneously, the latch of type L_2 is released to activate the Circuits 4 and 5. Once again, the combinational logic of these two circuits operate during ϕ_{n+3}, and we get the outputs Z_{n+1} and Z_{n+2} at the trailing edge of ϕ_{n+3}.

This complete block of architecture performs in a cascaded pipeline fashion, as the indices i_2, j_2 and the state S_{n+2} are fed back into the system at the end of ϕ_{n+2} (actually, i_{n+2} is fed back at the end of ϕ_{n+1} to allow for the increments at the trailing edge of ϕ_{n+2}). Hence, the operational gap between two iterations (e.g., n^{th} and $(n+2)^{th}$) of the system is two clock cycles (e.g., ϕ_n to ϕ_{n+2}), and we obtain two output bytes per iteration of the model.

Hence, the PRGA architecture proposed in Fig. 7 produces $2N$ bytes of output stream in N iterations, over $2N$ clock cycles. Note that the initial clock pulse ϕ_0 is an extra one, and the production of the output bytes lag the feedback cycle by one clock pulse in every iteration (e.g., ϕ_{n+3} in case of nth iteration). Hence, our model practically produces $2N$ output bytes in $2N$ clock cycles, that is "one byte per clock", after an initial lag of two clock cycles. □

4.1 Issues for the Circuit of KSA

Note that the general KSA routine runs for 256 iterations to produce the initial permutation of the S-box. Moreover, the steps of the KSA phase in RC4 are quite similar to the steps of PRGA, apart from the following:

- Calculation of j involves the key K along with S-box S and index i.
- Calculation of Z_1, Z_2 not required (actually, not recommended).

We propose the use of our PRGA architecture (Fig. 7) for the KSA round as well, with some modifications to the design, as follows.

K-register bank: We have to introduce a new register bank for key K. It will contain l number of 8-bit registers, where $8 \leq l \leq 15$ in practice.

K-register MUX: To read data $K[i_1 \bmod l]$ and $K[i_2 \bmod l]$ from the K-registers, we need to have two 16 to 1 multiplexer unit. The first l input lines of this MUX will be fed data from registers $K[0]$ to $K[l-1]$, and the rest $16-l$ inputs can be left floating (recall that $8 \leq l \leq 15$). The control lines of these MUX units will be $i_1 \bmod l$ and $i_2 \bmod l$ respectively, and hence the floating inputs will never be selected.

Modular Counters: To obtain $i_1 \bmod l$ and $i_2 \bmod l$, we have to incorporate two modular counters (modulo l) for the indices. These will be synchronous counters and the one for i_2 will have no clock input for the LSB position, similar to circuit in Fig. 1.

Extra 2-input Adders: Two 2-input parallel adders to be appended to Fig. 2 for adding $K[i_1 \bmod l]$ and $K[i_2 \bmod l]$ to j_1 and j_2 respectively.

No Outputs: Circuits of Fig. 4 and Fig. 5 have to be removed from the overall structure, so that no output byte is generated during KSA. If any such byte is generated, the key K may be compromised, and hence, the output module have to be disconnected during the KSA stage. One can do so by resetting these modules permanently throughout the KSA operation.

Using this modified hardware configuration, we can implement two rounds of KSA in 2 clock cycles, that is "one round per clock", after an initial lag of 1 cycle. Total time required for KSA will be $256 + 1 = 257$ clock cycles in this case.

4.2 Comparison with Existing Architecture

Combining our KSA and PRGA architectures, we can obtain $2N$ output-stream bytes in $2N + 258$ clock cycles, counting the initial delay of 2 cycles for PRGA. The best hardware implementation of RC4 till date is described in [6], which provides an output of N bytes in $3N + 768$ clock cycles. A formal comparison of the timings is shown in Table 3. One can easily observe that for large N, the throughput of our RC4 architecture is 3 times compared to that of the hardware configuration proposed in [6].

Table 3. Timing comparison of our hardware with that of [6]

Operations	Clock cycles needed for hardware in [6]	Clock cycles needed for our model
Per round of KSA	3	1
Complete KSA routine	$256 \times 3 = 768$	$256 + 1 = 257$
N output bytes from PRGA	$3N$	$N + 2$
N output bytes from RC4	$3N + 768$	$257 + (N + 2) = N + 259$
Per byte output from RC4	$3 + \frac{768}{N}$	$1 + \frac{259}{N}$

Another aspect to compare is the cost of the hardware configuration proposed in the two cases. Table 4 in Appendix C summarizes the hardware components required for our model. The reader may note that the hardware components used in the architecture proposed in [6, Figure 3] can be compared with the ones we use. The hardware of [6] uses three 256-byte RAM blocks for the S-box, while we use only one such 256-byte block built by master-slave J-K flip-flops. The number of registers for our K-register bank is the same as the ones used for the K-box in [6]. Both architectures use the same number of 8-bit registers in addition to these banks. Our architecture requires in excess only a constant number of combinational and sequential logic components compared to that in [6]. Thus, not only in terms of throughput, but also in terms of memory requirement, our architecture is better than that proposed in [6].

5 Conclusion

In this paper we try to answer the question of RC4 hardware efficiency in terms of the throughput, that is, the number of keystream bytes generated per clock cycle of the system. We have proposed a novel architecture for a generic RC4 hardware in this direction, using basic combinational and sequential logic components. Our architecture performs the KSA stage of RC4 at a rate of "one round per clock" and produces "one byte per clock" in case of the PRGA routine. To the best of our knowledge, this is 3 times faster than the current best hardware configuration for RC4.

Simulation of our architecture using VHDL and its implementation on an FPGA module are in progress. To get the best performance in terms of fast real-time clock cycles, one may also want to fabricate the architecture onto an ASIC board, or a processing chip. We would also like to add that this hardware model can generate even higher throughput provided one has enough memory space to perform the swap and read operations at the same time, that is, if two copies of the S array can be maintained.

Acknowledgements. The authors are grateful to the anonymous reviewers for their valuable comments and suggestions that helped in improving the quality of the paper.

References

1. Fluhrer, S.R., McGrew, D.A.: Statistical Analysis of the Alleged RC4 Keystream Generator. In: Schneier, B. (ed.) FSE 2000. LNCS, vol. 1978, pp. 19–30. Springer, Heidelberg (2001)
2. Fluhrer, S.R., Mantin, I., Shamir, A.: Weaknesses in the Key Scheduling Algorithm of RC4. In: Vaudenay, S., Youssef, A.M. (eds.) SAC 2001. LNCS, vol. 2259, pp. 1–24. Springer, Heidelberg (2001)
3. Galanis, M.D., Kitsos, P., Kostopoulos, G., Sklavos, N., Goutis, C.E.: Comparison of the Hardware Implementation of Stream Ciphers. Int. Arab. J. Inf. Tech. 2(4), 267–274 (2005)
4. Golic, J.: Linear statistical weakness of alleged RC4 keystream generator. In: Fumy, W. (ed.) EUROCRYPT 1997. LNCS, vol. 1233, pp. 226–238. Springer, Heidelberg (1997)
5. Hamalainen, P., Hannikainen, M., Hamalainen, T., Saarinen, J.: Hardware implementation of the improved WEP and RC4 encryption algorithms for wireless terminals. In: Proc. of Eur. Signal Processing Conf. pp. 2289–2292 (2000)
6. Kitsos, P., Kostopoulos, G., Sklavos, N., Koufopavlou, O.: Hardware Implementation of the RC4 stream Cipher. In: Proc. of 46th IEEE Midwest Symposium on Circuits & Systems 2003, Cairo, Egypt (2003), `http://dsmc.eap.gr/en/members/pkitsos/papers/Kitsos_c14.pdf`
7. Lee, J.-D., Fan, C.-P.: Efficient low-latency RC4 architecture designs for IEEE 802.11i WEP/TKIP. In: Proc. of Int. Symp. on Intelligent Signal Processing and Communication Systems ISPACS 2007, pp. 56–59 (2007)
8. Lynch, T., Swartzlander Jr., E.E.: A Spanning Tree Carry Lookahead Adder. IEEE Trans. on Computers 41(8), 931–939 (1992)
9. Mantin, I.: Predicting and Distinguishing Attacks on RC4 Keystream Generator. In: Cramer, R. (ed.) EUROCRYPT 2005. LNCS, vol. 3494, pp. 491–506. Springer, Heidelberg (2005)
10. Mantin, I.: A Practical Attack on the Fixed RC4 in the WEP Mode. In: Roy, B. (ed.) ASIACRYPT 2005. LNCS, vol. 3788, pp. 395–411. Springer, Heidelberg (2005)
11. Mantin, I., Shamir, A.: A Practical Attack on Broadcast RC4. In: Matsui, M. (ed.) FSE 2001. LNCS, vol. 2355, pp. 152–164. Springer, Heidelberg (2002)

12. Matthews Jr., D.P.: System and method for a fast hardware implementation of RC4. US Patent Number 6549622, Campbell, CA (April 2003), http://www.freepatentsonline.com/6549622.html
13. Maximov, A., Khovratovich, D.: New State Recovering Attack on RC4. In: Wagner, D. (ed.) CRYPTO 2008. LNCS, vol. 5157, pp. 297–316. Springer, Heidelberg (2008)
14. Mironov, I. (Not So) Random Shuffles of RC4. In: Yung, M. (ed.) CRYPTO 2002. LNCS, vol. 2442, pp. 304–319. Springer, Heidelberg (2002)
15. Paul, G., Maitra, S.: On Biases of Permutation and Keystream Bytes of RC4 towards the Secret Key. Cryptography and Communications - Discrete Structures, Boolean Functions and Sequences 1(2), 225–268 (2009)
16. Roos, A.: A class of weak keys in the RC4 stream cipher. Two posts in sci.crypt, message-id 43u1eh1j3@hermes.is.co.za, 44ebge llf@hermes.is.co.za (1995), http://marcel.wanda.ch/Archive/WeakKeys
17. Sinha, B.P., Srimani, P.K.: Fast Parallel Algorithms for Binary Multiplication and Their Implementation on Systolic Architectures. IEEE Trans. on Computers 38(3), 424–431 (1989)
18. Wagner, D.: My RC4 weak keys. Post in sci.crypt, message-id 447o1l$cbj@cnn.Princeton.EDU (1995), http://www.cs.berkeley.edu/~daw/my-posts/my-rc4-weak-keys
19. Performance results, http://www.ecrypt.eu.org/stream/perf/#results

Appendix A: Detailed Explanation for the Swapping Cases

Case 1: $i_2 \neq j_1$ and $j_2 \neq i_1$ and $j_2 \neq j_1$

These data transfers are symbolically represented by the following permutation on data in S_0.

$$\begin{pmatrix} i_2 & j_2 \\ j_2 & i_2 \end{pmatrix} \circ \begin{pmatrix} i_1 & j_1 \\ j_1 & i_1 \end{pmatrix}$$

This involves 4 simultaneous register to register transfers:

$$S_0[i_1] \rightarrow S_0[j_1], \quad S_0[j_1] \rightarrow S_0[i_1], \quad S_0[i_2] \rightarrow S_0[j_2], \quad S_0[j_2] \rightarrow S_0[i_2]$$

Case 2: $i_2 \neq j_1$ and $j_2 \neq i_1$ and $j_2 = j_1$

In this case the data transfers are represented by the following permutation on data in S_0.

$$\begin{pmatrix} i_2 & j_1 \\ j_1 & i_2 \end{pmatrix} \circ \begin{pmatrix} i_1 & j_1 \\ j_1 & i_1 \end{pmatrix}$$

This involves 3 simultaneous register to register transfers:

$$S_0[i_1] \rightarrow S_0[i_2], \quad S_0[i_2] \rightarrow S_0[j_1] = S_0[j_2], \quad S_0[j_1] \rightarrow S_0[i_1]$$

Case 3: $i_2 \neq j_1$ and $j_2 = i_1$ and $j_2 \neq j_1$

In this case the data transfers are represented by the following permutation on data in S_0.

$$\begin{pmatrix} i_2 & i_1 \\ i_1 & i_2 \end{pmatrix} \circ \begin{pmatrix} i_1 & j_1 \\ j_1 & i_1 \end{pmatrix}$$

This again involves 3 simultaneous register to register transfers:

$$S_0[i_1] \rightarrow S_0[j_1], \quad S_0[i_2] \rightarrow S_0[i_1] = S_0[j_2], \quad S_0[j_1] \rightarrow S_0[i_2]$$

Case 4: $i_2 \neq j_1$ **and** $j_2 = i_1$ **and** $j_2 = j_1$
In this case the data transfers are represented by the following permutation on data in S_0.

$$\begin{pmatrix} i_2 & i_1 \\ i_1 & i_2 \end{pmatrix} \circ \begin{pmatrix} i_1 & i_1 \\ i_1 & i_1 \end{pmatrix}$$

This involves 2 simultaneous register to register transfers:

$$S_0[i_1] \rightarrow S_0[i_2], \quad S_0[i_2] \rightarrow S_0[i_1] = S_0[j_1] = S_0[j_2]$$

Case 5: $i_2 = j_1$ **and** $j_2 \neq i_1$ **and** $j_2 \neq j_1$
In this case the data transfers are represented by the following permutation on data in S_0.

$$\begin{pmatrix} j_1 & j_2 \\ j_2 & j_1 \end{pmatrix} \circ \begin{pmatrix} i_1 & j_1 \\ j_1 & i_1 \end{pmatrix}$$

This involves 3 simultaneous register to register transfers:

$$S_0[i_1] \rightarrow S_0[j_2], \quad S_0[j_2] \rightarrow S_0[j_1] = S_0[i_2], \quad S_0[j_1] \rightarrow S_0[i_1]$$

Case 6: $i_2 = j_1$ **and** $j_2 \neq i_1$ **and** $j_2 = j_1$
In this case the data transfers are represented by the following permutation on data in S_0.

$$\begin{pmatrix} j_1 & j_1 \\ j_1 & j_1 \end{pmatrix} \circ \begin{pmatrix} i_1 & j_1 \\ j_1 & i_1 \end{pmatrix}$$

This involves 2 simultaneous register to register transfers:

$$S_0[i_1] \rightarrow S_0[j_1] = S_0[i_2] = S_0[j_2], \quad S_0[j_1] \rightarrow S_0[i_1]$$

Case 7: $i_2 = j_1$ **and** $j_2 = i_1$ **and** $j_2 \neq j_1$
In this case the data transfers are represented by the following permutation on data in S_0.

$$\begin{pmatrix} j_1 & i_1 \\ i_1 & j_1 \end{pmatrix} \circ \begin{pmatrix} i_1 & j_1 \\ j_1 & i_1 \end{pmatrix}$$

This is identity permutation, and does not involve any data transfer.

Case 8: $i_2 = j_1$ **and** $j_2 = i_1$ **and** $j_2 = j_1$
This case cannot occur, as it implies $i_1 = i_2$, which is impossible because $i_2 = i_0 + 2 = i_1 + 1$.

Appendix B: Master-Slave J-K Flip-Flops

The master-slave configuration is basically two J-K flip-flops connected together in series, as shown in Fig. 8. The input signals J and K are connected to the Master flip-flop which *locks* the input while the clock input is *high*. As the clock input of the Slave flip-flop is the inverse of the Master clock input, the outputs from the Master flip-flop are *seen* by the Slave only when the clock input goes

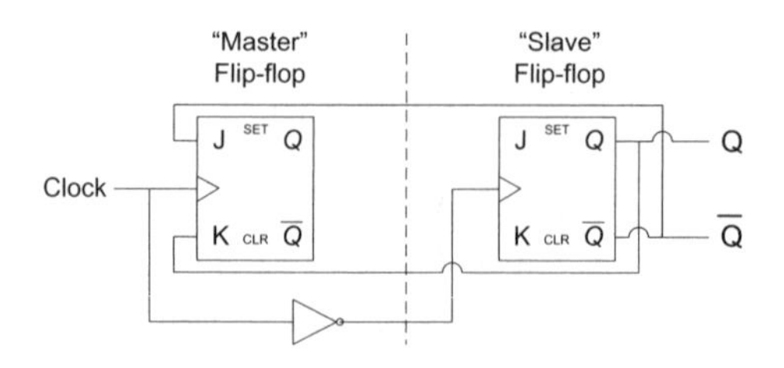

Fig. 8. Master-Slave J-K flip-flop

low. Therefore on the trailing edge of the clock pulse, the locked outputs of the Master flip-flop are fed through to the J-K inputs of the Slave flip-flop. The circuit accepts input at the Master J-K flip-flop when the clock signal is high, and passes it to the output on the trailing edge of the clock.

This enables us to make data interchange between two registers or cyclic data transfers among three registers, as needed by the swap operations, in just one clock. Recall that the registers in the S-register bank are made of these Master-Slave J-K flip-flops. We connect the registers participating in the swap operation using combinational logic circuits during the relevant clock cycle (ϕ_r, say), and wait for the clock pulse. At the leading edge of ϕ_r, the Master flip-flop of each register accepts the incoming new value, and it is passed on to the output through the Slave flip-flop only at the trailing edge of ϕ_r. Thus, the swap operation completes at the end of the cycle ϕ_r, and avoids all data collisions. This controlled flow of data helps perform the swap operations in a single clock-cycle.

Appendix C: Hardware Components

Table 4 lists all hardware components required for the RC4 architecture proposed in this paper. It counts all sequential and combinational components required for the circuit of PRGA (Fig. 7), as well as the extra components required for the proposed modifications to perform the KSA.

Table 4. Hardware components required to realize the proposed RC4 architecture

Component	Type	Number of pieces used	Module built using the components
Registers	8-bit (master-slave J-K)	256	S-register bank
	8-bit (normal)	l	K-register bank
	8-bit (parallel load)	2	Storage for j_1, j_2
Counters	8-bit (asynchronous)	2	Counters for i_1, i_2
	8-bit (modulo l)	2	Counters for i_1 mod l and i_2 mod l
Multiplexers	256 to 1	$4 + 2 = 6$	To read S-register values and Z_1, Z_2
	16 to 1	2	To read K-register values
	8 to 1	$4 + 1 = 5$	Fig. 3 and Fig. 5
	4 to 1	1	Fig. 4
	2 to 1	1	Fig. 2
Address Decoders	8 to 256	4	Fig. 3
Comparators	2 input	$1 + 3 + 2 = 4$	Fig. 2, Fig. 3 and Fig. 4
Parallel Adders	3 input	2	Fig. 2
	2 input	$1 + 1 + 8 + 2 = 12$	Fig. 2, Fig. 4, Fig. 5 and KSA extra adders
Latches	Edge-triggered	$4 + 8 = 12$	L_1 and L_2 latches in Fig. 7
Logic Gates	4 input OR	2	Fig. 3
	2 input AND	5	Fig. 3

Author Index